# Calculus
## of the
## Elementary
## Functions

# Calculus of the Elementary Functions

**MERRILL E. SHANKS**

**ROBERT GAMBILL**

*Purdue University*

**HOLT, RINEHART AND WINSTON, INC.**

*New York   Chicago   San Francisco   Atlanta
Dallas   Montreal   Toronto   London   Sydney*

# Preface

In recent years students often enter college with greater mathematical sophistication than in the past but with little, if any, greater manipulative skill. This text is written with the premise that many schools will find it desirable to devote a year to the development of students' computational power. Calculus has been the traditional instrument for this development and we think it will continue to be. While the student is acquiring skill in calculus per se, he improves at the same time his grasp of elementary algebra, trigonometry, and analytic geometry. Moreover, as power to apply calculus grows, and as he encounters many specific functions, he develops insight into the processes of analysis. We are convinced that at this level a grasp of theory must come mainly from problems that illuminate theory.

Therefore, in this book, we have tried to make the concepts and processes of calculus as transparent as possible. We want the student to develop his mathematical intuition and we have supplied an abundance of problems to practice on. We have tried to hold the text material to a minimum and have used pictures to convey ideas wherever possible to do so without distortion. Theorems too are held to a minimum, especially in the beginning, and some hard proofs are either omitted or sketched. Our point of view is that proofs that are straight calculation are fair game and should probably be required of the student. Whereas proofs that are more conceptual in nature should be seen but probably not required of the student.

The intent throughout is to present the theoretical structure as simply as possible without excessive concern about subtleties. Calculus, after all, is a certain kind of calculation; calculus is the solving of problems. And so the student must do many problems—about 1200 for the full year is minimal and about 1500 would seem about right.

The student should read the text with pencil and paper at hand. Not only may he need to fill in details of the text, but he may need to expand the Examples. The earlier Examples are usually fully expanded. Later Examples leave out steps that the student should be able to complete for himself.

We have tried to write a reasonably short book. Thick books discourage the reader and excess verbiage can obscure ideas. Moreover short exposition permits the teacher to expand on the text in his own style. Nevertheless, there is more than enough material here for a year's work even with well prepared students. The unstarred sections comprise a standard course. Selections can be made from the starred sections and problems to accommodate better students. These usually are either harder computation or more theoretical, and sometimes contain material given in advanced calculus.

Two other features are the exclusion of analytic geometry proper and the separation of differential and integral calculus. It seems to us that the monster texts that combine concepts of geometry, the derivative, and the integral into a unified whole present such a kaleidoscopic pattern that the student is often confused. In those books there may be one hundred or more pages devoted to pre-calculus topics including perhaps discussion of the real number system at a level for which the student sees no use. The pre-calculus background in this book is provided in Appendix A where the needed formulas and definitions are supplied for reference. The text proceeds at once with the calculus and soon the student is doing interesting problems. After all, there are but two basic techniques to be learned: differentiation and integration. Once these are mastered the student has powerful tools to attack problems which previously he could not touch. Too often students do not *feel* this gain in power. We think that in part this has been because they do not master one technique (differentiation) and learn to apply it before learning another.

Part I (Chapters 1 through 8) is devoted to the differential calculus of functions of one variable. The only mention of integration is a short section on antiderivatives in Chapter 1. This is for the benefit of students taking physics at the same time. However, if the instructor so desires, Chapter 9 on the definite integral can be taken up immediately following Chapter 4.

Part II (Chapters 9 through 13) is devoted to the integral calculus. It finishes with the theory of infinite series, which was treated in an informal intuitive way in Chapter 8.

Part III (Chapters 14 and 15) contains the standard topics of multivariate calculus. Although vectors are mentioned, no vector notation is used nor knowledge of vectors presupposed.

Rigorous proof has, on the whole, been relegated to the background and to later portions of the text. (Appendixes C and D are concerned with limits and continuity and some proofs of basic theorems.) We are much more concerned that a student have intuition about what is going on than that he remember proofs of analysis. But we are concerned that he understand the theorems and that he is able to verify the hypotheses in theorems and to apply them. Students arrive in college with such diverse backgrounds and attitudes that if one were to present calculus with complete rigor, one would have to provide the foundation for that rigor—and calculus proper would be delayed unduly. In the early chapters theorems about general functions are avoided where possible. In the early stages when one deals exclusively with the elementary functions general limit theorems are unnecessary, and to the student often seemingly irrelevant. A student needs only to "see" the particular limits that occur in treating elementary functions. The theorems that do occur are definitely utilized in problems—which is merely another way of saying that the text (except for historical remarks) is purposefully tied to expected student activity.

We have tried not to let the text get in the way of the students learning to calculate. The logical development is direct, and such that details and proofs of some theorems are easily supplied. We have tried to avoid the inclusion of anything that might have to be unlearned later. All that is here is usable.

*West Lafayette, Indiana*                                       Merrill E. Shanks
*January 1969*                                                  Robert Gambill

# Contents

*Part II* **INTEGRAL CALCULUS**                                        *215*

*PART I*

**DIFFERENTIAL CALCULUS**

Sir Isaac Newton
(1642–1727)
Born December 25, 1642 in Woolthorpe
Entered Trinity College
Cambridge 1661
Undergraduate degree January, 1664
At home 1665–1667
Fellow at Trinity 1667–1669
Professor of mathematics, Lucasian chair,
at Trinity 1669–1696
Publishing date: Philosophiae Naturalis
Principia Mathematica (Mathematical Principles
of Natural Philosophy) 1687
Warden of the London Mint, 1696
Master of the Mint 1699–1727

Newton is generally regarded as one of the great intellects of all time. His influence on the development of mathematics and physics was decisive. He entered Cambridge knowing almost no mathematics but advanced rapidly under the brilliant Isaac Barrow, his teacher.

The two years spent at home during a recurrence of the bubonic plague were fantastically productive. During this time: (1) He invented the calculus, which he called the method of *fluxions*, and had it in fairly complete form—for his own use. (2) He conceived the principle of universal gravitation and sketched its main outlines. (3) He discovered the decomposition of white light into a spectrum of colors and devised optical equipment.

On his return to Cambridge the pupil soon surpassed the teacher, and Barrow resigned his Lucasian chair of mathematics in favor of Newton. There a long and productive period ensued in which he published little material except at the urging of his friends. News of his activity was known in England mainly through letters and the words of friends, while the continent remained unaware of the method of fluxions.

*CHAPTER I*

## *THE DERIVATIVE*

### 1   *The Problem of Tangents*

In the seventeenth century mathematicians were concerned (among other things) with two major problems, "The Problem of Tangents" and "The Problem of Area," whose nature and significance will become apparent as the reader progresses with the text. The area problem will be introduced in Chapter 9 of Part II. Part I is devoted to the solution of the Problem of Tangents and its application to a variety of problems.

Actually there are two aspects to "The Problem of Tangents."

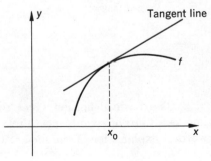

(1) Given a function $f$ what should we *mean* by a tangent line to the graph of $f$?

(2) Is there an easy way, a trick, for calculating what the tangent is?

In this chapter we are concerned only with question (1). Chapter 2 deals with and solves question (2). The reader may be surprised, as he proceeds, with the apparent simplicity of the solutions to both problems. But in the seventeenth century the problems were obscured by several

3

deficiencies. In the first place suitable notation (that marvelous shorthand that presents ideas concisely) was not well developed. Secondly, the real number system was not adequately formulated. (Not until the nineteenth century did Dedekind give a modern construction of the real numbers.) But the main obstacle to a solution to problem (1) was the fact that the central concept required the notion of *limit*— and limits were not properly understood. This, too, was to come later.

It is hard for the beginner to understand why calculus was not invented earlier. Indeed, Isaac Barrow (the teacher of Newton) and the great Pierre Fermat were aware of all the pieces of both problems, namely Tangents and Area. But the fact remains that it was left to the genius of Newton and Leibniz to show the way to handle both problems. [The historically interested reader is urged to consult the references at the end of the chapter.]

There *is* a question as to what the definition should be. Observe:

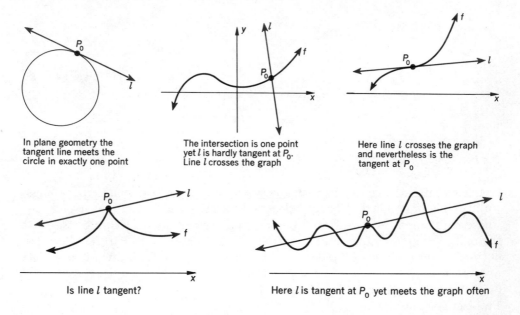

In plane geometry the tangent line meets the circle in exactly one point

The intersection is one point yet $l$ is hardly tangent at $P_0$. Line $l$ crosses the graph

Here line $l$ crosses the graph and nevertheless is the tangent at $P_0$

Is line $l$ tangent?

Here $l$ is tangent at $P_0$ yet meets the graph often

The basic definition gives the tangent line as a *limit* of secant lines. First we need a word of explanation of the idea of limit.

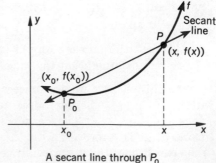

A secant line through $P_0$

We say that a function $f$ has a limit* $L$ as $x$ approaches $a$ if as $x$ gets closer to $a$ (that is, $|x - a|$ gets smaller, *but does not become zero*), then $f(x)$ gets closer to the number $L$ (that is, $|f(x) - L|$ gets arbitrarily small). This is expressed symbolically by

$$\lim_{x \to a} f(x) = L.$$

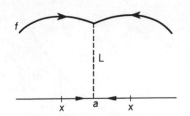

$f(x)$ approaches $L$ as $x$ approaches $a$, or $f(x) \to L$ as $x \to a$

**DEFINITION**　The tangent line to the graph of $f$ at $P_0 = (x_0, f(x_0))$ is the line through $P_0$ whose slope is the limit of the slopes of the secant lines through $(x_0, f(x_0))$ and $(x, f(x))$ as $x$ approaches $x_0$.

*Remark.* This definition does not permit vertical tangent lines, because in that case the limit of the slopes of the secant lines does not exist. Example 3 below illustrates this possibility.

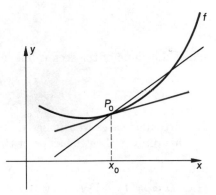

The tangent line is determined as soon as we know its slope, $m$, for then it is the line with slope $m$ passing through $(x_0, f(x_0))$.

The slope of a secant line is

$$\frac{f(x) - f(x_0)}{x - x_0}.$$

So as $x$ approaches $x_0$ this number should approach the slope of the tangent line. We formulate this in terms of "increments," a notation terminology that was devised by Leibniz.

$$\Delta x = x - x_0 = \text{increment}\dagger \text{ of } x;$$
$$\begin{aligned}
\Delta y &= y - y_0 \\
&= f(x) - f(x_0) \\
&= f(x_0 + \Delta x) - f(x_0) \\
&= \text{increment of } y.
\end{aligned}$$

---

\* We shall not try to be absolutely precise about limits in the text proper. Our concern is with an intuitive grasp of the concepts. Moreover, in the special cases that concern us, the required limit will be rather obvious because of the simplicity of the algebra involved.

A precise definition of limit and some theorems about limits can be found in Appendix C, Section 1.

† The symbol "$\Delta x$," read "delta-$x$," is a single quantity. Thus $\Delta x^2$ will mean $(\text{delta-}x)^2 = (\Delta x)^2$. $\Delta x$ can be either positive or negative. If $\Delta x$ is negative, the point $x$ lies "to the left" of $x_0$.

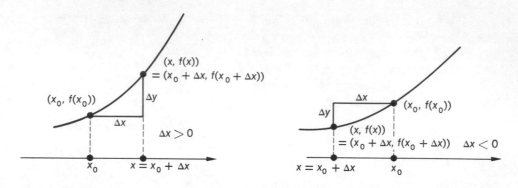

Then the slope of the secant line is

$$m_{\text{secant}} = \frac{\Delta y}{\Delta x} = \frac{f(x_0 + \Delta x) - f(x_0)}{\Delta x}.$$

In particular cases we can see that this slope approaches a limit as $\Delta x$ approaches 0. This limit is then the slope of the tangent line.

**Example 1**   If $y = f(x) = 2 - 2x - x^2$, then the slope of the tangent line at $(\frac{1}{2}, \frac{3}{4})$ is the limit of slopes of secant lines.

Here $x_0 = \frac{1}{2}$, $y_0 = \frac{3}{4}$,   and

$$\Delta y = f(x_0 + \Delta x) - f(x_0)$$
$$= f(\tfrac{1}{2} + \Delta x) - \tfrac{3}{4}$$
$$= 2 - 2(\tfrac{1}{2} + \Delta x) - (\tfrac{1}{2} + \Delta x)^2 - \tfrac{3}{4}$$
$$= -3\Delta x - \Delta x^2.$$

$$\frac{\Delta y}{\Delta x} = -3 - \Delta x.$$

And *now* it is easy to see what happens to $\Delta y / \Delta x$ as $\Delta x \to 0$. Observe that both $\Delta y$ and $\Delta x$ approach 0, so we could not have evaluated the limit simply by setting $\Delta x = 0$ and $\Delta y = 0$ for then we would have obtained the meaningless symbol $0/0$. Instead, we have so simplified the form of $\Delta y / \Delta x$ that we can *see* at once its limit as $\Delta x \to 0$. Clearly the limit is $-3$, and so

$$m = \text{slope of the tangent line at } (\tfrac{1}{2}, \tfrac{3}{4}) = -3,$$

and an equation of the tangent line is:

$$y - \tfrac{3}{4} = -3(x - \tfrac{1}{2}) \qquad \text{or} \qquad y + 3x - \tfrac{9}{4} = 0.$$

*Example 2*     In the Example 1 we found the slope of the tangent at a specific point. Usually we find the slope at an arbitrary point $(x, f(x))$. In this way we obtain a formula for the slope at any point. For example, if

$$y = f(x) = x + \frac{1}{x},$$

then     $\Delta y = f(x + \Delta x) - f(x) = \left(x + \Delta x + \dfrac{1}{x + \Delta x}\right) - \left(x + \dfrac{1}{x}\right)$

$$= \Delta x + \frac{1}{x + \Delta x} - \frac{1}{x} = \Delta x + \frac{-\Delta x}{x(x + \Delta x)}.$$

Hence,

$$\frac{\Delta y}{\Delta x} = 1 - \frac{1}{x(x + \Delta x)}.$$

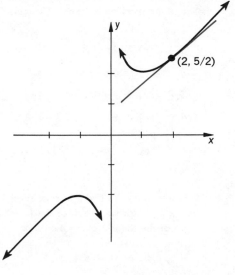

(2, 5/2)

Again, as in Example 1, we have so simplified $\Delta y/\Delta x$ that its limit as $\Delta x$ approaches 0 is clear. The slope of the tangent at $(x, x + 1/x) = m = 1 - 1/x^2$. For example, the slope at $(2, \frac{5}{2})$ is $m = \frac{3}{4}$, and an equation of the tangent line at this point is:

$$y - \tfrac{5}{2} = \tfrac{3}{4}(x - 2)$$

or

$$y - \tfrac{3}{4}x - 1 = 0.$$

*Example 3*     A secant line to $y = f(x) = x^{2/3}$ at $(0, 0)$ has slope

$$m = \frac{\Delta y}{\Delta x} = \frac{(\Delta x)^{2/3}}{\Delta x - 0} = (\Delta x)^{-1/3}.$$

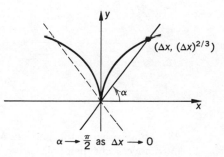

$(\Delta x, (\Delta x)^{2/3})$

$\alpha \longrightarrow \frac{\pi}{2}$ as $\Delta x \longrightarrow 0$

Now, as $\Delta x \to 0$ the slope does not approach a limiting value. Nevertheless, the inclination $\alpha$ of the secant line approaches $\pi/2$.

In such examples we agree that the tangent line at $(x_0, f(x_0))$ is the vertical line $x = x_0$. In this example $x_0 = 0$.

*Problems*

1.  Draw a large scale graph of $f(x) = x^2 - 2$ for $-2 \leq x \leq 2$. Compute slopes of secant lines through $(1, -1)$ for $\Delta x = 1, -1, 0.5, -0.5, 0.25, -0.25, 0.1, -0.1$. Draw each secant line and also the tangent line.

    [Slopes of secant: 3, 1, 2.5, 1.5, 2.25, 1.75, 2.1, 1.9]

2.  Sketch the graph of $y = f(x) = x^3 - 3x$ for $-2 \leq x \leq 2$. Compute slopes of secant lines through $(\frac{1}{2}, -\frac{11}{8})$ for $\Delta x = 1, 0.5, -0.5, 0.1, -0.1$, and draw each secant line and the tangent line.

    [Slopes of secant: 0.25, $-1.25$, $-2.75$, $-2.09$, $-2.39$]

In Problems 3 to 7 find the slope of the tangent line and an equation of the tangent line at the specified point as in Example 1. Then sketch each graph and the tangent line.

3.  $f(x) = 2x - 3$ ;   $(1, -1)$

4.  $f(x) = mx + b$;   $(x_0, y_0)$

5.  $f(x) = x^2$     ;   $(1, 1)$   [m = 2]

6.  $f(x) = x^2$     ;   $(-\frac{1}{2}, \frac{1}{4})$   [m = -1]

7.  $f(x) = \dfrac{1}{x}$     ;   $(1, 1)$

In Problems 8 to 12 find a formula for the slope of the tangent line at an arbitrary point $(x, f(x))$. Sketch the graph of the function and examine, by eye, the variation of the slope of the tangent along the curve.

8.  $\checkmark$ $f(x) = x^2 - x + 1$   [m = 2x - 1]

9.  $\checkmark$ $f(x) = -x^2 + x - 1$   [m = -2x + 1]

10.  $\checkmark$ $f(x) = 2x^3 - 3x^2 - 6x + 8$   [m = 6x^2 - 6x - 6]

11.  $\checkmark$ $f(x) = |x|$   $\begin{bmatrix} m = 1, & x > 0; \\ m = -1, x < 0 \end{bmatrix}$

12.  $f(x) = x - x^3$   [m = 1 - 3x^2]

*13.  Show that if the line $l: y = mx + b$ meets the parabola $y = x^2$ only in $(x_0, x_0^2)$ then $l$ must be tangent to the parabola.

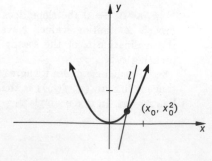

## 2  *The Derivative*

The slope of the tangent line to the graph of $f$ is called the *derivative* of $f$.

**DEFINITION**   The *derivative* of $f$ at $x_0$ (if it exists) is the limit of*

$$\frac{\Delta y}{\Delta x} = \frac{f(x_0 + \Delta x) - f(x_0)}{\Delta x}$$

as $\Delta x$ approaches zero. The derivative is denoted by†

$$\frac{dy}{dx} \qquad \text{or} \qquad f'(x_0).$$

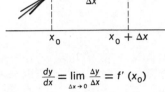

$$\frac{dy}{dx} = \lim_{\Delta x \to 0} \frac{\Delta y}{\Delta x} = f'(x_0)$$

$$= \lim_{\Delta x \to 0} \frac{f(x_0 + \Delta x) - f(x_0)}{\Delta x}$$

Using the notation for limits we have

$$\frac{dy}{dx} = f'(x_0) = \lim_{\Delta x \to 0} \frac{\Delta y}{\Delta x} = \lim_{\Delta x \to 0} \frac{f(x_0 + \Delta x) - f(x_0)}{\Delta x} \,.$$

*Example 1*   Find the derivative of $f$ at $x_0$ where $y = f(x) = x^2 + x$.

The process of applying the definition to find the derivative can be reduced to three steps:

(1)   $\Delta y = f(x_0 + \Delta x) - f(x_0)$

$= [(x_0 + \Delta x)^2 + (x_0 + \Delta x)] - [x_0^2 + x_0]$  $\left.\begin{array}{l}\text{This expression must}\\\text{be simplified as much}\\\text{as possible.}\end{array}\right.$

$= 2x_0 \Delta x + \Delta x^2 + \Delta x.$

(2)   $\dfrac{\Delta y}{\Delta x} = 2x_0 + \Delta x + 1.$

(3)   $\dfrac{dy}{dx} = f'(x_0) = 2x_0 + 1,$   after letting $\Delta x \to 0.$

---

* The ratio $\Delta y / \Delta x$ is often called *the difference quotient.*

† The symbol $dy/dx$ (which is the invention of Leibniz) represents a single quantity and is *not a fraction*. In Chapter 4, we shall give meanings to $dy$ and to $dx$, but not now.

The symbol "$dy/dx$" is read "dee y dee x," and "$f'(x_0)$" is read "f prime at $x_0$."

There is no need to keep the subscript on $x$ because $x_0$ can be any number. Thus,

$$\frac{dy}{dx} = f'(x) = 2x + 1.$$

It is clear that at each number $x$, at which there is a derivative, there is a number $f'(x)$. Hence from the given function $f$ a new one $f'$ has been obtained.

**DEFINITION**  If $f$ has a derivative at $x$, then $f$ is said to be *differentiable* at $x$.

$$\boxed{f \xrightarrow[\text{operator}]{\text{derivative}} f'}$$

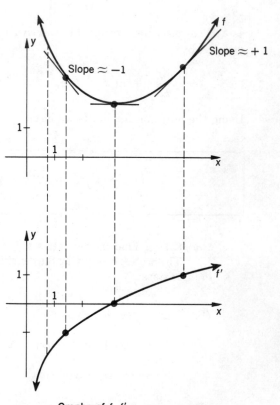

Slope $\approx -1$

Slope $\approx +1$

Graphs of $f$, $f'$

The *derived function* of $f$ is $f'$. The derivative is an *operator* that maps a function into its derived function. The word "operator" is used to indicate that we operate on a function $f$ to produce another function $f'$.

The derived function $f'$ can be sketched approximately from the graph of $f$ by estimating "by eye" the slope to the graph of $f$. Then the graphs of $f$ and $f'$ can be drawn one above the other. Though the graph of $f'$ will not be precise, still, $f'(x)$ should be positive where $f$ increases, negative where $f$ decreases, and zero where the tangent is horizontal.

*Example 2*    Find $f'(x)$ if $f(x) = \dfrac{1}{x+1}$.

$$y = f(x) = \frac{1}{x+1};$$

(1)     $\Delta y = f(x + \Delta x) - f(x)$

$$= \frac{1}{x + \Delta x + 1} - \frac{1}{x + 1}$$    $\begin{cases} \text{If we leave this alone} \\ \text{we will have trouble} \\ \text{with the limit.} \end{cases}$

$$= \frac{-\Delta x}{(x + \Delta x + 1)(x + 1)} \cdot$$    $\begin{cases} \text{This simplified form} \\ \text{for } \Delta y \text{ is much bet-} \\ \text{ter.} \end{cases}$

(2)     $\dfrac{\Delta y}{\Delta x} = \dfrac{-1}{(x + \Delta x + 1)(x + 1)} \cdot$

(3)     $\dfrac{dy}{dx} = f'(x) = \dfrac{-1}{(x + 1)^2},$    letting $\Delta x \to 0$.

**Example 3**    Sometimes one has to be ingenious, as this example will show. Find $f'(x)$ if $f(x) = \sqrt{x}$.

$$y = \sqrt{x}.$$

(1)     $\Delta y = f(x + \Delta x) - f(x) = \sqrt{x + \Delta x} - \sqrt{x}.$

This is a very inconvenient form for $\Delta y$. Let us see why.

$$\frac{\Delta y}{\Delta x} = \frac{\sqrt{x + \Delta x} - \sqrt{x}}{\Delta x}.$$

Now we are stuck. This limit as $\Delta x \to 0$ is not obvious. The trick we apply is to rationalize the numerator.

(2)     $\dfrac{\Delta y}{\Delta x} = \dfrac{\sqrt{x + \Delta x} - \sqrt{x}}{\Delta x} \dfrac{\sqrt{x + \Delta x} + \sqrt{x}}{\sqrt{x + \Delta x} + \sqrt{x}}$

$$= \frac{1}{\sqrt{x + \Delta x} + \sqrt{x}} \cdot$$

The limit is now easy.

(3)     $\dfrac{dy}{dx} = f'(x) = \dfrac{1}{\sqrt{x} + \sqrt{x}} = \dfrac{1}{2\sqrt{x}},$    letting $\Delta x \to 0$.

---

**Historical Remark**    Fermat had (in 1629) a method for finding slopes of tangent lines, but it did not involve limits (though it should have). He also had many of the formulas that are given in Chapter 2. In 1665 Isaac Barrow, also, could compute derivatives and apparently looked upon the derivative as a limit. But among mathematicians as a whole there was no agreement. Leibniz himself (about 1674) conceived of the derivative as the ratio of "infinitesimally small" quantities $dy$ and $dx$. The curse of infinitesimals took more than a century to eradicate. The concept of limit, which is the basis of a precise development, avoids the pitfall of infinitesimals. Newton himself apparently thought in terms of limits, but his language was not always free of uncertainty.

## Problems

In Problems 1 to 9 make a sketch of the graph of *f* on squared paper. Then sketch the graph of *f'* below it.

1.

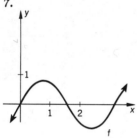

2.

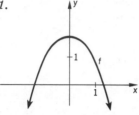

3.

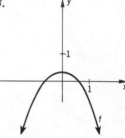

4.

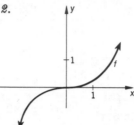

5.

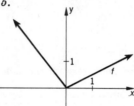

6.

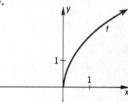

7.

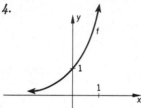

8.

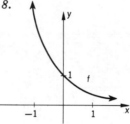

9.

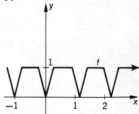

10.   In Problems 1 to 9 where is the derivative zero? positive? negative? non-existent?

11.   (a)   Sketch the graph of $F(x) = x - x^2$.
  (b)   Sketch the graph of $F'$ without doing any calculation.
  (c)   Find $F'(x)$ from the definition of derivative.
  (d)   Graph $F'$ as given by (c) and compare with (b).

12.   Carry out the same for $f(x) = x^3 - x$ as was done in Problem 11.

13.   (a)   Sketch the semi-circle given by $y = \sqrt{r^2 - x^2}$.
  (b)   Sketch, without computing, the graph of $dy/dx$.

(c)   Obtain $dy/dx$ from the definition of derivative.
(d)   Obtain $dy/dx$ from geometric considerations.
(e)   Graph (c) and (d) and compare with (b).

**In Problems 14 to 23 find $f'(x)$ from the definition of derivative.**

14. $f(x) = ax + b$ $\hspace{4cm}$ $[a]$

15. $f(x) = x - x^2 + 2$ $\hspace{3cm}$ $[1 - 2x]$

16. $f(x) = 3 + \dfrac{x}{x - 1}$ $\hspace{3cm}$ $\left[\dfrac{-1}{(x-1)^2}\right]$

17. $f(x) = Ax^2$ $\hspace{4cm}$ $[2Ax]$

18. $f(x) = Ax^3$ $\hspace{4cm}$ $[3Ax^2]$

19. $f(x) = \sqrt{x + 1}$ $\hspace{3cm}$ $[1/2\sqrt{x + 1}]$

20. $f(x) = 1/\sqrt{x + 1}$ $\hspace{2.5cm}$ $[-\frac{1}{2}(x + 1)^{-3/2}]$

21. $f(x) = 1/(x^2 + 1)$ $\hspace{2.5cm}$ $[-2x/(x^2 + 1)^2]$

22. $f(x) = \sqrt{x^2 + 1}$ $\hspace{3cm}$ $[x/\sqrt{x^2 + 1}]$

23. $f(x) = |x|$ $\hspace{2cm}$ $\left[\begin{array}{l} -1 \text{ if } x < 0, \quad 1 \text{ if } x > 0 \\ f'(0) \text{ does not exist} \end{array}\right]$

## 3   *The Fundamental Lemma*

The definition of derivative is*:

$$f'(x) = \lim_{\Delta x \to 0} \frac{f(x + \Delta x) - f(x)}{\Delta x}.$$

From the meaning of limit it follows that, when $\Delta x$ "gets small" (but not 0), then the difference

(1) $\hspace{3cm}$ $\dfrac{f(x + \Delta x) - f(x)}{\Delta x} - f'(x) = \eta$

must get small too. In other words, $\eta$ depends on $\Delta x$ (is a function of $\Delta x$) and approaches 0 as $\Delta x$ approaches 0. If we multiply both sides in (1) by $\Delta x$, then we obtain an important result:

---

* That we use the symbol "$\Delta x$" to denote an arbitrary increment in $x$ is irrelevant. Any symbol would do. Thus

$$f'(x) = \lim_{h \to 0} \frac{f(x + h) - f(x)}{h} = \lim_{w \to 0} \frac{f(x + w) - f(x)}{w}.$$

**FUNDAMENTAL LEMMA**   *If $f$ is dif-ferentiable at $x$, then*

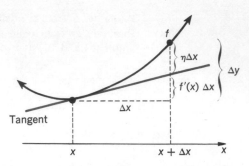

$$(2) \qquad \Delta y = f(x + \Delta x) - f(x)$$
$$= f'(x)\,\Delta x + \eta\,\Delta x,$$

*where $\eta$ is a function of $\Delta x$ and*

$$\lim_{\Delta x \to 0} \eta = 0.$$

A simple consequence of the Fundamental Lemma concerns the continuity of $f$. The definition of continuity is as follows:

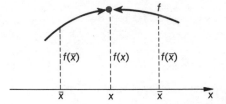

**DEFINITION**   A function $f$ is *continuous* at $x$ if $f$ is defined in an interval containing $x$, and $f(\bar{x})$ approaches $f(x)$ as $\bar{x}$ approaches $x$. In other words, if

$$\lim_{\bar{x} \to x} f(\bar{x}) = f(x).$$

Now, returning to the Fundamental Lemma and equation (2), if $\Delta x \to 0$, then $\bar{x} = x + \Delta x \to x$ and $f(\bar{x}) \to f(x)$. This proves the following corollary to the Fundamental Lemma.

**COROLLARY**   *If $f$ is differentiable at $x$ then $f$ is continuous at $x$.*

Thus, functions discontinuous at $x$ *cannot* have a derivative at $x$, and even continuous functions need not have a derivative at some particular number $x$, as the following graphs illustrate:

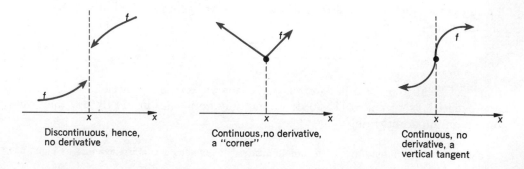

Discontinuous, hence, no derivative

Continuous, no derivative, a "corner"

Continuous, no derivative, a vertical tangent

There is an overabundance of symbols for derivatives. The most common ones are, if $y = f(x)$:

$$y', \quad \frac{dy}{dx}, \quad f'(x), \quad \frac{df}{dx}, \quad \frac{d}{dx}f(x), \quad Dy, \quad Df, \quad Df(x), \quad D_x f, \quad D_x y$$

with obvious modifications when other letters are used for the variables and functions.

When we wish to be sure we know where the derivative is to be evaluated we can write, for $f'(a)$,

$$\frac{dy}{dx}\bigg|_{x=a} \qquad \text{or} \qquad D_x y \bigg|_{x=a}.$$

The symbols "$d/dx$" and "$D_x$" are read as "*the derivative with respect to x.*" They have the advantage of indicating the independent variable. Thus,

$$\frac{d}{dx}(x^2 + 2ax + a^2) \qquad \text{or} \qquad D_x(x^2 + 2ax + a^2)$$

are quite clear, whereas the notation,

$$D(x^2 + 2ax + a^2)$$

may leave some doubt as to whether $x$ or $a$ represents the variable.

Leibniz' notation $dy/dx$ is particularly suggestive in applications because letters are often used to indicate the physical meaning:

$$\frac{dV}{dt} \qquad\qquad \frac{dp}{dv} \qquad\qquad \frac{dE}{dT}$$

derivative of            derivative of            derivative of
voltage with            pressure with            energy with
respect to time       respect to volume     respect to temperature

### Problems

In Problems 1 to 8 compute the indicated derivative by applying the definition of derivative.

1.  $\dfrac{d}{dt}(2t^2 - t + 7)$                                                    $[4t - 1]$

2.  $\dfrac{d}{du}\left(2u - \dfrac{1}{u}\right)$                                    $\left[2 + \dfrac{1}{u^2}\right]$

3.  $s = 16t^2 + 8t + 10; \quad ds/dt$                                       $[32t + 8]$

*optimal*

4. $\checkmark$ $D_x\left(ax + \dfrac{1}{ax}\right)$              $\left[a - \dfrac{1}{ax^2}\right]$

5. $\checkmark$ $\dfrac{d}{d\theta}(\sqrt{2\theta + 3})$              $\left[\dfrac{1}{\sqrt{2\theta + 3}}\right]$

6. $\checkmark$ $\dfrac{d}{du} u^3$                            $[3u^2]$

7. $\checkmark$ $\dfrac{d}{du} u^4$                            $[4u^3]$

*8. $\checkmark$ $\dfrac{d}{du} u^n$;   $n$ a positive integer              $[nu^{n-1}]$

9.   Do problems 1, 2, and 3 mentally.

In Problems 10 to 14 calculate the function $\eta$ as given by Equation (2). Verify that $\lim_{\Delta x \to 0} \eta = 0$.

10.   $f(x) = x^2 + 5$                            $[\eta = \Delta x]$

11.   $f(x) = x^3$                            $[\eta = 3x\,\Delta x + \Delta x^2]$

12.   $f(x) = x - (1/x)$                            $[\eta = -\Delta x/x^2\,(x + \Delta x)]$

13.   $S = \dfrac{1}{2t + 1}$              $\left[\eta = \dfrac{4\,\Delta t}{(2t + 1)^2\,(2\,(t + \Delta t) + 1)}\right]$

*14.   $u = \sqrt{2\theta}$              $\left[\eta = \dfrac{-\Delta\theta}{\sqrt{2\theta}\,(\sqrt{(\theta + \Delta\theta)} + \sqrt{\theta})^2}\right]$

## 4   *Rates; Velocity; Rate of Change*

Thus far the derivative has been interpreted as the slope of the tangent line. There is a second basic interpretation as a velocity, or rate of change.

If a point moves on a line, so that its coordinate $s$ is a function of time $t$, $s = f(t)$, then the *average velocity** in any time interval $[t, t + \Delta t]$ is

$$\text{average velocity} = v_{\text{av}} = \frac{\Delta s}{\Delta t} = \frac{f(t + \Delta t) - f(t)}{\Delta t}.$$

* Velocity can be positive, negative, or zero. Positive velocity means motion in the direction of increasing $s$, etc.

Our intuition views instantaneous velocity as being approximated by the average velocity over a small time interval, the approximation improving with smaller intervals. But this is the same as saying that the instantaneous velocity is the limit of the average velocity as $\Delta t$ approaches zero.

**DEFINITION**  The (*instantaneous\**) *velocity* at $t$ is defined to be

$$v = \frac{ds}{dt} = f'(t) = \lim_{\Delta t \to 0} \frac{f(t + \Delta t) - f(t)}{\Delta t}.$$

*Example 1*  A ball is shot vertically upward with an initial velocity of 64 feet per second (fps). It is shown in physics that, neglecting air resistance, the distance $s$ reached at time $t$ seconds later is

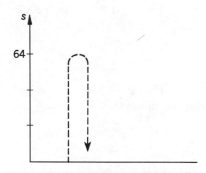

$$s = f(t) = 64t - 16t^2.$$

Then

$$\Delta s = \big[64(t + \Delta t) - 16(t + \Delta t)^2\big]$$
$$- \big[64t - 16t^2\big]$$
$$= 64\,\Delta t - 32t\,\Delta t - 16\,\Delta t^2.$$

Let us compute some average velocities from $t = 0$ to $t = \Delta t$. The table shows the result. Clearly $v_{av}$ is approaching $v = 64$ fps.

| $\Delta t$ | 2 | 1 | 0.5 | 0.25 | 0.1 | 0.01 | 0.001 |
|---|---|---|---|---|---|---|---|
| $\Delta s$ | 64 | 48 | 28 | 15 | 6.24 | 0.6384 | 0.063984 |
| $v_{av}$ | 32 | 48 | 56 | 60 | 62.4 | 63.84 | 63.984 |

In any time interval, $[t, t + \Delta t]$

$$v_{av} = \frac{\Delta s}{\Delta t} = 64 - 32t - 16\,\Delta t,$$

and

(1)  $$v = \frac{ds}{dt} = 64 - 32t.$$

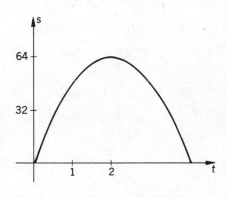

A graph of $s$ versus $t$ shows the motion. Clearly $v = 0$ at $t = 2$, both from the graph and from Equation (1). Before $t = 2$, $v$ is positive and the ball is rising. After $t = 2$, $v$ is negative and the ball is falling.

---

\* Usually one omits the adjective instantaneous. The word velocity alone always means the derivative.

This example illustrates an important connection between the behavior of a function and the sign of its derivative. First a definition:

**DEFINITION**   A function $f$ is said to be *increasing* on an interval $I$ if, for any two numbers $t_1$ and $t_2$ in $I$, $t_1 < t_2$ implies that $f(t_1) < f(t_2)$. A function $f$ is *decreasing* on $I$ if, for any two numbers $t_1$ and $t_2$ in $I$, $t_1 < t_2$ implies that $f(t_1) > f(t_2)$.

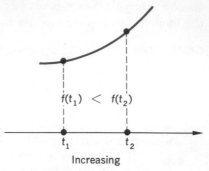

Increasing

The connection between function behavior and the derivative is given by the following theorem.

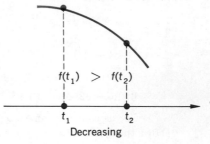

Decreasing

**THEOREM**   *If $f$ is differentiable in an interval $I$, and $f'(x) > 0$ on $I$, then $f$ is increasing on $I$. If $f'(x) < 0$ on $I$, then $f$ is decreasing on $I$.*

This theorem will be proved in Chapter 4. Meanwhile we will use it in problems.

Even when the independent variable is not a measure of time it is natural to speak of rate of change:

| *Average rate of change* of $f$ in the interval $[x, x + \Delta x]$ $= \dfrac{f(x + \Delta x) - f(x)}{\Delta x}$ | *Instantaneous rate of change* of $f$ at $x$ $= f'(x) = \lim_{\Delta x \to 0} \dfrac{f(x + \Delta x) - f(x)}{\Delta x}$ |
|---|---|

The velocity function $v = f'(t)$ also has a derivative which is called the *acceleration*:

$$\text{acceleration} = a = \frac{dv}{dt}.$$

In Example 1 the acceleration is $-32$ fps$^2$ $= -32$ feet per second per second. This is the constant *acceleration of gravity*.

*Problems*

1.  A point moves on a coordinate line so that

$$s = 4t^2 - 16t + 12$$

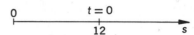

where $t$ is time in seconds.

(a)  What is the average velocity in the time interval $[1, 3]$? in the time interval $[1, 2]$? in the time interval $[1, 1 + \Delta t]$?
(b)  What is the velocity at $t = 1$?
(c)  Describe the motion of the particle. Draw a graph of $s$ versus $t$.

[(a) 0; $-4$; $-8 + 4\,\Delta t$.   (b) $-8$.
(c) $s$ has a minimum equal to $-4$ at $t = 2$; then $s$ increases.]

2.  If $F(x) = x^3 + 3x - 4$, what is the average rate of change of $F$ in the interval $[x, x + \Delta x]$? What is the instantaneous rate of change of $F$?

$$[3x^2 + 3x\,\Delta x + \Delta x^2 + 3; \quad 3x^2 + 3]$$

3.  A body falls from rest a distance $s = 16t^2$ where $s$ is in feet and $t$ is in seconds.

(a)  What is the average velocity from $t = 0$ to $t = 4$? from $t = 0$ to $t = 2$? from $t = 0$ to $t = 0.1$?
(b)  What is the average velocity from $t = 2$ to $t = 4$? from $t = 2$ to $t = 3$? from $t = 2$ to $t = 2.1$?
(c)  What is the instantaneous velocity at time $t$?
(d)  What is the acceleration?

4.  Apply the definition of derivative to show that

$$\frac{d}{dt}\,(at^3 + bt^2 + ct + d) = 3at^2 + 2bt + c.$$

In Problems 5 to 10 a particle moves on a line according to the given equation. Use the result of Problem 4 to find position, velocity and acceleration at the time indicated.

5.  $s = 2t^3 - 3t^2$    ; $t = 3$            $[s = 27; \quad v = 36; \quad a = 30]$

6.  $s = 2t - t^2$    ; $t = 1$            $[s = 1; \quad v = 0; \quad a = -2]$

7.  $s = 4 - 5t$    ; $t = 2$            $[s = -6; \quad v = -5; \quad a = 0]$

8.  $s = 2t^2 - 2t + 2$;  $t = 0$            $[s = 2; \quad v = -2; \quad a = 4]$

9.  $s = t^3 - 3t$    ; $t = 1$            $[s = -2; \quad v = 0; \quad a = 6]$

10. $s = t^2 - 24t$   ; $t = 0$            $[s = 0; \quad v = -24; \quad a = 2]$

In Problems 11 to 16 a particle moves on a line according to the given equation. For each problem answer these questions: (a) When is $s$ increasing? (b) When is the velocity 0? (c) What is the acceleration at any time? (d) Describe the motion.

11.  $s = t^2 - 2t - 12$            [(a) $t > 1$; (b) $t = 1$; (c) 2]

12.  $s = 8t - 16$            [(a) always; (b) never; (c) 0]

13.   $s = 12 - 3t$                     [(a) never; (b) never; (c) 0]

14.   $s = 16 + 20t - 16t^2$            [(a) $t < \frac{5}{8}$; (b) $t = \frac{5}{8}$; (c) $-32$]

15.   $s = 120 - 60t - t^3$             [(a) never; (b) never; (c) $-6t$]

16.   $s = 120 - 60t + t^3$             [(a) $t^2 > 20$; (b) $t^2 = 20$; (c) $6t$]

✓ 17.   The volume $V$ in cubic feet of water in a tank follows the equation $V = 3(t - 12)^2$ for $t$ in seconds with $t \geq 0$.

(a)   Does the volume decrease or increase?

(b)   What is the rate of change of $V$ at $t = 0$? $t = 5$? $t = 12$?

[(a) Decreases until $t = 12$; (b) $-72$;   $-42$;   0]

18.   A point moves so as to be at the tabulated positions at the various times, in seconds.

| $s$ | 4 | 19 | 24 | 19 | 4 | $-21$ |
|---|---|---|---|---|---|---|
| $t$ | 0 | 1 | 2 | 3 | 4 | 5 |

(a)   Plot $s$ against $t$ and connect the points by a smooth curve.

(b)   Estimate the velocity at $t = 1$ and $t = 2$.

(c)   Sketch a graph of $v$.

(d)   Sketch a graph of the acceleration.

## 5   Antiderivatives

When we differentiate a function $f$, we operate on the function to obtain a new function $f' = Df$. The derivative operator $D$ converts a function into its derivative. Now we ask whether, for a given function $g$, there is a function $G$ whose derivative is $g$.

$$f \xrightarrow{\quad D \quad} f'$$

$$?? \xrightarrow{\quad D \quad} g$$

$$DG = G' = g?$$

If there is such a function $G$, then $G$ is called an *antiderivative** of $g$.

**DEFINITION**     If $G$ is a function whose derivative is $g$, i.e., $G' = g$, then $G$ is called an *antiderivative* of $g$.

*Example 1*     For simple functions it is often easy to *guess* an antiderivative.

---

* Antiderivative is an appropriate name because we reverse, or undo, the process of differentiation. The process of finding antiderivatives will be taken up in detail in Chapters 10 and 11.

If $g(x) = 1$ then $G(x) = x$, or $G(x) = x + 1$, or $G(x) = x - 5$, or $G(x) = x + c$, where $c$ is any constant.

If $g(x) = x$, then $G(x) = \frac{1}{2}x^2$, or $G(x) = \frac{1}{2}x^2 + \frac{2}{3}$, or $G(x) = \frac{1}{2}x^2 + c$, where $c$ is any constant.

$$\frac{d}{dx}(x + c) = 1$$

$$\frac{d}{dx}\left(\tfrac{1}{2}x^2 + c\right) = x$$

Now we ask if we have found all the antiderivatives in Example 1. We infer that the answer is "yes" from the following theorem.

**THEOREM** *If, in an interval,*

$$\frac{d}{dx}G(x) = \frac{d}{dx}F(x),$$

*then, in that interval,*

$$G(x) = F(x) + \text{constant}.$$

A proof will be delayed until Chapter 4. It is clear from the theorem that, if we can guess *one* antiderivative, then we can get them *all* simply by adding a constant; that is, by adding an arbitrary constant function.

*Example 2* If we recall (Problem 4 of Section 4) that

$$\frac{d}{dx}(ax^3 + bx^2 + cx + d) = 3ax^2 + 2bx + c,$$

then certain antiderivatives can be obtained at once.

If $f'(x) = 2x$, then $f(x) = x^2 + c$: if $f'(x) = 3x^2 - x - 1$, then $f(x) = x^3 - \frac{1}{2}x^2 - x + c$, where $c$ is an arbitrary constant.

*Remark.* When we find antiderivatives, we are solving very simple *differential equations*. Thus we have for example:

**Differential Equation** $\quad \dfrac{dy}{dx} = -x^2 - 5x - \sqrt{2}$

**Solution** $\quad y = -\tfrac{1}{3}x^3 - \tfrac{5}{2}x^2 - \sqrt{2}x + c$

*Example 3* An important example occurs in physics. Near the surface of the earth* mass particles are accelerated downward by the *constant acceleration* of gravity $= g =$ approximately 32 ft/sec$^2$.

Then if $a = dv/dt = -g$ (negative because $s$ is measured positively upward), we have

$$v = ds/dt = -gt + c_1.$$

---

* As one recedes from the earth the attraction of gravity decreases, but near the earth—say, below 5000 ft—the variation in gravity is slight and would require a sensitive spring scale to detect it. On the moon $g$ is much less, about $\frac{1}{6}$ of its earth value.

We can find the constant $c_1$ if we know $v$ at some instant. Thus, if $v = v_0$ at $t = 0$, then $c_1 = v_0$ and

$$v = ds/dt = -gt + v_0.$$

Now a second antiderivative gives $s$

$$s = -\tfrac{1}{2}gt^2 + v_0t + c_2,$$

and the constant $c_2$ can be found if we know $s$ at some instant. If $s = s_0$ when $t = 0$, then

$$s = -\tfrac{1}{2}gt^2 + v_0t + s_0.$$

The path of the particle is completely determined by:

(1) the constant acceleration of gravity, $g$;
(2) the initial conditions $v = v_0$ and $s = s_0$ at $t = 0$.

***Example 4***     We can show graphically that some functions have antiderivatives. The figures below suggest how the graph of $f$ can be drawn from the graph of $f'$ and the knowledge of $f(x_0)$ for some number $x_0$.

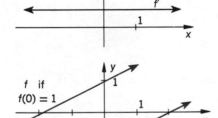
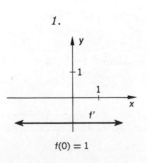
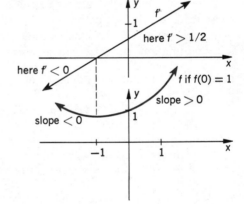

***Problems***

In Problems 1 to 6 make reasonable copies of the graphs of $f'$ on squared paper. Then draw the graph of $f$ which goes through the assigned point.

1.

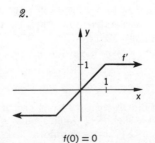

$f(0) = 1$

2.

$f(0) = 0$

3.

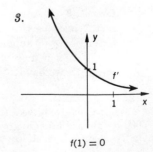

$f(1) = 0$

4.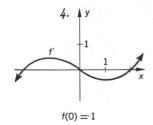

$f(0) = 1$

5.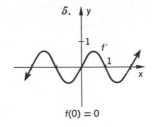

$f(0) = 0$

6.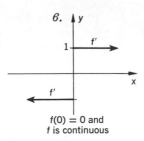

$f(0) = 0$ and
f is continuous

7.  If $f(2) = 2$ and $f'(x) = x - 2$, find $f(x)$.  $\left[\frac{1}{2}x^2 - 2x + 4\right]$

8.  If $F(-1) = 0$ and $F'(\theta) = \theta - \theta^2$, find $F(\theta)$.  $\left[\theta^2/2 - \theta^3/3 - 5/6\right]$

9.  If $\varphi(4) = -2$ and $\varphi'(t) = 1 + 2t^2$, find $\varphi(t)$.  $\left[t + 2t^3/3 - 146/3\right]$

10.  If $dy/dx = 2 - x$ and $y = 0$ at $x = 1$, find $y$.  $\left[2x - x^2/2 - 3/2\right]$

11.  If $d\theta/dt = -t^2$ and $\theta = 1$ at $t = 1$, find $\theta$.  $\left[(4 - t^3)/3\right]$

12.  A point moves on a line with constant acceleration $a = -4$. If the velocity is 2 at $t = 0$, find $v$ as a function of $t$.  $[v = -4t + 2]$

13.  A ball is dropped from the top of a building 120 ft high. Find a formula for the height $h$ above the ground in terms of the number of seconds of fall.

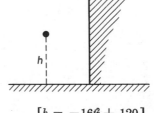

$[h = -16t^2 + 120]$

14.  Find $\dfrac{d}{dx}\left(\dfrac{x}{x-1}\right)$  and  $\dfrac{d}{dx}\left(\dfrac{1}{x-1}\right)$. What can you conclude?

$$\left[\frac{x}{x-1} - \frac{1}{x-1} = \text{constant}\right]$$

15.  Sketch the graph of $f(x) = \dfrac{x}{x-1} - \dfrac{1}{x-1}$.

16.  If $f'(x) = -1/x^2$ and $f(1) = 1$, find $f(x)$. Is $f$ uniquely determined?

$$\left[\begin{array}{l} f(x) = 1/x. \\ \text{For } x > 0 \ f \text{ is unique.} \\ \text{For } x < 0 \text{ an arbitrary} \\ \text{constant can be added.} \end{array}\right]$$

17.   An ideal spring, when stretched $(x > 0)$ or compressed $(x < 0)$ from its unstretched condition, exerts a restoring force $F$ proportional to the displacement: $F = kx$ where $k$ is a constant depending on the spring.

It is shown in physics that, if $W(x)$ is the work done in stretching the spring an amount $x$, then $dW/dx = F$. Obtain a formula for $W(x)$.

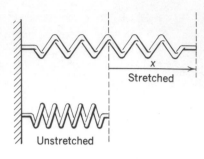

Stretched

Unstretched

$$[W(x) = \tfrac{1}{2}kx^2]$$

18.   The ends of the weightless lever pictured are held in place. The fulcrum $F$ starts at the right end and moves to the left with an acceleration $a = 2$ ft/sec². Sand starts to fall into the weightless container at the rate of $\frac{1}{3}$ lb/sec when the fulcrum starts moving. At what time will the lever be in balance if the ends are released?

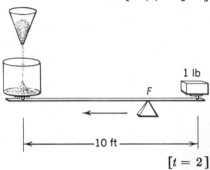

1 lb

$F$

10 ft

$$[t = 2]$$

## Review Problems

**In Problems 1 to 6 give the definition of the concept.**

1.   Tangent line to the graph of a function.

2.   The derivative of a function.

3.   Average velocity, (instantaneous) velocity.

4.   Function increasing on an interval.

5.   Average rate of change, (instantaneous) rate of change.

6.   Antiderivative of a function.

7.   If $f$ is differentiable at $x$ what can you say about the difference $(f(x + \Delta x) - f(x)) - f'(x)\,\Delta x$?

8.   Apply the definition of derivative to find the slope of the tangent line to $y = 1/(x + 1)$ at the point where $x = 1$.

9.   Find the equation of the tangent line to the graph of $y = \sqrt{x - 4}$ at the point where $x = 5$.

In Problems 10, 11, and 12 sketch a reasonable copy of the graphs on squared paper and then draw an approximation to the graphs of the derived functions.

10.

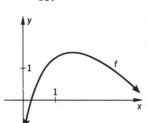

11.

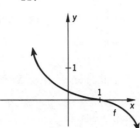

12.

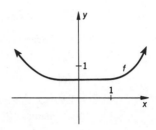

In Problems 13, 14, and 15 calculate the derivative by using the definition of derivative. Also calculate the function $\eta$ of the Fundamental Lemma.

13.  $\dfrac{d}{dx}\left(x^2 + \dfrac{1}{x}\right)$
$$\left[\; \eta = \Delta x + \frac{\Delta x}{x^2\,(x + \Delta x)} \;\right]$$

14.  $\dfrac{d}{dt}\sqrt{4t - 5}$
$$\left[\; \eta = \frac{-8\,\Delta t}{\sqrt{4t - 5}\,(\sqrt{4(t + \Delta t) - 5} + \sqrt{4t - 5})^2} \;\right]$$

15.  $\dfrac{d}{dt}\dfrac{1}{\sqrt{t}}$
$$\left[\; \eta = \frac{\Delta t\,(\sqrt{t + \Delta t} + 2\sqrt{t})}{2t\sqrt{t}\,\sqrt{t + \Delta t}\,(\sqrt{t + \Delta t} + \sqrt{t})^2} \;\right]$$

16.  A particle moves on a line according to the equation $s = 8t^2 - 32t - 8$.
   (a)   When is it moving in the direction of increasing $s$?
   (b)   What is its velocity at $t = 0$?          $[\,$(a) $t > 2$;   (b) $-32\,]$

17.  A ball is thrown upward with an initial velocity of 60 fps. How far above its point of release will it rise?          $[\,56\frac{1}{4}$ ft$\,]$

In Problems 18 and 19 graphs of $f'$ are given as well as a value of $f$ at some point. Copy the graph of $f'$ on squared paper and sketch the graph of $f$.

18.

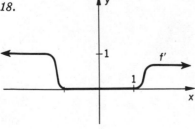

$f(0) = 1$

19.

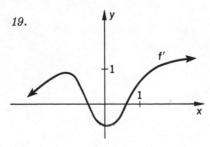

$f(0) = 0$

*Mathematical and Historical References*

1.  T. M. Apostal, *Calculus, Vol. I*. New York: Blaisdell Publishing Co., 1961, pp. 99–112 and 126–127.

2.  E. T. Bell, *Men of Mathematics*. New York: Simon and Schuster, 1937, pp. 56–130.

3.  C. B. Boyer, *The Concepts of the Calculus*. New York: Hafner Publishing Co., 1949, pp. 187–223.

4.  F. Cajori, *A History of Mathematics*. New York: The Macmillan Co., 1919.

5.  H. Eves, *An Introduction to the History of Mathematics*. New York: Holt, Rinehart and Winston, 1964, pp. 329–344.

6.  D. J. Struik, *A Concise History of Mathematics*. New York: Dover Publishing Co., 1948.

## CHAPTER 2

# THE TECHNIQUE OF DIFFERENTIATION

## 1  The Differentiation Formulas

In this chapter we derive formulas that enable one to find quickly the derivatives of simple functions *without having to apply the definition of derivative* directly. These formulas provide rules for differentiation and must be committed to memory as they are developed and used.

One of the accomplishments of Newton and Leibniz was the recognition that the problem of computing derivatives of elementary functions could be reduced to a small number of rules, or algorithms.* At the end of the chapter we shall look back on our accomplishment and say more precisely what we mean by "elementary functions."

In the meantime, here are the few formulas. *It is always assumed that the requisite derivatives exist.* In the formulas, $u$ and $v$ are functions of the variable $x$. The letters $a$ and $c$ represent constants, and $e$ is the base of natural logarithms, which will be defined later.

All of these formulas are rather easily proved by using the definition of derivative in Chapter 1. Most of the proofs are to be found in succeeding pages, but a few will be left as problems.

---

* An algorithm is a rule for calculation. The name derives from the 9th century Arab mathematician Al-Khowarizmi who wrote a book entitled *Hisâb al-jabr w'al-muqâbalah*. From the title comes our word "algebra."

I $\quad \dfrac{dc}{dx} = 0$

II $\quad \dfrac{d}{dx} x = 1 \qquad$ (The identity function has derivative 1.)

III $\quad \dfrac{d}{dx} cu = c \dfrac{du}{dx}$

IV $\quad \dfrac{d}{dx}(u \pm v) = \dfrac{du}{dx} \pm \dfrac{dv}{dx}$

V $\quad \dfrac{d}{dx} uv = u \dfrac{dv}{dx} + v \dfrac{du}{dx}$

VI $\quad \dfrac{d}{dx} u^n = nu^{n-1} \dfrac{du}{dx}, \quad$ and so in particular,

$\quad \dfrac{d}{dx} x^n = nx^{n-1}$

VII $\quad \dfrac{d}{dx} \dfrac{u}{v} = \dfrac{v \dfrac{du}{dx} - u \dfrac{dv}{dx}}{v^2}$

VIII $\quad \dfrac{d}{dx} \log_a u = \log_a e \dfrac{1}{u} \dfrac{du}{dx} = \dfrac{1}{\log a} \dfrac{1}{u} \dfrac{du}{dx}, \quad$ and so in particular,

$\quad \dfrac{d}{dx} \log u = \dfrac{1}{u} \dfrac{du}{dx}$

IX $\quad \dfrac{d}{dx} a^u = a^u \log a \dfrac{du}{dx}, \quad$ and so in particular,

$\quad \dfrac{d}{dx} e^u = e^u \dfrac{du}{dx}$

X $\quad \dfrac{d}{dx} \sin u = \cos u \dfrac{du}{dx}$

XI $\quad \dfrac{d}{dx} \cos u = -\sin u \dfrac{du}{dx}$

XII $\quad \dfrac{d}{dx} \tan u = \sec^2 u \, \dfrac{du}{dx}$

XIII $\quad \dfrac{d}{dx} \cot u = -\csc^2 u \, \dfrac{du}{dx}$

XIV $\quad \dfrac{d}{dx} \sec u = \sec u \tan u \, \dfrac{du}{dx}$

XV $\quad \dfrac{d}{dx} \csc u = -\csc u \cot u \, \dfrac{du}{dx}$

XVI $\quad \dfrac{d}{dx} \text{Arc} \sin u = \dfrac{1}{\sqrt{1 - u^2}} \dfrac{du}{dx}$

XVII $\quad \dfrac{d}{dx} \text{Arc} \cos u = \dfrac{-1}{\sqrt{1 - u^2}} \dfrac{du}{dx}$

XVIII $\quad \dfrac{d}{dx} \text{Arc} \tan u = \dfrac{1}{1 + u^2} \dfrac{du}{dx}$

XIX $\quad \dfrac{d}{dx} \text{Arc} \cot u = \dfrac{-1}{1 + u^2} \dfrac{du}{dx}$

XX $\quad$ **Chain Rule**    If $y$ is a function of the variable $u$, $y = y(u)$, and if $u$ is a function of the variable $x$, $u = u(x)$, then

$$\frac{dy}{dx} = \frac{dy}{du} \frac{du}{dx}$$

XXI $\quad$ If $y = y(x)$ has an inverse, $x = x(y)$, then

$$\frac{dy}{dx} = \frac{1}{dx/dy}$$

*Remark 1.* Besides the functions explicitly mentioned in this list, all sums, products, quotients, and compositions of these functions are functions whose derivatives can be found by applying the rules.

*Remark 2.* The problems of this chapter are selected for the development of *skill*. Accordingly, *all* students, from the best to the worst, should work at least two-thirds of the problems. By doing these problems one develops not only calculus skill but also *algebraic skill*, both of which are vital to further work. Students weak in algebra should do even more problems. The best students should do the special problems, whose numbers are marked with an asterisk.

**PROOF OF I**   Clearly, the graph of $y = c$ has slope 0. But also, formally,

$$\Delta y = 0$$

so

$$\frac{\Delta y}{\Delta x} = 0,$$

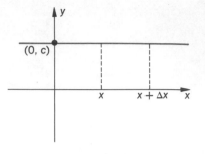

and so

$$\frac{dy}{dx} = \frac{dc}{dx} = \lim_{\Delta x \to 0} \frac{\Delta y}{\Delta x} = 0.$$

**PROOF OF II**   Clearly, the graph of $y = x$ has slope 1. Also, formally,

$$\Delta y = \Delta x$$

so

$$\frac{\Delta y}{\Delta x} = 1,$$

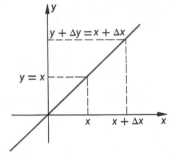

and so

$$\frac{dy}{dx} = \frac{d}{dx}x = \lim_{\Delta x \to 0} \frac{\Delta y}{\Delta x} = 1.$$

**PROOF OF III**   If $y(x) = cu(x)$, then

$$\Delta y = y(x + \Delta x) - y(x) = cu(x + \Delta x) - cu(x) = c\,\Delta u.$$

Then

$$\frac{\Delta y}{\Delta x} = c\,\frac{\Delta u}{\Delta x} \quad \text{and, letting } \Delta x \to 0, \quad \frac{dy}{dx} = c\,\frac{du}{dx}.$$

**PROOF OF IV**   If $y(x) = u(x) + v(x)$,

then $\Delta y = y(x + \Delta x) - y(x)$

$$= [u(x + \Delta x) + v(x + \Delta x)]$$
$$- [u(x) + v(x)]$$
$$= [u(x + \Delta x) - u(x)]$$
$$+ [v(x + \Delta x) - v(x)]$$
$$= \Delta u + \Delta v.$$

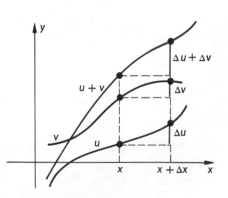

$$\frac{\Delta y}{\Delta x} = \frac{\Delta u}{\Delta x} + \frac{\Delta v}{\Delta x}.$$

Then, as $\Delta x \to 0$ the two terms on the right approach the derivatives of $u$ and $v$.

$$\frac{dy}{dx} = \frac{du}{dx} + \frac{dv}{dx}.$$

The proof for $y = u - v$ is analogous.

**PROOF OF VI (Special Case)**   This formula is valid for *any real number n.* Our proof here will be only for the case where $n$ is a positive integer—but we shall *use* the formula in its full generality. Proof for the general case comes in Section 4. If

$$y(x) = x^n, \text{ then}$$

$$\Delta y = (x + \Delta x)^n - x^n$$

$$= \left[ x^n + nx^{n-1}\Delta x + \frac{n(n-1)}{2} x^{n-2}\Delta x^2 + \cdots + \Delta x^n \right] - x^n,$$

by the binomial theorem. Therefore,

$$\Delta y = nx^{n-1}\Delta x + \frac{n(n-1)}{2} x^{n-2}\Delta x^2 + \cdots + \Delta x^n;$$

$$\frac{\Delta y}{\Delta x} = nx^{n-1} + \frac{n(n-1)}{2} x^{n-2}\Delta x + \cdots + \Delta x^{n-1}.$$

In this last expression all terms except the first have a factor $\Delta x$. Therefore, when $\Delta x \to 0$, each of these terms approaches 0 and so does their sum. Thus

$$\frac{d}{dx} x^n = \lim_{\Delta x \to 0} \frac{\Delta y}{\Delta x} = nx^{n-1}.$$

**Example 1**   $\dfrac{d}{dx} (3x^4) = 3 \dfrac{d}{dx} x^4$   (by III)

$$= 3 \cdot 4x^3 \quad \text{(by the special case of VI)}$$

$$= 12x^3.$$

**Example 2**   $\dfrac{d}{dx} (2x^{4/3} - 3x^{1/3}) = \dfrac{d}{dx} 2x^{4/3} - \dfrac{d}{dx} 3x^{1/3}$   (by IV)

$$= 2 \frac{d}{dx} x^{4/3} - 3 \frac{d}{dx} x^{1/3} \quad \text{(by III)}$$

$$= 2 \cdot \tfrac{4}{3} x^{1/3} - 3 \cdot \tfrac{1}{3} x^{-2/3} \quad \text{(by VI)}$$

$$= \tfrac{8}{3} x^{1/3} - x^{-2/3}.$$

**Problems**

In all problems the letters *a, b, c, d, n, m, α, β, γ,* represent constants. In Problems 1 to 31 you are to compute the indicated derivative.

1.   $y = 3x^5 - 2x^3$   ;   $Dy$   $[Dy = 3x^2(5x^2 - 2)]$

2.   $z = 3ax^2 + 2bx + c$   ;   $\dfrac{dz}{dx}$   $\left[ \dfrac{dz}{dx} = 6ax + 2b \right]$

3.   $y = (\sqrt{2})^3 - (\sqrt{2})^5$                    ;  $y'$                                    $[y' = 0]$

4.   $\dfrac{d}{dx}(x^{a+b})$                                           $\left[\dfrac{d}{dx}(x^{a+b}) = (a+b)x^{a+b-1}\right]$

5.   $f(y) = ay^n - by^m + c$          ;  $f'(y)$      $[f'(y) = nay^{n-1} - mby^{m-1}]$

6.   $\dfrac{d}{dx}7x^{-7}$                                             $\left[\dfrac{d}{dx}7x^{-7} = -49x^{-8}\right]$

7.   $f(x) = 8x^{5/2} - 4x^{3/2} - 2x^{1/2}$     ;  $f'(x)$   $[f'(x) = 20x^{3/2} - 6x^{1/2} - x^{-1/2}]$

8.   $y = \dfrac{ax+b}{c}$                         ;  $y'$                          $\left[y' = \dfrac{a}{c}\right]$

9.   $F(\theta) = 3\theta^{-2} - 2\theta^{-3}$             ;  $F'(\theta)$        $[F'(\theta) = 6\theta^{-4}(1-\theta)]$

10.  $f(\varphi) = \dfrac{1}{\varphi} - \dfrac{1}{\varphi^2}$                  ;  $f'(\varphi)$        $[f'(\varphi) = \varphi^{-2}(2\varphi^{-1} - 1)]$

11.  $y = 1 + x + \dfrac{x^2}{2!} + \dfrac{x^3}{3!} + \cdots + \dfrac{x^n}{n!}$ ;  $y'$    $\left[y' = 1 + x + \dfrac{x^2}{2!} + \right.$
$$\left. \cdots + \dfrac{x^{n-1}}{(n-1)!}\right]$$

12.  $\dfrac{d}{du}\left(\dfrac{u^a + u^b}{ab}\right)$                           $\left[\dfrac{d}{du}\left(\dfrac{u^a + u^b}{ab}\right) = \dfrac{u^{a-1}}{b} + \dfrac{u^{b-1}}{a}\right]$

13.  $y = a + \dfrac{b}{x}$                          ;  $y'$                        $\left[y' = -\dfrac{b}{x^2}\right]$

14.  $y = (\tfrac{7}{2}x^3 - 3x^2 + \tfrac{1}{3}x - 1) - (2x^4 + 2x^3 - \tfrac{1}{2}x^2 - \sqrt{2});$   $Dy$
$$[Dy = -8x^3 + \tfrac{9}{2}x^2 - 5x + \tfrac{1}{3}]$$

15.  $G(t) = \alpha + (\alpha + \beta)t + (\alpha + \beta + \gamma)t^2;$   $G'(2)$   $[G'(2) = 5\alpha + 5\beta + 4\gamma]$

16.  $H(x) = \dfrac{12}{x^{1/2}} + \dfrac{12}{x^{1/3}} - \dfrac{12}{x^{3/2}} - \dfrac{12}{x^{3/4}}$ ;  $H'(1)$                $[H'(1) = 17]$

17.  $\varphi(y) = 5y^{1/2} - \tfrac{1}{3}y^{3/2} - \tfrac{2}{3}y^{-1/2}$ ;  $\varphi'(2)$                $[\varphi'(2) = \tfrac{5}{6}\sqrt{2}]$

18.  $y(x) = \dfrac{ax^2 + bx + c}{x}$          ;  $y'(x)$                $\left[y'(x) = a - cx^{-2}\right]$

19. $z = \dfrac{(x+2)^2}{x}$     ; $z'$     $\left[ z' = \dfrac{x^2 - 4}{x^2} \right]$

20. $H(x) = \dfrac{(x-1)^3}{x^{1/2}}$     ; $H'(x)$     $[H'(x) = \tfrac{1}{2}x^{-3/2}(5x^3 - 9x^2 + 3x + 1)]$

21. $y = \tfrac{3}{5}x^{5/2} + 2x^{3/2} + 3x^{1/2}$     ; $y'$     $[y' = 3(x+1)^2/2\sqrt{x}]$

22. $F(t) = \dfrac{a^2 + 2abt + b^2t^2}{a^2 + 2ab + b^2}$     ; $F'(1)$     $\left[ F'(1) = \dfrac{2b}{a+b} \right]$

23. $\varphi(\theta) = \sqrt{3\theta^3} + \dfrac{3}{\theta^2}$     ; $\varphi'(\sqrt{3})$     $\left[ \varphi'(\sqrt{3}) = \dfrac{25}{\sqrt{3}} \right]$

24. $y = 2\sqrt{x}$     ; $Dy$     $[Dy = x^{-1/2}]$

25. $y = \sqrt{2x}$     ; $Dy$     $\left[ Dy = \dfrac{\sqrt{2}}{2} x^{-1/2} \right]$

26. $y = \dfrac{1}{\sqrt{2x}}$     ; $Dy$     $\left[ Dy = -\dfrac{\sqrt{2}}{4} x^{-3/2} \right]$

27. $y = \dfrac{1}{2\sqrt{x}}$     ; $Dy$     $[Dy = -\tfrac{1}{4}x^{-3/2}]$

28. $y = \dfrac{2}{\sqrt{x}}$     ; $Dy$     $[Dy = -x^{-3/2}]$

29. $y = \dfrac{x^2 - 1}{x + 1}$     ; $y'$     $[y' = 1]$

30. $y = \dfrac{a^2 + 2abx + b^2x^2}{a + bx}$     ; $y'$     $[y' = b]$

31. $y = \dfrac{x^3 - a^3}{x - a}$     ; $y'$     $[y' = 2x + a]$

32. Now re-do the preceding 31 problems mentally. In other words, by inspection of the function write down the final form for its derivative. You won't succeed completely, but it will improve your algebra and help you to see steps that can be left out.

33. State formulas I to VI in words.

In Problems 34 to 37, write out an equation of the tangent line to the graph of $y = f(x)$ at the indicated point.

34.   $y = a + \dfrac{b}{x}$ ;   $\left(a,\, a + \dfrac{b}{a}\right)$        $\left[\, y + \dfrac{b}{a^2} x - \dfrac{a^2 + 2b}{a} = 0 \,\right]$

35.   $y = \dfrac{ax + b}{c}$ ;   $\left(-\dfrac{b}{a},\, 0\right)$        $\left[\, y = \dfrac{ax + b}{c} \,\right]$

36.   $y = 3x^{-2} - 2x^{-3};\quad (2, \tfrac{1}{2})$        $[\,8y + 3x - 10 = 0\,]$

37.   $y = 3x^{-2} - 2x^{-3};\quad (\tfrac{1}{2}, -4)$        $[\,y - 48x + 28 = 0\,]$

38.   Find the point, in the first quadrant, on the parabola $y = x^2 + 4$, such that the tangent line at that point passes through the origin.

39.   Use the definition of derivative to prove the formula $D(au - bv) = aDu - bDv$.

40.   If $f(x) = \sum\limits_{k=0}^{30} (x^k - x^{k+1})$,   find $f'(x)$.        $[\,f'(x) = -31x^{30}\,]$

41.   It is true, (for the precise sense see Chapter 8), that

$$\sin x = x - \frac{x^3}{3!} + \frac{x^5}{5!} - \cdots + (-1)^n \frac{x^{2n+1}}{(2n+1)!} + \cdots;$$

$$\cos x = 1 - \frac{x^2}{2!} + \frac{x^4}{4!} - \cdots + (-1)^n \frac{x^{2n}}{(2n)!} + \cdots.$$

These are infinite series representations of $\sin x$ and $\cos x$.

Assuming that the derivatives of sin and cos can be obtained by differentiating the series term by term, find formulas for $D \sin x$ and $D \cos x$.

42.   Prove that if $f(x)$ is a polynomial then $f'(x)$ is also a polynomial. (See Appendix A for a definition of a polynomial.)

## 2   Derivatives of Products, Powers, and Quotients

A proof of formula V for the derivative of a product:

$$\frac{d}{dx} uv = u\frac{dv}{dx} + v\frac{du}{dx},$$

is left as an exercise for the reader. Observe that the formula says that the derivative of the product of two functions is *the first times the derivative of the second, plus the second times the derivative of the first*.

A proof for the general power formula, VI, when $n$ is a positive integer, is quite like the proof for the special case given in Section 1. In brief outline:

If $y = u^n$, then

$$\Delta y = y(x + \Delta x) - y(x) = [u(x + \Delta x)]^n - [u(x)]^n$$

$$= (u + \Delta u)^n - u^n \quad \text{(dropping the $x$ for brevity)}.$$

We now expand $(u + \Delta u)^n$ by the binomial theorem and obtain

$$\frac{\Delta y}{\Delta x} = nu^{n-1}\frac{\Delta u}{\Delta x} + \left[\frac{n(n-1)}{2}u^{n-2}\Delta u + \cdots + \Delta u^{n-1}\right]\frac{\Delta u}{\Delta x}.$$

As $\Delta x \to 0$ the term in square brackets approaches 0 and we obtain in the limit

$$\frac{dy}{dx} = nu^{n-1}\frac{du}{dx} + 0\frac{du}{dx}$$

$$= nu^{n-1}\frac{du}{dx}.$$

Observe that this formula for the derivative of a power of a *function* has the additional factor $du/dx$. The special case: $(d/dx)x^n$ comes from the general formula because $(d/dx)x = 1$.

A proof of the quotient formula, VII, will provide a model for the reader to follow in proving the product formula, V.

If $y = u/v$, then

$$\Delta y = y(x + \Delta x) - y(x) = \frac{u(x + \Delta x)}{v(x + \Delta x)} - \frac{u(x)}{v(x)}$$

$$= \frac{u + \Delta u}{v + \Delta v} - \frac{u}{v} = \frac{v\,\Delta u - u\,\Delta v}{v(v + \Delta v)};$$

$$\frac{\Delta y}{\Delta x} = \frac{v\dfrac{\Delta u}{\Delta x} - u\dfrac{\Delta v}{\Delta x}}{(v + \Delta v)v}.$$

Now as $\Delta x \to 0$, so does $\Delta v$; hence, the denominator approaches $v^2$. The two terms in the numerator approach $v\,du/dx$ and $u\,dv/dx$. Thus, in the limit we have the desired result:

$$\frac{d}{dx}\frac{u}{v} = \frac{v\dfrac{du}{dx} - u\dfrac{dv}{dx}}{v^2}.$$

In words the formula says that *the derivative of a quotient is equal to the denominator times the derivative of the numerator, minus the numerator times the derivative of the denominator, all divided by the denominator squared.*

**Example 1**   $\dfrac{d}{dx}(2x^2 - 7)^{3/2} = \frac{3}{2}(2x^2 - 7)^{1/2}\dfrac{d}{dx}(2x^2 - 7)$   (by VI)

$$= \tfrac{3}{2}(2x^2 - 7)^{1/2}(4x)$$

$$= 6x(2x^2 - 7)^{1/2}.$$

**Example 2**   If $y = (x^2 + 2)\sqrt{1 + x^2}$,   then

$$Dy = (x^2 + 2)\,D(1 + x^2)^{1/2} + (1 + x^2)^{1/2}\,D(x^2 + 2) \qquad \text{(by V)}$$

$$= (x^2 + 2)(\tfrac{1}{2})(1 + x^2)^{-1/2}\,D(1 + x^2) + (1 + x^2)^{1/2}(2x) \quad \text{(by VI)}$$

$$= (x^2 + 2)(\tfrac{1}{2})(1 + x^2)^{-1/2}(2x) + (1 + x^2)^{1/2}(2x)$$

$$= \frac{x(x^2 + 2)}{(1 + x^2)^{1/2}} + 2x(1 + x^2)^{1/2}$$

$$= \frac{3x^3 + 4x}{\sqrt{1 + x^2}}.$$

**Example 3**

$$\frac{d}{dx}\frac{x^2 + 2}{\sqrt{1 + x^2}} = \frac{(1 + x^2)^{1/2}\dfrac{d}{dx}(x^2 + 2) - (x^2 + 2)\dfrac{d}{dx}(1 + x^2)^{1/2}}{(1 + x^2)} \qquad \text{(by VII)}$$

$$= \frac{(1 + x^2)^{1/2}(2x) - (x^2 + 2)(\tfrac{1}{2})(1 + x^2)^{-1/2}(2x)}{(1 + x^2)} \qquad \text{(by VI)}$$

$$= \frac{x^3}{(1 + x^2)^{3/2}}.$$

*Remark.* One could also have treated this function as a product, rather than as a quotient and calculated $(d/dx)[(x^2 + 2)(1 + x^2)^{-1/2}]$.

### Problems

**In Problems 1 to 56 compute the indicated derivative. Part of the game is to get the derivative in the form specified.**

1.   $y = (a + bx^2)^3$ $\qquad\qquad$ ; $y'$ $\qquad\qquad\qquad$ $[y' = 6bx(a + bx^2)^2]$

2.   $y = (3 + x^3)(4 + x^2)$ $\qquad$ ; $y'$ $\qquad\qquad\qquad$ $[y' = 5x^4 + 12x^2 + 6x]$

3.   $f(\theta) = (a + \theta)^{1/2}(a - \theta)$ $\quad$ ; $f'(\theta)$ $\qquad\qquad$ $\left[f'(\theta) = \dfrac{-a - 3\theta}{2\sqrt{a + \theta}}\right]$

4.   $F(\theta) = (1 + a\theta)^n(1 + b\theta)^m$ $\quad$ ; $F'(\theta)$

$$[F'(\theta) = (1 + a\theta)^{n-1}(1 + b\theta)^{m-1}(na + mb + ab(n + m)\theta)]$$

5. $z = (a - x)(a^2 + ax + x^2)$ ; $\dfrac{dz}{dx}$ $\left[ \dfrac{dz}{dx} = -3x^2 \right]$

6. $W = (a + bx)(a^4 - a^3bx + a^2b^2x^2 - ab^3x^3 + b^4x^4)$; $W'$ $[W' = 5b^5x^4]$

7. $f(x) = \dfrac{a - x}{x}$ ; $f'(x)$ $\left[ f'(x) = -\dfrac{a}{x^2} \right]$

8. $f(x) = \dfrac{x}{a - x}$ ; $f'(x)$ $\left[ f'(x) = \dfrac{a}{(a - x)^2} \right]$

9. $y = \dfrac{x - a}{x + a}$ ; $Dy$ $\left[ Dy = \dfrac{2a}{(x + a)^2} \right]$

10. $y = \dfrac{x + a}{x - a}$ ; $Dy$ $\left[ Dy = \dfrac{-2a}{(x - a)^2} \right]$

11. $F(t) = \dfrac{1 + t}{1 + t^2}$ ; $F'(t)$ $\left[ F'(t) = \dfrac{1 - 2t - t^2}{(1 + t^2)^2} \right]$

12. $F(t) = \dfrac{t^2 + 2t + 1}{(t + 1)^2}$ ; $F'(t)$ $[F'(t) = 0]$

13. $T = (t^3 - 3t + 3)^3$ ; $\dfrac{dT}{dt}$ $\left[ \dfrac{dT}{dt} = 9(t^2 - 1)(t^3 - 3t + 3)^2 \right]$

14. $y = (3 + \frac{3}{2}x - \frac{1}{4}x^2)^{1/2}$ ; $y'$ $[y' = (3 - x)/2\sqrt{12 + 6x - x^2}]$

15. $\varphi = \dfrac{2\theta + 3}{3\theta + 2}$ ; $\varphi'$ $\left[ \varphi' = \dfrac{-5}{(3\theta + 2)^2} \right]$

16. $y = x\sqrt{a^2 - x^2}$ ; $y'$ $[y' = (a^2 - 2x^2)/\sqrt{a^2 - x^2}]$

17. $y = \dfrac{x^3}{(x^2 - 1)^{3/2}}$ $x^3(x^2 - 1)^{-\frac{3}{2}}$ ; $y'$ $\left[ y' = \dfrac{-3x^2}{(x^2 - 1)^{5/2}} \right]$

18. $y = \dfrac{1}{(a + x)^m (b + x)^n}$ ; $y'$ $\left[ y' = -\dfrac{(mb + na) + (m + n)x}{(a + x)^{m+1}(b + x)^{n+1}} \right]$

19. $f(x) = \dfrac{x}{1 - x} - \dfrac{1}{1 - x}$ ; $f'(x)$ $[f'(x) = 0]$

20. $f(x) = \dfrac{x}{(a^2 - x^2)^{3/2}}$ ; $f'(x)$ $\left[ f'(x) = \dfrac{a^2 + 2x^2}{(a^2 - x^2)^{5/2}} \right]$

21. $f(x) = (x + 1)\sqrt{x^2 - 2x + 2}$ ; $Df(x)$ $\left[ Df(x) = \dfrac{2x^2 - 2x + 1}{\sqrt{x^2 - 2x + 2}} \right]$

22. $\dfrac{d}{dx}(x^3 + 3x)^{5/3}$ $\left[ \dfrac{d}{dx}(x^3 + 3x)^{5/3} = 5(x^2 + 1)(x^3 + 3x)^{2/3} \right]$

23. $\varphi = (2x^3 + x^2 - 4x - 1)^4$ ; $\varphi'$

$[\varphi' = 8(3x^2 + x - 2)(2x^3 + x^2 - 4x - 1)^3]$

24. $H(t) = t\sqrt{9t^2 - 1}$ ; $H'(t)$ $\left[ H'(t) = \dfrac{18t^2 - 1}{\sqrt{9t^2 - 1}} \right]$

25. $F(x) = (a^{2/3} - x^{2/3})^{3/2}$ ; $F'(x)$ $[F'(x) = -x^{-1/3}(a^{2/3} - x^{2/3})^{1/2}]$

26. $\varphi(t) = \dfrac{2t(t - 2)^3 - 3t^2(t - 2)^2}{(t - 2)^6}$ ; $\varphi'(t)$ $\left[ \varphi'(t) = \dfrac{2(t^2 + 8t + 4)}{(t - 2)^5} \right]$

27. $f(x) = \dfrac{\sqrt{a^2 - x^2}}{x}$ ; $f'(x)$ $\left[ f'(x) = \dfrac{-a^2}{x^2\sqrt{a^2 - x^2}} \right]$

28. $y = \sqrt{1 + 2\sqrt{x}}$ ; $y'$ $[y' = 1/2\sqrt{x + 2x\sqrt{x}}]$

29. $\psi(\theta) = \dfrac{(\theta + 3)^2}{\theta + 2}$ ; $\psi'(\theta)$ $\left[ \psi'(\theta) = \dfrac{\theta^2 + 4\theta + 3}{(\theta + 2)^2} \right]$

30. $\lambda(x) = (2 + x)\sqrt{2 - x}$ ; $D\lambda(x)$ $\left[ D\lambda(x) = \dfrac{2 - 3x}{2\sqrt{2 - x}} \right]$

31. $\tau(\theta) = (1 + \theta^2)(1 + \theta^3)$ ; $\tau'(\theta)$ $[\tau'(\theta) = \theta(5\theta^3 + 3\theta + 2)]$

32. $G(t) = (\sqrt{a} - \sqrt{t})^2$ ; $G'(t)$ $[G'(t) = 1 - \sqrt{a/t}]$

33. $y = \dfrac{1 + x^2}{1 - x^2}$ ; $y'$ $\left[ y' = \dfrac{4x}{(1 - x^2)^2} \right]$

34. $w = (ax + b)^{n+1}$ ; $w'$ $[w' = a(n + 1)(ax + b)^n]$

35. $y = \dfrac{1}{x(x + 1)}$ ; $y'$ $\left[ y' = -\dfrac{2x + 1}{x^2(x + 1)^2} \right]$

36. $y = \dfrac{1}{x} - \dfrac{1}{x+1} + c$   ; $y'$   $\left[\, y' = -\dfrac{2x+1}{x^2(x+1)^2} \,\right]$

37. $y = \dfrac{1+x+x^2}{x}$   ; $Dy$   $\left[\, Dy = 1 - \dfrac{1}{x^2} \,\right]$

38. $f(t) = \sqrt{\dfrac{1+t^2}{1-t^2}}$   ; $f'(t)$   $\left[\, f'(t) = \dfrac{2t}{(1+t^2)^{1/2}(1-t^2)^{3/2}} \,\right]$

39. $f(t) = \sqrt{\dfrac{1-t^2}{1+t^2}}$   ; $f'(t)$   $\left[\, f'(t) = \dfrac{-2t}{(1+t^2)^{3/2}(1-t^2)^{1/2}} \,\right]$

40. $y = \dfrac{x^2}{(1-x^2)^{1/2}}$   ; $y'$   $\left[\, y' = \dfrac{2x - x^3}{(1-x^2)^{3/2}} \,\right]$

41. $z = \left(x + \dfrac{1}{x}\right)^2$   ; $\dfrac{dz}{dx}$   $\left[\, \dfrac{dz}{dx} = \dfrac{2(x^4 - 1)}{x^3} \,\right]$

42. $w(x) = \dfrac{x^a}{(1-x)^b}$   ; $Dw(x)$   $\left[\, Dw(x) = \dfrac{ax^{a-1} + (b-a)x^a}{(1-x)^{b+1}} \,\right]$

43. $y = (4x^3 - 3x^2 - \pi)^{3/2}$   ; $y'$   $[\, y' = 9(2x^2 - x)(4x^3 - 3x^2 - \pi)^{1/2} \,]$

44. $f(r) = (1+r)^2(r-1)^{-1/2}$   ; $f'(r)$   $\left[\, f'(r) = \dfrac{(3r - 5)(r+1)}{2(r-1)^{3/2}} \,\right]$

45. $\varphi(x) = \sqrt{x-1}\,\sqrt{x+1}$   ; $\varphi'(x)$   $[\, \varphi'(x) = x/\sqrt{x^2 - 1} \,]$

46. $y = (x-1)(x-2)(x-3)$   ; $y'$   $[\, y' = 3x^2 - 12x + 11 \,]$

47. $y = (x-1)^2(x-2)(x-3)$   ; $y'$   $[\, y' = (x-1)(4x^2 - 17x + 17) \,]$

48. $y = (x-1)^3(x-2)(x-3)$   ; $y'$   $[\, y' = (x-1)^2(5x^2 - 22x + 23) \,]$

49. $y = (x-1)^3(x-2)^2(x-3)$   ; $y'$

$[\, y' = (x-1)^2(x-2)(6x^2 - 26x + 26) \,]$

50. $y = (x^3+1)(x^2+1)\sqrt{x+1}$ ; $y'$

$[\, y' = (11x^5 + 10x^4 + 7x^3 + 11x^2 + 4x + 1)/2\sqrt{x+1} \,]$

51. $y = \sqrt{\dfrac{1-x}{1+x}}$   ; $y'$   $\left[\, y' = \dfrac{-1}{(1-x)^{1/2}(1+x)^{3/2}} \,\right]$

52. $y = \dfrac{x^5 - 5x + 7}{125}$ $\qquad ; \ y'$ $\qquad \left[ y' = \dfrac{x^4 - 1}{25} \right]$

53. $y = \sqrt{2x} + \dfrac{1}{\sqrt{2x}}$ $\qquad ; \ y'$ $\qquad \left[ y' = \dfrac{2x - 1}{(2x)^{3/2}} \right]$

54. $y = \sqrt{2x} + \sqrt{\dfrac{2}{x}}$ $\qquad ; \ y'$ $\qquad \left[ y' = \dfrac{x - 1}{\sqrt{2}\, x^{3/2}} \right]$

55. $y = \sqrt{\dfrac{x}{2}} + \sqrt{\dfrac{2}{x}}$ $\qquad ; \ y'$ $\qquad \left[ y' = \dfrac{x - 2}{(2x)^{3/2}} \right]$

\* 56. $y = \dfrac{1}{x + \sqrt{1 + x^2}}$ $\qquad ; \ y'$ $\qquad \left[ y' = \dfrac{x}{\sqrt{1 + x^2}} - 1 \right]$

*Hint:* Rationalize the denominator.

57. Now go back to Problem 1 and see how many you can do entirely mentally. If you cannot complete one mentally, write down as few steps as possible.

58. Sketch the graph of $y = \dfrac{1}{x\,(x + 1)} - \dfrac{1}{x} + \dfrac{1}{x + 1}$.

(See Problems 35, 36, and the theorem on page 21.)

59. Use VII to obtain $\dfrac{d}{dx} \dfrac{u}{c}$. Is this the easy way?

60. Use the definition of derivative to derive Formula V.

61. Prove that $D\,(uvw) = uv\,Dw + vw\,Du + uw\,Dv$.

\*62. Prove, directly from the definition of the derivative, that VI holds when $n$ is a negative integer, i.e., $Du^{-k} = -ku^{-k-1}\,Du$, for $k$ a positive integer.

\*63. Find the derivatives of the following functions wherever the derivatives exist.

(a) $f(x) = |\,x - 1\,|$ $\qquad \left[ \begin{array}{l} f'(x) = 1 \ \text{if } x > 1; \ = -1 \ \text{if } x < 1; \\ f'(1) \ \text{does not exist.} \end{array} \right.$

(b) $F(x) = |\,x^2 - 9\,|$ $\left[ \begin{array}{l} F'(x) = 2x \ \text{if } |\,x\,| > 3; \ = -2x \ \text{if } |\,x\,| < 3 \\ F'(\pm 3) \ \text{does not exist.} \end{array} \right.$

(c) $g(x) = |\,x - 2\,| - |\,x\,|$ $\quad [\,g'(x) = 0 \ \text{if } x > 2; \ = -2 \ \text{if } 0 < x < 2;$
$g'(x) = 0 \ \text{if } x < 0; \ g'(x) \ \text{does not exist at } x = 0, \ x = 2.\,]$

*64.* Show that if $y = -x + c\sqrt{1 + x^2}$, then $(1 + x^2)\dfrac{dy}{dx} - xy + 1 = 0$, where $c$ is any constant.

## 3 Derivatives of Logarithmic Functions

We now enlarge the class of functions we are able to differentiate by considering exponential and logarithmic functions.

We first consider the logarithm function. A graph of $\log_a u$ for $a > 1$ is shown in the figure.

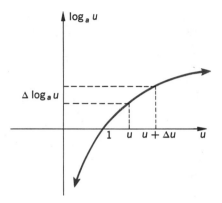

We apply the definition of the derivative to obtain formula VIII. If

$$y = \log_a u \qquad \text{and} \qquad u = u(x),$$

then

$$\Delta y = y(u + \Delta u) - y(u)$$

$$= \log_a (u + \Delta u) - \log_a u$$

$$= \frac{d}{du} (\log_a u)\, \Delta u + \eta\, \Delta u,$$

where $\eta \to 0$ as $\Delta u \to 0$ by the Fundamental Lemma of Chapter 1, assuming that $\log_a u$ is differentiable. Hence,

$$\frac{\Delta y}{\Delta x} = \frac{d}{du} (\log_a u)\frac{\Delta u}{\Delta x} + \eta\frac{\Delta u}{\Delta x}.$$

By letting $\Delta x \to 0$, we obtain

(1)
$$\frac{dy}{dx} = \frac{d}{du} (\log_a u)\frac{du}{dx} + 0\frac{du}{dx} = \frac{d}{du} (\log_a u)\frac{du}{dx},$$

because $\Delta u \to 0$ as $\Delta x \to 0$.

Thus, our problem has been reduced to finding $d/du\,(\log_a u)$. We now attack that problem:

$$\frac{d}{du} (\log_a u) = \lim_{\Delta u \to 0}\left[\frac{\log_a (u + \Delta u) - \log_a u}{\Delta u}\right].$$

This limit is awkward to evaluate as it stands. A trick will get us near the end:

(2)
$$\frac{\log_a (u + \Delta u) - \log_a u}{\Delta u} = \frac{1}{\Delta u}\log_a\left(\frac{u + \Delta u}{u}\right) = \frac{1}{\Delta u}\log_a\left(1 + \frac{\Delta u}{u}\right)$$

$$= \frac{1}{u}\frac{u}{\Delta u}\log_a\left(1 + \frac{\Delta u}{u}\right) = \frac{1}{u}\log_a\left(1 + \frac{\Delta u}{u}\right)^{u/\Delta u}.$$

Now as $\Delta u \to 0$, the first factor remains equal to $1/u$, and the problem becomes that of finding what happens to the second factor. In this second factor the number $\Delta u/u$ is small and approaches zero. Therefore, if we set $\Delta u/u = h$, we have to evaluate:

$$\lim_{h \to 0} \log_a (1 + h)^{1/h}.$$

To evaluate the above limit we first observe that, because the function $\log_a$ is continuous (see Appendix C, section 2), it is sufficient to evaluate the limit

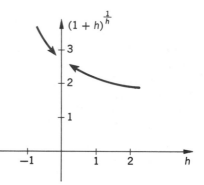

(3)    $$\lim_{h \to 0} (1 + h)^{1/h},$$

providing that it exists at all.

If we compute, and then graph, $(1 + h)^{1/h}$ for a few values of $h$, it becomes reasonable that the limit (3) exists and is a number between 2 and 3.

| $h$ | 1 | 1/2 | 1/4 | 1/16 | 1/128 | $-1/2$ | $-1/4$ | $-1/128$ |
|---|---|---|---|---|---|---|---|---|
| $(1 + h)^{1/h}$ | 2 | 2.25 | 2.44 | 2.65 | 2.70 | 4 | 3.16 | 2.72 |

This limit is the first really difficult one we have encountered. A proof that it exists* and is a number, denoted by $e$, that is greater than 2 can be found in Appendix D. So we shall accept these facts and proceed.

Thus,

$$\lim_{h \to 0} \log_a (1 + h)^{1/h} = \log_a \left[\lim_{h \to 0} (1 + h)^{1/h}\right] = \log_a e.$$

Now we obtain from (2)

$$\frac{d}{du} (\log_a u) = \frac{1}{u} \lim_{\Delta u \to 0} \log_a \left(1 + \frac{\Delta u}{u}\right)^{u/\Delta u} = \frac{1}{u} \log_a e.$$

Substituting this last result in equation (1), we obtain the desired formula,

VIII    $$\frac{d}{dx} \log_a u = \frac{1}{u} \frac{du}{dx} \log_a e.$$

---

* The limit (3) is not the most efficient way of computing $e$. In Chapter 8 we will compute $e$ by other methods, and we will find that $e = 2.71828\ldots$ to 5 decimal places.

In this formula there is an irritating numerical factor, $\log_a e$. The formula would be neater if this factor could be avoided. And so it can if we set $a = e$, for then the factor becomes equal to 1. In other words, as far as calculus is concerned, the natural, simplifying base to use for logarithms is this peculiar number $e$. We therefore make the following definition.

**DEFINITION**   *Logarithms to the base $e$ are called natural logarithms.* We shall denote the natural logarithm function by* $\log = \log_e$:

$$\log N = \log_e N.$$

Then the special case of VIII for the natural logarithm function is

$$\frac{d}{dx} \log u = \frac{1}{u} \frac{du}{dx}.$$

Actually, it is rare in mathematics, science, or engineering to use anything but natural logarithms. Logarithms to any other base can be converted to natural logarithms (see Appendix A), viz.,

$$\log_a N = \frac{1}{\log a} \log N.$$

A small table of natural logarithms is to be found in Appendix A.

*Example 1*   $\dfrac{d}{dx} \log (a^2 - x^2) = \dfrac{1}{a^2 - x^2} (-2x) = \dfrac{-2x}{a^2 - x^2}.$

*Example 2*   $\dfrac{d}{dx} \log \sqrt{a^2 - x^2} = \dfrac{d}{dx} \dfrac{1}{2} \log (a^2 - x^2).$

Observe that we first use a property of the log function. Now we differentiate:

$$\frac{d}{dx} \log \sqrt{a^2 - x^2} = \frac{1}{2} \frac{-2x}{a^2 - x^2} = \frac{-x}{a^2 - x^2}.$$

Naturally, we obtain the same result by using VIII directly:

$$\frac{d}{dx} \log \sqrt{a^2 - x^2} = \frac{1}{\sqrt{a^2 - x^2}} \frac{d}{dx} \sqrt{a^2 - x^2} = \frac{1}{\sqrt{a^2 - x^2}} \frac{-x}{\sqrt{a^2 - x^2}} = \frac{-x}{a^2 - x^2}.$$

*Problems*

**In Problems 1 to 39 differentiate the function.**

1.   $y = \log (x + 1)$                    $[y' = 1/(x + 1)]$

2.   $y = \log (ax + 1)$                   $[y' = a/(ax + 1)]$

---

* The natural logarithm is also denoted by ln.

3.   $\varphi(x) = \log (x^2 - x)$
$$\left[ \varphi'(x) = \frac{1}{x} + \frac{1}{x-1} \right]$$

4.   $f(x) = \log \dfrac{x}{x-1}$
$$\left[ f'(x) = \frac{1}{x} - \frac{1}{x-1} \right]$$

5.   $\lambda(x) = \log \dfrac{a-x}{a+x}$
$$\left[ D\lambda(x) = \frac{2a}{x^2 - a^2} \right]$$

6.   $y = \log e^x$
$$[y' = 1]$$

7.   $y = \log e^{2x}$
$$[y' = 2]$$

8.   $y = \log e^{x^2}$
$$[y' = 2x]$$

9.   $y = x \ln x$
$$[y' = 1 + \log x]$$

10.  $y = x \log x^2$
$$[y' = 2 + 2 \log x]$$

11.  $y = \dfrac{\log x}{x}$
$$\left[ y' = \frac{1 - \log x}{x^2} \right]$$

12.  $y = \dfrac{\log x}{\sqrt{x}}$
$$\left[ y' = \frac{1 - \log \sqrt{x}}{x \sqrt{x}} \right]$$

13.  $y = \log \dfrac{1 - x^2}{1 + x^2}$
$$\left[ \frac{dy}{dx} = \frac{4x}{x^4 - 1} \right]$$

14.  $z = \log^2 ax$   $z = (\log ax)(\log ax)$
$z' = \log ax \cdot \frac{1}{x} + \log ax \cdot \frac{1}{x}$
$$\left[ \frac{dz}{dx} = 2 (\log ax)/x \right]$$

15.  $\varphi(x) = \log (x^3 - 3x + 2)$
$$[\varphi'(x) = 3(x+1)/(x+2)(x-1)]$$

16.  $F(t) = \log \sqrt{\dfrac{1+t}{1-t}}$
$$\left[ F'(t) = \frac{1}{1 - t^2} \right]$$

17.  $f(\theta) = \log \log \theta$   $= \log u$   $u = \log \theta$
$$[Df(\theta) = 1/\theta \log \theta]$$

18.  $g(\theta) = \log \log^3 \theta$   $u = \log^3 \theta$
$$[Dg(\theta) = 3/\theta \log \theta]$$

19.  $h(\theta) = \log^3 (\log \theta)$
$$[Dh(\theta) = 3 (\log^2 \log \theta)/\theta \log \theta]$$

20.  $y = (a + b \log x)^c$
$$[y' = (b/x)c (a + b \log x)^{c-1}]$$

21.  $u = \log_a x^3$
$$[du/dx = (3/x) \log_a e]$$

22.  $y = e^{\log x}$   $\frac{1}{x} e^{\log x} = 1$
$$[y' = 1]$$

$\log x = x$

23. $y = \dfrac{1}{x \log x}$  $x \log x \; y = 1$  then differentiate  $\left[ y' = -\dfrac{1 + \log x}{x^2 \log^2 x} \right]$

24. $y = \log_a b^x$  $[y' = \log_a b]$

25. $y = \log_{10} (x^2 + 2x + 100)$  $[y' = 0.8686 \, (x + 1)/(x^2 + 2x + 100)]$

26. $v = \log (\sqrt{x^2 + 1} + x)$  $[v' = 1/\sqrt{1 + x^2}]$

27. $v = -\log (\sqrt{x^2 + 1} - x)$  $[v' = 1/\sqrt{x^2 + 1}]$

28. $v = \log (x + \sqrt{x^2 - 1})$  $[v' = 1/\sqrt{x^2 - 1}]$

29. $v = \log \dfrac{\sqrt{x^2 + 1} + x}{\sqrt{x^2 + 1} - x}$  $\left[ v' = \dfrac{2}{\sqrt{x^2 + 1}} \right]$

Suggestion for alternate solution: Rationalize the denominator.

30. $\psi(\theta) = \log \dfrac{1 - \theta}{1 + \theta}$  $\left[ \psi'(\theta) = \dfrac{-2}{1 - \theta^2} \right]$

31. $F(x) = \log \dfrac{1 + \sqrt{x}}{1 - \sqrt{x}}$  $\left[ F'(x) = \dfrac{1}{\sqrt{x}\,(1 - x)} \right]$

32. $y = \log |x|$  $\left[ y' = \dfrac{1}{x} \text{ for both } x > 0 \text{ and } x < 0 \right]$

33. $y = \log_2 x^2$  $[y' = (2/x) \log_2 e]$

34. $y = \log_{10} \sqrt{1 - 4x^2}$  $[y' = -1.7372x/(1 - 4x^2)]$

35. $y = \log \dfrac{1 + x + x^2}{1 - x + x^2}$  $\left[ \dfrac{dy}{dx} = \dfrac{2(1 - x^2)}{1 + x^2 + x^4} \right]$

36. $k(t) = \log \dfrac{t}{\sqrt{t^2 + 1} - t}$  $\left[ \dfrac{dk}{dt} = \dfrac{1}{t} + \dfrac{1}{\sqrt{t^2 + 1}} \right]$

37. $f(x) = \log \dfrac{\sqrt{1 + x} + \sqrt{1 - x}}{\sqrt{1 + x} - \sqrt{1 - x}}$  $\dfrac{\sqrt{1+x}\,+\,\sqrt{1-x}}{\sqrt{1+x}\,+\,\sqrt{1-x}}$  $\left[ f'(x) = \dfrac{-1}{x\sqrt{1 - x^2}} \right]$
rat. denominator

38. $F(x) = \log (\sqrt{1 + x^2} + \sqrt{1 - x^2})$  $\left[ F'(x) = \dfrac{1}{x} - \dfrac{1}{x\sqrt{1 - x^4}} \right]$

39. $u(y) = \tfrac{1}{6} \log \dfrac{(y + 1)^2}{y^2 - y + 1}$  $\left[ \dfrac{du}{dy} = \dfrac{1 - y}{2(y^3 + 1)} \right]$

40.   Now go back to Problem 1 and try to do as many as you can entirely in your head. For the others, write as few steps as you can.

41.   Verify the table, given on page 42, of values of $(1 + h)^{1/h}$.

42.   Show that

$$\lim_{h \to 0} \log_a (1 + h)^{1/h}$$

is the slope of the graph of $\log_a x$ at $x = 1$.

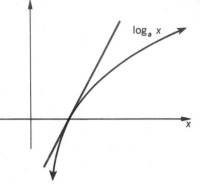

43.   Show that one does not need to remember $d/dx \log_a u$, only $d/dx \log u$, by changing $\log_a$ to a natural logarithm. (See Appendix A.)

## 4   *Exponential Functions; Logarithmic Differentiation*

The exponential function $\exp_a$ and the logarithmic function $\log_a$ are inverses of each other:

$$\exp_a x = a^x = y \quad \text{if and only if} \quad \log_a y = x.$$

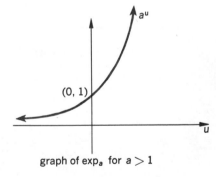

graph of $\exp_a$ for $a > 1$

If we use the standard notation, $f^{-1}$, for the inverse of a function $f$, then $\log_a = \exp_a^{-1}$, and $\exp_a = \log_a^{-1}$.

If the exponential function is differentiable,* a formula for its derivative is readily obtained from the derivative of $\log_a$. Thus, if $y = \exp_a u$ where $u$ is a function of $x$, then $y = a^u$, and

(1)                           $\log y = \log a^u = u \log a.$

Both sides of equation (1) are representations of the same function. The derivative of this function then is

$$\frac{1}{y}\frac{dy}{dx} = \frac{du}{dx} \log a \quad \text{(by VIII)}.$$

---

* We shall see in Section 7 that, in general, inverses of differentiable functions are also differentiable.

Consequently,

$$\frac{dy}{dx} = y \frac{du}{dx} \log a.$$

But $y = a^u$, and so we have the desired formula:

IX    $$\frac{d}{dx} a^u = a^u \frac{du}{dx} \log a.$$

If $a = e$, the awkward numerical factor $\log a$ becomes equal to 1, and we obtain the special case:

$$\frac{d}{dx} e^u = e^u \frac{du}{dx}.$$

The general formula IX is rarely needed. When it is needed, however, it can be derived from the special case by substituting $a = e^{\log a}$; then $a^u = (e^{\log a})^u = e^{u \log a}$.

**Example 1** *    $$\frac{d}{dx} \exp(x^2 - x + 1) = \frac{d}{dx} e^{x^2 - x + 1}$$

$$= e^{x^2 - x + 1} \frac{d}{dx}(x^2 - x + 1)$$

$$= (2x - 1) e^{x^2 - x + 1}.$$

**Example 2**    $$\frac{d}{dx} a^{\sqrt{1+x^2}} = a^{\sqrt{1+x^2}} \frac{d}{dx}(\sqrt{1 + x^2}) \log a$$

$$= a^{\sqrt{1+x^2}} \frac{x}{\sqrt{1 + x^2}} \log a.$$

This could have been done using only the special formula. Because $a = e^{\log a}$,

$$a^{\sqrt{1+x^2}} = (e^{\log a})^{\sqrt{1+x^2}} = e^{(\log a)\sqrt{1+x^2}}$$

$$\frac{d}{dx} a^{\sqrt{1+x^2}} = e^{\log a \sqrt{1+x^2}} \frac{d}{dx} (\log a)\sqrt{1 + x^2}$$

$$= e^{\log a \sqrt{1+x^2}} (\log a) \frac{x}{\sqrt{1 + x^2}}$$

$$= a^{\sqrt{1+x^2}} \frac{x}{\sqrt{1 + x^2}} \log a.$$

The method we used to obtain the formula for the derivative of $a^u$ was *first* to take the logarithm of both sides of $y = a^u$, and *then* to differentiate. This method can be applied to other functions and is referred to as the technique of **logarithmic differentiation**. Examples 3 to 6 will illustrate its use.

---

* The expression "$\exp(x)$" has the same meaning as "$e^x$."

In Section 2 we derived the power formula VI only for the case where $n$ is a positive integer. The next example uses the technique of logarithmic differentiation to establish VI with no restrictions on $n$.

**Example 3**    To differentiate $y = u^n$, where $u$ is a function of $x$, we use logarithmic differentiation:

$$\log y = \log u^n = n \log u;$$

$$\frac{1}{y}\frac{dy}{dx} = n\frac{1}{u}\frac{du}{dx};$$

$$\frac{dy}{dx} = n\frac{y}{u}\frac{du}{dx} = n\frac{u^n}{u}\frac{du}{dx}$$

$$= n\,u^{n-1}\frac{du}{dx}.$$

Thus, we have obtained the familiar power formula VI but *without any restrictions on the constant exponent n*. We now are on solid ground in our development of the basic algorithm for differentiation of the power function.

**Example 4**    If $y = u^v$ where both $u$ and $v$ are functions, and, of course, $u > 0$, then, using logarithmic differentiation,

$$\log y = \log u^v = v \log u;$$

$$\frac{1}{y}\frac{dy}{dx} = v\frac{d}{dx}\log u + (\log u)\frac{dv}{dx}$$

$$= v\frac{1}{u}\frac{du}{dx} + (\log u)\frac{dv}{dx};$$

$$\frac{dy}{dx} = y\left(\frac{v}{u}\frac{du}{dx} + \log u\frac{dv}{dx}\right).$$

$$\boxed{\frac{d}{dx}u^v = v\,u^{v-1}\frac{du}{dx} + u^v(\log u)\frac{dv}{dx}.}$$

This new formula, which combines features of VI and IX, could be added to our list of differentiation rules, but in practice it is equally simple just to use logarithmic differentiation, as in the next example.

**Example 5**    Differentiate $z = x^x$.

$$\log z = \log x^x = x \log x;$$

$$\frac{1}{z}\frac{dz}{dx} = x\frac{1}{x} + \log x;$$

$$\frac{dz}{dx} = z(1 + \log x) = x^x(1 + \log x).$$

*Example 6*    Sometimes logarithmic differentiation provides a method which involves simpler algebra than direct differentiation. For example,

$$y = \sqrt{\frac{(x^2 + 1)(x + 2)}{(x^2 + 2)(x + 3)}}$$

could be differentiated by using VI and VII and V. However, logarithmic differentiation is easier:

$$\log y = \tfrac{1}{2} \log (x^2 + 1) + \tfrac{1}{2} \log (x + 2) - \tfrac{1}{2} \log (x^2 + 2) - \tfrac{1}{2} \log (x + 3);$$

$$\frac{1}{y} \frac{dy}{dx} = \frac{x}{x^2 + 1} + \frac{1}{2} \frac{1}{(x + 2)} - \frac{x}{x^2 + 2} - \frac{1}{2} \frac{1}{x + 3};$$

$$\frac{dy}{dx} = y \left[ \frac{x}{x^2 + 1} + \frac{1}{2(x + 2)} - \frac{x}{(x^2 + 2)} - \frac{1}{2(x + 3)} \right],$$

and probably one would leave the derivative in this form.

*Remark.* We now have extended our repertory of functions that we can differentiate by simple rules. To differentiate $\log_a$ and $\exp_a$ we had but one difficult limit to evaluate, namely,

$$\lim_{h \to 0} (1 + h)^{1/h} = e.$$

### Problems

**Differentiate the following functions.**

1.  $f(t) = e^{2t}$  $\qquad\qquad\qquad\qquad\qquad\qquad$ $[\, f'(t) = 2e^{2t} \,]$

2.  $f(x) = b^{x+2}$  $\qquad\qquad\qquad\qquad\qquad$ $[\, f'(x) = b^{x+2} \log b \,]$

3.  $\varphi(\theta) = e^{a\theta + b}$  $\qquad\qquad\qquad\qquad\qquad$ $[\, \varphi'(\theta) = ae^{a\theta + b} \,]$

4.  $y = e^{a^2 + b^2 x^2}$  $\qquad\qquad\qquad\qquad$ $[\, y' = 2b^2 x e^{a^2 + b^2 x^2} \,]$

5.  $y = a^{\log_a x}$  $\qquad\qquad\qquad\qquad\qquad\qquad\qquad$ $[\, y' = 1 \,]$

6.  $y = e^{e^{2x}}$  $\qquad\qquad\qquad\qquad\qquad\qquad$ $[\, y' = 2e^{e^{2x} + 2x} \,]$

7.  $W = a^{t^3}$  $\qquad\qquad\qquad\qquad\qquad$ $[\, W' = 3t^2 (a^{t^3}) \log a \,]$

8.  $y = e^{\sqrt{2x}}$  $\qquad\qquad\qquad\qquad\qquad$ $\left[ \dfrac{dy}{dx} = \dfrac{e^{\sqrt{2x}}}{\sqrt{2x}} \right]$

9.  $F(x) = 2^{x^2 + 2x + 2}$  $\qquad\qquad$ $[\, F'(x) = 8(x + 1)2^{x^2 + 2x} \log 2 \,]$

10.  $\psi(t) = 3^{4 - t^2}$  $\qquad\qquad$ $[\, \psi'(t) = -162t(3^{-t^2}) \log 3 \,]$

11.  $k(x) = x^a e^{bx}$  $\qquad\qquad$ $[\, k'(x) = x^{a-1} e^{bx}(a + bx) \,]$

12.  $g(r) = e^r (r^2 + r + 1)$  $\qquad\qquad$ $\left[ \dfrac{dg}{dr} = e^r (r^2 + 3r + 2) \right]$

13.  $h(\theta) = 2^a e^{\theta/2}$  $\quad\quad\quad\quad\quad\quad\quad\quad\quad\quad\quad\quad\quad$ $[\,h'(\theta) = 2^{a-1} e^{\theta/2}\,]$

14.  $y = A e^{-1/x}$  $\quad\quad\quad\quad\quad\quad\quad\quad\quad\quad\quad\quad\quad$ $[\,y' = A e^{-1/x}/x^2\,]$

15.  $y = 10^{x^2+1}$  $\quad\quad\quad\quad\quad\quad\quad\quad\quad\quad\quad\quad\quad$ $[\,y' = 46.06x\,(10^{x^2})\,]$

16.  $y = \dfrac{a}{2}\,(e^{x/a} - e^{-x/a})$  $\quad\quad\quad\quad\quad\quad\quad\quad$ $[\,y' = \tfrac{1}{2}(e^{x/a} + e^{-x/a})\,]$

17.  $\checkmark$ $y = \dfrac{a}{2}\,(e^{x/a} + e^{-x/a})$  $\quad\quad\quad\quad\quad\quad\quad\quad$ $[\,y' = \tfrac{1}{2}(e^{x/a} - e^{-x/a})\,]$

18.  $y = x e^{-x}$  $\quad\quad\quad\quad\quad\quad\quad\quad\quad\quad\quad\quad\quad$ $[\,y' = e^{-x}(1 - x)\,]$

19.  $\checkmark$ $f(x) = \dfrac{e^x - e^{-x}}{e^x + e^{-x}}$  $\quad\quad\quad\quad\quad\quad\quad$ $\left[\, f'(x) = \dfrac{4}{(e^x + e^{-x})^2} \,\right]$

20.  $W = x^a a^x$  $\quad\quad\quad\quad\quad\quad\quad\quad\quad\quad\quad$ $\left[\, \dfrac{dw}{dx} = x^{a-1} a^x (a + \log a^x) \,\right]$

21.  $H(\theta) = e^\theta \log \theta$  $\quad\quad\quad\quad\quad\quad\quad\quad\quad$ $[\,H'(\theta) = e^\theta (1/\theta + \log \theta)\,]$

22.  $\varphi(t) = e^{\log t}$  $\quad\quad\quad\quad\quad\quad\quad\quad\quad\quad\quad\quad$ $[\,\varphi'(t) = 1\,]$

23.  $\varphi(t) = \log e^t$  $\quad\quad\quad\quad\quad\quad\quad\quad\quad\quad\quad\quad$ $[\,\varphi'(t) = 1\,]$

24.  $y = e^{-\log x^2}$  $\quad\quad\quad\quad\quad\quad\quad\quad\quad\quad\quad\quad$ $[\,y' = -2/x^3\,]$

25.  $y = \log e^{-x^2}$  $\quad\quad\quad\quad\quad\quad\quad\quad\quad\quad\quad\quad$ $[\,y' = -2x\,]$

26.  $y = e^{\log (1/x)}$  $\quad\quad\quad\quad\quad\quad\quad\quad\quad\quad\quad$ $[\,y' = -1/x^2\,]$

27.  $y = \log e^{1/x}$  $\quad\quad\quad\quad\quad\quad\quad\quad\quad\quad\quad\quad$ $[\,y' = -1/x^2\,]$

28.  $y = x^{1/x}$  $\quad\quad\quad\quad\quad\quad\quad\quad\quad\quad\quad\quad$ $[\,y' = x^{1/x}(1 - \log x)/x^2\,]$

29.  $\varphi(s) = s^{\log s}$  $\quad\quad\quad\quad\quad\quad\quad\quad\quad\quad\quad$ $[\,\varphi'(s) = \log s^2 \cdot s^{\log s - 1}\,]$

30.  $\varphi(s) = s^{1/\log s}$  $\quad\quad\quad\quad\quad\quad\quad\quad\quad\quad$ $[\,\varphi'(s) = 0\,]$

31.  $\checkmark$ $f(x) = (e^x)^x$  $\quad\quad\quad\quad\quad\quad\quad\quad\quad\quad$ $[\,f'(x) = 2x e^{x^2}\,]$

32.  $f(x) = e^{(x^x)}$  $\quad\quad\quad\quad\quad\quad\quad\quad\quad\quad$ $[\,f'(x) = x^x e^{x^x}(1 + \log x)\,]$

33.  $y = x^{(e^x)}$  $\quad\quad\quad\quad\quad\quad\quad\quad\quad$ $\left[\, y' = e^x x^{(e^x)}\left(\dfrac{1}{x} + \log x\right) \,\right]$

34.  $y = (x^e)^x$  $\quad\quad\quad\quad\quad\quad\quad\quad\quad\quad$ $[\,y' = e\,(x^{ex})(1 + \log x)\,]$

35.  $F(x) = x^{(x^x)}$  $\quad\quad$ $\left[\, F'(x) = x^x x^{(x^x)} \log x \left(1 + \log x + \dfrac{1}{x \log x}\right) \,\right]$

36.  $F(x) = (x^x)^x$  $\quad\quad\quad\quad\quad\quad\quad\quad\quad$ $[\,F'(x) = x^{x^2+1}(1 + \log x^2)\,]$

37.  $y = 10^{(10^x)}$  $\quad\quad\quad\quad\quad\quad\quad\quad\quad\quad$ $[\,y' = 10^x \cdot 10^{10^x} \log^2 10\,]$

*38.*   $y = e^{(e^{(e^x)})}$                                                     $[\,y' = e^x e^{(e^x)}\, e^{(e^{(e^x)})}\,]$

*39.*  $\sqrt{\phantom{}}$ $y = x^x + x^{1/x}$                                     $[\,y' = x^x (1 + \log x) + x^{(1/x - 2)}(1 - \log x)\,]$

*40.*   Now go back to Problem 1 and do as many as you can entirely in your head. Then do the others in as few steps as possible.

**In the following problems use logarithmic differentiation to find the derivative of the function.**

*41.*   $y = \sqrt{\dfrac{1 + x^2}{1 - x^2}}$                  $\left[\, y' = \dfrac{2xy}{1 - x^4} = \dfrac{2x}{(1 + x^2)^{1/2}(1 - x^2)^{3/2}} \,\right]$

*42.*   $f(x) = (x - 1)^2 (x - 2)(x - 3)$            $[\, f'(x) = 2(x - 1)(x - 2)(x - 3)$
$$+ (x - 1)^2 (x - 3) + (x - 1)^2 (x - 2)\,]$$

*43.*   $y = \dfrac{(x + 2)^2}{(x + 1)(x + 3)^3}$                 $\left[\, y' = \dfrac{-2(x + 2)(x^2 + 3x + 3)}{(x + 1)^2 (x + 3)^4} \,\right]$

*44.*   $y = \dfrac{(x + 1)^{3/2}(x - 1)^{3/4}}{(x + 2)^{1/3}}$           $\left[\, y' = \dfrac{(23x^2 + 45x - 14)(x + 1)^{1/2}}{12(x - 1)^{1/4}(x + 2)^{4/3}} \,\right]$

*45.*   $y = \sqrt{1 - x}\,\sqrt{1 - 2x}\,\sqrt{1 - 3x}$      $\left[\, y' = \dfrac{-(9x^2 - 11x + 3)}{\sqrt{1 - x}\,\sqrt{1 - 2x}\,\sqrt{1 - 3x}} \,\right]$

*46.*   $y = \dfrac{uv}{wz}$   $(u, v, w, z$ all functions of $x)$    $\left[\, y' = \dfrac{uv}{wz}\left(\dfrac{u'}{u} + \dfrac{v'}{v} - \dfrac{w'}{w} - \dfrac{z'}{z}\right) \,\right]$

*47.*   Graph the function $y = x^{1/\log x}$ (see Problem 30).

## 5   *Derivatives of the Trigonometric Functions*

We now extend our list of differentiation formulas to the trigonometric functions. As we derive the formula for the derivative of the sine function we will encounter a new limit. Once this limit is established, however, all the differentiation formulas are easily found.

First consider the graph of sin $u$.

Surprisingly, the graph of $D$ sin appears to be the same as the graph of cos. When the formula for the derivative is derived it will be seen that indeed that is the case.

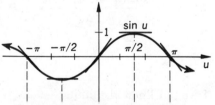

The graph of *D* sin is easily sketched, at least approximately

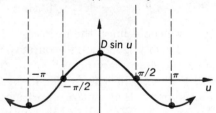

We apply the definition of the derivative to obtain formula X.
If

$$y = \sin u \qquad \text{and} \qquad u = u(x),$$

then

$$\Delta y = y(u + \Delta u) - y(u) = \sin(u + \Delta u) - \sin u$$

$$= \frac{d}{du}(\sin u)\,\Delta u + \eta\,\Delta u,$$

where $\eta \to 0$ as $\Delta u \to 0$, again by the Fundamental Lemma of Chapter 1.
Hence,

$$\frac{\Delta y}{\Delta x} = \frac{d}{du}(\sin u)\frac{\Delta u}{\Delta x} + \eta\frac{\Delta u}{\Delta x}.$$

By letting $\Delta x \to 0$ we obtain

$$(1) \qquad \frac{dy}{dx} = \frac{d}{du}(\sin u)\frac{du}{dx} + 0\frac{du}{dx} = \frac{d}{du}(\sin u)\frac{du}{dx},$$

because $\Delta u \to 0$ as $\Delta x \to 0$.

Thus, our problem has been reduced to finding $(d/du)(\sin u)$, a situation comparable to the derivation of formula VIII. We have

$$(2) \qquad \frac{d}{du}(\sin u) = \lim_{\Delta u \to 0} \frac{\sin(u + \Delta u) - \sin u}{\Delta u}.$$

This limit is difficult to evaluate as it stands. It is natural to use the addition formula for $\sin(u + \Delta u)$:

$$(3) \qquad \frac{\sin(u + \Delta u) - \sin u}{\Delta u} = \frac{\sin u \cos \Delta u + \cos u \sin \Delta u - \sin u}{\Delta u}$$

$$= \cos u\,\frac{\sin \Delta u}{\Delta u} + \sin u\,\frac{\cos \Delta u - 1}{\Delta u}.$$

If we but knew the limits:

$$\lim_{\Delta u \to 0}\frac{\sin \Delta u}{\Delta u} \quad \text{and} \quad \lim_{\Delta u \to 0}\frac{\cos \Delta u - 1}{\Delta u},$$

then we could finish the task.

From trigonometry, you may recall that for small $\theta$, $\sin \theta$ and $\theta$ are very nearly equal. (See the table on page 512.) Thus, when $\Delta u$ is small, $\sin \Delta u/\Delta u$ is very near 1. For a proof that $\sin \theta/\theta$ approaches 1 as $\theta \to 0$ consult Appendix D. We shall use that result here.

$$\lim_{\theta \to 0}\frac{\sin \theta}{\theta} = 1$$

(For a proof see Appendix D.)

The second of the two fractions can be handled by a simple trick.

$$\frac{\cos \Delta u - 1}{\Delta u} = \frac{(\cos \Delta u - 1)(\cos \Delta u + 1)}{\Delta u(\cos \Delta u + 1)} = \frac{-\sin^2 \Delta u}{\Delta u(\cos \Delta u + 1)}$$

$$(4) \qquad = \frac{\sin \Delta u}{\Delta u}\left(\frac{-\sin \Delta u}{\cos \Delta u + 1}\right).$$

Then, using (3) and (4) in equation (2), we have

$$\frac{d}{du}(\sin u) = \lim_{\Delta u \to 0} \frac{\sin(u + \Delta u) - \sin u}{\Delta u}$$

$$= \lim_{\Delta u \to 0}\left[\cos u \left(\frac{\sin \Delta u}{\Delta u}\right) + \sin u \left(\frac{\sin \Delta u}{\Delta u}\right)\left(\frac{-\sin \Delta u}{\cos \Delta u + 1}\right)\right]$$

$$= \cos u(1) - \sin u(1)\left(\frac{0}{1+1}\right) = \cos u.$$

Substituting this result in (1), we obtain formula X:

X $\quad \dfrac{d}{dx}(\sin u) = \cos u \dfrac{du}{dx}.$

The remaining formulas are now easy to derive by clever use of the formula for the sine.

If $y = \cos u$, then

$$y = \sin\left(\frac{\pi}{2} - u\right);$$

$$\frac{dy}{dx} = \cos\left(\frac{\pi}{2} - u\right)\frac{d}{dx}\left(\frac{\pi}{2} - u\right) = \sin u\left(-\frac{du}{dx}\right).$$

Therefore,

XI $\quad \dfrac{d}{dx}\cos u = -\sin u \dfrac{du}{dx}.$

The remaining formulas are even easier to derive. We simply express each of $\tan u$, $\cot u$, $\sec u$, and $\csc u$ in terms of $\sin u$ or $\cos u$. See Problem 62 below.

XII $\quad \dfrac{d}{dx}\tan u = \sec^2 u \dfrac{du}{dx}$

XIII $\quad \dfrac{d}{dx}\cot u = -\csc^2 u \dfrac{du}{dx}$

**XIV** $\quad \dfrac{d}{dx} \sec u = \sec u \tan u \dfrac{du}{dx}$

**XV** $\quad \dfrac{d}{dx} \csc u = -\csc u \cot u \dfrac{du}{dx}$

**Example 1** $\quad \dfrac{d}{dx} \sin (ax + b) = \cos (ax + b) \dfrac{d}{dx} (ax + b)$

$$= a \cos (ax + b).$$

**Example 2** $\quad \dfrac{d}{dx} \sqrt{\cos x^2} = \tfrac{1}{2} (\cos x^2)^{-1/2} \dfrac{d}{dx} (\cos x^2)$

$$= \tfrac{1}{2} (\cos x^2)^{-1/2} (-\sin x^2) \dfrac{d}{dx} x^2$$

$$= \dfrac{-x \sin x^2}{\sqrt{\cos x^2}}.$$

**Example 3** $\quad \dfrac{d}{d\theta} e^{\sec \sqrt{\theta}} = e^{\sec \sqrt{\theta}} \dfrac{d}{d\theta} \sec \sqrt{\theta}$

$$= e^{\sec \sqrt{\theta}} \sec \sqrt{\theta} \tan \sqrt{\theta} \dfrac{d}{d\theta} \sqrt{\theta}$$

$$= \dfrac{1}{2\sqrt{\theta}} e^{\sec \sqrt{\theta}} \sec \sqrt{\theta} \tan \sqrt{\theta}.$$

**Problems**

1. From the graph of the cosine function, sketch the graph of $D \cos$ by estimating slopes at several points.

2. From the graph of the tangent function, sketch the graph of $D \tan$. Then on the same sheet of graph paper, sketch the graphs of sec and $\sec^2$.

**Differentiate the following functions:**

3. $y = \cos 2x$  $\qquad\qquad\qquad\qquad\qquad [y' = -2 \sin 2x]$

4. $y = \cos x \sin x$  $\qquad\qquad\qquad\qquad\quad [Dy = \cos 2x]$

5. $y = \cos 2x \sin x$  $\qquad\qquad\quad [Dy = \cos 3x - \sin 2x \sin x]$

6. $\varphi = \sec 3\theta$  $\qquad\qquad\qquad\quad [d\varphi/d\theta = 3 \sec 3\theta \tan 3\theta]$

7. $r = a \sqrt{\cos 2\theta}$  $\qquad\qquad [dr/d\theta = -a \sin 2\theta / \sqrt{\cos 2\theta}]$

8. $\lambda(\theta) = \sin^2 3\theta + \cos^2 3\theta$ $\qquad\qquad [d\lambda/d\theta = 0]$

9. $x = a(\theta - \sin\theta)$ $\qquad\qquad [dx/d\theta = a(1 - \cos\theta)]$

10. $\tau(\theta) = \tan\sqrt{1-\theta}$ $\qquad [\tau'(\theta) = (-1/2\sqrt{1-\theta})\sec^2\sqrt{1-\theta}]$

11. $f(\theta) = \cot(a\theta^2 + 5)$ $\qquad [f'(\theta) = -2a\theta\csc^2(a\theta^2 + 5)]$

12. $F(x) = A\csc^2 3x$ $\qquad\qquad [F'(x) = -6A\csc^2 3x\cot 3x]$

13. $\mu(\theta) = \tan 2\theta - 2\theta$ $\qquad\qquad [\mu'(\theta) = 2\tan^2 2\theta]$

14. $y = \dfrac{1 - \tan 2x}{\sec 2x}$ $\qquad\qquad [y' = -2(\cos 2x + \sin 2x)]$

15. $y = (\csc 2x - \cot 2x)^2$ $\qquad\qquad [y' = (4\sin 2x)/(1 + \cos 2x)^2]$

16. $y = e^{a\theta}(a\sin b\theta - b\cos b\theta)$ $\qquad [y' = (a^2 + b^2)e^{a\theta}\sin b\theta]$

17. $\varphi(t) = \sqrt{\cos 2t + 1}$ $\qquad [\varphi'(t) = -\sin 2t(\cos 2t + 1)^{-1/2}]$

18. $y = \sin^4 x + \cos^4 x$ $\qquad\qquad [y' = -\sin 4x]$

19. $f(x) = \dfrac{\sin 2x}{x}$ $\qquad\qquad \left[ f'(x) = \dfrac{2x\cos 2x - \sin 2x}{x^2} \right]$

20. $f(x) = x\sin\dfrac{1}{x}$ $\qquad\qquad \left[ f'(x) = -\dfrac{1}{x}\cos\dfrac{1}{x} + \sin\dfrac{1}{x} \right]$

21. $f(x) = x^2\sin\dfrac{1}{x} = x^2\sin u \quad u = \frac{1}{x}$ $\qquad \left[ f'(x) = -\cos\dfrac{1}{x} + 2x\sin\dfrac{1}{x} \right]$

22. $G(\varphi) = \cos^3\varphi\sin\varphi$ $\qquad [G'(\varphi) = \cos^2\varphi(\cos^2\varphi - 3\sin^2\varphi)]$

23. $y = \cos^2\sqrt{x}$ $\qquad\qquad [y' = -(\cos\sqrt{x}\sin\sqrt{x})/\sqrt{x}]$

24. $y = \log\sin a\theta$ $\qquad\qquad [y' = a\cot a\theta]$

25. $y = \log\csc a\theta$ $\qquad\qquad [y' = -a\cot a\theta]$

26. $y = \log\cos a\theta$ $\qquad\qquad [y' = -a\tan a\theta]$

27. $y = \log\sec a\theta$ $\qquad\qquad [y' = a\tan a\theta]$

28. $y = \log\tan a\theta$ $\qquad [y' = a/\sin a\theta\cos a\theta = 2a\csc 2a\theta]$

29. $y = \log\cot a\theta$ $\qquad\qquad [y' = -a/\sin a\theta\cos a\theta]$

30. $\varphi(x) = \dfrac{\sin 2x}{1 + \cos 2x}$ $\qquad\qquad \left[ \varphi'(x) = \dfrac{2}{1 + \cos 2x} = \sec^2 x \right]$

31. $y = \sqrt{1 + \tan^2 3\theta}; \quad -\pi/2 < 3\theta < \pi/2$ $\qquad [y' = 3\sec 3\theta\tan 3\theta]$

32. $\alpha(x) = \tan\sqrt{x}$ $\qquad\qquad [\alpha'(x) = (\sec^2\sqrt{x})/2\sqrt{x}]$

33. $\beta(x) = \dfrac{\sec 2x}{\tan 2x + 1}$ $\qquad\left[\beta'(x) = \dfrac{2(\sin 2x - \cos 2x)}{(\sin 2x + \cos 2x)^2}\right]$

**(Do Problem 33 in two ways.)**

34. $h(\theta) = \log(\sec\theta + \tan\theta)$ $\qquad [h'(\theta) = \sec\theta]$

35. $h(\theta) = \log(\csc\theta + \cot\theta)$ $\qquad [h'(\theta) = -\csc\theta]$

36. $y = \dfrac{1 - \tan^2 x}{\tan x}$ $\qquad [y' = -\sec^2 x\,\csc^2 x]$

37. $y = \dfrac{a}{\cos 4x}$ $\qquad [y' = 4a\sec 4x\tan 4x]$

38. $y = \cos^2 ax$ $\qquad [y' = -2a\cos ax\sin ax]$

39. $y = -\sin^2 ax$ $\qquad [y' = -2a\cos ax\sin ax]$

40. $y = \cos^2 ax - \sin^2 ax$ $\qquad [y' = -2a\sin 2ax]$

41. $f(\theta) = \log\sqrt{\dfrac{1 + \sin\theta}{1 - \sin\theta}}$ $= \log\sqrt{\dfrac{1+\sin\theta}{1-\sin\theta}\cdot\dfrac{1+\sin\theta}{1+\sin\theta}} = \log\sqrt{(\sec\theta + \tan\theta)^2}$ $\qquad [f'(\theta) = \sec\theta]$

42. $F(\theta) = \log\tan\tfrac{1}{2}(\pi/2 + \theta)$ $\qquad [F'(\theta) = \sec\theta]$

43. $\alpha(\theta) = e^{\sin a\theta}$ $\qquad [d\alpha/d\theta = a\cos a\theta\, e^{\sin a\theta}]$

44. $\beta(\theta) = \sqrt{\sin\theta}$ $\qquad [D\beta(\theta) = \cos\theta/2\sqrt{\sin\theta}]$

45. $\gamma(\theta) = \sqrt{\sin\sqrt{\theta}}$ $\quad \sin\theta = u\left|\tfrac{1}{2}u^{-\frac{1}{2}}\times\right.$ $\sqrt{\theta} = \theta^{\frac{1}{2}}\left|\cos\sqrt{\theta}\cdot\tfrac{1}{2}\theta^{-\frac{1}{2}}\right.$ $\qquad\left[D\gamma(\theta) = \dfrac{\cos\sqrt{\theta}}{4\sqrt{\theta}\sqrt{\sin\sqrt{\theta}}}\right]$

46. $\delta(\theta) = \sin^\theta x$ $\qquad [D\delta(\theta) = (\sin^\theta x)(\log\sin x)]$

47. $Z = \sin(1/x^2)$ $= \sin u \quad u = x^{-2}$ $\qquad [DZ = -2\cos(1/x^2)/x^3]$

48. $w = \sin\log x$ $\qquad [Dw = (1/x)\cos\log x]$

49. $v = \log\sin x$ $\qquad [Dv = \cot x]$

50. $u = \log\sin^2 x$ $\qquad [Du = 2\cot x]$

51. $y = \log\sqrt{1 - 2\sin^2\theta}$ $\qquad [Dy = -\tan 2\theta]$

52. $f(\theta) = \theta + \cot\theta - \tfrac{1}{3}\cot^3\theta$ $\qquad [f'(\theta) = \cot^4\theta]$

53. $F(\theta) = \dfrac{\tan\theta - 1}{\sec\theta}$ $\qquad [F'(\theta) = \cos\theta + \sin\theta]$

54. $\zeta(\theta) = e^{\sin\theta}\cos\theta$ $\qquad [\zeta'(\theta) = e^{\sin\theta}(\cos^2\theta - \sin\theta)]$

55. $\tau(x) = \sec e^{ax}$ $\qquad [D\tau(x) = ae^{ax}\sec e^{ax}\tan e^{ax}]$

56. $y(\theta) = e^{a\theta}(a\sin\theta - \cos\theta)$ $\qquad [Dy(\theta) = (a^2 + 1)e^{a\theta}\sin\theta]$

**57.**  $y = x^{\sin x}$   $\qquad [Dy = x^{\sin x - 1} \sin x + \cos x \cdot x^{\sin x} \log x]$

**58.** ✓  $y = (\sin x)^x$   $\qquad [Dy = (\sin x)^x (\log \sin x + x \cot x)]$

**59.**  $y = (\sin x)^{\sin x}$   $\qquad [Dy = (\sin x)^{\sin x} \cos x (1 + \log \sin x)]$

**60.**  $F(x) = \dfrac{1}{\sqrt{b^2 - a^2}} \log \dfrac{\sqrt{b + a} + \sqrt{b - a} \tan (x/2)}{\sqrt{b + a} - \sqrt{b - a} \tan (x/2)}$

$$\left[ F'(x) = \frac{1}{a + b \cos x} \right]$$

**61.**   Now go back over Problems 3 to 60 and try to do them mentally.

**62.**   Derive formulas XII–XV.

**63.**   Examine a natural table of sines (a larger table than on page 512) to verify that $\sin \theta$ is very nearly $\theta$ for small $\theta$.

**\*64.**   Two legs, $a$ and $b$, of a triangle are of fixed length. The included angle, $\theta$, is allowed to vary. Find the rate of change of the third side $c$ with respect to $\theta$.

$$\left[ \frac{dc}{d\theta} = \frac{ab \sin \theta}{\sqrt{a^2 + b^2 - 2ab \cos \theta}} = \frac{ab \sin \theta}{c} \right]$$

**\*65.**   Differentiate the function represented on each side of the following identities. Obtain in this way either new identities or familiar or obvious ones.

(a)   $\sin^2 \theta + \cos^2 \theta = 1$   $\qquad\qquad\qquad\qquad\qquad [0 = 0]$

(b)   $\sec \theta = \dfrac{1}{\cos \theta}$   $\qquad\qquad\qquad\qquad \left[ \sec \theta \tan \theta = \dfrac{\sin \theta}{\cos^2 \theta} \right]$

(c)   $\tan \theta = \dfrac{\sin \theta}{\cos \theta}$   $\qquad\qquad\qquad\qquad \left[ \sec^2 \theta = \dfrac{\cos^2 \theta + \sin^2 \theta}{\cos^2 \theta} \right]$

(d)   $\sin 2\theta = 2 \sin \theta \cos \theta$   $\qquad\qquad [\cos 2\theta = \cos^2 \theta - \sin^2 \theta]$

(e)   $\sin (\theta + a) = \sin \theta \cos a + \cos \theta \sin a$

$$[\cos (\theta + a) = \cos \theta \cos a - \sin \theta \sin a]$$

(f)   $\tan (\theta + a) = \dfrac{\tan \theta + \tan a}{1 - \tan \theta \tan a}$

$$\left[ \sec^2 (\theta + a) = \frac{\sec^2 \theta (1 + \tan^2 a)}{(1 - \tan \theta \tan a)^2} \right]$$

(g)   $\sec^2 \theta + \csc^2 \theta = \csc^2 \theta \sec^2 \theta$

$$[\sec^2 \theta \tan \theta - \csc^2 \theta \cot \theta = \csc^2 \theta \sec^2 \theta \tan \theta - \sec^2 \theta \csc^2 \theta \cot \theta]$$

66.   Derive the formula for the derivative of cos $x$ from the derivative of sin $x$ and the identity $\sin^2 x + \cos^2 x = 1$.

*67.   Differentiate $\tan \theta$ with respect to $\sin \theta$, where $\theta$ is in the interval

$-\pi/2 < \theta < \pi/2$.                    $[d \tan \theta / d \sin \theta = 1/\cos^3 \theta]$

*68.   Differentiate $\sin \theta$ with respect to $\tan \theta$, where $\theta$ is in the interval

$-\pi/2 < \theta < \pi/2$.                    $[d \sin \theta / d \tan \theta = \cos^3 \theta]$

## 6   The Inverse Trigonometric Functions

In Section 4 we were able to obtain $D \exp_a$ because $\exp_a$ is the inverse of $\log_a$. The same idea applies to the inverse trigonometric functions.* First, one has to restrict the domain of the trigonometric functions in order that unique inverses exist. For the sine function this restricted domain is shown at the right.

If $y = \text{Arc sin } u$, where $u$ is some function of $x$, then

(1)                    $\sin y = u$.

Now differentiate the function in equation (1) to obtain

$$\cos y \frac{dy}{dx} = \frac{du}{dx},$$

(2)                    $$\frac{dy}{dx} = \frac{1}{\cos y} \frac{du}{dx}.$$

In many respects equation (2) is perfectly adequate, though somewhat unusual. What we desire, of course, is to express $1/\cos y$ in terms of $u$—but that is an easy exercise in trigonometry; all we must do is express $\cos y$ in terms of $\sin y = u$. We have

(3)   $\cos y = +\sqrt{1 - \sin^2 y} = +\sqrt{1 - u^2}$

because $-\pi/2 \leq y = \text{Arc sin } u \leq \pi/2$, and in this interval $\cos y$ is always positive or zero, so we must use the positive square root.

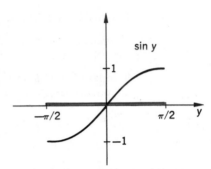

The graph of the sine function with domain restricted to $[-\frac{\pi}{2}, \frac{\pi}{2}]$

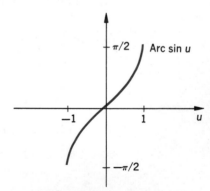

The principal inverse of the sine function $-\frac{\pi}{2} \leq \text{Arc sin } u \leq \frac{\pi}{2}$. Observe that the slope is always positive

* Because these functions are inverses of differentiable functions, they too are differentiable. See Section 7.

When (3) is used in (2) we obtain the desired formula

**XVI**
$$\frac{d}{dx} \text{Arc sin } u = \frac{1}{\sqrt{1 - u^2}} \frac{du}{dx}.$$

Observe that $u$ cannot be $\pm 1$ in this formula.
To obtain $D$ Arc cos $u$, a similar procedure is followed.
If $y = $ Arc cos $u$,

then
$$\cos y = u,$$

and
$$-\sin y \frac{dy}{dx} = \frac{du}{dx}.$$

One obtains

**XVII**
$$\frac{d}{dx} \text{Arc cos } u = \frac{-1}{\sqrt{1 - u^2}} \frac{du}{dx}.$$

The details are to be developed in Problem 2.

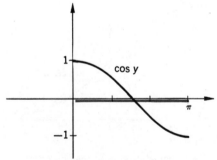

The graph of the cosine function with
domain restricted to $[0, \pi]$

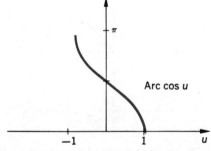

The principal inverse of the cosine function
$0 \leqslant$ Arc cos $u \leqslant \pi$. Observe that the slope
is always negative

The remaining formulas,

**XVIII**
$$\frac{d}{dx} \text{Arc tan } u = \frac{1}{1 + u^2} \frac{du}{dx},$$

**XIX**
$$\frac{d}{dx} \text{Arc cot } u = \frac{-1}{1 + u^2} \frac{du}{dx}$$

are even easier to derive because there is no problem of ambiguity in sign, as in equation (3). The principal inverses of the tangent and cotangent functions are shown in the figures below.

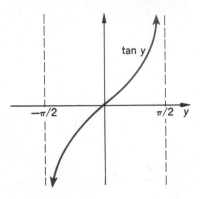

The graph of the tangent function
with domain restricted to $(-\pi/2, \pi/2)$

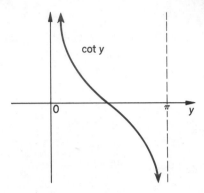

The graph of the cotangent function
with domain restricted to $(0, \pi)$

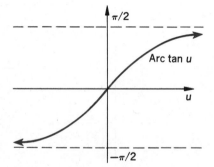

The principal inverse of the tangent
function $-\pi/2 <$ Arc tan $u < \pi/2$
observe that the slope is always
positive

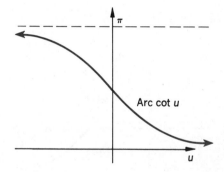

The principal inverse of the cotangent
function $0 <$ Arc cot $u < \pi$
observe that the slope is always negative

Naturally, there are also functions inverse to the secant and cosecant when their domains are suitably restricted, but on the whole they are seldom used, and, moreover, one can always avoid them. For example, if sec $y = u$, then cos $y = 1/u$, and $y =$ Arc cos $1/u$.

*Example 1*    $\dfrac{d}{d\theta}$ Arc sin $(2\theta - \theta^2) = \dfrac{1}{\sqrt{1 - (2\theta - \theta^2)^2}} \dfrac{d}{d\theta}(2\theta - \theta^2)$

$$= \frac{2(1 - \theta)}{\sqrt{1 - 4\theta^2 + 4\theta^3 - \theta^4}}.$$

**Example 2**   $\dfrac{d}{dx} \text{Arc tan} \dfrac{a}{x} = \dfrac{1}{1 + \dfrac{a^2}{x^2}} \dfrac{d}{dx} \left(\dfrac{a}{x}\right)$

$$= \dfrac{1}{1 + \dfrac{a^2}{x^2}} \dfrac{-a}{x^2} = \dfrac{-a}{a^2 + x^2} .$$

Observe that this is the same as $D$ Arc cot $(x/a)$.

### Problems

*1.* Verify from the graph that the formula for $D$ Arc tan $u$ appears to predict the correct slope.

*2.* Complete the proof of the formula for $D$ Arc cos $u$.

*3.* Derive the formula for $D$ Arc tan $u$.

*4.* Derive the formula for $D$ Arc cot $u$.

**Find the derivatives of the following functions.**

*5.* $y = \text{Arc sin} \dfrac{x}{a}, \quad a > 0$ $\qquad\qquad \left[ y' = \dfrac{1}{\sqrt{a^2 - x^2}} \right]$

*6.* $y = \text{Arc cos} \dfrac{x}{a}, \quad a > 0$ $\qquad\qquad \left[ y' = \dfrac{-1}{\sqrt{a^2 - x^2}} \right]$

*7.* $y = \text{Arc tan} \dfrac{x}{a}$ $\qquad\qquad \left[ y' = \dfrac{a}{a^2 + x^2} \right]$

*8.* $f(x) = \text{Arc cos } ax$ $\qquad\qquad [f'(x) = -a/\sqrt{1 - a^2 x^2}]$

*9.* $f(x) = \text{Arc tan } ax$ $\qquad\qquad [f'(x) = a/(1 + a^2 x^2)]$

*10.* $\varphi(x) = \text{Arc cos }(1/x)$ $\qquad\qquad [\varphi'(x) = 1/|x|\sqrt{x^2 - 1}]$

*11.* $f(x) = \text{Arc sec } x, \quad 0 < x < \pi/2$ $\qquad\qquad [f'(x) = 1/x\sqrt{x^2 - 1}]$

*12.* $\varphi(\theta) = \text{Arc sin }\sqrt{\theta}$ $\qquad\qquad [D\varphi(\theta) = 1/2\sqrt{\theta - \theta^2}]$

*13.* $\varphi(\theta) = \text{Arc tan }\sqrt{\theta}$ $\qquad\qquad [D\varphi = 1/2\sqrt{\theta}(1 + \theta)]$

*14.* $w = \text{Arc sin }\pi\theta^2$ $\qquad\qquad \left[ \dfrac{dw}{d\theta} = \dfrac{2\pi\theta}{\sqrt{1 - \pi^2\theta^4}} \right]$

*15.* $u = (1 + t^2) \text{Arc tan } t$ $\qquad\qquad \left[ \dfrac{du}{dt} = 1 + 2t \text{ Arc tan } t \right]$

16. $y = \text{Arc sin } \dfrac{a}{x}$
$$\left[\dfrac{dy}{dx} = \dfrac{-a}{|x|\sqrt{x^2 - a^2}}\right]$$

17. $y = \text{Arc csc } \dfrac{x}{a}, \quad a, x > 0$
$$\left[\dfrac{dy}{dx} = \dfrac{-a}{x\sqrt{x^2 - a^2}}\right]$$

18. $g(x) = \text{Arc cos } (2 - 3x)$
$$[Dg(x) = \sqrt{3}/\sqrt{-1 + 4x - 3x^2}]$$

19. $h(x) = \text{Arc sin } (ax + b)$
$$[Dh(x) = a/\sqrt{1 - b^2 - 2abx - a^2x^2}]$$

20. $y = \text{Arc cos } \dfrac{a}{\sqrt{a^2 + x^2}}$
$$\left[y' = \dfrac{ax}{|x|\,(a^2 + x^2)}\right]$$

21. $\psi(t) = \text{Arc cot } \dfrac{t}{t^2 - 1}$
$$\left[\psi'(t) = \dfrac{t^2 + 1}{t^4 - t^2 + 1}\right]$$

22. $z(t) = \text{Arc sin } \dfrac{t - 1}{2}$
$$\left[\dfrac{dz}{dt} = \dfrac{1}{\sqrt{3 + 2t - t^2}}\right]$$

23. $u(\theta) = \text{Arc tan } (\theta^2 - a^2)$
$$[du/d\theta = 2\theta/(1 + (\theta^2 - a^2)^2)]$$

24. $v(\theta) = \text{Arc sin } (2\theta - 1)$
$$[dv/d\theta = 1/\sqrt{\theta - \theta^2}]$$

25. $w(\theta) = \text{Arc cos } \sqrt{1 - \theta^2}$
$$[Dw(\theta) = \theta/|\theta|\sqrt{1 - \theta^2}]$$

26. $y = \text{Arc tan } \sqrt{x - 1}$
$$[y' = 1/2x\sqrt{x - 1}]$$

27. $T = \text{Arc sin } e^t$
$$\left[\dfrac{dT}{dt} = \dfrac{e^t}{\sqrt{1 - e^{2t}}} = \dfrac{1}{\sqrt{e^{-2t} - 1}}\right]$$

28. $y = \log \text{Arc tan } x$
$$\left[y' = \dfrac{1}{(1 + x^2)\,\text{Arc tan } x}\right]$$

29. $w = \text{Arc tan } \dfrac{ax + 1}{a - x}$
$$\left[\dfrac{dw}{dx} = \dfrac{1}{1 + x^2}\right]$$

30. $F(x) = \text{Arc sin } \sqrt{1 - x^2}, \quad x > 0$
$$[DF(x) = -1/\sqrt{1 - x^2}]$$

31. $\psi(\theta) = \text{Arc cot } (\theta/2)$
$$[D\psi(\theta) = -2/(4 + \theta^2)]$$

32. $X(\theta) = \text{Arc tan } \dfrac{\theta}{\sqrt{1 - \theta^2}}$
$$\left[\dfrac{dX}{d\theta} = \dfrac{1}{\sqrt{1 - \theta^2}}\right]$$

33. $y = \text{Arc sin } \dfrac{x - 1}{x + 1}, \quad x > 0$
$$\left[y' = \dfrac{1}{(x + 1)\sqrt{x}}\right]$$

34.  $y = (\cot\theta)\,\text{Arc}\cot\theta$  $\left[\,y' = (-\csc^2\theta)\,\text{Arc}\cot\theta - \dfrac{\cot\theta}{1+\theta^2}\,\right]$

35.  $\eta(\theta) = \theta\,\text{Arc}\tan 2\theta$  $\left[\,\dfrac{d\eta}{d\theta} = \text{Arc}\tan 2\theta + \dfrac{2\theta}{1+4\theta^2}\,\right]$

36.  $\zeta(\theta) = \text{Arc}\sin\log x$  $[\,D\zeta(\theta) = 1/x\,\sqrt{1-\log^2 x}\,]$

37.  $\varphi(\theta) = 2\,\text{Arc}\tan\sqrt{\dfrac{1-\cos\theta}{1+\cos\theta}},\quad 0<\theta<\pi$  $[\,\varphi'(\theta)=1\,]$

38.  $\gamma(t) = e^{\text{Arc}\sin t}$  $[\,D\gamma(t) = e^{\text{Arc}\sin t}/\sqrt{1-t^2}\,]$

39.  $\delta(t) = \text{Arc}\sin\sqrt{1-t}$  $[\,D\delta(t) = -1/2\,\sqrt{t-t^2}\,]$

40.  $\epsilon(\theta) = \text{Arc}\tan\tan\theta$  $[\,d\epsilon/d\theta = 1\,]$

41.  $\zeta(\theta) = \tan\text{Arc}\tan\theta$  $[\,D\zeta(\theta) = 1\,]$

42.  $\eta(\theta) = \text{Arc}\sin\sin\theta$  $\left[\,D\eta(\theta) = \dfrac{\cos\theta}{|\cos\theta|} = \pm 1\,\right]$

43.  $\lambda(\theta) = \text{Arc}\sin\sin\sqrt{\theta}$  $\left[\,D\lambda(\theta) = \dfrac{1}{2\sqrt{\theta}}\dfrac{\cos\sqrt{\theta}}{|\cos\sqrt{\theta}|}\,\right]$

44.  $\mu(\theta) = \text{Arc}\sin\sqrt{\sin\theta}$  $[\,D\mu(\theta) = \cos\theta/2\sqrt{\sin\theta-\sin^2\theta}\,]$

45.  $\nu(\theta) = \sin\text{Arc}\sin\sin\theta$  $[\,D\nu(\theta) = \cos\theta\,]$

46.  $\rho(x) = a\,\text{Arc}\cos\left(1-\dfrac{x}{a}\right) - \sqrt{2ax-x^2},\, a>0$  $\left[\,\rho'(x) = \dfrac{x}{\sqrt{2ax-x^2}}\,\right]$

47.  $\sigma(x) = \text{Arc}\tan\dfrac{x}{a} - \log\sqrt{\dfrac{x+a}{x-a}}$  $\left[\,D\sigma(x) = \dfrac{2ax^2}{a^4-x^4}\,\right]$

48.  $\tau(x) = \sqrt{a^2-x^2} + a\,\text{Arc}\sin(x/a),\quad a>0$
$[\,D\tau(x) = \sqrt{(a-x)/(a+x)}\,]$

49.  $f(\theta) = \theta^{\text{Arc}\sin\theta}$  $[\,f'(\theta) = \theta^{\text{Arc}\sin\theta}[\,(\text{Arc}\sin\theta)/\theta + (\log\theta)/\sqrt{1-\theta^2}\,]\,]$

50.  $F(\theta) = \sin\theta^{\text{Arc}\sin\theta}$  $\left[\,F'(\theta) = \sin\theta^{\text{Arc}\sin\theta}\left(\cot\theta\,\text{Arc}\sin\theta + \dfrac{\log\sin\theta}{\sqrt{1-\theta^2}}\right)\,\right]$

51.  $y = \text{Arc}\sin e^{\text{Arc}\tan x}$  $\left[\,y' = \dfrac{e^{\text{Arc}\tan x}}{(1+x^2)\sqrt{1-e^{2\,\text{Arc}\tan x}}}\,\right]$

**52.**   $y = (\text{Arc cos } x)^{\text{Arc sin } x}$

$$\left[ y' = \frac{(\text{Arc cos } x)^{\text{Arc sin } x}}{\sqrt{1 - x^2}} \left( \log \text{Arc cos } x - \frac{\text{Arc sin } x}{\text{Arc cos } x} \right) \right]$$

**53.**   Now return to Problem 5 and do as many as you can mentally, or by writing down as few steps as possible.

**\*54.**   According to the result of Problem 40, the function $f(x) = x - \text{Arc tan tan } x$ is a constant. Find $f(0)$ and $f(\pi)$. How do you explain these answers in view of the theorem on page 21?

**\*55.**   Differentiate each member of the following identities to obtain either new identities or familiar ones.

(a)   $\text{Arc cos } x + \text{Arc sin } x = \text{constant}$   (what constant?)

(b)   $\text{Arc tan } x + \text{Arc cot } x = \text{constant}$   (what constant?)

(c)   $\text{Arc tan tan } x = x + \text{constant}$   (what constant?)

(d)   $\tan (\text{Arc tan } \theta + \text{Arc tan } a) = \dfrac{\theta + a}{1 - a\theta}$

**\*56.**   Differentiate Arc tan $x$ with respect to Arc sin $x$.

$$\left[ \frac{d \text{ Arc tan } x}{d \text{ Arc sin } x} = \frac{\sqrt{1 - x^2}}{1 + x^2} \right]$$

**\*57.**   Differentiate $x^{\text{Arc sin } x}$ with respect to Arc sin $x$.

$$\left[ \frac{dx^{\text{Arc sin } x}}{d \text{ Arc sin } x} = x^{\text{Arc sin } x} \left( \log x + \frac{1}{x} \text{Arc sin } x \cos \text{Arc sin } x \right) \right]$$

# 7   *Elementary Functions; Chain Rule; Inverse Functions*

The functions that are called elementary are generated by combining a few very simple ones by a few rules of combination. Here are the simple functions:

### THE SIMPLEST ELEMENTARY FUNCTIONS

| *Constant* | *Identity* | *Powers* |
|---|---|---|
| $f(x) = c$ | $f(x) = x$ | $f(x) = x^a$ |

| *Exponential* | *Logarithmic* |
|---|---|
| $f(x) = \exp_a x = a^x$ | $f(x) = \log_a x$ |

| *Trigonometric* | *Inverse Trigonometric* |
|---|---|
| $f(x) = \sin x, \sec x$ | $f(x) = \text{Arc} \sin x$ |
| $= \cos x, \csc x$ | $= \text{Arc} \cos x$ |
| $= \tan x, \cot x$ | $= \text{Arc} \tan x, \text{Arc} \cot x$ |

These few simple functions can be combined only in the following ways, they can be repeatedly combined any finite number of times, and the resultant function will always be an elementary function.

### THE METHODS OF COMBINATION

| *Addition, Subtraction* | *Multiplication* | *Division* |
|---|---|---|
| $f(x) \pm g(x)$ | $f(x)g(x)$ | $\dfrac{f(x)}{g(x)}, g(x) \neq 0$ |

| *Composition\* of Two Functions* |
|---|
| $f(g(x))$ |

**DEFINITION**   The class of all functions obtained from the simplest elementary functions by the methods of combination constitute the *elementary functions*.

*Examples*   $x + \text{Arc} \sin x$ is a sum of two elementary functions. $\sqrt{1 + x^2}$ is the composition of $\sqrt{u}$ and $u = 1 + x^2$. Therefore both of these functions are elementary functions.

---

\* The composition of $f$ and $g$, or the composite of $f$ and $g$, is also denoted by $f \circ g$. Thus,
$$f \circ g(x) = f(g(x)).$$

The basic algorithms numbered I, II, VI, and VIII to XIX show that the derivatives of the simplest elementary functions are also elementary functions. For example,

$$D \text{ Arc tan } x = \frac{1}{1 + x^2},$$

and $1/(1 + x^2)$ is clearly an elementary function.

It is also true that the *derivative* of *any* elementary function (no matter how complexly it is made up from a long series of sums, products, quotients, and compositions) is also an elementary function. This important fact we state as a theorem.

**THEOREM**    *The derivative operator maps elementary functions into elementary functions. In other words, the set of elementary functions is closed under the operation of derivation.*

To prove this result all we need show is that, if functions $f$ and $g$ are elementary, and their derivatives $f'$ and $g'$ are also elementary, then the methods of combination give functions whose derivatives are also elementary. Part of this is already shown by formulas IV, V, VI, and VII. These show that sums, products, powers, and quotients of elementary functions produce functions with derivatives that are elementary functions.

It remains, therefore, to prove that the operation of composition also leads to functions that have derivatives that are elementary functions. A proof is supplied by formula XX. To prove XX recall how composition is defined:

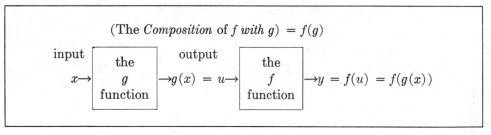

To prove XX, suppose that $y = f(u)$ and $u = g(x)$ and that $f$, $g$, $f'$, $g'$ are all elementary functions. We compute the derivative of $y = f(g(x))$.

$$\Delta y = f(u + \Delta u) - f(u)$$

$$= f'(u) \, \Delta u + \eta \, \Delta u,$$

where $\eta$ approaches zero as $\Delta u \to 0$, from the Fundamental Lemma (Chapter 1). Then we have

$$\frac{\Delta y}{\Delta x} = f'(u) \frac{\Delta u}{\Delta x} + \eta \frac{\Delta u}{\Delta x},$$

and when $\Delta x \to 0$, we obtain

$$\frac{dy}{dx} = f'(u) \frac{du}{dx} + 0 \frac{du}{dx},$$

because $\Delta u \to 0$ as $\Delta x \to 0$, and $\eta \to 0$ as $\Delta u \to 0$. Then

$$\frac{dy}{dx} = f'(u)\frac{du}{dx} = \frac{dy}{du}\frac{du}{dx}.$$

This gives the highly important *chain rule*:

**XX**    **(CHAIN RULE)**  $\dfrac{dy}{dx} = \dfrac{dy}{du}\dfrac{du}{dx}$

Observe the simplicity of the formula in this notation. The formula can also be expressed as:

$$\frac{d}{dx}f(g(x)) = f'(g(x))g'(x)$$

$$= (Df)(g(x))Dg(x).$$

From XX, in either of these last two forms, we see that the derivative of the composite $f(g(x))$ is an elementary function because both $f'(g(x))$ and $g'(x)$ are elementary.

*Remark.* The chain rule is used repeatedly in computing derivatives, but its full power will not be appreciated until techniques of integration are taken up in Chapter 10.

Formula XXI will be obtained by direct calculation. We assume that the inverse function exists, and so we will suppose that $f$ is increasing (if $f$ were decreasing the proof would look the same.)

The derivative $dx/dy$ of the inverse function is the limit, as $\Delta y \to 0$, of $\Delta x/\Delta y$. However,

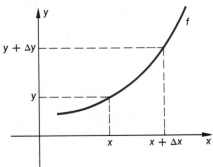

$$\frac{\Delta x}{\Delta y} = \frac{1}{\Delta y/\Delta x}.$$

Now, because $f$ is increasing,

$$\Delta y = f(x + \Delta x) - f(x)$$

is not zero, but $\Delta x$ and $\Delta y$ go to zero together. Therefore,

$$\lim_{\Delta y \to 0}\frac{\Delta x}{\Delta y} = \lim_{\Delta x \to 0}\frac{1}{\Delta y/\Delta x},$$

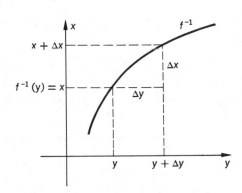

and we have the desired formula:

**XXI**    $\dfrac{dx}{dy} = \dfrac{1}{dy/dx}.$

*Remark 1.* Naturally this requires that $dy/dx \neq 0$.

*Remark 2.* In other notations we have

$$\frac{d}{dy} f^{-1}(y) = \frac{1}{f'(x)} = \frac{1}{f'(f^{-1}(y))}.$$

**Example 1**   If $y = 2u^2 - u + 1$   and   $u = x^2 + 1$, then

$$\frac{dy}{dx} = \frac{dy}{du}\frac{du}{dx} = (4u - 1)(2x).$$

This answer is often satisfactory in itself, but if one desires the derivative explicitly in terms of $x$, then

$$\frac{dy}{dx} = (4(x^2 + 1) - 1)(2x) = 2x(4x^2 + 3).$$

**Example 2**   Many problems that you have done using Algorithms I to XIX can also be done using XX. For example,

$$\frac{d}{dx} \log(x^2 - x + 1) = \frac{d}{du} \log u \frac{du}{dx}, \quad \text{where } u = x^2 - x + 1,$$

$$= \frac{1}{u}(2x - 1) = \frac{2x - 1}{x^2 - x + 1}.$$

$$\frac{d}{dx} \sqrt{x^2 - 4x + 9} = \frac{d}{du} \sqrt{u} \frac{du}{dx}, \quad \text{where } u = x^2 - 4x + 9$$

$$= \tfrac{1}{2}u^{-1/2}(2x - 4)$$

$$= \frac{x - 2}{\sqrt{x^2 - 4x + 9}}.$$

**Example 3**   If $y = x^3 + 3x + 3$, then

$$\frac{dx}{dy} = \frac{1}{dy/dx} = \frac{1}{3(x^2 + 1)}.$$

In this case the answer would be left in terms of $x$ because $x = f^{-1}(y)$ is not readily expressed.

But in a similar case we can get $dx/dy$ in terms of $y$: If $y = x^3 + 3$, then $x = (y - 3)^{1/3}$, and

$$\frac{dx}{dy} = \frac{1}{dy/dx} = \frac{1}{3x^2} = \frac{1}{3(y - 3)^{2/3}}.$$

*Problems*

In Problems 1 to 6 use the chain rule to find the derivative of the composite function.

1.  $y = 2u^3 - 3u^2 - 3u, \quad u = 2x^2 - 1$
$$\left[\frac{dy}{dx} = 12x\,(8x^4 - 12x^2 + 1)\right]$$

2.  $y = \dfrac{1}{2u-1} \quad, \quad u = \dfrac{1}{x+1}$
$$\left[\frac{dy}{dx} = \frac{2}{(1-x)^2}\right]$$

3.  $w = \text{Arc sin } (u + c), \quad u = \sin x$
$$\left[\frac{dw}{dx} = \frac{\cos x}{\sqrt{1 - (\sin x + c)^2}}\right]$$

4.  $w = \tan 2u \quad, \quad u = \text{Arc tan } \dfrac{x}{a}$
$$\left[\frac{dw}{dx} = \frac{2a}{a^2 + x^2}\sec^2 2u = \frac{2a\,(a^2 + x^2)}{(a^2 - x^2)^2}\right]$$

5.  $y = ue^{-u} \quad, \quad u = \log\,(x - 1)$
$$\left[\frac{dy}{dx} = \frac{1 - \log\,(x-1)}{(x-1)^2}\right]$$

6.  $y = \sec 2v \quad, \quad v = \text{Arc tan } t$
$$\left[\frac{dy}{dt} = \frac{2\sec 2v \tan 2v}{1 + t^2} = \frac{4t}{(1 - t^2)^2}\right]$$

In Problems 7 to 11 use Formula XXI to find the derivative of the inverse function. In all cases assume that the inverse does exist. The inequalities on x give a domain where this is the case.

7.  $y = \sqrt{2x + x^2} \quad, \quad x > 0$
$$\left[\frac{dx}{dy} = \frac{\sqrt{2x + x^2}}{1 + x} = \frac{y}{1 + x} = \frac{y}{\sqrt{1 + y^2}}\right]$$

8.  $y = \dfrac{x^2}{1 + x^2} \quad, \quad x > 0$
$$\left[\frac{dx}{dy} = \frac{(1 + x^2)^2}{2x} = \frac{1}{2\sqrt{y}\,(1 - y)^{3/2}}\right]$$

9.  $y = \tan x$
$$\left[\frac{dx}{dy} = \frac{1}{\sec^2 x} = \frac{1}{1 + y^2}\right]$$

10.  $y = \dfrac{x + 1}{x - 1} \quad, \quad x > 1$
$$\left[\frac{dx}{dy} = -\frac{(x - 1)^2}{2} = \frac{-2}{(y - 1)^2}\right]$$

11. $y = 6x - x^2$ , $x > 3$ $\qquad \left[ \dfrac{dx}{dy} = \dfrac{1}{2\,(3-x)} = \dfrac{-1}{2\sqrt{9-y}} \right]$

12. Interpret XXI in terms of slopes of $y = f(x)$ and $x = f^{-1}(y)$.

The following Problems 13 to 59 constitute a random list of elementary functions on which to practice. Observe that in each case the derivative is an elementary function.

13. $y = 3x^{3/4} - x^{3/2} + \sqrt{2}\,x^{-1/3} + \pi$ $\qquad \left[ y' = \tfrac{9}{4}x^{-1/4} - \tfrac{3}{2}x^{1/2} - \dfrac{\sqrt{2}}{3}x^{-4/3} \right]$

14. $f(x) = mx^{-n} + nx^{-m}$ $\qquad [\, f'(x) = -mn\,(x^{-n-1} + x^{-m-1})\,]$

15. $y = (x+a)^3/x$ $\qquad [\, y' = 2x + 3a - a^3/x^2 \,]$

16. $y = \dfrac{(x+a)^3}{x^3 + a^3}$ $\qquad \left[ y' = \dfrac{3a\,(x+a)\,(a-x)}{(x^2 - ax + a^2)^2} \right]$

17. $y = (3x)^{1/3} - (3x)^{-1/3}$ $\qquad [\, y' = (3x)^{-2/3}(1 + (3x)^{-2/3})\,]$

18. $f(x) = \dfrac{2x^2 + x - 6}{x + 2}$ $\qquad [\, Df(x) = 2 \,]$

19. $F(t) = (2 - t)(4 + 2t + t^2)$ $\qquad [\, DF(t) = -3t^2 \,]$

20. $y = (a^{3/2} - x^{3/2})^{2/3}$ $\qquad [\, y = -x^{1/2}/(a^{3/2} - x^{3/2})^{1/3} \,]$

21. $\varphi(\theta) = \theta/(\theta - a)$ $\qquad [\, \varphi'(\theta) = -a/(\theta - a)^2 \,]$

22. $\psi(x) = \dfrac{1}{x - \sqrt{x^2 - 1}}$ $\qquad \left[ \dfrac{d\psi}{dx} = 1 + \dfrac{x}{\sqrt{x^2 - 1}} \right]$

23. $g(x) = \sqrt{\dfrac{2x+1}{2x-1}}$ $\qquad \left[ Dg(x) = \dfrac{-2}{(2x+1)^{1/2}(2x-1)^{3/2}} \right]$

24. $h(x) = \log\sqrt{\dfrac{2x-1}{2x+1}}$ $\qquad \left[ h'(x) = \dfrac{2}{4x^2 - 1} \right]$

25. $\eta(t) = (3t^3 - 2t^2 + t - 1)^3$ $\qquad \left[ \dfrac{d\eta}{dt} = 3\,(9t^2 - 4t + 1)(3t^3 - 2t^2 + t - 1)^2 \right]$

26. $y = x^2 \log x$ $\qquad [\, y' = x + 2x \log x \,]$

27. $y = \log \dfrac{\sqrt[3]{a^3 - x^3}}{\sqrt{a^2 - x^2}}$ $\qquad \left[ y' = \dfrac{a^2 x}{(a-x)(a+x)(a^2 - ax + x^2)} \right]$

28. $F(x) = e^x \log \tan x$ $\left[ F'(x) = e^x \left( \log \tan x + \dfrac{1}{\sin x \cos x} \right) \right]$

29. $y = 3\sqrt{9 - x^2} + 2 \text{ Arc sin } (x/3)$ $[y' = (2 - 3x)/\sqrt{9 - x^2}]$

30. $y = \log (x + a + \sqrt{2ax + x^2})$ $[y' = 1/\sqrt{2ax + x^2}]$

31. $y = \sqrt{x^2 + a^2} + b \log (x + \sqrt{x^2 + a^2})$ $[y' = (x + b)/\sqrt{x^2 + a^2}]$

32. $B(t) = \text{Arc tan } \sqrt{t^2 - 1}$ $\left[ B'(t) = \dfrac{1}{t\sqrt{t^2 - 1}} \right]$

33. $T(\theta) = \text{Arc tan } \dfrac{a + \theta}{1 - a\theta}$ $\left[ T'(\theta) = \dfrac{1}{1 + \theta^2} \right]$

34. $w = \text{Arc tan } \dfrac{2 \tan \theta}{1 - \tan^2 \theta}$ $\left[ \dfrac{dw}{d\theta} = 2 \right]$

35. $\sigma(x) = e^{\log \sqrt{x}}$ $\left[ D\sigma(x) = \dfrac{1}{2\sqrt{x}} \right]$

36. $\tau(x) = e^{\sqrt{ax}}$ $\left[ D\tau(x) = \dfrac{ae^{\sqrt{ax}}}{2\sqrt{ax}} \right]$

37. $\lambda(t) = e^{\text{Arc sin } 2t}$ $\left[ \dfrac{d\lambda}{dt} = \dfrac{2}{\sqrt{1 - 4t^2}} e^{\text{Arc sin } 2t} \right]$

38. $f(\theta) = \text{Arc sin } (2 \sin \theta \cos \theta)$ $\left[ f'(\theta) = 2 \dfrac{\cos 2\theta}{|\cos 2\theta|} \right]$

39. $x = e^t (\sin 3t - 3 \cos 3t)$ $[dx/dt = 10e^t \sin 3t]$

40. $z = \sin^4 2\theta - \cos^4 2\theta$ $[dz/d\theta = 4 \sin 4\theta]$

41. $y = x^2 \sin (\pi/x)$ $[y' = -\pi \cos (\pi/x) + 2x \sin (\pi/x)]$

42. $z = \dfrac{1 - \cos ax}{\sin ax}$ $\left[ z' = \dfrac{a}{2} \sec^2 2ax \right]$

43. $w = \log \sqrt{(a + bx^c)^m}$ $\left[ Dw = \dfrac{mbcx^{c-1}}{2(a + bx^c)} \right]$

44. $k(x) = \text{Arc sin } (x/\sqrt{x^2 + 1})$ $[k'(x) = 1/(1 + x^2)]$

45.   $l(x) = \text{Arc cos } \sqrt{1 - x^2}$   $[l'(x) = x/|x|\sqrt{1-x^2}]$

46.   $M(x) = \text{Arc tan } \dfrac{x}{a} + \log \sqrt{\dfrac{x-a}{x+a}}$   $\left[ M'(x) = \dfrac{2ax^2}{x^4 - a^4} \right]$

47.   $N(x) = \log_{10}(2x^2 - 7x + 15)$   $\left[ N'(x) = 0.4343 \dfrac{4x-7}{2x^2 - 7x + 15} \right]$

48.   $w = \log(2x^3 \sqrt{4 - x^2})$   $\left[ w' = \dfrac{3}{x} - \dfrac{x}{4-x^2} = \dfrac{4(3-x^2)}{x(4-x^2)} \right]$

49.   $y = x\sqrt{a^2 - x^2} + a^2 \text{ Arc sin }(x/a)$   $[y' = 2\sqrt{a^2 - x^2}]$

50.   $f(x) = \text{Arc cot } \dfrac{1}{x} - \log \dfrac{1-x}{1+x}$   $\left[ f'(x) = \dfrac{3+x^2}{1-x^4} \right]$

51.   $z = x^3 \text{ Arc tan } x + \log \sqrt{1+x^2} - \tfrac{1}{2}x^2$   $[dz/dx = 3x^2 \text{ Arc tan } x]$

52.   $w = x + \sqrt{1-x^2} \text{ Arc cos } x$   $\left[ \dfrac{dw}{dx} = -\dfrac{x \text{ Arc cos } x}{\sqrt{1-x^2}} \right]$

53.   $u(x) = \text{Arc tan } \dfrac{b \sin x}{a + b \cos x}$   $\left[ \dfrac{du}{dx} = \dfrac{b(b + a \cos x)}{a^2 + 2ab \cos x + b^2} \right]$

54.   $v(x) = \log(\sec ax + \tan ax)$   $[Dv(x) = a \sec ax]$

55.   $w(\theta) = \text{Arc sin } \sqrt{\tan \theta}$   $[Dw(\theta) = \sec^2 \theta / 2\sqrt{\tan \theta - \tan^2 \theta}]$

56.   $R(x) = xe^{-1/x}$   $[DR(x) = (1 + 1/x)e^{-1/x}]$

57.   $S(x) = (e^x)^x$   $[S'(x) = 2xe^{x^2}]$

58.   $f(t) = t\sqrt{4 - 3t^2}$   $[f'(t) = (4 - 6t^2)/\sqrt{4 - 3t^2}]$

59.   $y = \dfrac{\sin ax}{x}$   $\left[ \dfrac{dy}{dx} = \dfrac{ax \cos ax - \sin ax}{x^2} \right]$

60.   Now try to do Problems 13 to 60 mentally. At this time you should succeed pretty well at this mental exercise.

**Find the following derivatives.**

61.   $\dfrac{d}{dx}\left(\dfrac{2}{x}\right)^{x/2}$   $\left[ \dfrac{1}{2}\left(\dfrac{2}{x}\right)^{x/2}\left(\log \dfrac{2}{x} - 1\right) \right]$

62. $\dfrac{d}{dx}\left(\dfrac{x}{2}\right)^{2/x}$ $\qquad\qquad\qquad\left[\dfrac{2}{x^2}\left(\dfrac{x}{2}\right)^{2/x}\left(1-\log\dfrac{x}{2}\right)\right]$

63. $\dfrac{d}{dx}\,x^{\log x}$ $\qquad\qquad\qquad\qquad\qquad\left[\dfrac{2}{x}\,x^{\log x}\log x\right]$

*64.  Find the derivative of Arc tan $x/\sqrt{1-x^2}$ with respect to Arc cos $(2x^2-1)$. (Assume $x>0$.) $\qquad\qquad\qquad\qquad\qquad\qquad\qquad [-\tfrac{1}{2}]$

*65.  Find the derivative of Arc cos $x$ with respect to Arc sin $x$. (Assume $x>0$.) $\qquad\qquad\qquad\qquad\qquad\qquad\qquad\qquad\qquad [-1]$

## CHAPTER 3

## 1 *Functions Defined Implicitly*

We have been differentiating elementary functions that are given by rather simple formulas:

$$(1) \qquad\qquad y = f(x),$$

where $f$ is given explicitly in terms of the simplest elementary functions.

Suppose now that one did not, as in (1), have a formula for $y$ explicitly in terms of $x$ so that one had to solve for $y$. For example,

$$3x^2 + 2\sqrt{x} - 7 - y = 0 \qquad \text{instead of} \qquad y = 3x^2 + 2\sqrt{x} - 7, \quad \text{or}$$

$$x^2 e^x - \sin y = 5 \qquad \text{instead of} \qquad y = \text{Arc} \sin (x^2 e^x - 5), \quad \text{or}$$

$$x - y^2 = 0 \qquad \text{instead of} \qquad y = \sqrt{x} \ \text{ or } \ y = -\sqrt{x}, \quad \text{or}$$

$$x^2 + y^2 = 9 \qquad \text{instead of} \qquad y = \pm\sqrt{9 - x^2}.$$

Equations such as these, which we may suggest symbolically by

$$(2) \qquad\qquad F(x, y) = 0,$$

are said "to define* $y$ implicitly as a function of $x$." In the examples that we will encounter this will actually be the case (often evidently so), and so we shall *assume* that equation (2) determines one, or more, differentiable functions $y(x)$. The process of finding $y'(x)$ without actually solving for $y(x)$ is called *implicit differentiation*.

**Example 1**    If $x^2 + y^2 = 9$ determines a differentiable function $y(x)$, then

(3) $$x^2 + [y(x)]^2 = 9$$

must be an identity in $x$. Hence $x^2 + [y(x)]^2$ is the constant function 9, even though it may not look like it.

In this case, of course, it is easy to solve for $y(x)$. We get either $y(x) = \sqrt{9 - x^2}$ or $y(x) = -\sqrt{9 - x^2}$, and

$$x^2 + [y(x)]^2 = x^2 + [\pm\sqrt{9 - x^2}]^2 = x^2 + (9 - x^2) = 9.$$

Suppose we differentiate the constant function in the left member of (3). We get

$$\frac{d}{dx}(x^2 + y^2) = \frac{d}{dx} 9$$

$$2x + 2yy' = 0.$$

Hence, $$y' = -\frac{x}{y}.$$

This is a perfectly good equation for finding the slope of the tangent line at a point $(x, y)$ on the circle. Note that this is the same answer, though differently expressed, as we get by first solving for $y$:

$$y = \pm\sqrt{9 - x^2}.$$

$$y' = \pm\tfrac{1}{2}(9 - x^2)^{-1/2}(-2x) = \frac{-x}{\pm\sqrt{9 - x^2}}$$

$$= -\frac{x}{y}.$$

**Example 2**    Find $dy/dx$ if $\sin y = x + y$.

Observe that, unlike Example 1, there is no easy way to solve for $y$. Differentiating each side of the equation, just as it stands, and remembering that $y$ is a differentiable function of $x$, we get

---

* A more careful formulation of what this means, and when one can be sure that there is a function defined, will be discussed in Chapter 14.

That there may be no function determined is evident from the example: $x^2 + y^2 + 1 = 0$. Clearly no real function $y = f(x)$ can satisfy this equation.

$$\frac{d}{dx}\sin y = \frac{d}{dx}(x+y)$$

$$\cos y \, y' = 1 + y'$$

$$y' = \frac{1}{\cos y - 1}.$$

**Example 3**    Find an equation of the tangent line to the graph of $x^3 + y^3 - 3x^2y^2 + 1 = 0$ at the point $(1,1)$.

Differentiating each side of the equation, just as it stands, and remembering that $y$ is a function of $x$, we get

$$3x^2 + 3y^2y' - 6xy^2 - 6x^2yy' = 0,$$

whence

(4)
$$y' = \frac{2xy^2 - x^2}{y^2 - 2x^2y}.$$

Then the slope of the tangent line at $(1,1)$ is obtained from (4) with $x = 1$, $y = 1$; $m = -1$. An equation of the tangent line is $y - 1 = -(x - 1)$.

### Problems

In Problems 1 to 10 find $y'$ by implicit differentiation.

1.    $2x^2 - 5y^2 = 10$ $\qquad\qquad\qquad\qquad\qquad$ $[y' = 2x/5y]$

2.    $x^3 + x^2y + xy^2 + y^3 = 4$ $\qquad\qquad$ $\left[y' = -\dfrac{3x^2 + 2xy + y^2}{x^2 + 2xy + 3y^2}\right]$

3.    $x^3 + y^3 - 3axy = 0$ $\qquad\qquad\qquad$ $\left[y' = \dfrac{ay - x^2}{y^2 - ax}\right]$

4.    $x^{2/3} + y^{2/3} = a^{2/3}$ $\qquad\qquad\qquad$ $[y' = -y^{1/3}/x^{1/3}]$

5.    $\tan y - y = \sin x + x$ $\qquad\qquad$ $[y' = \cot^2 y \,(\cos x + 1)]$

6.    $\log(x+y) = xe^y$ $\qquad\qquad\qquad$ $\left[y' = \dfrac{(x+y)e^y - 1}{1 - x(x+y)e^y}\right]$

7.    $\sin xy = y$ $\qquad\qquad\qquad\qquad$ $\left[y' = \dfrac{y\cos xy}{1 - x\cos xy}\right]$

8.    $x^y = y^x$ $\qquad\qquad\qquad\qquad$ $\left[y' = \dfrac{(x\log y - y)y}{(y\log x - x)x}\right]$

9.  $y - \text{Arc tan}\, \dfrac{x}{y} = 0$
$$\left[ y' = \frac{y}{x^2 + y^2 + x} \right]$$

10.  $\sin^2 y = a^2 \cos 2x$
$$\left[ y' = \frac{-a^2 \sin 2x}{\sin y \cos y} \right]$$

In Problems 11 to 18

(a) find $y'$ by implicit differentiation.

(b) Then solve for $y(x)$ and find $y'$ explicitly in terms of $x$.

(c) Obtain the solution in (b) by substituting $y(x)$ in the solution of part (a).

11.  $x^2 + xy - y^2 = 5$
$$\left[ \text{(a)}\ y' = \frac{2x + y}{2y - x} \right]$$

12.  $x^2 + 2x + xy = 3$
$$\left[ \text{(a)}\ y' = -\frac{2x + 2 + y}{x} \right]$$

13.  $y^2 + y = \log x$
$$\left[ \text{(b)}\ y' = \frac{\pm 1}{x\sqrt{1 + 4 \log x}} \right]$$

14.  $x^{1/2} + y^{1/2} = a^{1/2}$
$$\left[ \text{(b)}\ y' = -\frac{a^{1/2} - x^{1/2}}{x^{1/2}} \right]$$

15.  $\sin y = x^2 + x$
$$\left[ \text{(b)}\ y' = \frac{\pm (2x + 1)}{\sqrt{1 - (x^2 + x)^2}} \right]$$

16.  $\log \sin \dfrac{y}{a} = \sqrt{x}$
$$\left[ \text{(b)}\ y' = \frac{\pm a e^{\sqrt{x}}}{2\sqrt{x}\ \sqrt{1 - e^{2\sqrt{x}}}} \right]$$

17.  $y^2 (2a - x) = x^3$   (Cissoid)
$$\left[ \text{(b)}\ y' = \frac{\pm (3a - x) x^{1/2}}{(2a - x)^{3/2}} \right]$$

18.  $(y - x^2)^2 = x^5$
$$\left[ \text{(a)}\ y' = \frac{5x^4 + 4x (y - x^2)}{2 (y - x^2)} \right]$$

In Problems 19 to 21 find an equation of the tangent line at the indicated point.

19.  $x^2 + xy - y^2 = 5$ ; $(3, 4)$    $[y - 2x + 2 = 0]$

20.  $\log (x + y) = xe^y$ ; $(0, 1)$    $[(1 - e)x + y = 1]$

*21.* $x^3 + x^2y + xy^2 + y^3 = 5$;  $(\frac{1}{2}, \frac{3}{2})$ $\qquad$ $[17y + 9x - 30 = 0]$

*22.* Obtain the formula for $Dx^{p/q}$, where $p$ and $q$ are positive integers, by considering $y = x^{p/q}$, as defined implicitly by $y^q = x^p$.

*23.* Obtain the formula for differentiating $u = f^{-1}(x)$ from implicit differentiation of $f(u) = x$.

*24.* Suppose $\qquad y = \sqrt{\sin x + \sqrt{\sin x + \sqrt{\sin x + \sqrt{\cdots \text{ etc.}}}}}$

where the square roots continue indefinitely. Suppose (take for granted) that this defines, in some sense, $y$ as a differentiable function of $x$. Show that

$$y' = \frac{\cos x}{2y - 1}.$$

*25.* Suppose $\qquad y = x + \cfrac{1}{x + \cfrac{1}{x + \cfrac{1}{x + \text{etc.}}}}$

where the "infinite continued fraction" keeps on forever. Furthermore, suppose that this defines $y$ as a differentiable function. Show that

$$y' = \frac{y}{2y - x} = \frac{y^2}{1 + y^2}.$$

## 2  *Algebraic Functions*

These are the functions that one gets from constant functions and the identity function by algebraic operations. To begin with, we have:

| POLYNOMIALS | and | RATIONAL FUNCTIONS |
|---|---|---|
| $f(x) = a_0x^n + a_1x^{n-1} + \cdots + a_n$ | | $f(x) = \dfrac{\text{a polynomial in } x}{\text{a polynomial in } x}$ |
| *e.g.,*  $f(x) = 3x^4 - 2x^3 + 7x + 2$ | | *e.g.,*  $f(x) = \dfrac{x^2 + x - 7}{x^3 - 2}$ |

Then if one permits square roots, cube roots, etc., one gets more* functions, as shown below:

$$f(x) = \sqrt{x}; \qquad\qquad f(x) = (x^2 + 1)^{1/4} + x;$$

$$f(x) = x^{2/3} + x^{-1/2}; \qquad\qquad f(x) = (x^4 + 2x^2 + 1)^{1/3} + x^{2/3}.$$

But these are not all. Any polynomial equation in $x$ and $y$ for example,

$$(1) \qquad (x^2 + 3x - 1)y^9 - (\sqrt{2}x^5 - x^3 + x + 2)y^7 + x^2y^3 + x^4y - 1 = 0$$

defines $y$ implicitly as a differentiable *algebraic* function of $x$. Indeed, because equation (1) is of degree nine in $y$ there will be in general nine choices for $y$ for each $x$, possibly some complex. The various kinds of algebraic functions are as follows, where $A(x)$, $B(x)$, $A_0(x)$, $\cdots$, $A_n(x)$ are *polynomials* in $x$:

polynomials or

rational integral functions $\qquad :\qquad y - A(x) = 0$

rational (fractional) functions $\qquad :\qquad A(x)y - B(x) = 0$

general algebraic function (irrational): $\qquad A_0(x)y^n + \cdots + A_n(x) = 0$

*Remark.* It is established in advanced algebra that sums, products, and quotients of algebraic functions are also algebraic. This is not a trivial result to prove, but, granting that it is true, one can then show that the derivative of an algebraic function is algebraic.

**THEOREM**    *If $y(x)$ is an algebraic function then $y'(x)$ is an algebraic function.*

We shall illustrate the theorem by an example. The argument is the same in general. If $y(x)$ is defined implicitly by

$$(x^2 + 1)y^4 - (7x - 2)y^2 + x^2 - x + 1 = 0,$$

then $\qquad 4(x^2 + 1)y^3y' + 2xy^4 - 2(7x - 2)yy' - 7y^2 + 2x - 1 = 0;$

$$y'(x) = -\frac{2xy^4 - 7y^2 + 2x - 1}{4(x^2 + 1)y^3 - 2(7x - 2)y}.$$

In this formula, $y'$ is obtained from algebraic functions (constants, $x$, and $y$) by the operations of addition, multiplication, and division. By the remark above, $y'$ is algebraic.

---

* It is not immediately evident that $f(x) = \sqrt{x}$, for example, is not a polynomial, or even a rational function, even though it does not look like it. That it is not a polynomial or rational function can be proved using the Fundamental Theorem of algebra. See Appendix A.

That not all functions are algebraic is true but not obvious. In fact, a proof is quite difficult. The nonalgebraic functions are called *transcendental*. The simplest transcendental functions are the exponential and trigonometric functions and their inverses.

The differentiable functions can be classified in a way suggested by the chart at the right.

| All differentiable functions | |
|---|---|
| Algebraic functions | Nonalgebraic (transcendental) functions |
| Rational functions | the simplest transcendental functions: exponential logarithmic, trigonometric, inverse trigonometric |
| Rational Integral functions (polynomials) | |

## Problems

In Problems 1 to 4 an algebraic function can be found explicitly. Find $y'$ in two ways.

1.  $(x^2 + 1)y^2 + 2xy + 1 - x = 0$
$$\left[ y' = \frac{1 - 2y - 2xy^2}{2(x^2 + 1)y + 2x} \right]$$

2.  $y^{2/3} + 2xy^{1/3} + x^2 - 1 = 0$
$$\left[ y' = -3(x \pm 1)^2 \right]$$

3.  $y^3 - 3y^2x + 3yx^2 - x^3 = x^2 + 1$
$$\left[ y' = 1 + \frac{2x}{3}(x^2 + 1)^{-2/3} \right]$$

4.  $y^n - x^n = a^n$, $n$ a positive integer
$$\left[ y' = \left(\frac{x}{y}\right)^{n-1} \right]$$

5.  Can a transcendental function have an algebraic function for its derivative? Why?

6.  If $f$ and $g$ are functions such that $f' = g'$ in an interval, and $g$ is algebraic, what is $f$? Why?

*7.  Explain why the sine function is not a polynomial function.

*8.  Explain why the sine function is not a rational function.

## 3 Higher Order Derivatives

The elementary functions are differentiable (except possibly at isolated points) and have derivatives that are also elementary functions. Hence the derivatives too are differentiable. Then $f'$ becomes the *first* derivative of $f$, $Df' = f''$ the *second* derivative, etc.

$$
\begin{aligned}
f(x) &= 2x^5 + x^2 \\
Df(x) = f'(x) &= 10x^4 + 2x \\
Df'(x) = f''(x) &= 40x^3 + 2 \\
Df''(x) = f'''(x) &= 120x^2 \\
&\text{etc.}
\end{aligned}
$$

There are many notations for the successive derivatives of $y = f(x)$. The more common ones are as follows:

$$\frac{dy}{dx}, \frac{d}{dx}\frac{dy}{dx} = \frac{d^2y}{dx^2}, \cdots, \frac{d}{dx}\frac{d^{n-1}y}{dx^{n-1}} = \frac{d^ny}{dx^n} = n\text{th derivative}$$

$$y', y'', y''', y^{iv}, \cdots, y^{(n)}$$

$$\frac{d}{dx}f(x), \frac{d^2}{dx^2}f(x), \cdots, \frac{d^n}{dx^n}f(x)$$

$$f'(x), f''(x), \cdots, f^{(n)}(x)*$$

$$Dy, D^2y, \cdots, D^ny$$

**Example 1** The derivative of a polynomial of degree $n$ is a polynomial of degree $n-1$:

$$\frac{d}{dx}(a_0x^n + a_1x^{n-1} + \cdots + a_n) = na_0x^{n-1} + (n-1)a_1x^{n-2} + \cdots + a_{n-1}.$$

The $n$th derivative is

$$\frac{d^n}{dx^n}(a_0x^n + \cdots + a_n) = n!a_0.$$

The $(n+1)$th derivative is 0.

**Example 2** If $y = A \sin kx$,
$$y' = kA \cos kx;$$
$$y'' = -k^2A \sin kx;$$
$$\cdots$$
$$y^{(2n)} = (-1)^nk^{2n}A \sin kx;$$
$$y^{(2n+1)} = (-1)^nk^{2n+1}A \cos kx.$$

---

* The notation $f$, $f'$, $f''$, $\cdots$, $f^{(n)}$ was devised by the great French analyst J. L. Lagrange (1736–1813). (See Chapter 10 of *Men of Mathematics* by E. T. Bell.) This notation is preferable to the $d^ny/dx^n$ notation because it makes it clear that the derivatives are functions.

*Example 3*    If a function $y$ is defined implicitly, its higher derivatives can also be found without solving for $y$. The chain rule applies.

If $y(x)$ is defined by $b^2x^2 + a^2y^2 = a^2b^2$, then

(1) $$2b^2x + 2a^2yy' = 0,$$

$$y' = -\frac{b^2x}{a^2y}, \quad \text{and}$$

$$\frac{d^2y}{dx^2} = y'' = Dy' = \frac{a^2y(-b^2) + (b^2x)a^2y'}{(a^2y)^2} = \frac{-a^2b^2y + a^2b^2x\left(-\dfrac{b^2x}{a^2y}\right)}{a^4y^2}$$

$$= \frac{-a^2b^2y^2 - b^4x^2}{a^4y^3} = \frac{-b^2(a^2y^2 + b^2x^2)}{a^4y^3}$$

$$= \frac{-b^2a^2b^2}{a^4y^3}, \quad \text{because } a^2y^2 + b^2x^2 = a^2b^2.$$

So $$y'' = \frac{-b^4}{a^2y^3}.$$

The second derivative can also be found by implicit differentiation from equation (1), instead of solving (1) for $y'$ and then differentiating. Thus, from

$$2b^2x + 2a^2yy' = 0,$$

we get $$b^2 + a^2yy'' + a^2(y')^2 = 0.$$

Then $$b^2 + a^2yy'' + a^2\left(-\frac{b^2x}{a^2y}\right)^2 = 0,$$

and, solving for $y''$, the same result is obtained.

The third order derivative can be obtained by differentiating $y''$.

### Problems

**In Problems 1 to 18 find the indicated derivative.**

1.  $f(x) = 2x^3 - \sqrt{6}x^2 + \sqrt{2}x - 1; \quad f''$ $\qquad$ $[12x - 2\sqrt{6}]$

2.  $y = \dfrac{x}{x-1}$ $\qquad ; \quad y''$ $\qquad$ $[2(x-1)^{-3}]$

3.  $\varphi(t) = \log(1 + at)$ $\qquad ; \quad \varphi'''(t)$ $\qquad$ $[2a^3(1 + at)^{-3}]$

4.  $z = A \sin kx$ $\qquad ; \quad D^2z$ $\qquad$ $[-k^2z]$

5.  $f(\theta) = \sin a\theta$ $\qquad ; \quad f^{(2n)}(\theta)$ $\qquad$ $[(-1)^n a^{2n} f(\theta)]$

6.  $y = (x^2 + 1) \operatorname{Arc} \tan x$ $\qquad ; \quad y'''$ $\qquad$ $\left[\dfrac{4}{(x^2+1)^2}\right]$

7.  $y(t) = e^{at}(A \sin bt + B \cos bt)$ ; $y''(t)$

$$[e^{at}[(a^2A - 2abB - b^2A) \sin bt + (a^2B + 2abA - b^2B) \cos bt]]$$

8.  $G(x) = \dfrac{x^3}{1-x}$        ; $D^4G(x)$           $[24/(1-x)^5]$

*9.  $F(x) = x^k \log x$       ; $F^{(k+1)}(x)$           $[k!/x]$

*Hint:* Use induction on the positive integer $k$.

10.  $y = \dfrac{1-x}{1+x}$        ; $D^n y$           $\left[\dfrac{2(-1)^n n!}{(1+x)^{n+1}}\right]$

11.  $x^2 - y^2 = 8$        ; $y''$           $[-8/y^3]$

12.  $y^2 + xy + x^2 = 5$     ; $D^2 y$        $[-30(2y+x)^{-3}]$

13.  $2x^2 + 3y^2 = 6$      ; $y'''$        $[-8x/(3y^5)]$

14.  $e^y + y = x$        ; $y'''$      $[e^y(e^y + 1)^{-5}(2e^y - 1)]$

15.  $u = \sin(u + x)$      ; $D^2 u(x)$      $\left[\dfrac{-\sin(u+x)}{(1 - \cos(u+x))^3}\right]$

16.  $\log(x - y) = x + y$     ; $y''$      $[-4(x-y)(1+x-y)^{-3}]$

17.  $y = a + bxe^y$       ; $y''$       $\left[\dfrac{2b^2e^{-y} - b^3x}{(e^{-y} - bx)^3}\right]$

18.  $e^{x+y} = y^2$        ; $y''$        $\left[\dfrac{2y}{(2-y)^3}\right]$

19.  Show that all derivatives of $Ae^{ax} \cos bx + Be^{ax} \sin bx$ are of the same form, namely $c_1 e^{ax} \cos bx + c_2 e^{ax} \sin bx$ for some constants $c_1, c_2$.

20.  If $f$ is any $n$-fold differentiable function then $D^n(e^{ax}f) = e^{ax}(D^n f + na D^{n-1}f + \cdots + \binom{n}{r}a^r D^{n-r}f + \cdots + a^n f)$, where $\binom{n}{r}$ is the binomial coefficient, $\binom{n}{r} = n!/r!(n-r)!$. Verify this formula for $n = 1, 2, 3, 4$.

*21.  Establish the general formula of Problem 20 by mathematical induction.

22.  The $n$th derivative of a product is given by Leibniz' formula:

$$D^n(uv) = u^{(n)}v + nu^{(n-1)}v' + \cdots + \binom{n}{r}u^{(n-r)}v^{(r)} + \cdots + v^{(n)}u$$

$$= \sum_{r=0}^{n} \binom{n}{r}u^{(n-r)}v^{(r)}.$$

Verify the formula for $n = 1, 2, 3, 4$.

*23.  Establish the general formula of Problem 22 by mathematical induction.

## *4   *Multiple Roots of Polynomial Equations*†

A polynomial equation of degree $n$,

$$(1) \qquad a_0 x^n + a_1 x^{n-1} + \cdots + a_n = 0, \quad a_0 \neq 0,$$

has multiple roots if the associated polynomial,

$$(2) \qquad \begin{aligned} P(x) &= a_0 x^n + a_1 x^{n-1} + \cdots + a_n \\ &= a_0 (x - r_1)(x - r_2) \cdots (x - r_n), \end{aligned}$$

when factored, has repeated factors.

If there are only $k < n$ distinct roots $r_1, r_2, \cdots, r_k$, then the polynomial $P(x)$ factors as

$$P(x) = a_0 (x - r_1)^{p_1} \cdots (x - r_k)^{p_k}$$

where the integer exponents $p_1, \cdots, p_k$ are such that

---

$$x^4 + 7x^3 + 18x^2 + 20x + 8 = 0$$

factors as

$$(x + 2)^3(x + 1) = 0$$

Thus, $-2$ is a triple root and $-1$ a simple root.

---

$$p_1 + p_2 + \cdots + p_k = n.$$

The numbers $p_1, \cdots, p_k$ are called the *multiplicities* of the roots $r_1, \cdots, r_k$.

It is possible to discover multiple roots of even quite high degree equations because of the following theorem which concerns a polynomial $P(x)$ and its derivative $P'(x)$. This theorem is valid even for complex roots, so to be precise we need to know that the formula for the derivative of a polynomial in a complex variable $x$ is exactly the same as that for a polynomial in a real variable. We shall not stop to prove this, but simply remark that the proof of the formula for the derivative of a polynomial

$$P(x) = a_0 x^n + a_1 x^{n-1} + \cdots + a_{n-1} x + a_n$$

is valid even if $x$ and $\Delta x$ are complex numbers; thus,

$$P'(x) = \lim_{\Delta x \to 0} \frac{P(x + \Delta x) - P(x)}{\Delta x} = a_0 n x^{n-1} + a_1 (n - 1) x^{n-2} + \cdots + a_{n-1}.$$

**THEOREM**   *If the polynomial equation $P(x) = 0$ has a multiple root $r$ of multiplicity $m > 1$, then $P(x)$ and $P'(x)$ have a common factor of $(x - r)^{m-1}$. If $P(x) = 0$ has no multiple roots, then $P(x)$ and $P'(x)$ have no common factor of degree greater than or equal to $1$.*

**PROOF**   Suppose $r$ is a multiple root, of multiplicity $m > 1$, of $P(x) = 0$. Then

$$P(x) = (x - r)^m Q(x),$$

---

† See Appendix A for elementary facts about polynomial equations.

where $Q(x)$ is a polynomial of degree $m$ less than the degree of $P(x)$.

$$P'(x) = (x - r)^{m-1}[mQ(x) + (x - r)Q'(x)].$$

Clearly $(x - r)^{m-1}$ divides both $P(x)$ and $P'(x)$.

Now suppose that $P(x)$ has no multiple roots. Then

$$P(x) = a_0(x - r_1)(x - r_2) \cdots (x - r_n),$$

where the $r_i$, $i = 1, \cdots, n$ are distinct, and

$$\begin{aligned}
(3) \qquad P'(x) = {} & a_0(x - r_1) \cdots (x - r_{n-1}) \\
& + a_0(x - r_1) \cdots (x - r_{n-2})(x - r_n) + \cdots \\
& + a_0(x - r_2) \cdots (x - r_n)
\end{aligned}$$

Since $P'(r_i) \neq 0$, as one sees from (3), it follows from the remainder theorem that $(x - r_i)$ is not a factor of $P'(x)$, for $i = 1, \cdots, n$. Thus, $P(x)$ and $P'(x)$ have no common factors.

**COROLLARY 1**   *The greatest common divisor of $P(x)$ and $P'(x)$ will determine all the multiple roots of $P(x) = 0$.*

**COROLLARY 2**   *If $P(x)$ has a root $r$ of multiplicity $m$, then $P(x)$ and $P^{(m-1)}(x)$ have a common factor $x - r$.*

> *Example 1*   Find the multiple roots, if any, of $x^3 - 3x^2 + 4 = 0$.
>
> Here,   $P(x) = x^3 - 3x^2 + 4$
>
> $\qquad P'(x) = 3(x^2 - 2x) = 3x(x - 2).$
>
> Since 0 is not a root of $P(x) = 0$, the only possible common factor of $P'(x)$ and $P(x)$ is $x - 2$. This is easily checked. $P(2) = 0$, whence, by the remainder theorem, $x - 2$ is a factor of $P(x)$. It must also be a root of multiplicity 2, since $(x - 2)$ occurs to the power 1 in $P'(x)$.
>
> $\qquad P(x) = x^3 - 3x^2 + 4 = (x - 2)^2(x + 1).$
>
> The roots are 2, 2, −1.

*Remark.* Often, by examining $P(x)$ and $P'(x)$, one can see by inspection that certain factors of $P'(x)$ cannot be factors of $P(x)$. An examination of possibilities can greatly simplify the work.

Moreover, as soon as one multiple root has been found the degree of $P(x)$ can be reduced by dividing out the multiple factor.

If two polynomials are each completely factored, then one can find the factors common to both. The *greatest common divisor* (gcd) of two polynomials is the polynomial of highest degree that will divide both. It is unique up to a constant factor. Thus, if the polynomials are completely factored the gcd is easily found by inspection. Since factoring polynomials is difficult, we need a way of finding the gcd that does not require factoring the polynomials. There is a simple direct technique for finding the gcd of two polynomials, called the Euclidean algorithm.* The next example will illustrate its use.

**Example 2**    Find multiple roots, if any, and solve the equation $x^4 + 4x^3 - 16x - 16 = 0$.

Although one can guess the roots in this particular problem, we proceed with the method.

$$P(x) = x^4 + 4x^3 - 16x - 16;$$

$$P'(x) = 4x^3 + 12x^2 - 16 = 4(x^3 + 3x^2 - 4).$$

To find the gcd of $P(x)$ and $P'(x)$ we may disregard the numerical factor 4 and use the Euclidean algorithm.

$$x^4 + 4x^3 - 16x - 16 = (x + 1)(x^3 + 3x^2 - 4) + (-3x^2 - 12x - 12)$$

$$x^3 + 3x^2 - 4 = (-\tfrac{1}{3}x + \tfrac{1}{3})(-3x^2 - 12x - 12) + 0.$$

Therefore, the gcd is $-3x^2 - 12x - 12$, or, disregarding the numerical factor $-3$, $x^2 + 4x + 4 = (x + 2)^2$.

We conclude therefore that $-2$ is a root of multiplicity 3. The remaining root is easily found to be 2.

Other methods will give the same result without use of the Euclidean algorithm. Thus, it is evident by inspection that $P'(1) = 0$. Hence, $(x - 1)$ is a factor of $P'(x)$, $P'(x) = (x - 1)(x^2 + 4x + 4) = (x - 1)(x + 2)^2$. Since $x - 1$ is not a factor of $P(x)$, the only chance is that $(x + 2)^2$ divides $P(x)$, and, as is easily checked, it does.

Another method might be to look for triple roots by inspecting $P''(x) = 12x^2 + 24x = 12x(x + 2)$. Since $x$ is not a factor of $P(x)$, the only possibility is that $x + 2$ divides $P(x)$, and it does, so $(x + 2)^3$ divides $P(x)$.

### Problems

**In Problems 1 to 9 find multiple roots, if any. Then find all roots.**

    *1.*   $x^3 + 5x^2 + 3x - 9 = 0$                                    $[-3, -3, 1]$

    *2.*   $x^4 + 4x^3 + 2x^2 - 8x - 8 = 0$                     $[-2, -2, \pm\sqrt{2}]$

---

\* The algorithm was invented by Euclid to find the gcd of two *integers*. A quite analogous process works for polynomials. To illustrate its use for integers we calculate the gcd of 966 and 437.

$$\text{larger number} = 966 = 2(437) + 92: \quad \text{the gcd also divides 92.}$$

$$437 = 4(92) + 69: \quad \text{the gcd also divides 69.}$$

$$92 = 1(69) + 23: \quad \text{the gcd also divides 23.}$$

$$69 = 3(23) + 0: \quad \text{therefore, the gcd divides 23.}$$

On the other hand, 23 divides 69, 92, 437, and 966. Therefore, 23 is the gcd.

3.  $x^4 - 8x^3 + 24x^2 - 32x + 16 = 0$ $\qquad\qquad$ $[2, 2, 2, 2]$

4.  $x^4 - 4x^3 + 16x - 16 = 0$ $\qquad\qquad$ $[2, 2, 2, -2]$

5.  $x^4 - 4x^2 + 4 = 0$ $\qquad\qquad$ $[-\sqrt{2}, -\sqrt{2}, \sqrt{2}, \sqrt{2}]$

6.  $x^4 - 4x^3 + 4x^2 - 1 = 0$ $\qquad\qquad$ $[1, 1, 1 \pm \sqrt{2}]$

7.  $x^6 + x^4 - x^2 - 1 = 0$ $\qquad\qquad$ $[i, i, -i, -i, 1, -1]$

8.  $x^6 - 6x^4 + 12x^2 - 8 = 0$ $\qquad\qquad$ $[\sqrt{2}, \sqrt{2}, \sqrt{2}, -\sqrt{2}, -\sqrt{2}, -\sqrt{2}]$

9.  $x^6 - 3x^4 + 4 = 0$ $\qquad\qquad$ $[\sqrt{2}, \sqrt{2}, -\sqrt{2}, -\sqrt{2}, i, -i]$

10.  $x^4 - 2x^3 + x^2 - 12x + 20 = 0$ $\qquad\qquad$ $[2, 2, -1 + 4i, -1 - 4i]$

In Problems 11 to 14 show that there are no multiple roots.

11.  $2x^3 - 3x^2 - 12x + 12 = 0$

12.  $x^n - a^n = 0$

13.  $x^4 + x^2 + 1 = 0$

14.  $x^4 - 4x^2 + 4x - 4 = 0$

*15.  Why, for equation (3), page 85, is $P'(r_i) \neq 0$ for $i = 1, \cdots, n$?

## CHAPTER 4

## APPLICATIONS
## OF THE DERIVATIVE
## TO GRAPHING

1 *The Slope of a Curve*

From Chapter 1 we know that the derivative $f'(x_0)$ is the slope of the tangent line at the point $P_0 = (x_0, f(x_0))$ on the graph of $f$. We shall also use the expression:

"the slope of the graph at $P_0$"

for the slope of the tangent line.

If $f'(x_0) > 0$, then the tangent line goes "uphill" from left to right, and the inclination $\theta$ of the tangent line is an acute angle.

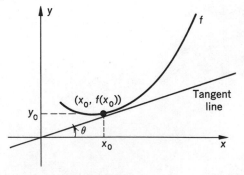

If $f'(x_0) < 0$, then the tangent line goes "downhill," and $\theta$ is an obtuse angle.

Horizontal tangents have $\theta = 0$ and slope 0.

Not all points on the graph have tangent lines.

Points where there is a tangent line that is vertical are points where there is no derivative. Other points with no derivative may also exist.

These various kinds of points of tangency are illustrated in the following figure.

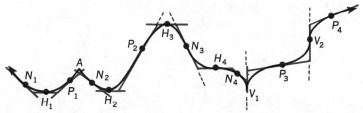

A = no tangent; $N_i$ = negative slope; $P_i$ = positive slope; $H_i$ = horizontal tangent; $V_i$ = vertical tangent

Drawing tangent lines to the graph at well selected points can aid tremendously in sketching the graph. Thus, if the graph of the function sketched above is replaced by segments of the tangent lines at the points $N_i$, $P_i$, $H_i$, $V_i$, one gets:

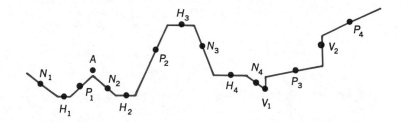

To sketch the graph of a function, it is obviously useful to know where the function increases or decreases. The criterion we use is the following, which we will prove in Section 5.

A function $f$ is *increasing in intervals where the derivative $f'(x)$ is positive (and decreasing where $f'(x)$ is negative)*.

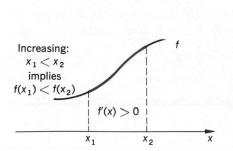

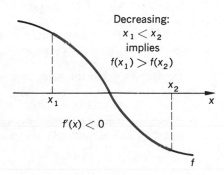

*Example 1*

(a) Sketch the graph of the polynomial $f(x) = \frac{1}{6}(2x^3 - 3x^2 - 12x + 12)$ and the tangent lines at the points where $x = -2, -1, 0, 1, 2, 3$.

(b) Approximate the graph of $f$ by segments of the tangent lines at those points.

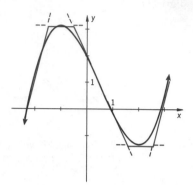

(c) Where is the tangent horizontal?

(d) Where is the slope positive? negative?

(e) Sketch the graph of $f$.

The derivative is

(1) $$f'(x) = x^2 - x - 2 = (x - 2)(x + 1).$$

The following table gives points in (a) with the corresponding slopes.

| $x$ | $-2$ | $-1$ | $0$ | $1$ | $2$ | $3$ |
|---|---|---|---|---|---|---|
| $f(x)$ | $\frac{4}{3}$ | $\frac{19}{6}$ | $2$ | $-\frac{1}{6}$ | $-\frac{4}{3}$ | $\frac{1}{2}$ |
| $f'(x)$ | $4$ | $0$ | $-2$ | $-2$ | $0$ | $4$ |

The segment approximation of $f$ is drawn in solid colored lines in the illustration.

Clearly, from (1), the tangent is horizontal only at $x = 2, -1$.

The sign of $f'(x)$ is determined by examining the signs of the factors.

$$\begin{cases} f'(x) = (x - 2)(x + 1) \\ x < -1, f'(x) = (-)(-) > 0 \\ -1 < x < 2, f'(x) = (-)(+) < 0 \\ x > 2, f'(x) = (+)(+) > 0 \end{cases}$$

The graph of $f$ fits the tangent lines remarkably well. Observe that $f$ is increasing where $f'(x) > 0$ (and decreasing where $f'(x) < 0$).

*Remark.* Obviously this example is selected for simplicity. The points are perfectly chosen, and the derivative is pleasantly factorable. Let us consider another example.

*Example 2*    For the function given by the polynomial

$$f(x) = \frac{1}{12}(3x^4 + 8x^3 - 6x^2 - 24x + 12),$$

sketch the tangent lines to the graph at the points where $x = -3, -2, -\frac{3}{2}, -1, 0, 1, 2$, and then proceed as in Example 1.

The derivative is

$$f'(x) = x^3 + 2x^2 - x - 2.$$

The values of $f(x)$ and $f'(x)$ at the specified values of $x$ are computed and shown in the table.

The points where the tangent is horizontal and where the slope is positive or negative require that we know the factors of $f'(x)$. In general factoring a cubic polynomial (and even more so, higher degree polynomials) can be a hard problem in itself. However, for polynomials with integer coefficients, any factors $x - r$ with $r$ rational can be found by *guessing*. (See Appendix A.)

| $x$ | $f(x)$ | $f'(x)$ |
|-----|--------|---------|
| $-3$ | 4.8 | $-8$ |
| $-2$ | 1.7 | 0 |
| $-1.5$ | 1.9 | 0.6 |
| $-1$ | 2.1 | 0 |
| 0 | 1 | $-2$ |
| 1 | $-0.6$ | 0 |
| 2 | 4.3 | 12 |

In this example, the possible linear factors with integer coefficients are $x \pm 1$ and $x \pm 2$.

For $(x - 1)$, we find that $f'(1) = 0$ and $f'(x) = (x - 1)(x^2 + 3x + 2)$. Now, the remaining quadratic factor is easily factored:

$$f'(x) = (x - 1)(x + 1)(x + 2).$$

The intervals where the slope is positive or negative are shown below. The required tangent line approximations to $f$, and the graph of $f$ are shown in the figure.

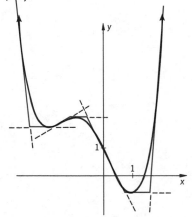

$$\left\{ \begin{array}{l} f'(x) = (x - 1)(x + 1)(x + 2) \\ \text{If } x < -2, f'(x) = (-)(-)(-) < 0 \\ \text{If } -2 < x < -1, f'(x) = (-)(-)(+) > 0 \\ \text{If } -1 < x < 1, f'(x) = (-)(+)(+) < 0 \\ \text{If } x > 1, f'(x) = (+)(+)(+) > 0 \end{array} \right\}$$

*Remark.* Clearly the problem is specially contrived so we could factor $f'(x)$ easily. Also, the points selected for drawing the tangent lines were carefully selected. In some problems you must make this choice yourself.

### Problems

In Problems 1 to 4, (a) draw the tangent lines to the graph of f at the specified values of x; (b) approximate the graph of f by segments of these tangent lines; (c) find where the slope is zero, positive, and negative; (d) sketch the graph of f.

1. $f(x) = \frac{1}{6}x^3 - \frac{1}{2}x + 1$ at $x = -2, -1, 0, 1, 2$

2. $f(x) = \frac{1}{2}(x^3 - 2x^2)$ at $x = -1, 0, 1, 2$

3.  $f(x) = \dfrac{x}{2}(x-2)^2$ at $x = -1, 0, 1, 2, 3$

4.  $f(x) = \dfrac{x^2}{48}(3x^2 - 4x - 36)$ at $x = -3, -2, -1, 0, 1, 3, 4$

In Problems 5 to 14 find where the slope is zero, positive, or negative. Then sketch the graph and sketch the tangent lines at a few points on the graph.

5.  $f(x) = \frac{1}{12}(x^3 - 6x^2 + 10)$
    $$\left[\begin{array}{l} f'(x) > 0 \text{ if } x > 4 \text{ or } x < 0 \\ f'(x) < 0 \text{ if } 0 < x < 4 \end{array}\right]$$

6.  $f(x) = \frac{1}{4}(x^3 + 3x^2 - 9x - 12)$
    $$\left[\begin{array}{l} f'(x) > 0 \text{ if } x > 1 \text{ or } x < -3 \\ f'(x) < 0 \text{ if } -3 < x < 1 \end{array}\right]$$

7.  $f(x) = (x-2)^2(x-1)$
    $$\left[\begin{array}{l} f'(x) < 0 \text{ if } \frac{4}{3} < x < 2 \\ f'(x) > 0 \text{ if } x > 2 \text{ or } x < \frac{4}{3} \end{array}\right]$$

8.  $f(x) = \frac{1}{3}(x^3 + 2)$
    $$[f'(x) > 0 \text{ if } x \neq 0]$$

9.  $f(x) = (2x+1)(x-1)^2$
    $$\left[\begin{array}{l} f'(x) > 0 \text{ if } x > 1 \text{ or } x < 0 \\ f'(x) < 0 \text{ if } 0 < x < 1 \end{array}\right]$$

10. $f(x) = (2x+1)^2(x-1)^2$
    $$\left[\begin{array}{l} f'(x) > 0 \text{ if } x > 1 \text{ or } -\frac{1}{2} < x < \frac{1}{4} \\ f'(x) < 0 \text{ if } \frac{1}{4} < x < 1 \text{ or } x < -\frac{1}{2} \end{array}\right]$$

11. $f(x) = (x+1)(x-1)^3$
    $$\left[\begin{array}{l} f'(x) > 0 \text{ if } x > -\frac{1}{2} \text{ and } x \neq 0 \\ f'(x) < 0 \text{ if } x < -\frac{1}{2} \end{array}\right]$$

12. $f(x) = \dfrac{x+1}{x-1}$
    $$[f'(x) < 0 \text{ for all } x \neq 1]$$

13. $f(x) = \dfrac{(x+1)^2}{4(x-1)}$
    $$\left[\begin{array}{l} f'(x) > 0 \text{ if } x > 3 \text{ or } x < -1 \\ f'(x) < 0 \text{ if } -1 < x < 3 \text{ and } x \neq 1 \end{array}\right]$$

14. $f(x) = \dfrac{1}{x^2+1}$
    $$[f'(x) > 0 \text{ if } x < 0; f'(x) < 0 \text{ if } x > 0]$$

15. Show, by considering $f(x) - f(0)$, that $f$ given by $f(x) = (x-1)e^x + 1$ is positive if $x$ is positive.

16. Show that $x - \log(1+x)$ is positive if $x$ is positive.

*17. Show that $\sqrt{x} - \log(1+x)$ is positive if $x$ is positive.

18. If $k$ is a positive constant, show that $(1/k)(1 - e^{-2kx}) - 2xe^{-2kx}$ is positive if $x$ is positive.

## 2  *Tangent and Normal Lines*

Knowing the slope $f'(x_0)$ of the tangent line at the point $(x_0, y_0)$ on the graph of a function, we can write an equation of the line:

$$y - y_0 = f'(x_0)(x - x_0).$$

Moreover, the *normal* line can also be found. It is the line perpendicular to the tangent line at the point $(x_0, y_0)$ and, hence, has the equation

$$y - y_0 = -\frac{1}{f'(x_0)}(x - x_0),$$

if $f'(x_0) \neq 0$.

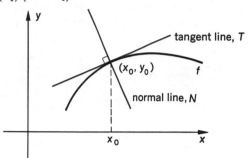

These remarks also apply to functions defined implicitly.

*Example 1*   Find equations of the tangent and normal lines at a point $(x_0, y_0)$ on the circle

$$x^2 + y^2 = a^2.$$

To find the slope we differentiate both members of the equation, remembering that $y$ is a function of $x$.

$$2x + 2yy' = 0$$

whence $y' = -x/y$ and the slope $m = -x_0/y_0$ if $y_0 \neq 0$.

Observe that we could have predicted this without calculus, using only elementary geometry.

An equation of the tangent line is

$$y - y_0 = -\frac{x_0}{y_0}(x - x_0) \qquad \text{or} \qquad yy_0 + xx_0 = x_0^2 + y_0^2 .$$

However, $x_0^2 + y_0^2 = a^2$ because $(x_0, y_0)$ is on the circle. Therefore an equation of the tangent line is

(T) $$yy_0 + xx_0 = a^2.$$

An equation of the normal line is

$$y - y_0 = \frac{y_0}{x_0}(x - x_0) \quad \text{(here } x_0 \neq 0\text{), or}$$

(N) $$x_0 y - y_0 x = 0.$$

Observe that although equation (T) was derived under the assumption that $y_0 \neq 0$, and equation (N) under the assumption that $x_0 \neq 0$, both are valid in all cases. For example, if $y_0 = 0$, then $x_0 = \pm a$, and the tangent line is given by $x = \pm a$.

*Example 2*    Where is the tangent line to the ellipse $3x^2 + 4y^2 = 12$ parallel to the line $2y - x = 2$?

The slope of the given line is $\frac{1}{2}$. The slope at $(x, y)$ on the ellipse is given by $6x + 8yy' = 0$ or

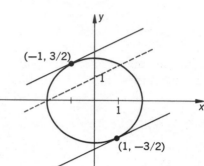

$$y' = -\frac{3x}{4y}$$

If $y'$ is to be $\frac{1}{2}$, then we have

$$-\frac{3x}{4y} = \frac{1}{2}.$$

To find points on the ellipse with slope $\frac{1}{2}$, this last equation must be solved simultaneously with the equation of the ellipse $3x^2 + 4y^2 = 12$. The solution of this pair of simultaneous equations yields the two points $(-1, \frac{3}{2})$ and $(1, -\frac{3}{2})$.

### Problems

1.  Find equations of the tangent and normal to the ellipse $2x^2 + y^2 = 11$ at the point $(1, 3)$.                    $[$ (T) $2x + 3y = 11$; (N) $3x - 2y = -3]$

2.  Find equations of the tangent and normal to the ellipse $b^2x^2 + a^2y^2 = a^2b^2$ at the point $(x_0, y_0)$.
    $$\left[ \begin{array}{l} \text{(T)}\ b^2xx_0 + a^2yy_0 = a^2b^2 \\ \text{(N)}\ a^2xy_0 - b^2yx_0 = x_0y_0(a^2 - b^2) \end{array} \right]$$

3.  Use the result of Problem 2 to show that if the tangent meets the $x$-axis in a point $(x_1, 0)$, and the normal meets the $x$-axis in a point $(x_2, 0)$, then $x_1x_2 = a^2 - b^2$.

4.  At what points is the tangent line to the parabola $y^2 - 6x = 0$ parallel to the line $2x + y = 7$?                                   $[(\frac{3}{8}, -\frac{3}{2})]$

5.  At what points of the hyperbola $2x^2 - y^2 = 4$ is the normal line parallel to the line $x + 2y = 3$?                       $[(2, 2)$ and $(-2, -2)]$

6.  Show that the circles $x^2 + y^2 - 2ax = 0$ and $x^2 + y^2 - 2by = 0$, where $a, b > 0$, intersect at right angles at a point in the first quadrant.
    $$\left[ \text{point} = \left( \frac{2ab^2}{a^2 + b^2}, \frac{2a^2b}{a^2 + b^2} \right); \text{slopes} = \frac{a^2 - b^2}{2ab}, \frac{2ab}{b^2 - a^2} \right]$$

7.  What is the angle of intersection of the curves $y = 1 - x^2$ and $y = 1 + x$?
$$[\,45° \text{ at } (0, 1), \text{ Arc tan } \tfrac{1}{3} \text{ at } (-1, 0)\,]$$

8.  Find an equation of the tangent line to the parabola $y^2 = 2px$ at the point $(x_0, y_0)$. Show that the tangent line intersects the axis of the parabola at $(-x_0, 0)$.

9.  Find an equation of the normal to the parabola $y^2 = 2px$ at the point $(x_0, y_0)$. Show that the normal intersects the $x$-axis at the point $(x_0 + p, 0)$.

10.  Where on the curve $y^3 = x^2$ is the slope equal to $-1$?  $[\,(-8/27, 4/9)\,]$

11.  Where does the graph of $y = \sin x$ have slope 1?  $[\,\text{at } 2\pi k,\ k \text{ any integer}\,]$

12.  Where is the sine function increasing?  $[\,-\pi/2 + 2\pi k < x < \pi/2 + 2\pi k\,]$

13.  Find the angle(s) of intersection of the parabolas $4y + 4 = 3x^2$ and $2y - 6 = x^2$.
$$[\,\text{Arc tan } 2/25 \text{ at } (\pm 4, 11)\,]$$

14.  Where, except at the origin, are the tangents to the folium of Descartes, $x^3 + y^3 - 3axy = 0$ (see Appendix A) parallel to the $x$-axis?  $[\,(\sqrt[3]{2}a, \sqrt[3]{4}a)\,]$

15.  At what angles does the line $3x - 2y + 5 = 0$ intersect the parabola $2(y - 2)^2 = 3x + 5$? Draw a sketch.
$$[\,\text{Arc tan } 18 \text{ at } (-1, 1);\ \text{Arc tan } 18/25 \text{ at } (1, 4)\,]$$

16.  Find equations of the tangent and normal to the graph of $y = x \log x^2$ at $x = 3$.

$$\left[ \begin{array}{l} \text{(T) } y - 2(1 + \log 3)x + 6 = 0 \\ \text{(N) } 2(1 + \log 3)y + x - 3 - 12\log 3 - 12(\log 3)^2 = 0 \end{array} \right]$$

17.  For what number $x_0 > 0$ does the tangent line to the graph of $y = xe^{-x}$ intersect the $x$-axis at $5x_0$?  $[\,x_0 = \tfrac{5}{4}\,]$

*18.  Find an equation of the normal $N$ to the parabola $y^2 = 2px$ at the point $P_0 = (x_0, y_0)$.

Show that the line through $P_0$ and the focus makes the same angle with $N$ as does the line through $P_0$ parallel to the axis.

This proves the "focusing property" of the parabola. A horizontal beam of light would be reflected to concentrate all rays at the focus.

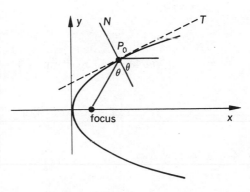

*19. A focusing property similar to that of the parabola (see Problem 18) holds for the ellipse.

Find an equation of the normal, $N$, at $P_0 = (x_0, y_0)$. Then show that the angles between $N$ and $P_0F_1$ and $P_0F_2$ are the same, where $F_1$ and $F_2$ are the foci.

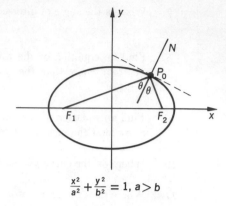

$$\frac{x^2}{a^2} + \frac{y^2}{b^2} = 1, a > b$$

*20. Establish a property of the hyperbola, $b^2x^2 - a^2y^2 = a^2b^2$, analogous to the property of Problem 19 for the ellipse.

## 3 Maxima and Minima

**DEFINITION** A function $f$ has an *absolute maximum* at $x_0$ if, for every $x$ in the domain of $f$,

(1) $$f(x_0) \geqq f(x).$$

The number $f(x_0)$ is the *absolute maximum* of $f$. (A similar statement, with the inequality sign reversed, defines an absolute minimum.)

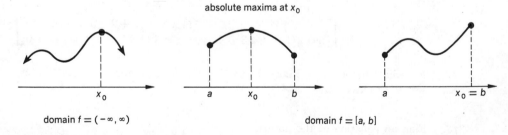

absolute maxima at $x_0$

domain $f = (-\infty, \infty)$        domain $f = [a, b]$

A function can be increasing everywhere on its domain and yet have no maximum. For example,

$$f(x) = 1 - \frac{1}{x} \quad \text{on } (0, \infty)$$

has no maximum. Yet $f(x)$ increases toward 1 as $x \to \infty$.

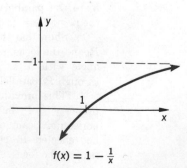

$f(x) = 1 - \frac{1}{x}$

**DEFINITION**    A function $f$ has a *relative maximum* (or *relative minimum*) at $x_0$ if there is an interval $I$, centered at $x_0$, on which $f$ has an absolute maximum (or absolute minimum) at $x_0$. The number $f(x_0)$ is a *relative maximum* (or *relative minimum*) for $f$.

relative maxima at $x_0$

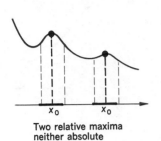

Two relative maxima
neither absolute

One relative maximum
that is absolute

One relative maximum
not absolute

*Remark.* The modifiers "absolute" or "relative" are often omitted if no confusion can arise.

In many problems and examples, we will need to specify both of the numbers $x_0$ and $f(x_0)$. For the sake of brevity, we shall often use the expression: "maximum at $(x_0, f(x_0))$" instead of the longer phrase "$f$ has a maximum $f(x_0)$ at $x_0$".

**THEOREM**    *If $f$ has a relative maximum (or minimum) at $x_0$, and if $f'(x_0)$ exists, then $f'(x_0) = 0$.*

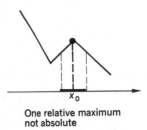

**PROOF**    We examine the sign of the difference quotient, once with $\Delta x > 0$, and once with $\Delta x < 0$, where $x_0 + \Delta x$ is always in the interval $I$ where $f$ has a maximum. We have

$$\Delta y = f(x_0 + \Delta x) - f(x_0) \leqq 0$$

because $f$ has a maximum at $x_0$. If $\Delta x > 0$, then $\Delta y/\Delta x \leqq 0$. If $\Delta x < 0$, then $\Delta y/\Delta x \geqq 0$. So, in the limit, as $\Delta x \to 0$:

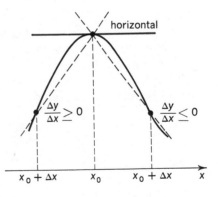

$$\frac{dy}{dx} = f'(x_0) \leqq 0 \qquad \text{and} \qquad \frac{dy}{dx} = f'(x_0) \geqq 0.$$

Therefore, $0 \leqq f'(x_0) \leqq 0$ and $f'(x_0) = 0$.

A similar proof works at a relative minimum.

*Remark.* One can have maxima even though the derivative does not exist, as is suggested by the figure at the right.

From the fact that, if $f'(x) > 0$ in an interval, then $f$ is increasing in that interval

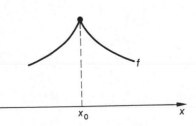

[decreasing if $f'(x) < 0$], one can prove the following test for relative maxima and minima.

**TEST FOR MAXIMA (MINIMA)**   *If $f$ is differentiable in some interval $I$ with midpoint $x_0$, and if for $x$ in $I$,*

$$f'(x) > 0 \quad \text{for } x < x_0,$$
$$f'(x) < 0 \quad \text{for } x > x_0,$$

*then $f$ has a relative maximum at $x_0$.*

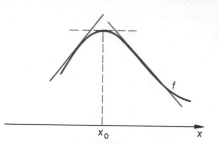

A similar test holds for a relative minimum; that is, if $f'(x) < 0$ for $x < x_0$, and $f'(x) > 0$ for $x > x_0$, then $f$ has a relative minimum at $x_0$.

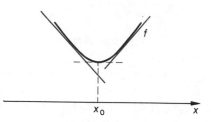

Therefore, we can find relative maxima or minima (at least some of them) by finding where $f'(x) = 0$, and then applying the above test.

The numbers $x$ where either $f'(x)$ does not exist, or where $f'(x) = 0$ are called *critical values*. If the domain of the function is an interval, and if the function has a maximum or minimum *within* the interval, then the maximum or minimum must occur at a critical value. In this case one searches among the critical values for maxima and minima. However, the function can have a maximum or minimum at an end of an interval without the end of the interval being a critical value.

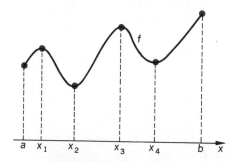

Domain of $f = [a, b]$
$x_1, x_2, x_3, x_4$, are critical values
relative maximum at $x_1, x_3$
relative minimum at $x_4$
absolute minimum at $x_2$
absolute maximum at $b$,
$b$ is not a critical value

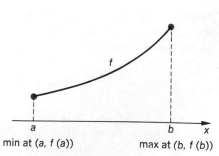

min at $(a, f(a))$                    max at $(b, f(b))$

Domain of $f = [a, b]$
Neither the maximum nor
the minimum is at $a$
critical value

*Remark* 1. If $f'(x_0) = 0$, but $f'(x)$ has the same sign on either side of $x_0$, then $f(x_0)$ is neither a maximum nor a minimum.

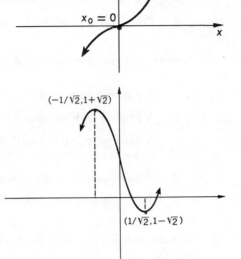

*Remark* 2. The test can also apply to functions continuous in the interval $I$ but not differentiable at $x_0$. See Example 2.

**Example 1**    Test the function $f(x) = 2x^3 - 3x + 1$ for relative maxima and minima. Then sketch its graph.

$$f(x) = 2x^3 - 3x + 1;$$

$$f'(x) = 6x^2 - 3 = 3(2x^2 - 1)$$

$$= 3(\sqrt{2}x - 1)(\sqrt{2}x + 1).$$

Thus, $x = 1/\sqrt{2}$ and $x = -1/\sqrt{2}$ are critical values. Now we examine the signs of $f'(x)$ in the appropriate intervals:

$$f(x) = 2x^3 - 3x + 1$$

If $x < -1/\sqrt{2}, f'(x) > 0.$
If $-1/\sqrt{2} < x < 1/\sqrt{2}, f'(x) < 0.$  $\Big\}$  $f$ has a relative maximum at $x = -1/\sqrt{2}$

If $-1/\sqrt{2} < x < 1/\sqrt{2}, f'(x) < 0.$
If $x > 1/\sqrt{2}, f'(x) > 0.$  $\Big\}$  $f$ has a relative minimum at $x = 1/\sqrt{2}$

The minimum and maximum values are $f(\pm 1/\sqrt{2}) = 1 \mp \sqrt{2}.$

**Example 2**    Test the function $f(x) = 1 + 2(x - 1)^{2/3}$ for relative maxima and minima. Then sketch its graph.

$$f(x) = 1 + 2(x - 1)^{2/3};$$

$$f'(x) = \tfrac{4}{3}(x - 1)^{-1/3} = \frac{4}{3(x - 1)^{1/3}},$$

if $x \neq 1$. From this equation we see that the only critical value is $x = 1$. If we examine the variation in sign of $f'(x)$ near $x = 1$, we get:

$$f'(x) < 0 \text{ if } x < 1$$

and

$$f'(x) > 0 \text{ if } x > 1.$$

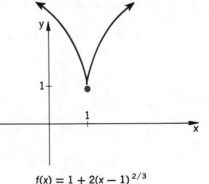

$$f(x) = 1 + 2(x - 1)^{2/3}$$

Thus, $f(1) = 1$ is a relative (in fact, an absolute) minimum, even though $f'(1)$ does not exist.

### Problems

1. State the definition of an absolute minimum.

2. Prove that, if $f$ has a relative minimum at $x_0$, and if $f'(x_0)$ exists, then $f'(x_0) = 0$.

**In Problems 3 to 7 sketch graphs of functions $f$ with the stated properties.**

3. A relative minimum at 2 but not an absolute minimum.

4. A function with its domain the interval $[0, 2]$ and an absolute maximum at 1.

5. A function with domain $[0, 2]$ with a relative maximum at 1 and an absolute maximum at 2. Is $f'(2) = 0$?

6. A function with a relative minimum at 0, but with $f'(0)$ not existing.

7. A differentiable function with neither a maximum nor a minimum.

**In Problems 8 to 34 examine the functions for maxima and minima. Then sketch the graphs.**

8. $f(x) = 3x - 1$, on $(-\infty, +\infty)$      [no maxima or minima]

9. $f(x) = 3x - 1$, on $[-1, 2]$      [max at $(2, 5)$; min at $(-1, -4)$]

10. $f(x) = 1 - x - x^2$      [max at $(-1/2, 5/4)$]

11. $f(x) = x^3 + x^2 - x - 1$      [min at $(1/3, -32/27)$; max at $(-1, 0)$]

12. $f(x) = x^3 + x^2 + x - 1$      [no max or min; $f$ is always increasing]

13. $f(x) = x^2 + x + 1$      [min at $(-\frac{1}{2}, \frac{3}{4})$]

Hence, conclude that $x^2 + x + 1 > 0$ for all $x$.

14. $f(x) = \dfrac{8}{x^2 + 4}$      [max at $(0, 2)$]

15. $f(x) = \sin \pi x$      [max at $\frac{1}{2} + 2k$; min at $-\frac{1}{2} + 2k$, $k$ an integer]

16. $f(x) = \sin^2 \pi x$      [max at $(\frac{1}{2} + k, 1)$; min at $(0 + k, 0)$, $k$ an integer]

17. $f(x) = x^2(x - 1)^2$      [min at $(0, 0)$; max at $(\frac{1}{2}, \frac{1}{16})$; min at $(1, 0)$]

18. $f(x) = x^3(x - 1)^2$      $\begin{bmatrix} \text{neither at } (0, 0); \text{ max at } (3/5, 108/3125) \\ \text{min at } (1, 0) \end{bmatrix}$

19. $f(x) = \frac{1}{4}x^4 + \frac{1}{3}x^3 - \frac{1}{2}x^2 - x + 1$      $\begin{bmatrix} \text{neither at } (-1, 17/12) \\ \text{min at } (1, 1/12) \end{bmatrix}$

20.  $f(x) = 1 + (x - 1)^{1/3}$                    [neither at $(1, 1)$]

21.  $f(x) = x^{1/3}(x - 2)^{1/3}$     $\begin{bmatrix} \text{neither at } (0, 0); \text{min at } (1, -1) \\ \text{neither at } (2, 0) \end{bmatrix}$

22.  $f(x) = \dfrac{x^2 - x - 1}{x^2 - x + 1}$     [min at $(1/2, -5/3)$; $f(x) \to 1$ as $x \to \pm\infty$]

23.  $f(x) = e^{-x^2}$     [max at $(0, 1)$; $f(x) \to 0$ as $x \to \pm\infty$]

24.  $f(x) = \dfrac{x}{x^2 + 1}$     $\begin{bmatrix} \text{max at } (1, \tfrac{1}{2}); \text{min at } (-1, -\tfrac{1}{2}) \\ f(x) \to 0 \text{ as } x \to \pm\infty \end{bmatrix}$

25.  $f(x) = \dfrac{x}{x^2 + 1}$,   on $[0, 1)$     [min at $(0, 0)$; no max]

26.  $f(x) = \dfrac{x^2}{x^2 + 1}$     [min at $(0, 0)$; $f(x) \to 1$ as $x \to \pm\infty$]

27.  $f(x) = \dfrac{(2 - x)^3}{2(1 - x)}$     [min at $(1/2, 27/8)$]

28.  $f(x) = x + 2\sqrt{2 - x}$     [max at $(1, 3)$]

29.  $f(x) = x + 2\sqrt{2 - x}$,   on $[-1, 1]$     $\begin{bmatrix} \text{min at } (-1, 2\sqrt{3} - 1) \\ \text{max at } (1, 3) \end{bmatrix}$

30.  $f(x) = x^2(x - 1)^2$,   on $[0, 2]$     $\begin{bmatrix} \text{min at } (0, 0); \text{max at } (1/2, 1/16) \\ \text{min at } (1, 0); \text{max at } (2, 4) \end{bmatrix}$

31.  $f(x) = xe^{-x}$,   on $x \geqq 0$     [min at $(0, 0)$; max at $(1, 1/e)$]

32.  $f(x) = x^2 e^{-x}$     [min at $(0, 0)$; max at $(2, 4/e^2)$]

33.  $f(x) = e^{-x} \sin x$,   on $x > 0$     $\begin{bmatrix} \text{max at } \left( (8k + 1)\dfrac{\pi}{4}, \dfrac{\sqrt{2}}{2e^{(8k+1)\pi/4}} \right) \\ \\ \text{min at } \left( (8k + 5)\dfrac{\pi}{4}, \dfrac{-\sqrt{2}}{2e^{(8k+5)\pi/4}} \right) \\ \\ k = 0, 1, 2, 3 \cdots \end{bmatrix}$

34.  $f(x) = x + \sin x$     [no max or min]

## 4   *The Second Derivative Test; Concavity; Flex Points*

There is a simple way to tell when the graph of a function bends upward or downward. First we need a definition.

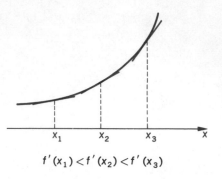

**DEFINITION\***   A function $f$, or its graph, is *concave up* on an interval $I$ if $f'$ is an increasing function on $I$. It is *concave down* on $I$ if $f'$ is decreasing on $I$.

$$f'(x_1) < f'(x_2) < f'(x_3)$$

The concavity of the graph is often called the *direction of bending*; that is either up or down.

There is a simple sufficient condition for concavity on an interval which we state as a theorem.

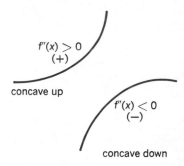

$f''(x) > 0$
$(+)$
concave up

$f''(x) < 0$
$(-)$
concave down

**THEOREM**   *If $f''$ exists on an interval $I$ and if $f''(x) > 0$ on $I$ then $f$ is concave up on $I$; if $f''(x) < 0$ on $I$ then $f$ is concave down on $I$.*

**PROOF**   If $f''(x) > 0$ on $I$, then $f'$ is increasing on $I$, and so $f$ is *concave up*.

*Remark* 1. The direction of bending does not predict whether the original function $f$ is increasing, or not increasing—as the figures below suggest.

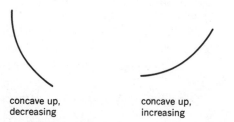

concave up,
decreasing

concave up,
increasing

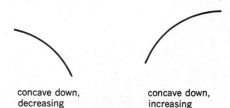

concave down,
decreasing

concave down,
increasing

---

\*There is a more general definition that does not require that $f$ be differentiable. This definition is as follows: The graph of $f$ is *concave up* if, for each pair of points on the graph, the chord connecting the points *lies above* the graph. It can be shown (after the Mean Value Theorem in Section 5) that the text definition implies this one.

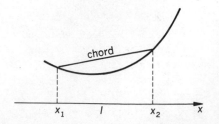

chord

$x_1$    $I$    $x_2$    $x$

A point at which the direction of bending changes is of interest. This suggests the following definition.

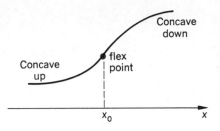

**DEFINITION**    The point $(x_0, f(x_0))$ is a *point of inflection* (or *flex point*) if the graph of $f$ reverses its direction of bending at $(x_0, f(x_0))$.

*Remark 2.* If the second derivative $f''$ exists and is continuous, then at a flex point $(x_0, f(x_0))$ it is necessary that $f''(x_0) = 0$, because $f''(x)$ must change sign as $x$ increases through $x_0$. See Example 1.

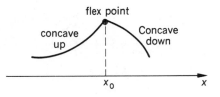

*Remark 3.* The second derivative can fail to exist at $x_0$ and still have $(x_0, f(x_0))$ as a flex point.

*Remark 4.* $(x_0, f(x_0))$ is *not always* a flex point when $f''(x_0) = 0$. For example, if $f(x) = x^4$, then $f''(0) = 0$, but $(0, 0)$ is not a flex point.

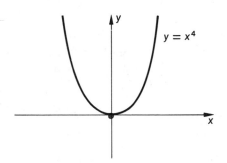

As a corollary to the theorem on the direction of bending, we have the following:

## SECOND DERIVATIVE TEST FOR MAXIMA AND MINIMA    *If $f'(x_0) = 0$, and if $f''(x) < 0$ in an interval centered at $x_0$, then $f$ has a relative maximum at $x_0$.*

*If $f''(x) > 0$ in an interval centered at $x_0$, then $f$ has a relative minimum at $x_0$.*

We give the proof for a relative maximum at $x_0$. Thus, if $f''(x) < 0$, then $f'$ is decreasing in the interval. But $f'(x_0) = 0$. Therefore, $f'(x) > 0$ if $x < x_0$, and $f'(x) < 0$ if $x > x_0$. This makes $f(x_0)$ a relative maximum.

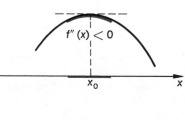

In practice one usually has only to examine the sign of $f''(x_0)$. This is valid if $f''$ is continuous at $x_0$, for then $f''(x)$ will have the same sign as $f''(x_0)$ in some interval containing $x_0$. Thus, *in this case* the test reads as follows:

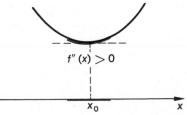

*If $f'(x_0) = 0$ and $f''(x_0) > 0$, then $f$ has a relative minimum at $x_0$.*

*If $f'(x_0) = 0$ and $f''(x_0) < 0$, then $f$ has a relative maximum at $x_0$.*

**Example 1**   Examine $f$ for maxima, minima, flex points, and where concave up or down. Then sketch the graph.

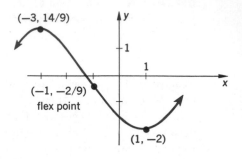

$f(x) = \frac{1}{9}(x^3 + 3x^2 - 9x - 13)$

$f'(x) = \frac{1}{3}(x^2 + 2x - 3)$

$\quad = \frac{1}{3}(x + 3)(x - 1).$

Therefore, $-3, 1$ are critical values since $f'(x) = 0$ there.

$$f''(x) = \frac{2}{3}(x + 1)$$

and the point $(-1, -2/9)$ is a flex point because $f''(-1) = 0$, and the direction of bending changes at $(-1, -2/9)$. The graph is concave up for $x > -1$, and concave down for $x < -1$.

At the critical values:

$$f''(-3) = -\frac{4}{3} < 0,$$

so $x = -3$ gives a maximum $= f(-3) = 14/9$.

$$f''(1) = \frac{4}{3} > 0,$$

so $x = 1$ gives a minimum $= f(1) = -2$.

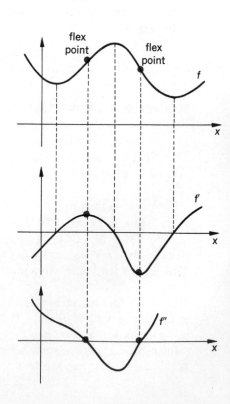

**Example 2**   For a given function $f$, the figure shows the result of a free-hand sketch of $f'$ and $f''$, starting with the graph of $f$.

Observe that the flex points occur at a maximum and a minimum of $f'$, and at zeros of $f''$.

**Example 3**    Examine $F$ for maxima, minima, and direction of bending. Then sketch the graph.

$$F(x) = \frac{1}{1 - x^2}.$$

Observe that $F$ is not defined at $\pm 1$, but that $F(x)$ becomes large in absolute value for $x$ near $\pm 1$.

$$F'(x) = \frac{2x}{(1 - x^2)^2}$$

so that 0 is a critical value and $(0, 1)$ is a critical point.

$$F''(x) = \frac{2 + 6x^2}{(1 - x^2)^3}.$$

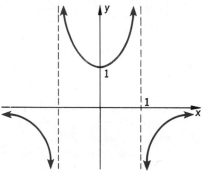

Since the numerator in $F''(x)$ is always positive, the sign of $F''(x)$ depends only on the denominator. Thus, if $-1 < x < 1$, then $F''(x) > 0$, and the graph is concave up. In particular $F''(0) > 0$, so $F(0) = 1$ is a relative minimum.

If either $x < -1$ or $x > 1$, then $F''(x) < 0$, and the graph is concave down.

If there is doubt about the sketch of a graph, simply *plot a few points.* In this case it helps to observe that

$$F(x) \rightarrow 0 \qquad \text{as} \quad x \rightarrow \pm \infty;$$
$$F(x) \rightarrow -\infty \quad \text{as} \quad x \rightarrow 1 \text{ from the right, or } -1 \text{ from the left;}$$
$$F(x) \rightarrow +\infty \quad \text{as} \quad x \rightarrow 1 \text{ from the left, or } -1 \text{ from the right.}$$

**Example 4**    Examine the function $g$ for maxima, minima, and direction of bending. Then sketch the graph.

$$\left. \begin{aligned} g(x) &= x \log x; \\[1em] g'(x) &= 1 + \log x. \end{aligned} \right\} \text{Note that } x \text{ must be positive.}$$

Then $g'(x) = 0$ if $\log x = -1$ or $x = e^{-1}$. Thus $x = e^{-1}$ is the only critical value.

$$g''(x) = \frac{1}{x},$$

and since $x$ is positive, $g''(x)$ is positive. Hence, the graph is concave up everywhere. In particular $g''(e^{-1}) > 0$, so $g(e^{-1}) = -e^{-1}$ is a minimum.

Clearly $g(x) \rightarrow \infty$ as $x \rightarrow \infty$. Also, $g(1) = 0$ and $g'(1) = 1$. These facts help in sketching the graph. What happens to $g(x) = x \log x$ as $x$ approaches 0 is not at all obvious. This particular limit happens to be zero:

$$\lim_{x \to 0} x \log x = 0.$$

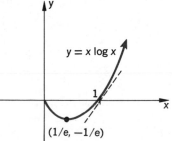

This fact has been used in drawing the graph. We shall see in Chapter 7 how to prove that this is so.

*Example 5*     Examine the function $f(x) = x^n e^{-\alpha x}$ ($n$ a positive integer, $\alpha > 0$) for maxima, minima, and direction of bending. Then sketch the graph. What happens to $f(x)$ as $x \to \infty$? (This important function occurs in many applications.)

$$f(x) = x^n e^{-\alpha x};$$

$$f'(x) = x^{n-1} e^{-\alpha x} (n - \alpha x).$$

$$f'(x) = 0 \text{ if } x = 0 \text{ (and } n > 1) \text{ or } x = n/\alpha.$$

Examining the first derivative near the critical values, we see that $f(n/\alpha) = (n/\alpha)^n e^{-n}$ is a maximum for all $n \geq 1$, and that $f(0) = 0$ is a minimum, if $n$ is even. If $n$ is odd, $f(0) = 0$ is neither a maximum nor a minimum.

For the second derivative we have

$$f''(x) = x^{n-2} e^{-\alpha x} [(\alpha x - n)^2 - n], \quad (n \geq 2).$$

$$f''(x) = 0 \quad \text{if } x = 0 \text{ (and } n > 2) \text{ or } x = (n + \sqrt{n})/\alpha \text{ or } x = (n - \sqrt{n})/\alpha.$$

Examining the second derivative we see that;

(a) The points where $x = (n \pm \sqrt{n})/\alpha$ are flex points, if $n \geq 2$. The point where $x = 2/\alpha$ is a flex point, if $n = 1$.

(b) The point $(0, 0)$ is a flex point, if $n$ is odd and $n \geq 3$. For all other values of $n$ the point $(0, 0)$ is not a flex point.

(c) If $n = 1$, the graph is concave down for $-\infty < x < 2/\alpha$ and concave up for $2/\alpha < x < \infty$.

(d) If $n$ is even, the graph is concave up for $-\infty < x < (n - \sqrt{n})/\alpha$ and $(n + \sqrt{n})/\alpha < x < \infty$; and concave down for $(n - \sqrt{n})/\alpha < x < (n + \sqrt{n})/\alpha$.

(e) If $n$ is odd, and $n \geq 3$, the graph is concave down for $-\infty < x < 0$ and $(n - \sqrt{n})/\alpha < x < (n + \sqrt{n})/\alpha$, and concave up for $0 < x < (n - \sqrt{n})/\alpha$ and $(n + \sqrt{n})/\alpha < x < \infty$.

We now show that $f(x) \to 0$ as $x \to \infty$ for all $n$. Observe that $f(x) > 0$ for $x > 0$ and that $f(x)$ is decreasing (although concave up) for $x > (n + \sqrt{n})/\alpha$.

Denote by $f_n(x)$ the function $f_n(x) = x^n e^{-\alpha x}$, $n = 0, 1, 2, \cdots$. Then $f_n(x) = x f_{n-1}(x)$, $n = 1, 2, 3, \cdots$. If $f_{n-1}(x)$ does *not* $\to 0$ as $x \to \infty$, then $f_n(x) \to \infty$ as $x \to \infty$, which is impossible.

The figures show the graphs of $y = x^n e^{-\alpha x}$ for $n = 1, 2, 3$.

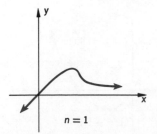

$n = 1$

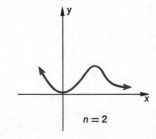

$n = 2$

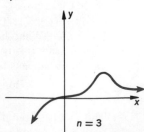

$n = 3$

*Problems*

In Problems 1 to 4 copy an approximation to the graph of *f* on squared paper, and then sketch free-hand the graphs of *f'* and *f''*, one above the other, as in Example 2.

**1.**

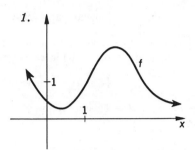

**2.**

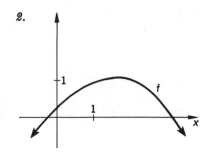

**3.**

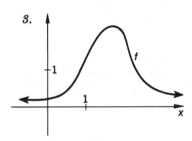

**4.**

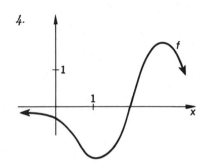

In Problems 5 to 20 examine the graphs for flex points and direction of bending.

5. $f(x) = 1/x$ — [concave up, $x > 0$; concave down, $x < 0$]

6. $f(x) = 1/x^2$ — [concave up everywhere]

7. $g(x) = \dfrac{1}{x^2 + 1}$ — $\begin{bmatrix} \text{concave up, } |x| > 1/\sqrt{3} \\ \text{concave down, } |x| < 1/\sqrt{3} \\ \text{flex points } (\pm 1/\sqrt{3}, 3/4) \end{bmatrix}$

8. $h(x) = x^3 + ax + b$ — $\begin{bmatrix} \text{concave up, } x > 0; \text{ concave down, } x < 0 \\ \text{flex point } (0, b) \end{bmatrix}$

9. $y(x) = \sin \pi x$ — $\begin{bmatrix} \text{concave up, } -1 + 2k < x < 2k \\ \text{concave down, } 2k < x < 1 + 2k \\ \text{flex points } (k, 0), k \text{ an integer} \end{bmatrix}$

10. $y(x) = e^{-x}$ — [concave up everywhere]

11.   $Z(x) = \log x$                                          [concave down everywhere]

12.   $T(x) = \tan x$

$$\begin{bmatrix} \text{concave up, } k\pi < x < \pi/2 + k\pi \\ \text{concave down, } k\pi - \pi/2 < x < k\pi \\ \text{flex points } (k\pi, 0), k \text{ an integer} \end{bmatrix}$$

13.   $f(x) = x^4$                                              [concave up everywhere]

14.   $f(x) = x^n$,   $n$ an integer $> 2$

$$\begin{bmatrix} \text{concave up everywhere if } n \text{ is even;} \\ \text{if } n \text{ is odd, then } (0, 0) \text{ is a flex point} \end{bmatrix}$$

15.   $\varphi(x) = 5 - 2x + x^2$                               [concave up everywhere]

16.   $\psi(x) = 3 + (x + 1)^{1/3}$    $\begin{bmatrix} \text{concave down, } x > -1; \text{ concave up, } x < -1 \\ \text{flex point } (-1, 3) \end{bmatrix}$

17.   $\lambda(x) = 3 + (x + 1)^{2/3}$    [concave down everywhere, except at $(-1, 3)$]

18.   $C(x) = e^{ax} + e^{-ax}$                                 [concave up everywhere]

√19.   $G(x) = \frac{1}{6}(x^4 - 4x^3 - 6x^2 + 6x + 7)$

$$\begin{bmatrix} \text{concave up, } x > 1 + \sqrt{2} \text{ or } x < 1 - \sqrt{2} \\ \text{concave down, } 1 - \sqrt{2} < x < 1 + \sqrt{2} \\ \text{flex points } (1 \pm \sqrt{2}, -8/3 \mp (7/3)\sqrt{2}) \end{bmatrix}$$

20.   $F(x) = \frac{3}{10}x^5 - 3x^4 + 11x^3 - 18x^2 + 18x + 18$

$$\begin{bmatrix} \text{concave up, } x > 3 \text{ or } 1 < x < 2 \\ \text{concave down, } 2 < x < 3 \text{ or } x < 1 \\ \text{flex points } (1, 263/10), (2, 158/5), (3, 369/10) \end{bmatrix}$$

In Problems 21 to 35 test for relative maxima and minima using the second derivative test. Then sketch the graph.

21.   $\varphi(x) = x^3 - 3x^2 - 9x - 1$                        [max at $(-1, 4)$; min at $(3, -28)$]

22.   $f(x) = \dfrac{x}{x^2 + 1}$                               [max at $(1, \frac{1}{2})$; min at $(-1, -\frac{1}{2})$]

23.   $f(x) = \dfrac{x^2}{x^2 + 1}$                             [min at $(0, 0)$]

24.   $\psi(x) = 3x^3 - 36x - 3$                               [max at $(-2, 45)$; min at $(2, -51)$]

25.   $G(x) = \dfrac{2}{x} + x$                                [max at $(-\sqrt{2}, -2\sqrt{2})$; min at $(\sqrt{2}, 2\sqrt{2})$]

26. $f(x) = \dfrac{4}{x^2} + x$ [min at $(2, 3)$]

*27. $f(x) = x^{1/x}$ [max at $(e, e^{1/e})$]

28. $F(x) = x^x$ [min at $(e^{-1}, e^{-e^{-1}})$]

29. $y(x) = x^5 - 20x^2 + 4$ $\left[\begin{array}{l}\text{max at } (0, 4); \text{min at } (2, -44) \\ \text{flex point } (\sqrt[3]{2}, -18\sqrt[3]{4} + 4)\end{array}\right]$

30. $Y(x) = x/\log x$ [min at $(e, e)$]

31. $h(x) = 3x^5 - 65x^3 + 540x$ $\left[\begin{array}{l}\text{max at } (-3, -594), (2, 656) \\ \text{min at } (-2, -656), (3, 594)\end{array}\right]$

32. $Y(x) = x^4 - 2x^2 + 5$ [max at $(0, 5)$; min at $(\pm 1, 4)$]

33. $F(x) = x(x - 2)^2 + 1$ [max at $(2/3, 59/27)$; min at $(2, 1)$]

34. $F(x) = x(x - 2)^3 + 1$ $\left[\begin{array}{l}\text{min at } (1/2, -11/16) \\ \text{flex points } (2, 1), (1, 0)\end{array}\right]$

35. $f(x) = 2x^3 - 3x^2 - 6x + 3$ $\left[\begin{array}{l}\text{max at } \left(\dfrac{1 - \sqrt{5}}{2}, \dfrac{-1 + 5\sqrt{2}}{2}\right) \\ \text{min at } \left(\dfrac{1 + \sqrt{5}}{2}, \dfrac{-1 - 5\sqrt{2}}{2}\right)\end{array}\right]$

In Problems 36 to 44 find maxima, minima, and direction of bending. Then sketch the graph.

36. $f(x) = \frac{1}{20}(2x^3 - 3x^2 - 36x + 28)$ $\left[\begin{array}{l}\text{max at } \left(-2, \dfrac{18}{5}\right) \\ \text{min at } \left(3, \dfrac{-53}{20}\right)\end{array}\right]$

37. $y(x) = \dfrac{8a^3}{x^2 + 4a^2}$ [max at $(0, 2a)$; flex points $(\pm 2a, a)$]

38. $F(x) = \sin^2 \pi x$ $\left[\begin{array}{l}\text{max at } (k/2, 1), k \text{ odd} \\ \text{min at } (k/2, 0), k \text{ even} \\ \text{flex points } (1/4 + k/2, 1/2)\end{array}\right]$

39. $f(x) = a - (x - b)^{2/3}$ [max at $(b, a)$; otherwise concave up]

40. $f(x) = a - (x - b)^{1/3}$ [flex point $(b, a)$]

41. $f(x) = a - (x - b)^{1/2}$ [max at $(b, a)$; otherwise concave up]

42.   $\varphi(x) = \log(1 + x^2)$    [min at $(0, 0)$; flex points $(\pm 1, \log 2)$]

43.   $F(x) = \dfrac{96}{x^2} + \dfrac{24}{x}$    [max at $(-8, -3/2)$; flex point $(-12, -4/3)$]

44.   $y(x) = \frac{1}{4}(x^4 - 6x^2 + 6)$    $\left[\begin{array}{l} \text{max at } (0, 3/2); \text{ min at } (\pm\sqrt{3}, -3/4) \\ \text{flex points } (\pm 1, 1/4) \end{array}\right]$

In Problems 45 to 57 sketch the graph, showing maxima or minima, or flex points as seems desirable to draw a good picture.

45.   $\varphi(x) = x^2(x - 1)^2$    $\left[ \text{flex points } \left(\dfrac{5 \pm \sqrt{5}}{10}, \dfrac{4}{100}\right) \right]$

46.   $\varphi(x) = x^3(x - 1)^2$    $\left[ \text{flex points } (0, 0), \left(\dfrac{6 \pm \sqrt{6}}{10}, \dfrac{21(20 \pm \sqrt{6})}{25000}\right) \right]$

47.   $f(x) = 2x \log x$    [no flex point]

48.   $\lambda(x) = x \log x^2$    [flex point $(0, 0)$]

49.   $f(x) = x^2 \log x$    [flex point $(e^{-3/2}, -\frac{3}{2}e^{-3})$]

50.   $y(x) = (x + 1)^{2/3}(x - 1)^2$    $\left[\begin{array}{l} \text{max at } (-\frac{1}{2}, \frac{9}{8}2^{1/3}) \\ \text{min at } (-1, 0), (1, 0) \end{array}\right]$

51.   $y(x) = x - \sin x$    $\left[\begin{array}{l} \text{no max or min} \\ \text{flex points } (k\pi, k\pi), k \text{ an integer} \end{array}\right]$

52.   $\varphi(x) = 80 - 9x^2 - x^3$    $\left[\begin{array}{l} \text{max at } (0, 80); \text{ min at } (-6, -28) \\ \text{flex point } (-3, 26) \end{array}\right]$

*53.   $f(x) = x \cos x, \quad -\dfrac{\pi}{2} \leqq x \leqq \dfrac{\pi}{2}$    $\left[\begin{array}{l} \text{min at } (-0.86, -0.56) \\ \text{max at } (0.86, 0.56) \\ \text{flex point } (0, 0) \end{array}\right]$

54.   $\Omega(x) = \dfrac{1}{x^2} - 2x$    $\left[\begin{array}{l} \text{no max; min at } (-1, 3) \\ \text{concave up everywhere} \end{array}\right]$

55.   $\omega(x) = \dfrac{\pi}{2} - x + \tan x, \quad -\dfrac{\pi}{2} < x < \dfrac{\pi}{2}$    $\left[ \text{flex point } \left(0, \dfrac{\pi}{2}\right) \right]$

56.   $S(x) = 2\sin x + \sin 2x, \quad 0 \leqq x \leqq 2\pi$    $\left[\begin{array}{l} \text{max at } (\pi/3, 3\sqrt{3}/2) \\ \text{min at } (5\pi/3, -3\sqrt{3}/2) \\ \text{one flex point at } (\pi, 0) \end{array}\right]$

*57.  $y(x) = \cos x\,(1 + \sin x)$,  $-\pi \leqq x \leqq \pi$

$$\left[\begin{array}{l} \text{max at } (\pi/6,\ 3\sqrt{3}/4);\ \text{min at } (5\pi/6,\ -3\sqrt{3}/4) \\ \text{flex points } (\pm\pi/2,\ 0),\ (\text{Arc sin } (-1/4),\ 3\sqrt{15}/16),\ (\pi - \text{Arc sin } 1/4,\ -3\sqrt{15}/16) \end{array}\right]$$

## 5  *The Mean Value Theorem*

We shall use the Mean Value Theorem in this section, although its proof is delayed until Chapter 7. The Theorem has a simple geometric interpretation indicated in the figure. If $f$ is "smooth enough" on the interval $[a, b]$, then there is at least one point $(c,\ f(c))$, with $a < c < b$ at which the tangent line is parallel to the chord between $(a,\ f(a))$, and $(b,\ f(b))$.

The algebraic statement, and clarification of what "smooth enough" means is as follows.

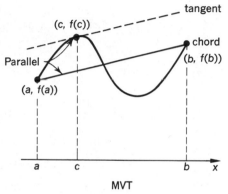

MVT

## MEAN VALUE THEOREM OF DIFFERENTIAL CALCULUS* (MVT)     *If $f$ is differentiable in the open interval $(a, b)$ and continuous at $a$ and $b$, then there is a number $c$, with $a < c < b$, such that*

$$\frac{f(b) - f(a)}{b - a} = f'(c).$$

*Remark.* The smoothness requirements for $f$ cannot be relaxed as the figures below show.

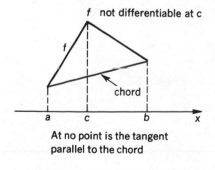

At no point is the tangent parallel to the chord

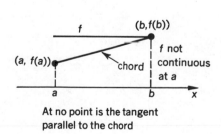

At no point is the tangent parallel to the chord

---

* There is also a Mean Value Theorem of Integral Calculus. When there can be no confusion, we refer simply to the Mean Value Theorem (MVT).

As a first application of the MVT we prove what we have used repeatedly in graphing:

**THEOREM 1** *If f is differentiable in an interval I, and $f'(x) > 0$ on I, then f is increasing on I.*

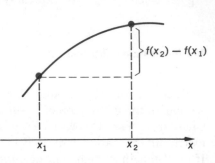

**PROOF**   Suppose $x_1$, $x_2$ are two points of I with $x_1 < x_2$. By the MVT

$$f(x_2) - f(x_1) = f'(c)(x_2 - x_1),$$

$$\text{where } x_1 < c < x_2.$$

But $f'(c) > 0$ and $x_2 - x_1 > 0$, and so $f(x_2) - f(x_1) > 0$.

When $x = t$ and measures time, and $f(t)$ measures distance $s$, then

$$\frac{\Delta y}{\Delta x} = \frac{\Delta s}{\Delta t} = \frac{f(t_2) - f(t_1)}{t_2 - t_1} = \text{average velocity from } t_1 \text{ to } t_2.$$

The MVT asserts that at some instant $t_0$, $t_1 < t_0 < t_2$ the velocity $f'(t_0)$ is equal to the average velocity. See Problems 18 and 19 below.

As a second application of the MVT we prove a theorem used in Chapter 1.

**THEOREM 2**   *If $f' = g'$ on an interval I then $f = g + constant$.*

**PROOF**   Consider the function

$$h = f - g.$$

Then $h' = f' - g' = 0$ on I. By the MVT, for some $c$ between $x_1$ and $x_2$,

$$h(x_2) - h(x_1) = h'(c)(x_2 - x_1).$$

But $h'(c) = 0$. Therefore,

$$h(x_2) - h(x_1) = 0.$$

In other words, $h$ is a constant function,

$$h(x) = \text{constant} = C.$$

Then

$$h(x) = f(x) - g(x) = C,$$

and

$$f(x) = g(x) + C.$$

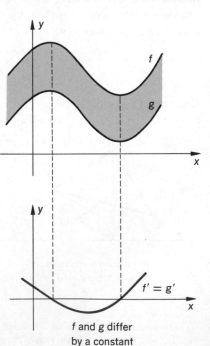

f and g differ by a constant

*Example 1*　If $f(x) = 1/x + x$, find all numbers $c$ such that

$$f(3) - f(\tfrac{1}{2}) = f'(c)(3 - \tfrac{1}{2}).$$

We first find $f'$:

$$f'(x) = -\frac{1}{x^2} + 1.$$

Then, by the MVT

$$f(3) - f\left(\frac{1}{2}\right) = \frac{10}{3} - \frac{5}{2}$$

$$= \left(-\frac{1}{c^2} + 1\right)\left(3 - \frac{1}{2}\right)$$

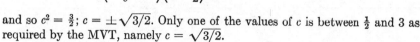

and so $c^2 = \tfrac{3}{2}$; $c = \pm\sqrt{3/2}$. Only one of the values of $c$ is between $\tfrac{1}{2}$ and 3 as required by the MVT, namely $c = \sqrt{3/2}$.

　　We could not, for this function $f$, apply the MVT to a case where $a < 0$ and $b > 0$ because $f$ is not even defined at $x = 0$.

*Example 2*　The MVT can be used to approximate $f(b)$ when $f(a)$ is known. Thus, to compute $\sqrt{26}$, we remember that $\sqrt{25} = 5$ and consider the function $f$,

$$f(x) = \sqrt{x}.$$

Then, $f(26) - f(25) = f'(c)(26 - 25)$, where $25 < c < 26$. So $\sqrt{26} - \sqrt{25} = 1/2\sqrt{c}$. Now, although we do not know $\sqrt{c}$ exactly we do know it is near 5 and, therefore,

$$\sqrt{26} - 5 \approx \tfrac{1}{10}\qquad \text{The right side is clearly a bit too large.}$$

$$\sqrt{26} \approx 5 + \tfrac{1}{10} = 5.1$$

## Problems

In Problems 1 to 9 illustrate the MVT for the indicated $a$ and $b$ if the theorem applies. That is, make a tracing, or rough sketch, of each graph and then find graphically the number, or numbers $c$, predicted by the MVT.

*1.*

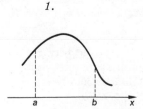

*2.*

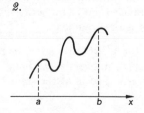

*3.*

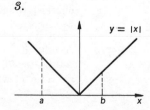

4.

5.
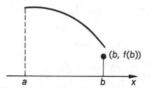

6.
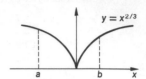
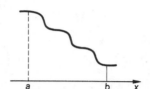

$y = x^{2/3}$

7.
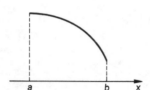

8.

(b, f(b))

9.
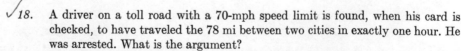

In Problems 10 to 17 verify that the MVT applies to the function in the interval $[a, b]$ and then find all numbers $c$ predicted by the MVT. In each problem draw a suitable picture.

10. $f(x) = x^2 - 3x - 2$ ; $a = 0, b = 3$ $\qquad$ $[c = \frac{3}{2}]$

11. $f(x) = x^2 - 3x - 2$ ; $a = -2, b = 2$ $\qquad$ $[c = 0]$

12. $f(x) = \dfrac{x + 1}{x - 1}$ ; $a = 2, b = 4$ $\qquad$ $[c = 1 + \sqrt{3}]$

13. $f(x) = \sqrt{x - 2}$ ; $a = 2, b = 6$ $\qquad$ $[c = 3]$

14. $f(x) = \sqrt{9 - x^2}$ ; $a = -3, b = 0$ $\qquad$ $[c = -3/\sqrt{2}]$

15. $f(x) = \sqrt{x^2 + 9}$ ; $a = 2\sqrt{10}, b = 4\sqrt{10}$ $\qquad$ $[c = 9]$

16. $f(x) = \dfrac{x^2 + 6x + 5}{x + 1}$ ; $a = 1, b = 7$ $\qquad$ $[c \text{ arbitrary}]$

17. $f(x) = \dfrac{x^2 + 6x + 5}{x - 1}$ ; $a = 2, b = 7$ $\qquad$ $[c = 1 + \sqrt{6}]$

18. A driver on a toll road with a 70-mph speed limit is found, when his card is checked, to have traveled the 78 mi between two cities in exactly one hour. He was arrested. What is the argument?

*19. Another driver on the same toll road as in Problem 18 traveled between two toll booths 70 mi apart in exactly one hour. He stopped at the first booth to get a ticket which he turned in at the second booth. He too could be arrested. Why?

20.  Use the MVT to approximate $\sqrt{101}$; $\sqrt{99}$                    [10.05; 9.95]

21.  Use the MVT to approximate $\sqrt{168}$                    [$12\frac{25}{26}$]

22.  (a)  Use the MVT to approximate log $(1 + h)$ for small $|h|$.
     (b)  Do the same for $\log_{10} (1 + h)$                    [(a) $h$; (b) $0.43h$]

## 6  Differentials

In this section we give a meaning to the symbols "$dx$" and "$dy$." The notation is an invention of Leibniz,* but our interpretation is modern.

We have seen in Section 3 of Chapter 1 that (the Fundamental Lemma)

$$(1) \qquad \Delta y = f(x + \Delta x) - f(x)$$

$$= f'(x)\, \Delta x + \eta\, \Delta x$$

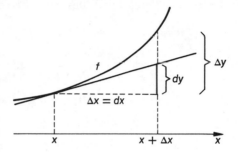

where $\eta \to 0$ as $\Delta x \to 0$. Thus, $\eta\, \Delta x$ is "small of second order" when $\Delta x$ is small, because it is a product of two small numbers. Therefore, the first term in (1) should be a good approximation to $\Delta y$ when $\Delta x$ is small†:

$$(2) \qquad \Delta y \approx f'(x)\, \Delta x.$$

Obviously the approximation is good when the curve "hugs the tangent line" rather closely.

Observe that in equation (2) $\Delta x$ can be any number whatever. This equation motivates our definition of the differential.

**DEFINITION**     If $f$ is differentiable, then the *differential* of $f$ at $x$ is

$$(3) \qquad\qquad\qquad df(x) = f'(x)\, dx$$

where *dx is any real number*.‡ If we set $y = f(x)$, then

$$(4) \qquad\qquad\qquad dy = f'(x)\, dx = y'\, dx.$$

---

* Leibniz thought of $dx$ and $dy$ as "infinitely small differences." But a number is either zero, or not zero, and so he was in error with respect to the way he conceived his calculus. The differential was fundamental for him, and he called his method of calculating differences *"the differential calculus."*

Throughout the seventeenth century, and later, too, one finds the statement, "the derivative is the ratio of infinitesimals." The modern treatment, as given in the text, avoids this by making $dx$ and $dy$ numbers.

† We shall see in Chapter 7 how to judge the "goodness" of this approximation. For the present we shall be content to observe how it works in examples.

‡ Thus, $x$ and $dx$ are quite unrelated. Hence, $df(x)$ is a function of the two variables $x$ and $dx$.

Therefore, if $dx \neq 0$, we get the usual notation by dividing each member of (4) by $dx$:

$$\frac{dy}{dx} = y',$$

so in all cases the derivative is the ratio of differentials. Because $f'(x)$ is, in equation (3), the coefficient of $dx$, the derivative $f'(x)$ is occasionally called the *differential coefficient*.

The differential can be used to approximate the increment $\Delta y$. Thus, $\Delta y = f(x + \Delta x) - f(x) \approx dy = f'(x)\, dx$ (if $dx = \Delta x$). The approximation is easy to compute because the differential is a linear function of $dx$ (it is proportional to $dx$). Examples 2 and 3 will illustrate the concept.

*Example 1*    The differential of $y = (x - x^2)$ is $dy = (1 - 2x)\, dx$.

If $x = 1$ and $dx = 0.5$,    then $dy = -0.5$.

If $x = 1$ and $dx = 0.01$,    then $dy = -0.01$.

If $x = 2$ and $dx = -0.1$,    then $dy = 0.3$.

In each case $dy/dx = f'(x)$.

*Example 2*    Compute, using differentials, an approximation to $\sqrt{16.5}$.
Let $y = f(x) = \sqrt{x}$. Then, using $x = 16$, $\Delta x = dx = 0.5$, we obtain

$$\Delta y = f(x + \Delta x) - f(x) = \sqrt{16.5} - \sqrt{16};$$

$$dy = f'(x)\, dx = \frac{1}{2\sqrt{x}}\, dx = \frac{1}{2(4)}\, (0.5) = 0.0625.$$

Because $dy \approx \Delta y$, we obtain

$$\sqrt{16.5} = y + \Delta y \approx y + dy = \sqrt{16} + 0.0625 = 4.0625.$$

The value of $\sqrt{16.5}$, correct to 4 decimal places, is 4.0620.

*Example 3*    Compute, using differentials, an approximation to $\cos 29°$.
Let $y = f(x) = \cos x$. Then, using $x = \pi/6$, $\Delta x = dx = -\pi/180$ (because 1 deg $= \pi/180$ radians), we obtain

$$\Delta y = f(x + \Delta x) - f(x) = \cos (\pi/6 - \pi/180) - \cos \pi/6;$$

$$dy = f'(x)\, dx = -\sin x \, dx = -(\tfrac{1}{2})(-\pi/180) \approx 0.0087.$$

Because $dy \approx \Delta y$ we obtain, since $\cos \pi/6 = \sqrt{3}/2 \approx 0.8860$,

$$\cos 29° = y + \Delta y \approx y + dy = 0.8660 + 0.0087 = 0.8747.$$

The value of $\cos 29°$, correct to five decimal places, is 0.87462.

*Example 4*    Differentials can be computed when functions are defined implicitly. For example, if $y$ is a function of $x$ such that

$$2x^2 + y^2 = 11,$$

then the left side is a constant function, and its differential is

(1)         $4x\,dx + 2yy'\,dx = 0.$

But $y'\,dx = dy$, whence (1) becomes*

(2)         $4x\,dx + 2y\,dy = 0.$

Then         $dy = -\dfrac{2x}{y}\,dx.$

If $(x, y) = (-1, 3)$ and $dx = 0.4$, then the change in $y$ is approximately

$$dy = \tfrac{2}{3}(0.4) \approx 0.27.$$

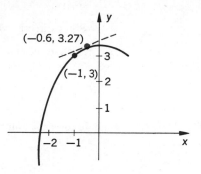

Therefore, the point on the curve near $(-1, 3)$ at which $x = -1 + 0.4 = -0.6$ is approximately $(-0.6, 3.27)$.

### Problems

In Problems 1 to 5 sketch the graph, look up $\Delta y$, and compute $dy$ for the indicated values.

1.  $y = x^2$, $x = 2$       , $dx = 0.2$        $[\Delta y = 0.84,\ dy = 0.8]$

2.  $y = \sqrt{x}$, $x = 9$     , $dx = 0.1$      $[\Delta y = 0.01662,\ dy = 0.01667]$

3.  $y = \sin x$, $x = 0$    , $dx = 0.1$        $[\Delta y = 0.0998,\ dy = 0.1]$

4.  $y = \sin 2x$, $x = \pi/6$,  $dx = 0.01$     $[\Delta y = 0.00985,\ dy = 0.01]$

5.  $y = \log x$, $x = 1$    , $dx = 0.01$       $[\Delta y = 0.00995,\ dy = 0.01]$

In Problems 6 to 13 compute approximations.

6.  $\sqrt{25.1}$           $[5.01]$          10.  $\sqrt[7]{126}$           $[1.9955]$

7.  $\sqrt{99}$             $[9.95]$          11.  $\cos 61°$              $[0.4849]$

8.  $(1.05)^{-3}$           $[0.85]$          12.  $\tan 46°$              $[1.035]$

9.  $\sqrt[3]{214}$         $[5.981]$         13.  Arc tan 0.99           $[0.7804]$

14.  Show that $h$ is a reasonable approximation to $\log(1 + h)$ if $|h|$ is small.

*15.  Show that $\dfrac{h}{1 + h} < \log(1 + h) < h$, if $h > 0.$

---

* Observe that in (2), $dx$ and $dy$ are "on equal footing." That is, we would have obtained equation (2) if we had begun by considering $x$ as a function of $y$.

**16.** (a) Show that $1 + nh$ is a reasonable approximation to $(1 + h)^n$ if $|h|$ is small.

(b) Show directly what the error is in case $n = -1$.

(c) Show directly what the error is in case $n = \frac{1}{2}$.

$$\left[ \text{(b) error} = \frac{h^2}{1 + h} ; \quad \text{(c) error} = \frac{-\frac{1}{4}h^2}{\sqrt{1 + h} + 1 + \frac{1}{2}h} \right]$$

**17.** Find $dy$ for $(x, y)$ on the folium of Descartes, $x^3 + y^3 - 3axy = 0$.

$$\left[ dy = \frac{-x^2 + ay}{y^2 - ax} dx \right]$$

**18.** Find $dy$ for $(x, y)$ on the parabola $x^{1/2} + y^{1/2} = a^{1/2}$. $\quad [dy = (-y^{1/2}/x^{1/2}) \, dx]$

**19.** The area of a circle of radius $r$ is $A = \pi r^2$. Find $dA$ and interpret it geometrically. $\quad [dA = 2\pi r \, dr = (\text{circumference}) \, dr]$

**20.** The volume of a sphere of radius $r$ is $V = \frac{4}{3}\pi r^3$. Find $dV$ and interpret it geometrically.

**21.** An orange with a diameter of 3 in. has a skin $\frac{1}{8}$ in. thick.

(a) What is the volume of the skin?

(b) What percentage of the total volume is skin? $\quad [\text{(a) } 9\pi/8 \quad \text{(b) } 25\%]$

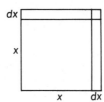

**22.** A square of side $x$ is increased to a square of side $x + dx$. Show geometrically the difference between $\Delta A$ and $dA$ where $A$ is the area.

**\*23.** Newton's method for approximating to the roots of an equation $f(x) = 0$ can be viewed in terms of differentials.

Suppose (refer to the figure) that $x_0$ is an initial approximation to a root of $f(x) = 0$, and that the (nearby) root is at $\bar{x}$.

The tangent line at $(x_0, f(x_0))$ meets the $x$-axis at $x_1$, which is often a better approximation to $\bar{x}$. To find $x_1$, we calculate $dx = x_1 - x_0$:

$$dy = f(x_0) = -f'(x_0) \, dx,$$

so $dx = -f(x_0)/f'(x_0)$. Then $x_1 = x_0 + dx$, or

$$x_1 = x_0 - \frac{f(x_0)}{f'(x_0)}.$$

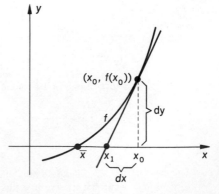

Repetition of the process gives successive approximations $x_2, x_3, \cdots$, which may approach $\bar{x}$. The successive formulas are:

$$x_2 = x_1 - \frac{f(x_1)}{f'(x_1)}, \quad x_3 = x_2 - \frac{f(x_2)}{f'(x_2)}, \quad \text{etc.}$$

(a)  Use Newton's method to approximate the root of $x^3 - 3x + 1 = 0$, which is between 0 and 1.

(b)  Use Newton's method to solve the equation $\cos x - x = 0$, starting with $x_0 = \pi/4$.

## CHAPTER 5

APPLICATIONS:
MAXIMA AND MINIMA
RATES

### 1  Problems of Maxima and Minima

In many situations in science, engineering, or economics one wishes to find the greatest (or least) value that some quantity can attain, while constrained in some way. Some possibilities are maximum profit, minimum cost, minimum energy, and least stress.

Such problems differ from the graphing problems of Chapter 4 in this respect: *One must choose a formula for the problem before he can proceed.* Once this is done the methods of Chapter 4 apply.

The procedure can be outlined as follows:

(1) Draw a diagram if appropriate.

(2) Choose a variable to use in representing the desired quantity $Q$. Then label your diagram appropriately. In some problems this involves choosing a convenient coordinate system.

(3) Express $Q$ in terms of your chosen variable, and note the domain of the function $Q$.

(4) Differentiate $Q$ with respect to your variable to find the critical values.

(5) Check among the critical values *and* the end points of intervals of definition, if appropriate, to determine the maximum or minimum values.

*Example 1*    From a square sheet of metal $a$ feet on a side an open-top box is to be made by cutting squares from each corner and folding up the sides. What are the dimensions of the box of maximum volume?

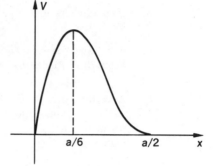

Let $x$ be the height of the box ($0 \leq x \leq a/2$), then

$$V = \text{volume of box} = x(a - 2x)^2.$$

Clearly $V = 0$ when $x = 0$ or $x = a/2$. For intermediate values, $V$ is positive, and we expect a maximum somewhere. In this problem the graph of $V$ is easily sketched, so we will have no doubt in choosing the proper critical value:

$$\frac{dV}{dx} = a^2 - 8ax + 12x^2 = 0,$$

$$(a - 2x)(a - 6x) = 0;$$

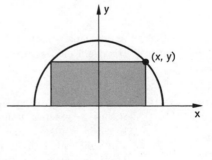

whence $x = a/6$ gives the maximum. This can also be seen by noting either that

(1) $dV/dx$ changes sign from $+$ to $-$ as $x$ goes through $a/6$, or that

(2) $d^2V/dx^2 = -8a + 24x = -4a < 0$   at $x = a/6$.

The dimensions of the box are $a/6$, $2a/3$, $2a/3$.


*Example 2*    Find the dimensions of the rectangle of maximum area that can be inscribed in a semicircle of radius $r$.

Choose axes as shown in the figure. Then $x^2 + y^2 = r^2$, and $y = \sqrt{r^2 - x^2}$. The area $A$ of the rectangle is

$$A = 2xy = 2x\sqrt{r^2 - x^2}.$$

$$\frac{dA}{dx} = 2\sqrt{r^2 - x^2} + 2x\,\frac{-x}{\sqrt{r^2 - x^2}}$$

$$= \frac{2(r^2 - 2x^2)}{\sqrt{r^2 - x^2}}$$

Then $dA/dx = 0$ if $r^2 - 2x^2 = 0$ or $x = r/\sqrt{2}$. The dimensions are $r\sqrt{2}$ and $r/\sqrt{2}$.

Clearly we have a maximum. (Notice the sign change of $dA/dx$ at $x = r/\sqrt{2}$.)

*Problems*

1. A man with 300 yd of fencing wishes to enclose a rectangular area as large as possible along the bank of a straight river. What dimensions should he use?    [75 yd × 150 yd]

2. One number is 12 more than another. What numbers give the least product?
[−6, 6]

3. What positive number plus its reciprocal gives the least sum?    [1]

4. Two numbers add to 12. Find them if twice one of them plus their product is a maximum.    [7, 5]

5. A rectangle is inscribed in a semicircle of radius *a*. What dimensions give maximum perimeter?    $[4a/\sqrt{5}, a/\sqrt{5}]$

6. A rectangle is inscribed in a semicircle of radius *a*. What dimensions give maximum perimeter, not counting the part along the diameter?
$[\sqrt{2}a, a/\sqrt{2}]$

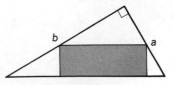

7. Two houses are 300 yd and 500 yd from a straight power line and are 800 yards apart measured along the power line. Where should they attach to the power line to make the total length of cable a minimum?
[300 yards from *A*]

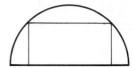

8. A rectangle is inscribed as shown, in a right triangle. Find the rectangle of maximum area.    [Area = *ab*/4]

9. A rectangle is inscribed, as shown, in a right triangle. Find the rectangle of maximum area.    [Area = *ab*/4]

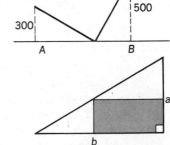

10. A man is in a rowboat 3 mi from shore and wishes to reach a point *B* 5 mi along the beach in the least time. If he rows at 2 mph and walks at 4 mph, (a) where should he land? (b) Does it matter how far along the beach he has to go? (c) Where should he land if *B* is 1 mile along the beach?

(a) $\sqrt{3}$ mi from *A*
(b) No, if *B* is farther than $\sqrt{3}$ from *A*

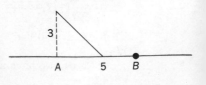
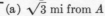

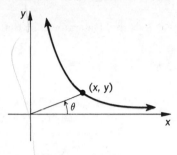

11. The radius vector to a point on the curve $y = 3\sqrt{3}/x, x > 0$, makes an angle $\theta$ with the positive $x$-axis. When is the rate of change of $\theta$ with respect to $x$ a minimum? (Note that $d\theta/dx < 0$.)     [at $x = \sqrt{3}, d\theta/dx = -\frac{1}{2}$]

12. What point on the parabola $y^2 = 2px$ is nearest the point $(a, 0)$, where $a > p > 0$.     $[(a - p, \sqrt{2p(a - p)})]$

*13. Solve Problem 12 if $0 \leqq a \leqq p$.

14. What is the least distance from a point on the line $ax + by + c = 0$ to the origin?     $[|c|/\sqrt{a^2 + b^2}]$

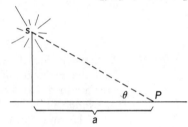

15. Radiation intensity on the ground at $P$ is directly proportional to $\sin \theta$ and inversely proportional to the square of the distance from the source $S$. What height for the source gives maximum intensity at $P$?     $[a/\sqrt{2}]$

16. A box with an open top is to be made from a 7 in. $\times$ 15 in. piece of sheet metal by cutting squares from the corners and folding. What size square should be cut out in order that the volume be a maximum?     $[1\frac{1}{2}$ in.$]$

17. What width $w$ gives maximum area to the trapezoid?     [12]

Problems 18 to 21 refer to the figure at the right, below. A point $P$ is at distances $a$ and $b$ from two perpendicular lines.

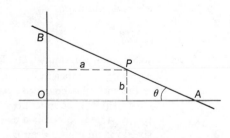

18. Determine $\theta$ so that $|AB|$ is a minimum.     [Arc tan $(b/a)^{1/3}$]

19. Determine $\theta$ so that $|OA| + |OB|$ is a minimum.     [Arc tan $(b/a)^{1/2}$]

20. Determine $\theta$ so that $|OA| \cdot |OB|$ is a minimum.     [Arc tan $b/a$]

*21. Determine $\theta$ so that $|OA| + |OB| + |AB|$ is a minimum.

$$\left[ \text{Arc tan } \frac{b + \sqrt{2ab}}{a + \sqrt{2ab}} \right]$$

*22. Angle $A$ and side $a$ of triangle are given. Prove that the area is a maximum when the triangle is isosceles.

(*Hint:* Use $\theta$ as your variable.)

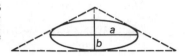

23. An isosceles triangle is circumscribed about an ellipse as pictured. Show that the area of the triangle is a minimum when its altitude is $3b$.

24. A rectangle has dimensions $a$ and $b$. Find the rectangle of maximum area circumscribed about it. (*Hint:* Use $\theta$ as your variable.)
      [A square of side $(a+b)/\sqrt{2}$]

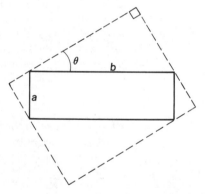

25. A cylindrical boiler is to contain 1000 cu ft. What are the most economical dimensions?
      [radius $= 10/\sqrt[3]{2\pi}$]

26. Referring to the figure, when is $x^2 + y^2$ a minimum?          [$z = c/2$]

27. For the figure of Problem 26, when is $x + y$ a minimum?
      [$z = ac/(a+b)$]

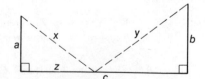

28. For a given subsonic airplane at a given altitude the drag force $D$ varies jointly as the drag coefficient $C_D$ and the square of the speed $v$. The drag coefficient $C_D$ is the sum of a constant (parasite) drag coefficient $C_{D_0}$ and an induced drag coefficient $C_{D_i}$ (due to lift): $C_D = C_{D_0} + C_{D_i}$. The induced drag is given by $C_{D_i} = A/v^4$ where $A$ is a constant depending on the airplane. Show that at the speed for minimum drag $C_{D_i} = C_{D_0}$.

29. The power required for level flight of the airplane of Problem 28 is $P = Dv$. Show that at the speed for minimum power $C_{D_i} = 3C_{D_0}$.

30. A wall 16 ft high is 8 ft from a house. What is the shortest ladder to reach the house from the ground outside the wall?   (Hint: Use $\theta$.)

$$[\text{length} = 8(1 + 2^{2/3})^{3/2} \approx 33.3 \text{ ft}]$$

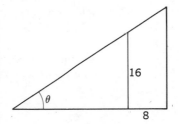

31. A box with square base and open top is to have a surface area of 60 sq ft. What dimensions give maximum volume?   $[2\sqrt{5} \times \sqrt{5}]$

32. Find the dimensions of the rectangle of maximum perimeter that can be inscribed in an ellipse of semi-axes $a$ and $b$.   $[2a^2/\sqrt{a^2 + b^2}, 2b^2/\sqrt{a^2 + b^2}]$

33. A wall mural $a$ ft high has its lower edge $b$ ft above an observer's eye. How far from the wall gives the best view (that is, the maximum angle subtended)?   $[\sqrt{b(a + b)}]$

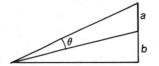

34. Find the narrowest width for $b$ in order that the beam of length $c$ can be gotten round the corner. Neglect the thickness of the beam.

$$[b = (c^{2/3} - a^{2/3})^{3/2}]$$

(Hint: Use $\theta$.)

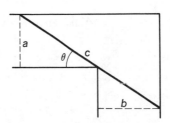

35. Find the cone of maximum volume that can be made by cutting a sector from a circle of radius $a$ and glueing the cut edges together.   $[r = \sqrt{\tfrac{2}{3}}a]$

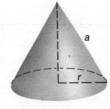

36. A corner of a rectangular strip of paper is folded over to reach the edge at $A$. Find $x$ so the length of the fold $l$ is a minimum. $[\frac{3}{4}a]$

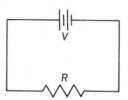

37. For the figure of Problem 36, what should $x$ be in order that the area of $\triangle BCD$ be a minimum? $[\frac{2}{3}a]$

38. The power output in watts of a battery in the simple circuit pictured is $P = V^2R/(r+R)^2$, where $V$ is the voltage of the battery, $r$ its internal resistance in ohms, and $R$ is the external resistance. Show that maximum power occurs when $R = r$.

39. Radiant heat from a point source varies inversely as the square of the distance and directly as the intensity of the source. If two sources at $O_1$ and $O_2$ a distance $a$ apart have intensities $c_1$ and $c_2$, what point between them is coolest?

[at a distance from $O_1 = ac_1^{1/3}/(c_1^{1/3} + c_2^{1/3})$]

40. A shotputter releases the shot at height $h$, angle of inclination $\alpha$, and initial velocity $v_0$. The range $R$ is defined by the formula

$$-gR^2 + v_0^2 R \sin 2\alpha + 2hv_0^2 \cos^2 \alpha = 0.$$

What angle $\alpha$ gives the maximum range?

$$\left[ \tan 2\alpha = \frac{v_0}{gh} \sqrt{2gh + v_0} \right]$$

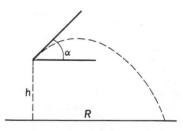

## 2  Problems with Auxiliary Variables

In Section 1 the quantity to be maximized, or minimized, was expressed as a function of a single variable. Occasionally this is inconvenient, or even impossible. In these situations the quantity $Q$ is expressed in terms of several variables, $x, y, \cdots$ which are in turn related to one another by equations that determine all but one of them *implicitly* as functions of the remaining one. Consider some examples.

*Example 1*      Find the cylinder of maximum volume that can
be inscribed in a sphere of radius $r$.

With the notation of the figure the volume of the cylinder is

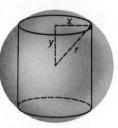

(1) $$V = 2\pi x^2 y.$$

But $x$ and $y$ are connected by the equation

(2) $$x^2 + y^2 = r^2.$$

To use the method of Section 1 we would solve equation
(2) for $x$ or $y$, and then substitute in (1). In this case one
could do this easily. However, it is not necessary to do so. To
find the maximum of $V$ we differentiate (1), remembering that *y is a function of x
implicitly defined by equation (2)*. From equation (1)

(3) $$\frac{dV}{dx} = 2\pi x^2 y' + 4\pi xy = 0,$$

where we have set the derivative equal to zero because we seek the critical points.
We also differentiate the equation relating $x$ and $y$,

$$2x + 2yy' = 0 \quad \text{from (2)}.$$

From this last equation we can (in all cases) solve for $y'$ in terms of $x$ and $y$.
This gives

$$y' = -\frac{x}{y}.$$

Now this formula for $y'$ is used in equation (3);

$$\frac{dV}{dx} = 2\pi x^2 \left(-\frac{x}{y}\right) + 4\pi xy = 0, \quad \text{or}$$

(4) $$2\pi x \left(\frac{-x^2 + 2y^2}{y}\right) = 0.$$

Thus, either $x = 0$, or $-x^2 + 2y^2 = 0$. Clearly, $x = 0$ gives minimum volume.
The equation $-x^2 + 2y^2 = 0$ yields

(5) $$y^2 = x^2/2 \quad \text{or} \quad y = x/\sqrt{2},$$

whence, for maximum volume,

(6) $$\text{altitude} = 2y = \sqrt{2}x = \sqrt{2} \text{ times the radius}.$$

That we actually have found a maximum is clear from equation (4)
because $dV/dx$ changes sign from $+$ to $-$ at $y = x/\sqrt{2}$.

The solution as given in (6) does not give the altitude and radius of the
cylinder, but only the relationship between them. This is often the way answers
will be left. However, from either (5) or (6) the altitude and radius at maximum
volume can be found by using (2). One obtains

$$\text{radius} = x = \sqrt{\tfrac{2}{3}}r, \quad \text{altitude} = 2y = 2r/\sqrt{3}.$$

*Example 2*    Of all right circular cones of given volume $V$ what proportions give minimum lateral surface area?

The lateral surface area of the cone is

$$S = \pi r \sqrt{r^2 + h^2}.$$

The variables $h$ and $r$ are connected by

(1) $$V = \tfrac{1}{3}\pi r^2 h = \text{constant}.$$

Then, at the minimum area,

(2) $$\frac{dS}{dr} = \pi \sqrt{r^2 + h^2} + \pi r \frac{r + h \dfrac{dh}{dr}}{\sqrt{r^2 + h^2}} = 0.$$

Because the volume is constant, we obtain, upon differentiating (1),

(3) $$\frac{1}{3}\pi \left( 2rh + r^2 \frac{dh}{dr} \right) = 0.$$

If we solve for $dh/dr$ in (3) and use the result in (2), we obtain

(4) $$\frac{dS}{dr} = \frac{\pi}{\sqrt{r^2 + h^2}} \left( r^2 + h^2 + r^2 + rh\left( \frac{-2h}{r} \right) \right) = 0,$$

whence $\qquad h = \sqrt{2}\,r.$

This gives the proportions of the cone, as required. The actual dimensions in terms of $V$, if desired, could be found from $V = \pi r^2 h/3$; one gets $h = \sqrt[3]{6V/\pi}$.

That a minimum has been found is fairly clear for geometric reasons, but it can be seen also from (4) that

$$\frac{dS}{dr} = \frac{\pi}{\sqrt{r^2 + h^2}}\,(2r^2 - h^2),$$

and the derivative changes sign from $-$ to $+$ at $r = h/\sqrt{2}$.

## Problems*

In these problems it is not required that auxiliary variables be used. In most cases it will be convenient to use them.

1.    Find the cylindrical can with open top that has least total surface area for a given volume.    [radius = height]

---

* A large number of these problems, as well as those of the preceding list, are quite old. Some are to be found, for example, in Todhunters' *Differential Calculus*, 5th edition, 1871. The well-known "wineglass" problem, number 32 on this list, is from Granville's *Elements of Differential and Integral Calculus*, 1904, revised 1911.

2.  What are the proportions of the cone of given volume that has the minimum total surface area (bottom plus lateral surface)?      [altitude $= 2\sqrt{2}$ radius]

3.  Find the dimensions of the rectangle of maximum area that can be inscribed as shown in an isosceles triangle.
    [base $b$, altitude $a/2$]

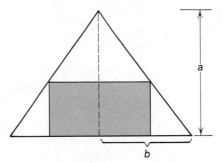

4.  The product of two positive numbers is a given number $A$. How is the first related to the second if the sum of the first plus twice the second is a minimum?
    [first $= 2$ times second]

5.  Find the dimensions of the rectangle of maximum area that can be inscribed in the ellipse $b^2x^2 + a^2y^2 = a^2b^2$.      [$\sqrt{2}a \times \sqrt{2}b$]

6.  What circular sector with a given perimeter has greatest area?
    [arc $=$ twice radius]

7.  The stiffness of a beam is jointly proportional to its width and the cube of its depth. Find the stiffest beam that can be cut from a log of diameter $a$.
    [width $= a/2$, depth $= \sqrt{3}a/2$]

8.  A gas tank of volume $V$ is to be made in the shape of a cylinder surmounted by a hemisphere. What should be its proportions for minimum material?
    [radius $=$ height]

9.  A tank is to be made as in Problem 8 but with a fixed total surface area. What should its proportions be for maximum volume?      [radius $=$ height]

10. Find the dimensions of the cone of minimum volume that can be circumscribed about a sphere of radius $a$.      [height $= 4a$, radius $= \sqrt{2}a$]

11. A rectangular region of given area $A$ is to be fenced in, using a wall for one of the sides. What should its proportions be to require the least amount of fencing?
    [side parallel to wall $=$ twice the other sides]

12. A wire of length $c$ is cut in two. One piece is bent to form a square and the other piece to form a circle. (a) How should it be cut to enclose the minimum area? (b) maximum area?      [(a) perimeter of square $= 4c/(4 + \pi)$]

13. Find the right triangle of greatest area that has a hypotenuse of given length $c$.
[isosceles]

14. Find the isosceles triangle of least area that can be circumscribed about a circle of radius $a$.
[altitude $= 3a$]

15. Find the isosceles triangle of greatest area that can be inscribed in a circle of radius $a$.
[equilateral]

16. A water tank is to have a square base and open top and contain 1000 gal. If the base is twice as costly as the sides, what proportions give minimum material cost?
[depth $=$ side of base]

17. A page of a book is to contain 24 sq in. of print. If margins at the top and bottom of the page are $1\frac{1}{2}$ in. and at the sides 1 in., what is the size of the page of least area?
[6 in. $\times$ 9 in.]

18. At what point does the tangent line to $y = \sqrt{x + 1}$ make, with the axes, a right triangle of least area?
[$(-2/3, 1/\sqrt{3})$]

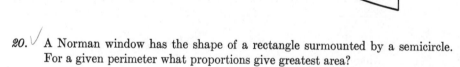

19. At what point $P$ does the rectangle with a vertex at $P$ have maximum area?
[$(a/3, \sqrt{2ap/3})$]

20. A Norman window has the shape of a rectangle surmounted by a semicircle. For a given perimeter what proportions give greatest area?
[radius $=$ height of rectangle]

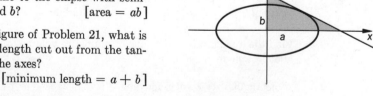

21. What is the minimum area of the triangle formed by the axes and the tangent line to the ellipse with semi-axes $a$ and $b$?
[area $= ab$]

22. For the figure of Problem 21, what is the least length cut out from the tangent by the axes?
[minimum length $= a + b$]

*23. A strip of sheet metal of width $c$ is to be bent to form a circular trough. What should the angle $\theta$ be for maximum carrying capacity?
[$\theta = \pi$]

24. What are the dimensions of the right circular cylinder of maximum lateral surface area that can be inscribed in a sphere of radius $a$?

$$[\text{altitude} = \sqrt{2}a, \text{radius} = a/\sqrt{2}]$$

*25. What are the dimensions of the right circular cylinder of maximum total surface area that can be inscribed in a sphere of radius $a$?

$$\left[\text{radius} = \left(1 + \frac{1}{\sqrt{5}}\right)^{1/2} a/\sqrt{2}, \text{altitude} = \sqrt{2}\left(1 - \frac{1}{\sqrt{5}}\right)^{1/2} a\right]$$

26. Find the cylinder of greatest lateral surface area that can be inscribed in the right circular cone of height $b$ and radius of base $a$.    $[\text{height} = (b/a)\,\text{radius}]$

27. Find the cylinder of greatest total surface area that can be inscribed in the cone of Problem 26. Consider two cases: (a) $b > 2a$, (b) $b \leq 2a$.

$$\left[\begin{array}{l} \text{(a) radius} = \dfrac{ab}{2(b-a)}, \text{altitude} = \dfrac{b(b-2a)}{2(b-a)} \\[2ex] \text{(b) maximum for height} = 0 \end{array}\right]$$

28. Find the cone of maximum volume that can be inscribed in a sphere of radius $a$.

$$[\text{height} = \tfrac{4}{3}a]$$

29. Find the cone of maximum lateral surface area that can be inscribed in a sphere of radius $a$.    $[\text{height} = \tfrac{4}{3}a]$

*30. Find the cone of maximum total surface area that can be inscribed in a sphere of radius $a$.

$$\left[\text{height} = \frac{a}{16}(23 - \sqrt{17})\right]$$

31. In a cone of height $b$ and radius of base $a$ another cone is inscribed upside down. Find the dimensions of the inscribed cone of maximum volume.

$$[\text{radius} = \tfrac{2}{3}a, \text{height} = \tfrac{1}{3}b]$$

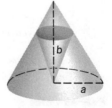

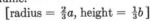

*32. Into a full conical wineglass of depth $a$ and generating angle $\alpha$ there is carefully dropped a sphere of such a size as to cause the greatest overflow. Show that the radius of the sphere is $a \sin\alpha/(\sin\alpha + \cos 2\alpha)$.

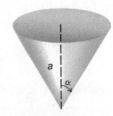

33. An observer at $P$ views the extended radius of the circle, where $b > a + l$. Show that the angle $\theta$ subtended at $P$ is a maximum when $\cos \varphi = (a + l)/b$. Interpret this result geometrically.

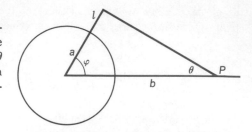

*34. Fermat's principle in optics implies that the path of a ray of light between points $A$ and $B$, in substances with different velocities of light, is such that the time of travel is a minimum. Show that the ray crosses the interface at a point $C$ such that $\sin \theta_1 / \sin \theta_2 = v_1 / v_2$.

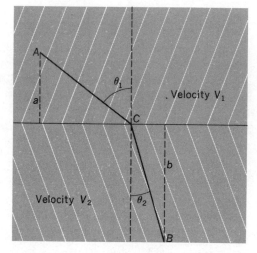

*35. An equilateral triangle is circumscribed about an arbitrary triangle as shown. Show that the maximum area it can have is

$$\frac{1}{\sqrt{3}}\left(b^2 + c^2 - 2bc \cos\left(A + \frac{\pi}{3}\right)\right).$$

(*Hint:* Use $\theta$ and $\varphi$. The problem can be done without calculus.)

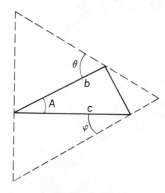

## 3   Related Rates

Suppose that variables $x$ and $y$ are connected by some equation that determines one of them implicitly as a function of the other:

$$f(x, y) = 0, \quad \text{for example,} \quad x - 2xy + y^2 - 5 = 0.$$

Now suppose that one of them, say $x$, is a function of the time $t$, $x = x(t)$. Then $y$ becomes a function of $t$ and we can compute $dy/dt$ by implicit differentiation.

*Notational Remark.* We shall sometimes use the "dot notation" devised by Newton. (See the historical remark on page 150.) This brief, convenient shorthand is often used in physics and engineering for derivatives with respect to time.

$$\frac{dx}{dt} = \dot{x}, \quad \frac{dy}{dt} = \dot{y}, \quad \frac{du}{dt} = \dot{u}, \quad \cdots.$$

Second and third derivatives become

$$\frac{d^2x}{dt^2} = \ddot{x}, \quad \frac{d^3x}{dt^3} = \dddot{x},$$

and so on.

**Example 1**    A point moves on the hyperbola $y^2 - 2x^2 = 2$. What is the rate of change of $y$ when $dx/dt = \frac{1}{2}$ and $x = 1, y = -2$?

In the figure it is clear that the point is moving in the direction of the arrow, so we expect to find that $\dot{y} < 0$.

From the equation of the hyperbola we get, on differentiating,

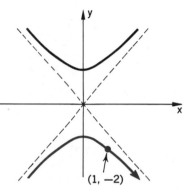

$$2y\frac{dy}{dt} - 4x\frac{dx}{dt} = 0 \quad \text{or} \quad 2y\dot{y} - 4x\dot{x} = 0,$$

whence, at an arbitrary instant,

$$\dot{y} = \frac{2x\dot{x}}{y},$$

and, at the instant of concern,

$$\dot{y} = \frac{2(1)\left(\frac{1}{2}\right)}{-2} = -\frac{1}{2}.$$

**Example 2**    If dry sand spills to form a conical pile with height equal to the radius of the base, how fast is the height increasing when the height is 3 ft and sand is being added at the rate of 2 cu ft/min?

The volume $V$ is given by

$$V = \tfrac{1}{3}\pi r^2 h = \tfrac{1}{3}\pi h^3$$

whence    $$\frac{dV}{dt} = \pi h^2 \frac{dh}{dt}.$$

At the moment of concern,

$$\dot{V} = 2 \quad \text{and} \quad h = 3,$$

$$\dot{V} = 2 = \pi h^2 \dot{h} = \pi 3^2 \dot{h},$$

whence    $\dot{h} = 2/9\pi$ ft/min.    or    $8/3\pi$ in./min.

*Example 3* In some problems there may be several variables changing with time.

Airplane $A$ is 100 mi east of an airport and flying east at 300 mph. Airplane $B$ is 50 mi south of the airport and flying north at 200 mph. They are at the same altitude. How fast is the distance between them changing?

With the notation of the figure,

$$z = \sqrt{x^2 + y^2},$$

and at the moment of interest,

$$x = 100, \quad y = 50, \quad \dot{x} = 300, \quad \dot{y} = -200,$$

with $\dot{y}$ negative because $y$ is decreasing. Then

$$\dot{z} = \frac{1}{2\sqrt{x^2 + y^2}}(2x\dot{x} + 2y\dot{y}) \quad \text{(at any time)}$$

$$= \frac{1}{\sqrt{100^2 + 50^2}}(100\,(300) + 50\,(-200)) \quad \text{(at the given instant)}$$

$$= 80\sqrt{5} \text{ mph} \approx 180 \text{ mph}.$$

## Problems

**In Problems 1 to 6 compute either $\dot{x}$ or $\dot{y}$ at the instant indicated.**

1. $x^2 = 2py, \quad \dot{y} = -2, \quad x = p$                      $[\dot{x} = -2]$

2. $x^3 + y^3 - 3xy = 0, \quad \dot{x} = -1, \quad x = \frac{3}{2}, \quad y = \frac{3}{2}$        $[\dot{y} = 1]$

3. $b^2x^2 + a^2y^2 = a^2b^2, \quad \dot{x} = \frac{1}{2}, \quad x = a/\sqrt{2}, \quad y > 0$     $[\dot{y} = -b/2a]$

4. $y = mx + b, \quad \dot{y} = c$                               $[\dot{x} = c/m]$

5. $y^2 = x^2(x + 1), \quad \dot{x} = a, \quad x = 1, \quad y = -\sqrt{2}$     $[\dot{y} = -5\sqrt{2}a/4]$

6. $xe^{-x} - y = 0, \quad \dot{y} = -e^{-1}, \quad x = 2$                  $[\dot{x} = e]$

7. Where on the ellipse $x^2 + 2y^2 = 6$ is $\dot{x} = -\dot{y}$ as the point $(x, y)$ moves?
                                               $[(2, 1) \text{ or } (-2, -1)]$

8. On the parabola $y^2 = 2px$ where is $\dot{x} = \dot{y}$? What is the slope there?
                                          $[(p/2, p), dy/dx = 1]$

9. Where on the graph of $y = x^3 + 3x^2 - 9x + 6$ is $\dot{y} = 0$, regardless of $\dot{x}$?
                                          $[(-3, 33), (1, 1)]$

10. A 24-ft ladder leans against a high wall. If the foot of the ladder is pulled away from the base of the wall at the rate of 6 ft/min, how fast is the top moving when the foot is 8 ft from the base of the wall?     $[\text{descending } 3/\sqrt{2} \text{ ft/min}]$

11.   A boy $4\frac{1}{2}$ ft tall walks toward a light 10 ft above the ground at the rate of 6 fps. How fast is his shadow changing in length?        [decreasing at 54/11 fps]

12.   In Problem 11 how fast is the shadow of his head moving?
                                                    [toward the light at 120/11 fps]

*13. ✓ A walk is perpendicular to a long wall, and a man strolls along it away from the wall at the rate of 3 fps. There is a light 8 ft from the walk and 24 ft from the wall. How fast is his shadow moving along the wall when he is 20 ft from the wall?                                                        [36 fps]

*14.   A lighthouse is $\frac{1}{2}$ mi out from a long straight line of cliffs. If the lighthouse beam rotates at the rate of 3 rpm, how fast is the light spot on the cliff moving at a point 1 mi along the shore from the point nearest the lighthouse?    [$15\pi$ miles/min]

15.   A ball is thrown vertically upward and reaches a height of 100 ft. If the angle of elevation of the sun is 60 deg, how fast is the shadow of the ball moving 2 sec after it begins to fall?                                    [$64/\sqrt{3}$ fps]

16. ✓ At a given instant the legs of a right triangle are 5 cm and 12 cm long. If the short leg is increasing at the rate of 1 cm/sec and the long leg is decreasing at the rate of 2 cm/sec, how fast is the area changing?        [1 cm²/sec]

17.   In Problem 16 how fast is the hypotenuse changing?        [$-19/13$ cm/sec]

*18.   An overpass is at right angles to the main road and 20 ft above it. A car on the main road is 150 ft from the overpass and traveling toward it at 50 mph. A car on the overpass is 100 ft from the main road and traveling 60 mph away from it. How fast are they separating? (Disregard the dimensions of the cars.)
                                        [approaching at approximately 12 fps]

19.   Two rail tracks cross at an angle of 60 deg. A train along one track travels at 70 mph and reaches the intersection 2 hr after a train on the other track, traveling 50 mph has passed. How fast were they separating 1 hr after the slow train passed the intersection? (Assume that the line between the trains is across the 60 deg angle.)                    [approaching at approximately 39 mph]

20. ✓ A man on a pier pulls in a rope attached to a small boat at the rate of 1 fps. If his hands are 10 ft above the place where the rope is attached, how fast is the boat approaching the pier when there is 20 ft of rope out?
                                                [approximately 1.2 fps]

21.   The dimensions of a box are 10 in., 12 in., and 16 in. If the shorter sides are decreasing at the rate of 0.2 in. per min, and the longest side is increasing at the rate of 0.3 in. per min, how fast is the volume changing?
                                    [decreasing at approximately 34 cu in./min]

22.   A trough 10 ft long has a cross section that is an isosceles triangle 3 ft deep and 8 ft across. If water flows in at the rate of 2 cu ft/min, how fast is the surface rising when the water is 2 ft deep?                            [0.9 in./min]

23. ✓ A plane flying at 1 mi altitude is 2 mi distant from an observer, measured along the ground, and flying directly away from the observer at 400 mph. How fast is the angle of elevation changing?        [decreasing at approximately 1.3 deg/sec]

*24.* A satellite shortly after launch is 50 mi high and 25 mi down range. If it is traveling at 2 mi/sec at an angle of 30 deg with the horizon, how fast is the angle of elevation at the launch site changing?

[decreasing approximately 1.13 deg/sec]

*25.* Water is flowing into a conical tank 12 ft across and 12 ft deep. If the water is rising at the rate of 1 in./min when the water is 6 ft deep, what is the rate of flow?          [$3\pi/4$ cu ft/min]

*26.* A cylindrical reservoir is 10 ft high and 10 ft in diameter. Water flows in at the rate of $1\frac{1}{2}$ cu ft/sec. How fast is the water level rising?          [approximately 0.23 in./sec]

*27.* A small spherical balloon is being inflated at the rate of 1 cu ft/min. At what rate is the diameter increasing 2 sec after inflation begins?

[approximately 4 ft/min]

*28.* How fast is the surface area of the balloon of Problem 27 increasing?

[approximately 1.6 ft²/min]

*29.* A cylindrical tank with axis horizontal has a diameter of 6 ft and a length of 15 ft. It holds oil to a depth of 4 ft when a leak starts to drain off the oil at the rate of 10 cu ft/min. How fast is the level falling?

[approximately 1.4 in./min]

*30.* A hemispherical bowl is 2 ft in diameter. Liquid is poured in at the rate of 200 cu in./min. How fast is the surface rising just as it overflows?

[approximately 0.44 in./min]

*31.* Water flows out of a spherical reservoir 20 ft in diameter. If the water level is falling at the rate of 1 in./min when the water is 15 ft deep, how fast is the volume decreasing?          [19.6 cu ft/min]

*32.* A ball of radius $a$ rests in a hemispherical bowl of radius $2a$ containing water. Show that $\dot{V} = 2\pi a h \dot{h}$.

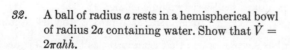

*33.* A ball falls from a height of 80 ft. How fast is the angle of elevation changing for an observer whose eye is 6 ft above ground level, and who is 60 ft from the point below the ball, when the ball is 26 ft from the ground? [−56.1 deg/sec]

## CHAPTER 6

# PLANE CURVES
# AND MOTION ON THEM

## 1  Parametric Equations

In this section and the next we shall explain what we mean by the word "curve." In other words we provide a description of those sets of points that we choose to call curves. In the meantime we shall study plane curves defined by *parametric equations* and investigate tangent and normal lines to such curves.

**DEFINITION**   *A plane parameterized curve* is a continuous image in the plane of an interval of real numbers.

This means that the coordinates of a point $(x, y)$ on the curve are continuous functions as follows:

$$(1) \qquad \begin{array}{c} x = \varphi(t) \\ y = \psi(t) \end{array} \Bigg\} \quad a \leq t \leq b,$$

Interval *I* is bent and stretched to give the curve.

where $\varphi$ and $\psi$ are continuous functions on the interval $[a, b]$. The variable $t$ is called a *parameter*, and equations (1) are called *parametric equations*.

137

In examples and problems $\varphi$ and $\psi$ will usually be differentiable functions in the interval $I$, and, in these cases, will actually be elementary functions. Such curves will be called *smooth curves*.

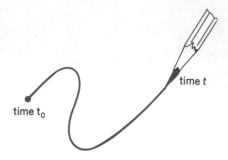

In time interval $[t_0, t]$ the pencil has traced out the curve

Sometimes the interval $I$ is the union of a finite number of closed intervals (end to end) in each of which the curve is smooth. Such curves are called *piecewise-smooth* curves.

*Remark 1.* In some problems and examples we will have parameterizations that are more general than the definition allows. For example (see Example 2), the equations

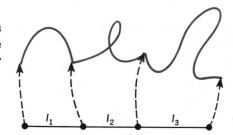

(2) $$x = t, \qquad y = \frac{1}{t}$$

give, in any closed interval not containing $t = 0$, a smooth parameterized curve. If we adhere strictly to the definition of a parameterized curve, (1), then we must be sure that the domain of the functions $\varphi$ and $\psi$ is a closed interval $I$. However, we will make free use of equations such as (2) without indicating an interval $I$, and we will call the set of points defined by such equations a parameterized curve.

If the parameter $t$ is eliminated between equations (1), then a rectangular equation for the curve results (see Examples 1 and 2). Often the parameterized curve will not be the graph of $y$ as a function of $x$ (or $x$ as a function of $y$). Nevertheless when this is the case, we can find the slope according to the following theorem.

**THEOREM**   *Consider the smooth curve given by*

$$x = \varphi(t), \qquad y = \psi(t).$$

*If $\varphi'(t_0) \neq 0$, say $\varphi'(t_0) > 0$, and if $\varphi'$ is continuous, then $x = \varphi(t)$ is increasing on an interval $I$ containing $t_0$ in its interior, and $\varphi(t)$ maps $I$ on an interval $I_1$ containing $x_0$. Then on $I_1$, $y$ is a function of $x$, and*

(3) $\quad \dfrac{dy}{dx} = \dfrac{dy/dt}{dx/dt}$, $\qquad$ *wherever* $\dfrac{dx}{dt} \neq 0$.

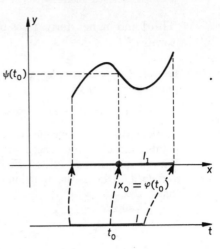

**PROOF** $\quad$ Because $\varphi'(t_0) > 0$, $\varphi'(t) > 0$ on an interval $I$ containing $t_0$. Then $x = \varphi(t)$ is increasing on $I$ and maps $I$ on an interval $I_1$ of the $x$-axis. For each $x$ in $I_1$ there is a unique $t$ in $I$. Thus, $y = \psi(t)$ is a function of $x$ for $x$ in $I_1$. We compute $dy/dx$ directly.

In the interval from $t$ to $t + \Delta t$ the increments of $x$ and $y$ are

$$\Delta y = \psi(t + \Delta t) - \psi(t);$$

$$\Delta x = \varphi(t + \Delta t) - \varphi(t).$$

Then

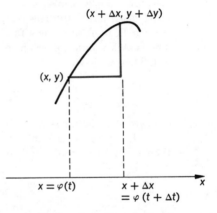

$$\frac{\Delta y}{\Delta x} = \frac{\Delta y/\Delta t}{\Delta x/\Delta t},$$

and $\Delta x \neq 0$ because $t$ and $t + \Delta t$ correspond to different values of $x$. As $\Delta t \to 0$ we get the desired result,

$$y' = \frac{dy}{dx} = \frac{dy/dt}{dx/dt} = \frac{\dot{y}}{\dot{x}} = \frac{\psi'(t)}{\varphi'(t)}.$$

Equation (3) gives $dy/dx$ in terms of $t$. To obtain $dy/dx$ in terms of $x$ requires that $x = \varphi(t)$ be solved for $t$ in terms of $x$ and substituted in (3).

*Remark 2.* If $\dot{x} = \varphi'(t) = 0$ at some point, and $\dot{y} = \psi'(t) \neq 0$, the roles of $x$ and $y$ can be reversed because $x$ is a function of $y$.

Higher derivatives are found using the same theorem. For example, to find $y''$ proceed as follows.

From equation (3) $y'$ is a function of $t$, say $y' = \lambda(t)$. Then, applying the Theorem,

(4) $$\frac{d^2y}{dx^2} = \frac{d}{dx} y' = \frac{dy'/dt}{dx/dt} = \frac{\lambda'(t)}{\varphi'(t)}.$$

Third and higher derivatives use the same method because equation (4) gives $y''$ in terms of $t$.

**Example 1**   Sketch the graph of the parameterized curve

$$x = t^2 - 1, \quad y = 2t - 2, \quad -\infty < t < \infty.$$

Mark on the graph the values of $t$ for a few points, and draw the tangent line there. What happens as $t \to \pm\infty$? Find $dy/dx$ and $dx/dy$. Discuss the direction of bending.

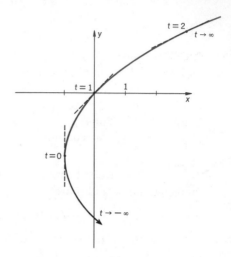

We have

$$\dot{x} = 2t, \qquad \dot{y} = 2.$$

Because $\dot{x} = 2t$ vanishes at $t = 0$, we must, according to the Theorem, restrict $t$, say $t > 0$, to obtain $y$ as a function of $x$. The sketch shows the graph for $t > 0$ in red. Then we have

(5)
$$\frac{dy}{dx} = \frac{dy/dt}{dx/dt} = \frac{2}{2t} = \frac{1}{t}.$$

Observe that the tangent line becomes vertical as $t \to 0$.

If we reverse the roles of $x$ and $y$ and note that $\dot{y} = 2$, so $\dot{y}$ is never zero, then $x$ is a function of $y$ and

$$\frac{dx}{dy} = \frac{\dot{x}}{\dot{y}} = \frac{2t}{2} = t,$$

as, of course, it should be from (5). Note that (5) is also valid for $t < 0$.

The second derivative $d^2y/dx^2$ is found from (5):

$$\frac{d^2y}{dx^2} = \frac{d}{dx}y' = \frac{dy'/dt}{dx/dt} = \frac{-1/t^2}{2t} = -\frac{1}{2t^3}.$$

Thus, the graph of $y$ as a function of $x$ is concave down for $t > 0$ and concave up for $t < 0$.

If we consider $x$ as a function of $y$, then

$$\frac{d^2x}{dy^2} = \frac{dx'}{dy} = \frac{dx'/dt}{dy/dt} = \frac{1}{2}.$$

Thus, the graph of $x$ as a function of $y$ is concave up everywhere.

Note that $d^2y/dx^2$ and $d^2x/dy^2$ are *not* reciprocals of each other.

In this example a rectangular equation for the curve is easily obtained: $t = (y + 2)/2$ and so $4(x + 1) = (y + 2)^2$.

*Example 2*  Sketch the graph of the parameterized curve

$$x = t, \quad y = \frac{1}{t}, \quad t \neq 0.$$

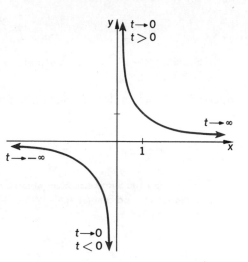

Find $y'$ and $y''$ in terms of $t$ and discuss the direction of bending.

This example is one to which Remark 1 applies. We have

$$\dot{x} = 1, \qquad \dot{y} = -\frac{1}{t^2}.$$

Because $\dot{x}$ is never zero, $y$ is a function of $x$, and

$$y' = \frac{dy}{dx} = \frac{-1/t^2}{1} = -\frac{1}{t^2}.$$

The slope is always negative, and so $y$ is a decreasing function of $x$ on any interval not containing 0. For the second derivative we have

$$\frac{d^2y}{dx^2} = \frac{dy'}{dx} = \frac{dy'/dt}{dx/dt} = \frac{2/t^3}{1} = \frac{2}{t^3}.$$

The curve is concave up for $t > 0$ and concave down for $t < 0$.

*Example 3*  Sketch the graph of

$$x = t^3 - 3t - 2, \quad y = t^2 - t - 2,$$
$$-\infty < t < \infty.$$

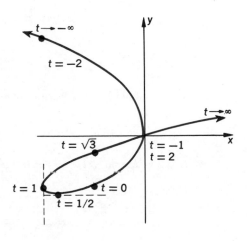

This one is harder. In the first place, the parameter is harder to eliminate, and even when one does (the result is $y^3 = x^2 - 3xy$), the resultant cubic equation in $x$ and $y$ is not recognizable, and is as hard to graph as the parametric equations.

We simply watch the behavior of $x$, $y$, $y'$, and $y''$, and plot enough points.

$$\dot{x} = 3(t^2 - 1), \qquad \dot{y} = 2t - 1,$$

$$y' = \frac{dy}{dx} = \frac{2t - 1}{3(t^2 - 1)}.$$

Thus, we conclude that:

    (i)    $x$ has a maximum at $t = -1$ and a minimum at $t = 1$;

    (ii)   $y$ has a minimum at $t = \frac{1}{2}$;

(iii) $\dfrac{d^2y}{dx^2} = \dfrac{dy'}{dx} = \dfrac{dy'/dt}{dx/dt} = -\dfrac{2}{9}\dfrac{(t^2-t+1)}{(t^2-1)^3}$ .

So, since $t^2 - t + 1$ is always positive, the curve is concave down for $t^2 > 1$.

*Reminder.* When in doubt, plot more points.

### Problems

In Problems 1 to 8 find equations of the tangent and normal at the point corresponding to the selected parameter value. Where is the tangent horizontal? vertical?

1. $x = 2t, \quad y = 1 + \dfrac{1}{2t}; \quad t = 1$ $\qquad$ $[\text{T}: x + 4y = 8; \text{N}: 8x - 2y = 13]$

2. $x = 2t, \quad y = 2t + 1; \quad t = -1$ $\qquad$ $[\text{T}: y = x + 1; \text{N}: x + y + 3 = 0]$

3. $x = 2s^2, \quad y = 2s + 1; \quad s = -\tfrac{1}{2}$ $\qquad$ $\begin{bmatrix} \text{T}: 2y + x = 1; \text{N}: x - 2y = \tfrac{1}{2} \\ \text{vert. tan at } (0, 1) \end{bmatrix}$

4. $x = e^t, \quad y = 2e^{-t}; \quad t = \log 2$ $\qquad$ $[\text{T}: 2y + x = 4; \text{N}: 2x - y = 3]$

5. $x = \sin\theta, \quad y = \sin 2\theta; \quad \theta = \dfrac{\pi}{6}$ $\qquad$ $\begin{bmatrix} \text{T}: 4x - 2\sqrt{3}y = -1 \\ \text{N}: 2\sqrt{3}x + 4y = 3\sqrt{3} \\ \text{vert. tan at } (\pm 1, 0) \\ \text{horiz. tan at } (\pm 1/\sqrt{2}, \pm 1) \end{bmatrix}$

6. $x = 2\sin\theta, \quad y = 2\cos\theta; \quad \theta = \dfrac{\pi}{6}$ $\qquad$ $\begin{bmatrix} \text{T}: \sqrt{3}y + x = 4 \\ \text{N}: \sqrt{3}x - y = 0 \\ \text{vert. tan at } (\pm 2, 0) \\ \text{horiz. tan at } (0, \pm 2) \end{bmatrix}$

7. $x = \text{Arc}\sin\sqrt{t}, \quad y = t^2 - 2t; \quad t = \tfrac{1}{2}$ $\qquad$ $\begin{bmatrix} \text{T}: 4x + 4y = \pi - 3 \\ \text{N}: 4x - 4y = \pi + 3 \\ \text{horiz. tan at } (0, 0), (\pi/2, -1) \end{bmatrix}$

8. $x = t^3 - 3t, \quad y = t^2 - 2t; \quad t = -\tfrac{1}{2}$ $\qquad$ $\begin{bmatrix} \text{T}: 16x - 12y = 7 \\ \text{N}: 24x + 32y = 73 \\ \text{vert. tan at } (2, 3) \end{bmatrix}$

In Problems 9 to 13 find $y'$, $y''$, $y'''$ in terms of the parameter.

9. $x = e^{2t}, \quad y = e^t - e^{-t}$ $\qquad$ $[y''' = \tfrac{3}{8}e^{-5t} + \tfrac{15}{8}e^{-7t}]$

10. $x = t^2 - t, \quad y = t - 3$ $\qquad\qquad\qquad\qquad\qquad [y''' = 12/(2t-1)^5]$

11. $x = t - 3, \quad y = t^2 - t$ $\qquad\qquad\qquad\qquad\qquad\qquad [y''' = 0]$

12. $x = t^2 + 1, \quad y = t^3 + 3t$ $\qquad\qquad\qquad\quad [y''' = 3(3-t^2)/8t^5]$

13. $x = a(\cos\theta + \theta\sin\theta),$
$\quad\ y = a(\sin\theta - \theta\cos\theta)$ $\qquad\qquad \left[ y''' = \dfrac{3\theta\sin\theta - \cos\theta}{a^2\theta^3\cos^5\theta} \right]$

In Problems 14 to 30 find in what parameter intervals (a) y is increasing, (b) x is increasing, (c) the graph of y versus x is concave up, or (d) the graph of x versus y is concave up. Use this information to sketch the curve, and indicate on the graph the variation of the parameter along the curve. Where convenient find a rectangular equation.

14. $x = 2t, \quad y = t^2$ $\qquad\qquad\qquad\qquad$ [(c) concave up everywhere]

15. $x = t^2, \quad y = t^3$ $\qquad\qquad\qquad\qquad$ [(d) concave down everywhere]

16. $x = t^2 - 1, \quad y = 2t + 1$ $\qquad\qquad\qquad\qquad$ [(c) $t < 0$; (d) all $t$]

17. $x = 2t^3 - 3t^2, \quad y = t^2 - 1$ $\qquad\qquad$ [(c) $0 < t < 1$; (d) $t > 0$]

18. $x = 1 + \cos^2\theta, \quad y = 1 + \sin\theta$ $\qquad \left[ \begin{array}{l} \text{(c) } \pi + 2\pi k < \theta < 2\pi + 2\pi k \\ \text{(d) never} \end{array} \right]$

19. $x = \log t, \quad y = t^2 + 1$ $\qquad\qquad\qquad\qquad$ [(c) always; (d) never]

20. $x = s^2 - s + 1, \quad y = 2 - s$ $\qquad \left[ y' = \dfrac{-1}{(2s-1)}, y'' = \dfrac{2}{(2s-1)^3} \right]$

21. $x = t^2 - 2t, \quad y = t^2 - t$ $\qquad \left[ y' = \dfrac{2t-1}{2(t-1)}, y'' = \dfrac{-1}{4(t-1)^3} \right]$

22. $x = \cos\theta, \quad y = \cos 2\theta$ $\qquad\qquad\qquad$ [$y' = 4\cos\theta, y'' = 4$]

23. $x = 2 - e^{2t}, \quad y = e^{-t} - 2$ $\qquad\qquad$ [$y' = \frac{1}{2}e^{-3t}, y'' = \frac{3}{4}e^{-5t}$]

24. $x = \sin^2 t, \quad y = \cos^2 t$ $\qquad\qquad\qquad$ [$y' = -1, y'' = 0$]

25. $x = a\tan\theta, \quad y = b\sec\theta, \quad -\dfrac{\pi}{2} < \theta < \dfrac{\pi}{2}$ $\quad \left[ y' = \dfrac{b}{a}\sin\theta, y'' = \dfrac{b}{a^2}\cos^3\theta \right]$

26. $x = \sin^4\theta, \quad y = \cos^4\theta$ $\qquad \left[ y' = -\tan^2\theta, y'' = \dfrac{-1}{2\sin^2\theta\cos^4\theta} \right]$

*27. $x = \dfrac{3t}{1+t^3}, \quad y = \dfrac{3t^2}{1+t^3}$ $\qquad \left[ y' = \dfrac{t(2-t^3)}{1-2t^3}, y'' = \dfrac{2(t^3+1)^4}{3(1-2t^3)^3} \right]$

(the folium of Descartes) See Remark 1.

*28.   $x = t^3 - 3t$,  $y = t^2 - 1$
$$\left[ y' = \frac{2t}{3\,(t^2 - 1)},\, y'' = \frac{-2\,(t^2 + 1)}{9\,(t^2 - 1)^3} \right]$$

*29.   $x = (t + 1)^2\,(t - 1)$,  $y = (t + 1)\,(t - 1)^2$
$$\left[ y' = \frac{(t - 1)\,(3t + 1)}{(t + 1)\,(3t - 1)},\, y'' = \frac{4\,(3t^2 + 1)}{(t + 1)^3\,(3t - 1)^3} \right]$$

30.   $x = a \tan \theta$,  $y = b \cos^2 \theta$;  $-\dfrac{\pi}{2} < \theta < \dfrac{\pi}{2}$
$$\left[ y' = -\frac{2b}{a} \cos^3 \theta \sin \theta,\, y'' = \frac{2b}{a^2} \cos^4 \theta\,(3 - 4 \cos^2 \theta) \right]$$

31.   Show that the curves $x = 2t - 2$, $y = 2 - 2t^2$, and $x = s - 4$, $y = 2 - (s^2/2)$ intersect at right angles. Sketch the figure.

*32.   Derive equation (3) by using the chain rule and the formula for the derivative of an inverse function.

## 2   *Plane Curves*

Section 1 was concerned with parameterized curves, that is, sets "traversed" in a certain way. Now we consider what we choose to call, simply, "curve" without any qualifying adjective.

**DEFINITION**   A plane curve is a subset of the plane that can be a parameterized curve. That is, some parameter interval can be mapped continuously onto the subset.

*Remark 1.* A curve can be parameterized in many ways. See Examples 1 and 2.

*Remark 2.* An important parameter choice for applications is $t$, the elapsed time from some instant.

*Remark 3.* The graph of $y = f(x)$, where $f$ is a continuous function, is a curve because, if we choose the parameter $t = x$, then
$$x = t \qquad \text{and} \qquad y = f(t).$$

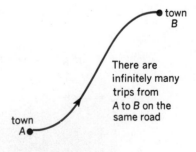

*Example 1*   A particularly important parameterization of a line is that for which the parameter $s$ is the directed distance from a given point, $(x_0, y_0)$ on the line.

Examination of the figure gives, if $(x, y) \neq (x_0, y_0)$,

$$\frac{x - x_0}{s} = \cos \alpha, \qquad \frac{y - y_0}{s} = \cos \beta,$$

where $\alpha$ and $\beta$ are the direction angles (see Appendix A) of the directed line. The choice of which ray from $(x_0, y_0)$ is to correspond to positive distance from $(x_0, y_0)$ gives the positive direction on the line. Then

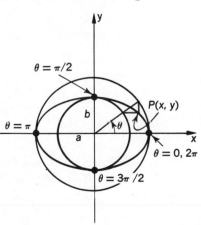

(1) $\quad x = x_0 + s \cos \alpha, \quad y = y_0 + s \cos \beta.$

A rectangular equation is obtained by eliminating $s$,

$$(x - x_0) \cos \beta = (y - y_0) \cos \alpha.$$

Our formula for slope $[y' = (dy/ds)/(dx/ds)]$ becomes rather trivial here: $y' = \cos \beta / \cos \alpha$, which is clear geometrically. However, the derivatives of equation (1) express an important relationship between the coordinates, distance along the curve, and direction cosines. (See Section 4.)

(2) $$\frac{dx}{ds} = \cos \alpha, \qquad \frac{dy}{ds} = \cos \beta.$$

There are other parameterizations. Thus, if we set $s = t^3$ in (1), we get

$$x = x_0 + t^3 \cos \alpha, \quad y = y_0 + t^3 \cos \beta, \quad -\infty < t < \infty.$$

Or, if we set $s = \tan \theta$, we get

$$x = x_0 + \cos \alpha \tan \theta, \quad y = y_0 + \cos \beta \tan \theta, \quad -\pi/2 < \theta < \pi/2.$$

**Example 2**    An ellipse is a curve because it can be parameterized. One way is to use the angle $\theta$ of the figure as parameter. Then the coordinates of $P$ are given by

$$x = a \cos \theta, \quad y = b \sin \theta, \quad -\infty < \theta < \infty,$$

where $a$ and $b$ are the semi-axes of the ellipse. That the point $P = (x, y)$ is on the ellipse follows from the equation

$$\frac{x^2}{a^2} + \frac{y^2}{b^2} = \cos^2 \theta + \sin^2 \theta = 1.$$

Parameter variation along the ellipse is shown in the figure. Derivatives are as follows,

$$\dot{x} = -a \sin \theta, \qquad \dot{y} = b \cos \theta;$$

$$y' = -\frac{b}{a} \cot \theta = -\frac{b^2 x}{a^2 y};$$

$$y'' = \frac{-b(-\csc^2 \theta)/a}{-a \sin \theta} = \frac{-b}{a^2} \csc^3 \theta = \frac{-b^4}{a^2 y^3}.$$

Another parameterization is obtained by using the slope $m$ of a line through $(-a, 0)$. The line is

$$y = m(x + a),$$

and solving for the intersection (other than $(-a, 0)$) of this line with the ellipse $b^2x^2 + a^2y^2 = a^2b^2$ gives

$$x = \frac{a(b^2 - a^2m^2)}{b^2 + a^2m^2}, \quad y = \frac{2ab^2m}{b^2 + a^2m^2},$$

$$-\infty < m < \infty.$$

Observe that one point is missed as $m$ varies, namely the point $(-a, 0)$. Derivatives are

$$\dot{x} = -\frac{4a^3b^2m}{(b^2 + a^2m^2)^2}, \quad \dot{y} = \frac{2ab^2(b^2 - a^2m^2)}{(b^2 + a^2m^2)^2};$$

$$y' = -\frac{b^2 - a^2m^2}{2a^2m} = -\frac{b^2x}{a^2y}.$$

### Problems

In Problems 1 to 6 parameterize the curve using the suggested parameter. Show on a graph the parameter variation along the curve. Then compute $y'$ and $y''$ in terms of the parameter.

1. The circle $x^2 + y^2 = a^2$; parameter $= \theta =$ the measure of the angle between the positive $x$-axis and the radius to $(x, y)$.    $[x = a\cos\theta, y = a\sin\theta]$

2. The circle $x^2 + y^2 = a^2$; parameter $= m =$ the slope of a line through $(0, a)$.
$$\left[ x = -\frac{2am}{1 + m^2}, y = \frac{a(1 - m^2)}{1 + m^2} \right]$$

3. The parabola $y^2 = 2px$; parameter $= m =$ the slope of a line through the origin.*
$$[x = 2p/m^2, y = 2p/m]$$

4. The folium of Descartes, $x^3 + y^3 - 3axy = 0$; parameter $= m =$ the slope of a line through the origin.*
$$\left[ x = \frac{3am}{1 + m^3}, y = \frac{3am^2}{1 + m^3} \right]$$

---

* Problems 3, 4, and 6 exhibit parameterizations in the sense of Remark 1, page 138.

*5. The cycloid, generated by a point on a circle of radius $a$ rolling on the $x$-axis; parameter $= \theta =$ the angle shown.
$$\begin{bmatrix} x = a\,(\theta - \sin \theta), \\ y = a\,(1 - \cos \theta) \end{bmatrix}$$

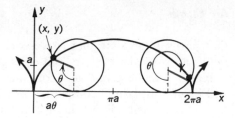

*6. The conchoid of Nicomedes, $x^2 y^2 = (a^2 - y^2)(b + y)^2$; parameter $= \theta =$ the angle shown.*
$$[x = b \tan \theta + a \sin \theta,\ y = a \cos \theta]$$

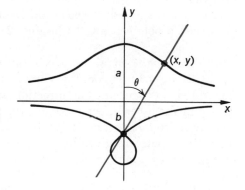

7. Find a rectangular equation for the curve:
$$x = e^t + 1, \quad y = e^{-t} - 1, \quad -\infty < t < \infty.$$

Does the curve exhaust the graph of the rectangular equation? Find $y'$, $y''$ both from the parametric equations and from the rectangular equation.
$$[(x-1)(y+1) = 1,\ \text{No},\ y' = -e^{-2t},\ y'' = 2e^{-3t}]$$

8. Show that the curve given by
$$x = \sin^2 \theta - 1, \quad y = \sin \theta - 1, \quad -\infty < \theta < \infty,$$
is part of a parabola. Sketch a graph.

9. What is the curve given by
$$x = 2 - t^2, \quad y = t^2 - 1, \quad -\infty < t < \infty?$$
Interpret $t$ geometrically.
$$\begin{bmatrix} \text{The ray pointing left from } (2, -1) \text{ on the line } x + y = 1; \\ t = \pm\sqrt{s/\sqrt{2}}, \text{ where } s \text{ is the distance from } (2, -1). \end{bmatrix}$$

*10. Show that the curve represented by
$$x = a_1 t^2 + b_1 t + c_1, \quad y = a_2 t^2 + b_2 t + c_2$$
is a parabola if $a_1 b_2 - a_2 b_1 \neq 0$.

## 3 Arc Length

One naturally expects that curves will have length. A definition of arc length as a limit of lengths of inscribed polygons will be given, and formulas for arc length will be obtained, in Chapter 12.

At present all we need are formulas for the derivative of arc length. We derive such formulas on the basis of a plausible geometric assumption below. A precise proof must await Chapter 12.

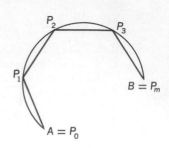

Arc length between $A$ and $B$

$$= \lim \sum_{i=1}^{m} \left| P_{i-1} P_i \right|$$

as the length of the longest side $\to 0$

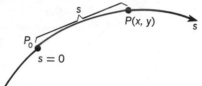

Arc length on a curve is measured from some fixed point $P_0$ on the curve. It will be measured positively in one direction and negatively in the other direction. Thus, the length of the curve from $P_0$ to a point $P(x, y)$ is a function of $P$. If the curve is given by $y = f(x)$, then the arc length is a function of $x$.

Referring to the figures, $\eta$ is the function of the Fundamental Lemma (Chapter 1) that $\to 0$ as $\Delta x \to 0$. Then it is geometrically plausible that, if $dx = \Delta x > 0$ and $s$ increases with $x$,

$$\sqrt{\Delta x^2 + \Delta y^2} \leqq \Delta s \leqq \Delta l + |\eta| \, \Delta x.$$

Because

$$\Delta l = \sqrt{dx^2 + dy^2} = \sqrt{1 + (dy/dx)^2} \, dx,$$

we have

$$\sqrt{1 + \left(\frac{\Delta y}{\Delta x}\right)^2} \leqq \frac{\Delta s}{\Delta x} \leqq \sqrt{1 + \left(\frac{dy}{dx}\right)^2} + |\eta|,$$

and now, as $\Delta x \to 0$, so does $\eta$, and we get

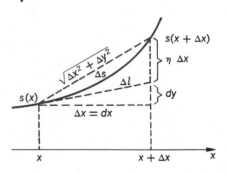

$$(1) \qquad \frac{ds}{dx} = \sqrt{1 + \left(\frac{dy}{dx}\right)^2}.$$

The negative root must be used in (1) if $s$ decreases as $x$ increases.

If $y$ had been the independent variable, we would have obtained

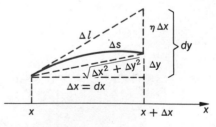

$$(2) \qquad \frac{ds}{dy} = \pm \sqrt{1 + \left(\frac{dx}{dy}\right)^2}.$$

If the curve had been given in parametric form, $x = x(t)$, $y = y(t)$, we would have obtained

(3)
$$\frac{ds}{dt} = \pm \sqrt{\left(\frac{dx}{dt}\right)^2 + \left(\frac{dy}{dt}\right)^2}.$$

In terms of differentials, equations (1), (2), and (3) imply that

$$ds^2 = dx^2 + dy^2.$$

*Example*   Suppose that arc length on the right half of the hyperbola $4x^2 - y^2 = 8$ is measured positively downward. Then

$$\left(\frac{ds}{dx}\right)^2 = 1 + \frac{16x^2}{y^2} = \frac{y^2 + 16x^2}{y^2}.$$

$$\frac{ds}{dx} = - \frac{\sqrt{20x^2 - 8}}{y},$$

where the minus sign must be chosen because of the direction of increasing $s$. Observe that, if $y > 0$ and $dx > 0$, then $ds < 0$, which agrees with the figure.

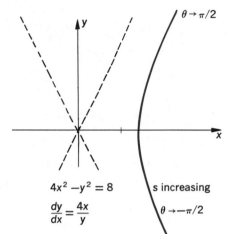

$4x^2 - y^2 = 8$

$\dfrac{dy}{dx} = \dfrac{4x}{y}$

$s$ increasing

If the following parametric equations are used for the hyperbola:

$$x = \sqrt{2}\sec\theta, \quad y = 2\sqrt{2}\tan\theta, \quad -\frac{\pi}{2} < \theta < \frac{\pi}{2};$$

then
$$dx = \sqrt{2}\sec\theta\tan\theta\, d\theta; \qquad dy = 2\sqrt{2}\sec^2\theta\, d\theta;$$

$$ds^2 = (2\sec^2\theta\tan^2\theta + 8\sec^4\theta)\, d\theta^2$$

$$= 2\sec^2\theta\,(5\tan^2\theta + 4)\, d\theta^2,$$

whence
$$ds = -\sqrt{2}\sec\theta\sqrt{5\tan^2\theta + 4}\, d\theta,$$

where we must choose the minus sign because $s$ decreases as $\theta$ increases.

### Problems

In Problems 1 to 4 express $ds$ in terms of either $x$ and $dx$, or $y$ and $dy$.

1.   $y^2 = -2x$;   $s$ increases as $y$ increases          $\left[\sqrt{1 + y^2}\, dy\right]$

2.   $x^2 = 2y$;   $s$ increases as $x$ increases          $\left[\sqrt{1 + x^2}\, dx\right]$

3.   The upper semicircle of $x^2 + y^2 = a^2$;   $s$ increases as one moves counterclockwise

$$\left[-\frac{a}{\sqrt{a^2 - x^2}}\, dx\right]$$

4.  $y = e^{ax}$; $s$ decreases as $x$ increases $\qquad$ $\left[ -\sqrt{1 + a^2 e^{2ax}}\, dx \right]$

**In Problems 5 to 14 express ds in any convenient way.**

5.  $x^2 + y^2 = a^2$; $s$ decreases as one moves counterclockwise $\qquad$ $[ds = (a/y)\, dx]$

6.  $x = a \cos\theta$, $y = b \sin\theta$; $s$ increases with $\theta$
$$[ds = \sqrt{a^2 \sin^2\theta + b^2 \cos^2\theta}\, d\theta]$$

7.  $xy = a^2$, $x < 0$; $s$ increases with $x$
$$\left[ ds = \frac{\sqrt{a^4 + x^4}}{x^2}\, dx \right]$$

8.  $2x^2 + 3y^2 = 18$
$$\left[ ds = \pm \frac{\sqrt{54 - 2x^2}}{3y}\, dx \right]$$

9.  $x = 3t$, $y = 2t - t^2$ $\qquad$ $[ds = \pm\sqrt{4t^2 - 8t + 13}\, dt]$

10.  $x = e^\theta \sin\theta$, $y = e^\theta \cos\theta$; $s$ decreases with increasing $\theta$ $\qquad$ $[ds = -\sqrt{2}e^\theta\, d\theta]$

11.  $x = t^2$, $y = t^3$; $s$ increases with $t$ and $t < 0$ $\qquad$ $[ds = -t\sqrt{4 + 9t^2}\, dt]$

12.  $3x^2 - 4y^2 + 24 = 0$, lower half; $s$ increases with $x$
$$[ds = (-\sqrt{3}/4y)\sqrt{7x^2 + 32}\, dx]$$

13.  $x = a \cos^3\theta$, $y = a \sin^3\theta$, $a > 0$; $s$ increases with $\theta$ $\qquad$ $\left[ ds = \frac{3a}{2}\left|\sin 2\theta\right| d\theta \right]$

14.  $y = \frac{1}{4}(2 \log x - x^2)$; $s$ decreases with increasing $x$ $\qquad$ $\left[ ds = -\frac{(x^2 + 1)}{2x}\, dx \right]$

*15.  Use the answer to Problem 13 to show that the length of the hypocycloid of four cusps is $6a$.

*16.  Use the answer to Problem 10 to show that the length of the spiral between $\theta = 0$ and $\theta = \pi/2$ is $\sqrt{2}[\exp(\pi/\sqrt{2}) - 1]$.

## 4   Motion on a Curve; Velocity; Acceleration

The most important parameter for applications is a measure $t$, of time.* For example, to follow the motion of a satellite one must be able to calculate its position in space at each instant.

---

\* **Historical Remark.** Newton made continuous motion basic to his calculus. He conceived of $x$ and $y$ on the graph of $y = f(x)$ as "generating the graph by their motion." He called $x$ and $y$ *fluents*, or flowing quantities. Their "rates of generation" (which we must regard as derivatives with respect to time) Newton called their *fluxions*—which he denoted by $\dot{x}$ and $\dot{y}$. In this way the derivative became, using Leibniz notation, $dy/dx = \dot{y}/\dot{x}$.

Therefore, if we have motion along a plane curve

$$x = \varphi(t), \qquad y = \psi(t),$$

then the derivatives $\dot{x} = \varphi'(t)$ and $\dot{y} = \psi'(t)$ are velocities in the $x$ and $y$ directions. They are the *components of velocity*:

$$v_x = \frac{dx}{dt}; \qquad v_y = \frac{dy}{dt}.$$

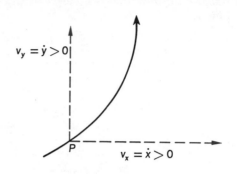

The components of velocity at $P$ can be represented by arrows drawn at $P$ of lengths $|v_x|$ and $|v_y|$ and directed parallel to the axes according to the signs of $v_x$ and $v_y$. If $v_x > 0$, the arrow points toward increasing $x$. If $v_y > 0$, the arrow points toward increasing $y$.

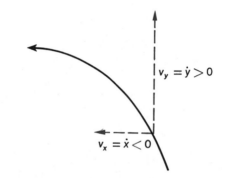

The *velocity vector* **v** is the resultant of the component velocities (and is represented by the arrow that is the diagonal of the rectangle of the figure).* It is directed along the tangent line because $y' = \dot{y}/\dot{x}$, and it has magnitude

$$|\mathbf{v}| = v = \sqrt{v_x^2 + v_y^2}.$$

The magnitude $v$ is called the *speed*.

If $s$ increases in the direction of motion along the curve, then

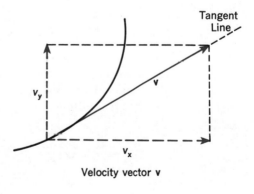

Velocity vector **v**

$$v = \sqrt{\left(\frac{dx}{dt}\right)^2 + \left(\frac{dy}{dt}\right)^2} = \frac{ds}{dt}.$$

Thus, $ds/dt$ is simply the speed along the curve.

---

* The vector **v** represents the displacement of a point in unit time if it were moving with constant velocity **v**.

*Remark.* If $\tau$ is the inclination of the tangent line, then

$$\frac{dx}{dt} = v_x = v \cos \tau = \frac{ds}{dt} \cos \tau,$$

$$\frac{dy}{dt} = v_y = v \sin \tau = \frac{ds}{dt} \sin \tau,$$

or

$$\begin{cases} dx = ds \cos \tau, \\ dy = ds \sin \tau. \end{cases}$$

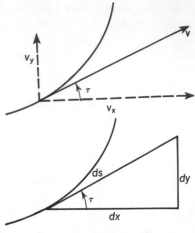

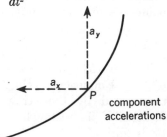

Mnemonic device

If we differentiate the components of velocity, we obtain the *components of acceleration*:

$$a_x = \frac{dv_x}{dt} = \frac{d^2x}{dt^2}, \qquad a_y = \frac{dv_y}{dt} = \frac{d^2y}{dt^2}.$$

The components of acceleration at $P$ can be represented by arrows drawn at $P$ of lengths $|a_x|$ and $|a_y|$ and directed parallel to the axes and in the direction of increasing coordinates if $a_x$ and $a_y$ are positive.

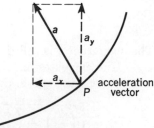

component accelerations

The resultant of these components (represented by the diagonal arrow) is the *acceleration vector*, **a**, which has magnitude

$$|a| = a = \sqrt{a_x^2 + a_y^2}.$$

acceleration vector

The relations of coordinates, component velocities, and accelerations can be summarized by this diagram*:

$$(x, y) \xrightarrow{D_t} (\dot{x}, \dot{y}) = (v_x, v_y) \xrightarrow{D_t} (\dot{v}_x, \dot{v}_y) = (a_x, a_y)$$

*Remark.* The acceleration vector need not point either along the curve or normal to it. The examples below will illustrate this. In Section 7 the components of **a** along the tangent and along the normal will be found.

**Example 1**    A point moves on the straight line $y = \frac{1}{2}x + 1$ in the direction of increasing $x$ at the constant speed of 2 units per second. Sketch the path and the velocity and acceleration vectors at $(-1, \frac{1}{2})$.

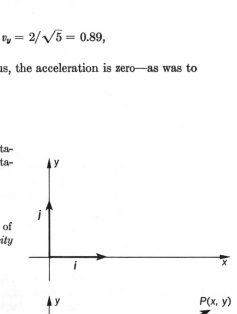

Since $v = 2$ is given, we have

$$v_x = 2 \cos \tau, \qquad v_y = 2 \sin \tau,$$

where $\tau$ is the inclination of the line. And since slope $= \tan \tau = \frac{1}{2}$,

$$\cos \tau = 2/\sqrt{5} \qquad \sin \tau = 1/\sqrt{5},$$

and

$$v_x = 4/\sqrt{5} = 1.79 \qquad v_y = 2/\sqrt{5} = 0.89,$$

regardless of where one is on the line. Thus, the acceleration is zero—as was to be expected.

---

* The concepts are neatly expressed in vector notation, which is not assumed here. However, the notation is simple enough, and is as follows:

The *position vector* to the point $P$ is

$$\mathbf{r} = x\mathbf{i} + y\mathbf{j}$$

where $\mathbf{i}$ and $\mathbf{j}$ are unit vectors (that is, vectors of length one), along the axes as shown. The *velocity vector* is

$$\mathbf{v} = d\mathbf{r}/dt = \dot{x}\mathbf{i} + \dot{y}\mathbf{j} = v_x\mathbf{i} + v_y\mathbf{j}.$$

The *acceleration vector* is

$$\mathbf{a} = d\mathbf{v}/dt = \dot{v}_x\mathbf{i} + \dot{v}_y\mathbf{j} = a_x\mathbf{i} + a_y\mathbf{j}.$$

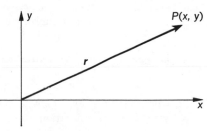

Vector notation and its general use is a product of the late nineteenth century. There are several styles of notation. The one used here was devised by Josiah W. Gibbs (1839–1903) professor of mathematical physics at Yale.

In the text we have simply written out the components of **v** and **a**. In the nineteenth century this practice was referred to as "Cartesian methods."

*Example 2*    A point moves on a circle of radius $r$ with constant angular velocity $\omega$. Find the velocity and acceleration vectors and draw them at some point.

Choose axes as indicated in the figure and assume the motion is counter-clockwise. Then, measuring time from an instant when the point is at $(r, 0)$, we have the equations of motion:

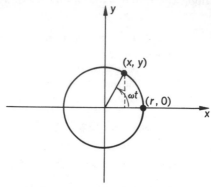

$$x = r \cos \omega t,$$

$$y = r \sin \omega t,$$

$$v_x = \dot{x} = -\omega r \sin \omega t,$$

$$v_y = \dot{y} = \omega r \cos \omega t,$$

$$a_x = \dot{v}_x = -\omega^2 r \cos \omega t,$$

$$a_y = \dot{v}_y = -\omega^2 r \sin \omega t.$$

From these expressions we see that the velocity vector is always directed along the tangent, and that the acceleration vector is always perpendicular to the velocity and points toward the center of the circle.

For the speed $v$ we get

$$v = \sqrt{\omega^2 r^2 \sin^2 \omega t + \omega^2 r^2 \cos^2 \omega t} = \omega r,$$

which was to be expected; $v$ is constant.

For the magnitude of the acceleration $a$, we get

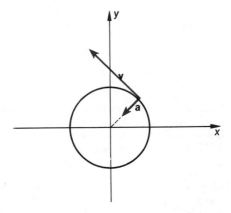

$$a = \sqrt{\omega^4 r^2 \cos^2 \omega t + \omega^4 r^2 \sin^2 \omega t} = \omega^2 r,$$

and we find the surprising (perhaps) result that, even though the speed $v$ is constant, the acceleration is not zero.

The magnitude of the acceleration can also be written as

$$a = \omega^2 r = \frac{\omega^2 r^2}{r} = \frac{v^2}{r}.$$

We shall see in Section 7 that on any curve there is an analogous formula for the normal component of the acceleration.

A satellite in uniform circular motion around the earth is accelerated by gravity toward the center of the earth.

*Example 3*    A point moves according to

$$x = 2t^2 - 1, \qquad y = 2 - t,$$

where $t$ is time. Find $v_x$, $v_y$, and $v$ and $a_x$, $a_y$, and $a$. Then sketch the path, with velocity and acceleration vectors at the point where $t = -1$.

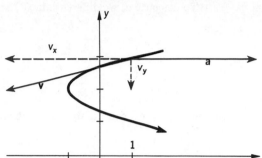

$$\begin{cases} v_x = 4t, \quad v_y = -1, \\[2mm] v = \sqrt{16t^2 + 1}. \end{cases}$$

$$\begin{cases} a_x = 4, \quad a_y = 0, \\[2mm] a = 4. \end{cases}$$

The path is the parabola

$$2(y - 2)^2 = (x + 1).$$

At $t = -1$ the point is at $(1, 3)$ and

$$v_x = -4, \qquad v_y = -1,$$

$$a_x = 4, \qquad a_y = 0.$$

*Example 4*    A point moves downward on the parabola $y^2 = 2x$ with constant speed $v = 3$. Find the components of velocity and acceleration in terms of $x$. Then sketch the curve and draw the velocity and acceleration vectors at $(2, -2)$.

Because the point is on the parabola $y^2 = 2x$, we have

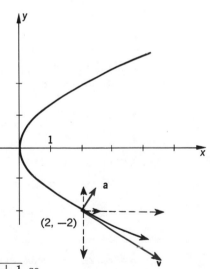

$$(1) \qquad\qquad y\dot{y} = \dot{x}.$$

Because $v = 3$, we also have

$$(2) \qquad \dot{x}^2 + \dot{y}^2 = v^2 = 9.$$

Solving equations (1) and (2) for $\dot{x}$ and $\dot{y}$ will give the component velocities. We get

$$\dot{y}^2 = \frac{9}{y^2 + 1} \qquad \text{and} \qquad \dot{y} = \frac{-3}{\sqrt{y^2 + 1}}$$

because the point moves downward. Hence

$$(3) \qquad \dot{y} = v_y = \frac{-3}{\sqrt{2x + 1}}.$$

From equation (1) we get $\dot{x} = -3y/\sqrt{2x + 1}$, so

$$(4) \qquad\qquad \dot{x} = v_x = \begin{cases} \dfrac{-3\sqrt{2x}}{\sqrt{2x + 1}} & \text{if } y \geqq 0, \\[4mm] \dfrac{3\sqrt{2x}}{\sqrt{2x + 1}} & \text{if } y < 0. \end{cases}$$

To compute $a_y$ we differentiate (3) and get

$$\ddot{y} = a_y = \frac{3\dot{x}}{(2x+1)^{3/2}}$$

$$= \frac{\mp 9\sqrt{2x}}{(2x+1)^2} \quad (- \text{ if } y \geqq 0, + \text{ if } y < 0).$$

To compute $a_x$ we can differentiate either (1) or (4). Equation (1) is easier. We have

$$y\ddot{y} + \dot{y}^2 = \ddot{x}$$

or

$$\ddot{x} = \mp \frac{9y\sqrt{2x}}{(2x+1)^2} + \frac{9}{2x+1}$$

$$= -\frac{18x}{(2x+1)^2} + \frac{9}{2x+1} = \frac{9}{(2x+1)^2}.$$

At the point $(2, -2)$

$$v_x = 6/\sqrt{5}, \qquad v_y = -3/\sqrt{5};$$

$$a_x = 9/25, \qquad a_y = 18/25.$$

## Problems

In Problems 1 to 14 find $v_x$, $v_y$, $v$, $a_x$, $a_y$, and $a$. Then sketch the path, showing the velocity and acceleration vectors at the given time.

1.  $x = 2t,\ \ y = t^2;\ \ t = \frac{1}{2}$    $[v_x = 2,\ v_y = 2t,\ a_x = 0,\ a_y = 2\,]$

2.  $x = t,\ \ y = t^3;\ \ t = 1$    $[v_x = 1,\ v_y = 3t^2,\ a_x = 0,\ a_y = 6t\,]$

3.  $x = t^2,\ \ y = t^3;\ \ t = 1$    $[v_x = 2t,\ v_y = 3t^2,\ a_x = 2,\ a_y = 6t\,]$

4.  $x = 2t,\ \ y = e^t;\ \ t = 1$    $[v_x = 2,\ v_y = e^t,\ a_x = 0,\ a_y = e^t\,]$

5.  $x = \cos t,\ \ y = 2\sin t;\ \ t = \dfrac{\pi}{2}$    $\begin{bmatrix} v_x = -\sin t,\ v_y = 2\cos t \\ a_x = -\cos t,\ a_y = -2\sin t \end{bmatrix}$

6.  $x = \cos 2t,\ \ y = \sin t;\ \ t = \pi$    $\begin{bmatrix} v_x = -2\sin 2t,\ v_y = \cos t \\ a_x = -4\cos 2t,\ a_y = -\sin t \end{bmatrix}$

7.  $x = 2t,\ \ y = \log t;\ \ t = 1$    $[v_x = 2,\ v_y = 1/t,\ a_x = 0,\ a_y = -1/t^2\,]$

8.  $x = \sin^2 t,\ \ y = \sqrt{2p}\sin t;\ \ t = 0$    $\begin{bmatrix} v_x = \sin 2t,\ v_y = \sqrt{2p}\cos t \\ a_x = 2\cos 2t,\ a_y = -\sqrt{2p}\sin t \end{bmatrix}$

9.  $x = 2 - 3\sin t,\ \ y = 1 + 2\cos t;\ \ t = \dfrac{\pi}{2}$    $\begin{bmatrix} v_x = -3\cos t,\ v_y = -2\sin t \\ a_x = 3\sin t,\ a_y = -2\cos t \end{bmatrix}$

10. $x = t, \quad y = t \log t; \quad t = e^{-1}$  $\left[\begin{array}{l} v_x = 1, v_y = 1 + \log t \\ a_x = 0, a_y = 1/t \end{array}\right]$

*11. $x = \dfrac{3t}{1 + t^3}, \quad y = \dfrac{3t^2}{1 + t^3} \quad ; \quad t = 1$ (folium of Descartes)

$$\left[ v_x = \frac{3(1 - 2t^3)}{(1 + t^3)^2}, v_y = \frac{3(2t - t^4)}{(1 + t^3)^2}; a_x = \frac{18(t^5 - 2t^2)}{(1 + t^3)^3}, a_y = \frac{6(1 - 7t^3 + t^6)}{(1 + t^3)^3} \right]$$

*12. $x = b(\sin^2 t - \cos^2 t), \quad y = b \tan t (\sin^2 t - \cos^2 t), \quad t = 0$ (strophoid)
$$\left[\begin{array}{l} v_x = 2b \sin 2t, v_y = b(\tan^2 t - 1) + 2b \tan t \sin 2t \\ a_x = 4b \cos 2t, a_y = 2b \sec^2 t (\tan t + \sin 2t) + 4b \tan t \cos 2t \end{array}\right]$$

13. $x = \frac{1}{2} \cos t^2, \quad y = \frac{1}{2} \sin t^2; \quad t = 0$ and $t = \sqrt{\pi}/2$ $\left[\begin{array}{l} \text{at } t = \sqrt{\pi}/2 \\ a_x \approx -1.82, a_y \approx -0.42 \end{array}\right]$

14. $x = b(\pi t - \sin \pi t), \quad y = b(1 - \cos \pi t); \quad t = \frac{1}{2}$ (cycloid). Show that the magnitude of the acceleration is constant.
$$[v_x = \pi b(1 - \cos \pi t), v_y = \pi b \sin \pi t; a_x = \pi^2 b \sin \pi t, a_y = \pi^2 b \cos \pi t]$$

15. A point moves on the parabola $x^2 = 4y$ so that $v_y = \text{constant} = \frac{1}{2}$ unit per second. Find the components of velocity and acceleration at $(-2, 1)$. Then sketch the curve and find $\mathbf{v}$ and $\mathbf{a}$ at $(-2, 1)$.
$$[v_x = -\tfrac{1}{2}, v_y = \tfrac{1}{2}, a_x = \tfrac{1}{8}, a_y = 0]$$

16. A point moves on the cubic $x^3 = 4y$ so that $v_y = \text{constant} = 2$ units per second. Find the components of velocity and acceleration at $(2, 2)$. Then sketch the curve and find $\mathbf{v}$ and $\mathbf{a}$ at $(2, 2)$.   $[v_x = \tfrac{2}{3}, v_y = 2, a_x = -\tfrac{4}{9}, a_y = 0]$

17. A point moves on $y = e^x$ in the direction of increasing $x$ with constant speed $= c$. Express $v_x, v_y, a_x,$ and $a_y$ in terms of $x$.
$$\left[\begin{array}{l} v_x = c(1 + e^{2x})^{-1/2}, v_y = ce^x(1 + e^{2x})^{-1/2} \\ a_x = -c^2 e^{2x}/(1 + e^{2x})^2, a_y = c^2 e^x/(1 + e^{2x})^2 \end{array}\right]$$

18. A point moves upward on the right half of the hyperbola $4x^2 - 3y^2 = 24$ at the constant speed $v = 6$ units per second. Find the components of velocity at $(3, -2)$. Then sketch the curve and the velocity vector at $(3, -2)$.
$$[v_x = -6/\sqrt{5}, v_y = 12/\sqrt{5}]$$

19. A point moves on part of a parabola given by $x^{1/2} - y^{1/2} = b^{1/2}$ with $v_x = \text{constant} = c$. Find $v_y, a_x,$ and $a_y$ in terms of $x$.
$$[v_y = c(1 - \sqrt{b/x}), a_x = 0, a_y = c^2 b^{1/2}/2x^{3/2}]$$

**20.** In the absence of wind resistance a projectile is subject only to the acceleration of gravity. Then

$$a_x = 0, \quad a_y = -g = -32 \text{ ft/sec}^2,$$

if axes are chosen as in the figure.

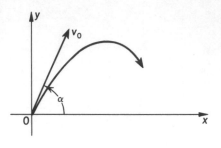

If a projectile is shot at an initial speed of $v_0$ fps at an angle $\alpha$ with the horizontal, (a) find $v_x$ and $v_y$ in terms of $t$, (b) find $x$ and $y$ in terms of $t$; (c) then find where the projectile lands.

**21.** If a bead slides without friction down a wire under the influence of gravity, then its speed is

$$v = \sqrt{2gh} = 8\sqrt{h},$$

where $h$ is the height through which it has fallen.

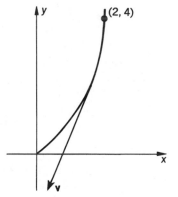

Suppose a bead slides down the parabola $y = x^2$ starting from the point $(2, 4)$. Find $v_x$ and $v_y$ in terms of $x$.

$$\left[ v_x = \frac{-8\sqrt{4 - x^2}}{\sqrt{1 + 4x^2}}, v_y = \frac{-16x\sqrt{4 - x^2}}{\sqrt{1 + 4x^2}} \right]$$

## 5   Polar Coordinates*

Suppose a curve is given in polar coordinates by

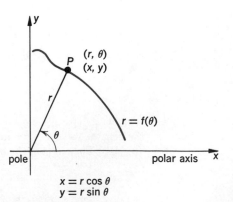

$$(1) \qquad r = f(\theta)$$

and suppose the polar-coordinate system is related to a rectangular coordinate system in the standard manner shown in the figure. Then

$$(2) \qquad x = r \cos \theta, \qquad y = r \sin \theta$$

are the equations connecting the rectan-

---

* Graphs of a number of curves in polar coordinates are to be found in Appendix A.

gular coordinates $(x, y)$ and the polar coordinates* $(r, \theta)$.

In order to find the inclination $\tau$, of the tangent line to a curve described in polar coordinates, it is convenient to introduce an auxiliary angle $\psi$ (see the figures), which is defined to be the angle between the extended radius vector at $P$ and the tangent to the curve at $P$, and it is measured positively counterclockwise.

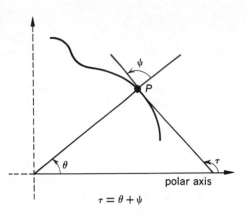

$$\tau = \theta + \psi$$

There is always a relationship of the form

$$\psi = \tau - \theta + n\pi$$

(where $n$ is an integer) between $\tau$, $\theta$, and $\psi$. The integer $n$ will depend on the particular curve under consideration and also (as is shown by the figures) on where point $P$ is on the curve. In any case, $\tan \psi$ can be computed directly from the equation of the curve $r = f(\theta)$.

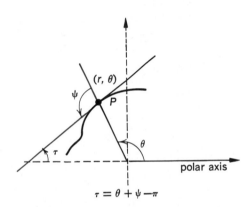

$$\tau = \theta + \psi - \pi$$

**THEOREM**    $\tan \psi = \dfrac{r}{dr/d\theta}$ .

**PROOF**    Though tricky, a proof is made by direct computation. We have

$$\tan \psi = \tan (\tau - \theta + n\pi) = \tan (\tau - \theta)$$

because the tangent function has period $\pi$. Then

(3)
$$\tan \psi = \tan (\tau - \theta) = \frac{\tan \tau - \tan \theta}{1 + \tan \tau \tan \theta} .$$

Now

(4)
$$\tan \tau = \frac{dy}{dx} = \frac{dy/d\theta}{dx/d\theta} ,$$

---

* **Historical Remark.** The first systematic use of polar coordinates was made by Jakob Bernoulli (1654–1705) who used them to discuss spirals. He was professor of mathematics in Basle from 1687 till his death. Bernoulli was extremely active in developing and promoting the Leibnizian calculus. He, more than anyone else in those early years after Leibniz' invention, was influential in making the power of Leibniz' notation known in Europe, which he did by extending and refining what Leibniz had begun. He made significant discoveries in geometry, probability, and the calculus of variations.

and from (2),

(5) $$\frac{dx}{d\theta} = -r\sin\theta + \frac{dr}{d\theta}\cos\theta, \quad \frac{dy}{d\theta} = r\cos\theta + \frac{dr}{d\theta}\sin\theta.$$

Using (4) and (5) and the identity $\tan\theta = \sin\theta/\cos\theta$ in (3), we obtain, after some algebraic simplification, the desired result:

$$\tan\psi = \frac{r}{dr/d\theta}.$$

**Example 1**   Sketch a graph of $r = 2/(1 - \cos\theta)$. Find $\psi$ as a function of $\theta$ and sketch the angle $\psi$ for $\theta = \pi/2$ and $5\pi/3$.

The graph is a parabola with focus at the pole.

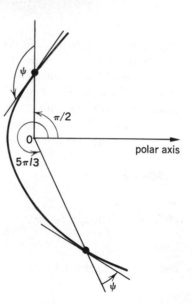

$$\tan\psi = \frac{r}{dr/d\theta} = \frac{2/(1-\cos\theta)}{-2\sin\theta/(1-\cos\theta)^2}$$

$$= \frac{\cos\theta - 1}{\sin\theta} = -\tan\frac{\theta}{2}$$

$$= \tan\left(-\frac{\theta}{2}\right).$$

Therefore

$$\psi = -\frac{\theta}{2} + n\pi$$

for some integer $n$. From the figure we see that $n = 1$.

At $\theta = \pi/2$,   $\psi = -\pi/4 + \pi = 3\pi/4$.

At $\theta = 5\pi/3$,   $\psi = -5\pi/6 + \pi = \pi/6$.

Equations (5) can also be used to find the derivative of arc length with respect to $\theta$.

$$\left(\frac{ds}{d\theta}\right)^2 = \left(\frac{dx}{d\theta}\right)^2 + \left(\frac{dy}{d\theta}\right)^2 = (-r\sin\theta + \dot{r}\cos\theta)^2 + (r\cos\theta + \dot{r}\sin\theta)^2$$

$$= r^2 + \dot{r}^2 = r^2 + \left(\frac{dr}{d\theta}\right)^2.$$

Thus,

(6) $$ds^2 = r^2\,d\theta^2 + dr^2.$$

The formulas for tan $\psi$ and $ds^2$ can be recalled at once from the figure at the right. The curved "right triangle," with sides equal to the increments shown, suggests the correct relations between the differentials that give the formulas for tan $\psi$ and $ds^2$. Naturally, this figure does not constitute a proof. One also gets the useful relations

(7)     $\sin \psi = r\dfrac{d\theta}{ds}$     $\cos \psi = \dfrac{dr}{ds}$,

if $s$ increases with $\theta$.

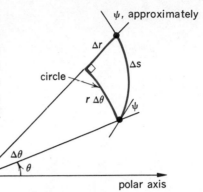

*Remark.* Curves in polar coordinates can also be defined implicitly. For example, the lemniscate of Bernoulli has the equation $r^2 = a^2 \cos 2\theta$. In such cases tan $\psi$ can be computed by implicit differentiation.

The "differential triangle"

$\tan \psi = r\dfrac{d\theta}{dr}$

$ds^2 = dr^2 + r^2 d\theta^2$

$\sin \psi = r\dfrac{d\theta}{ds}, \cos \psi = \dfrac{dr}{ds}$

**Example 2**    Find the inclination of the tangent line to the spiral $r = \theta$ at the point where $\theta = \pi/2$.
   From the figure,

$\tau = \theta + \psi = (\pi/2) + \psi$,   whence

$\tan \tau = \tan \left(\dfrac{\pi}{2} + \psi\right) = -\cot \psi.$

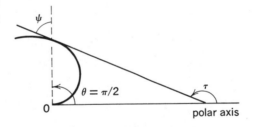

Then   $\tan \psi = \dfrac{r}{dr/d\theta} = \dfrac{\theta}{1} = \dfrac{\pi}{2}$,

and              $\tan \tau = -\cot \psi = -2/\pi \approx -0.6366.$

$\tau \approx 147.5°.$

### Problems

In Problems 1 to 7 sketch the curve. Find $\psi$ as a function of $\theta$ and sketch the angle $\psi$ for the indicated values of $\theta$.

1.   $r = e^{a\theta}$,   $a > 0$;   $\theta = \pi/4, 2\pi/3$ (This is called the equiangular spiral.)
                                              [$\psi = $ Arc tan $1/a$]

2.   $r\theta = a$,   $a > 0$;   $\theta = 1$              [$\psi = -$Arc tan $\theta + \pi$]

3. $r = a \cos \theta$; $\theta = \dfrac{\pi}{6}$
$$\left[\begin{matrix} \psi = \theta + \dfrac{\pi}{2}, \, 0 < \theta < \dfrac{\pi}{2} \\[2mm] \psi = \theta - \dfrac{\pi}{2}, \, \dfrac{\pi}{2} \le \theta \le \pi \end{matrix}\right]$$

4. $r = 1 + 2 \cos \theta$; $\theta = \dfrac{11\pi}{6}$
$$\left[ \tan \psi = \dfrac{1 + 2 \cos \theta}{-2 \sin \theta} \right]$$

5. $r = a\theta^2$, $a > 0$; $\theta = 2\pi$
$$[\, \psi = \text{Arc} \tan \theta/2 \,]$$

6. $r(1 + \cos \theta) = a$, $a > 0$; $\theta = \dfrac{\pi}{2}, \dfrac{3\pi}{2}$
$$\left[\begin{matrix} \psi = \dfrac{\pi}{2} - \dfrac{\theta}{2}, \, 0 \le \theta < \pi \\[2mm] \psi = \dfrac{3\pi}{2} - \dfrac{\theta}{2}, \, \pi < \theta < 2\pi \end{matrix}\right]$$

7. $r = \cos 2\theta$; $\theta = \pi/2$
$$[\, \tan \psi = -\tfrac{1}{2} \cot 2\theta \,]$$

**In Problems 8 to 13 find $ds/d\theta$. Assume that $s$ increases with $\theta$.**

8. $r = e^{a\theta}$
$$[\, ds/d\theta = \sqrt{1 + a^2 e^{a\theta}} \,]$$

9. $r = 1 + 2 \cos \theta$
$$[\, ds/d\theta = \sqrt{5 + 4 \cos \theta} \,]$$

10. $r = a \cos \theta$
$$[\, ds/d\theta = |\, a \,| \,]$$

11. $r(1 + \cos \theta) = a$
$$[\, ds/d\theta = \sqrt{2} \,|\, a \,|/(1 + \cos \theta)^{3/2} \,]$$

12. $r^2 = a^2 \cos 2\theta$
$$[\, ds/d\theta = |\, a \,| \sqrt{\sec 2\theta} \,]$$

13. $2r^3 = 3\theta^2$
$$[\, ds/d\theta = \sqrt{(3r^3 + 2)/3r} \,]$$

14. Draw a figure to show that the angle $\varphi$ between two polar curves at a point of intersection is given by

$$\tan \varphi = \pm \frac{\tan \psi_1 - \tan \psi_2}{1 + \tan \psi_1 \tan \psi_2}.$$

What condition implies that they meet at right angles?

15. Find the angle between the straight line $r \sin \theta = 2$ and the circle $r = 4 \sin \theta$ at their points of intersection. $[\pi/2]$

16. At what angles do the circle $r = \cos \theta$ and the rose $r = \cos 2\theta$ intersect?
$$\left[\begin{matrix} \pi/4 \text{ at the pole; } 0 \text{ at } (a, 0); \\ \text{Arc} \tan 3\sqrt{3}/5 \text{ at } (\tfrac{1}{2}, \pm\pi/3). \end{matrix}\right]$$

17. Show that the parabolas $r = a/(1 + \cos \theta)$ and $r = b/(1 - \cos \theta)$ intersect at right angles, where $a, b > 0$.

18. Find the angle between the parabola $r = a/(1 + \cos \theta)$ and the line $r \sin \theta = a$.

In Problems 19 to 22 draw a figure and find the points of intersection of the polar curves. Then show that they intersect at right angles at some points.

19.　$r = a\theta,\quad r = a/\theta$

20.　$r = a(1 - \cos\theta),\quad r(1 - \cos\theta) = a$

21.　$r^2 = a^2 \sin 2\theta,\quad r^2 = a^2 \cos 2\theta$

22.　$r = a\sec^2(\theta/2),\quad r = b\csc^2(\theta/2)$

In Problems 23 to 27 find the slope of the tangent line at the specified point.

23.　$r = a\cos 3\theta;\quad$ pole $\hfill [1/\sqrt{3},\ -1/\sqrt{3},\ \text{undefined}\,]$

24.　$r = a\sin 3\theta;\quad \theta = \pi/6 \hfill [-\sqrt{3}\,]$

25.　$r = a\theta;\quad \theta = 1 \hfill [-5.77\,]$

26.　$r\cos^2\theta = a;\quad r = 2a \hfill [3]$

27.　$r = e^{a\theta};\quad \theta = \pi/4 \hfill [(a+1)/(a-1)\,]$

28.　Show that the inclination $\tau$ of the tangent line to the parabola $r = a\sec^2(\theta/2)$ satisfies the equation $\tau + \psi = \pi$.

29.　Show that the inclination $\tau$ of the tangent to the curve $r = a\sin^3(\theta/3)$ satisfies $\tan\tau = \tan 4\psi$.

*30.　Derive equations (7) from the identity $\sin^2\psi = \tan^2\psi/(1 + \tan^2\psi)$ and the formula for $\tan\psi$.

31.　Finish the proof of the theorem on page 159 by completing the algebraic simplification.

*32.　Find the points $(r,\theta)$ where the graph of $r = \log\theta$ intersects itself. Then find the angle at which the curve meets itself at the point nearest the pole.
$$[\text{at intersections } \theta = (2n+1)\pi + \sqrt{(2n+1)^2\pi^2 + 4}\,)/2\,]$$

## 6　Motion in Polar Coordinates

Suppose that a point moves along the polar curve $r = f(\theta)$.

　　In this section we find the components of velocity and acceleration in the directions of increasing $r$ and increasing $\theta$. These components are called the *radial* and the *transverse* components.

　　The positive directions for these components are shown in the figure.

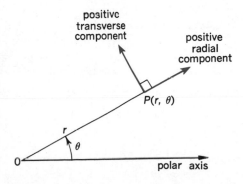

The components of velocity, $v_r$ and $v_\theta$, in the directions of increasing $r$ and increasing $\theta$ are easily obtained because the velocity is tangent to the curve. From the figure at the right,

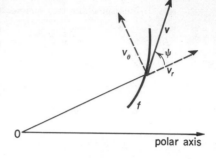

$v_r = v \cos \psi = $ the radial component of velocity,

$v_\theta = v \sin \psi$
   $= $ the transverse component of velocity.

But $v = ds/dt$, and from equations (7) of Section 5 we have

$$\cos \psi = \frac{dr}{ds} \quad \text{and} \quad \sin \psi = r\frac{d\theta}{ds},$$

whence

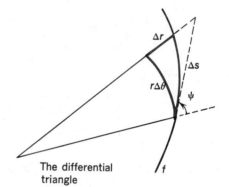

(1) $\quad v_r = \dfrac{dr}{dt} = \dot{r} \qquad v_\theta = r\dfrac{d\theta}{dt} = r\dot{\theta}.$

Of course, the speed is then

(2) $\quad \left(\dfrac{ds}{dt}\right)^2 = v^2 = \dot{r}^2 + r^2\dot{\theta}^2.$

The differential triangle

These velocity components are easily remembered by using the differential triangle.

**Example 1**   Find $v_r$, $v_\theta$, and $v$ at the point where $\theta = \pi/6$ on the cardioid $r = a(1 + \cos \theta)$, if at that point $\dot{\theta} = 1$ radian/sec.

Sketch the path and draw the velocity vector at the point where $\theta = \pi/6$.

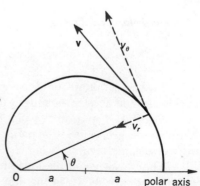

$v_r = \dot{r} = -a \sin \theta\, \dot{\theta} = -\dfrac{a}{2},$

$v_\theta = r\dot{\theta} = a(1 + \cos \theta)\dot{\theta} = a\left(\dfrac{2 + \sqrt{3}}{2}\right) = 1.87a,$

$v = \sqrt{\dfrac{a^2}{4} + \dfrac{(2 + \sqrt{3})^2}{4}a^2} = \sqrt{2 + \sqrt{3}}\,a$

$\quad = 1.93a.$

To obtain the radial and transverse components of the acceleration in polar coordinates we must work harder since the radial and transverse directions are different at different points.

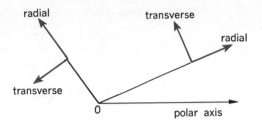

There are two steps to the derivation. First, the polar components of acceleration are expressed in terms of the rectangular components of acceleration. This is done by obtaining the projections of $a_x$ and $a_y$ on the radius vector and on a perpendicular to it. The projections are easily calculated by considering the two rectangles pictured with the common diagonal $\mathbf{a}$. We obtain

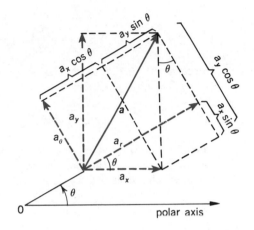

$$(3) \quad \begin{cases} a_r = a_x \cos \theta + a_y \sin \theta, \\ \\ a_\theta = -a_x \sin \theta + a_y \cos \theta. \end{cases}$$

Second, $a_x$ and $a_y$ are computed in terms of $r$ and $\theta$ and their derivatives

From

$$x = r \cos \theta, \qquad y = r \sin \theta$$

we obtain

$$\dot{x} = \dot{r} \cos \theta - r \sin \theta\, \dot{\theta}, \qquad \dot{y} = \dot{r} \sin \theta + r \cos \theta\, \dot{\theta},$$

so that finally,

$$(4) \quad \begin{cases} a_x = \ddot{x} = -(r\ddot{\theta} + 2\dot{r}\dot{\theta}) \sin \theta + (\ddot{r} - r\dot{\theta}^2) \cos \theta, \\ \\ a_y = \ddot{y} = (r\ddot{\theta} + 2\dot{r}\dot{\theta}) \cos \theta + (\ddot{r} - r\dot{\theta}^2) \sin \theta. \end{cases}$$

Now, substituting (4) in (3), we obtain, after a little algebra, the desired result:

$$(5) \quad \begin{cases} a_r = \ddot{r} - r\dot{\theta}^2, \\ \\ a_\theta = r\ddot{\theta} + 2\dot{r}\dot{\theta} = \dfrac{1}{r}\dfrac{d}{dt}(r^2\dot{\theta}). \end{cases}$$

**Example 2**   Find $a_r$, $a_\theta$, and $a$, if a point moves on the parabola $r = b \sec^2 (\theta/2)$, $b > 0$, at a constant speed $v = ds/dt = cb$, and $s$ increases with $\theta$. Sketch the curve and draw the acceleration at $\theta = \pi/2$ if $c = 2$.

The constant speed permits the calculation of $\dot\theta$ from

(6) $\qquad v = \dfrac{ds}{dt} = \dfrac{ds}{d\theta}\dfrac{d\theta}{dt} :$

$$\frac{ds}{d\theta} = \sqrt{\left(\frac{dr}{d\theta}\right)^2 + r^2} = \sqrt{b^2 \sec^4 \frac{\theta}{2} \tan^2 \frac{\theta}{2} + b^2 \sec^4 \frac{\theta}{2}} = b \sec^3 \frac{\theta}{2},$$

$$-\pi < \theta < \pi.$$

From (6) we then get

$$cb = b\left(\sec^3 \frac{\theta}{2}\right)\dot\theta, \qquad \text{so} \qquad \dot\theta = c \cos^3 \frac{\theta}{2}.$$

Because $\dot r = (dr/d\theta)\dot\theta$, we obtain $\dot r = bc \sin(\theta/2)$.

Then, the derivatives of $\dot\theta$ and $\dot r$ are found to be

$$\ddot\theta = -\frac{3}{2} c^2 \cos^5 \frac{\theta}{2} \sin \frac{\theta}{2}, \qquad \ddot r = \frac{bc^2}{2} \cos^4 \frac{\theta}{2}.$$

Finally, substitution of $r$, $\dot r$, $\dot\theta$, $\ddot r$, and $\ddot\theta$ in equations (5) gives the components of acceleration, which reduce to

$$a_r = -\frac{bc^2}{2} \cos^4 \frac{\theta}{2},$$

$$a_\theta = \frac{bc^2}{2} \cos^3 \frac{\theta}{2} \cdot \sin \frac{\theta}{2}.$$

The magnitude of the acceleration is

$$a = \frac{bc^2}{2} \cos^3 \frac{\theta}{2}.$$

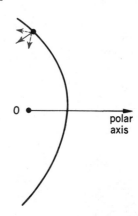

O

polar axis

### Problems

1.  A point moves on the parabola $r = b \sec^2 (\theta/2)$, $b > 0$, so that $\dot\theta = \text{constant} = c > 0$. Compute $v_r$, $v_\theta$, $a_r$, and $a_\theta$. Then sketch the path and draw the acceleration at $\theta = \pi/2$, if $c = 1$.

$$\left[ a_r = bc^2 \sec^2 \frac{\theta}{2}\left(\frac{3}{2}\sec^2 \frac{\theta}{2} - 2\right), \ a_\theta = 2bc^2 \sec^2 \frac{\theta}{2}\tan \frac{\theta}{2} \right]$$

2.  A point moves counterclockwise round the circle $r = b$ at the constant angular velocity $\dot\theta = \omega$. Compute $a_r$ and $a_\theta$. $\qquad [a_r = -b\omega^2, \ a_\theta = 0]$

3.  A point moves on the spiral $r = 2\theta$ so that $\dot\theta = \text{constant} = c$. Find $a_r$ and $a_\theta$. Then sketch the path and draw the acceleration at $\theta = (\pi/2)$, if $c = 1$. $\qquad [a_r = -2\theta c^2, \ a_\theta = 4c^2]$

4. A point moves on the spiral $r = 2\theta$ with constant speed $v = c$ in the direction of increasing $\theta$. Find $a_r$ and $a_\theta$. Then sketch the path and draw the acceleration at $\theta = \pi/2$, if $c = 1$.

$$\left[ a_r = -\frac{c^2\theta(2+\theta^2)}{2(1+\theta^2)^2}, a_\theta = \frac{c^2(2+\theta^2)}{2(1+\theta^2)^2} \right]$$

5. A point moves on the logarithmic spiral $r = e^\theta$ so that $\dot\theta = $ constant $= c$. Find $a_r$ and $a_\theta$. Then sketch the path and draw the acceleration at $\theta = \pi/4$, if $c = 1$.

$$[a_r = 0, a_\theta = 2c^2 e^\theta]$$

6. A point moves on the logarithmic spiral $r = e^\theta$ with constant speed $v = c$ in the direction of increasing $\theta$. Find $a_r$ and $a_\theta$. Then sketch the path and draw the acceleration at $\theta = \pi/4$, if $c = 1$.

$$\left[ a_r = -\frac{c^2}{2} e^{-\theta}, a_\theta = \frac{c^2}{2} e^{-\theta} \right]$$

7. A point moves on the limacon $r = b + c\cos\theta$, where $b > c > 0$ so that $\dot\theta = $ constant. Where is the speed least? Where is the magnitude of the acceleration least?

$$[v_{min} = (b-c)|\dot\theta| \text{ at } \theta = \pi, a_{min} = |b - 2c||\dot\theta^2| \text{ at } \theta = \pi]$$

8. A point moves on the cardioid $r = b(1 + \cos\theta)$. Show that the speed is $\sqrt{2br\dot\theta^2}$.

9. A point moves on the cardioid $r = b(1 + \cos\theta)$ with constant speed $v = c$. Find $a_r$ and $a_\theta$.

$$[a_r = -3c^2/4b, a_\theta = (-3c^2\sin\theta)/4r]$$

10. A point moves on the lemniscate $r^2 = b^2\cos 2\theta$ with constant speed $v = c$ in the direction of increasing $\theta$, $-\pi/4 \leq \theta \leq \pi/4$. Find $a_r$ and $a_\theta$. Where is $a$ least?

$$\left[ a_r = -\frac{3c^2}{b}(\cos 2\theta)^{3/2}, a_\theta = \frac{-3c^2}{b}(\cos 2\theta)^{1/2}\sin 2\theta \right]$$

11. A point moves clockwise on the circle $r = b\sin\theta$ with constant speed $v = c$. Find $a_r$ and $a_\theta$. Then sketch the path and draw the acceleration if $b = 2$ and $c = 1$.

$$\left[ a_r = -\frac{2c^2}{b}\sin\theta, a_\theta = \frac{2c^2}{b}\cos\theta \right]$$

12. Obtain equations (4) by differentiating $\dot x$ and $\dot y$.

13. Obtain equations (5) from equations (4).

# 7 Curvature

In addition to arc length, curves have another intrinsic property called the *curvature*. The curvature gives a measure of the amount, or rate, of bending along the curve. We shall see in Section 8 that this new concept has an important application to motion on a curve.

**DEFINITION** The *curvature*, $K$, of a curve at a point of the curve is (see the figure)

(1)
$$K = \frac{d\tau}{ds},$$

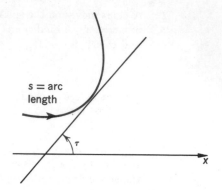

s = arc
length

the rate of turning of the tangent line with respect to arc length.

The sign of $K$ will depend on which direction along the curve corresponds to increasing $s$. $K$ will be positive if the curve "bends left" as $s$ increases since then $\Delta\tau > 0$ for $\Delta s > 0$.

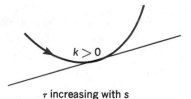

$k > 0$

$\tau$ increasing with s

*Example 1* For circles the curvature is computed directly from (1) and the definition of derivative:

$$\frac{d\tau}{ds} = \lim_{\Delta s \to 0} \frac{\Delta\tau}{\Delta s} = \lim_{\Delta\tau \to 0} \frac{\Delta\tau}{R\,\Delta\tau} = \frac{1}{R}.$$

This agrees with our intuition: Small circles have large curvature. Moreover, this suggests another definition.

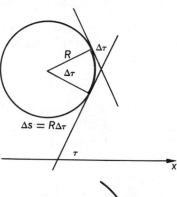

$\Delta s = R\Delta\tau$

**DEFINITION** *The radius of curvature is*

$$R = \frac{1}{|K|}.$$

The *circle of curvature* at the point $P$ on a curve is the circle of radius $R$ that is tangent to the tangent line at $P$. Its center is on the concave side of the curve, and is called the *center of curvature*.

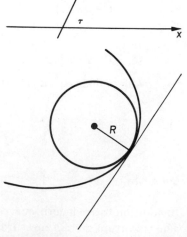

R

In most cases we shall compute curvature directly from the definition, (1). However, it should be mentioned that there are formulas that permit easy calculation of curvature in terms of the function defining the curve. See Problems 13, 14, and 15.

**Example 2**  Find the curvature of the parabola $y^2 = 2px$, $p > 0$.

Before beginning, observe that from the graph we would expect maximum curvature at the origin.

The slope of the tangent is $\tan \tau = dy/dx = p/y$. Thus,

$$\tau = \text{Arc} \tan \frac{dy}{dx} = \text{Arc} \tan \frac{p}{y}, \quad \text{if } y > 0.$$

Then

$$K = \frac{d\tau}{ds} = \frac{d\tau}{dy}\frac{dy}{ds}.$$

If we assume that $s$ increases with $y$, then

$$\frac{d\tau}{dy} = \frac{y^{-p/2}}{1 + (p^2/y^2)} = \frac{-p}{y^2 + p^2},$$

$$\frac{dy}{ds} = \frac{dy}{\sqrt{dx^2 + dy^2}} = \frac{1}{\sqrt{(dx/dy)^2 + 1}} = \frac{1}{\sqrt{(y^2/p^2) + 1}},$$

and

$$K = -\frac{p}{y^2 + p^2}\frac{p}{\sqrt{y^2 + p^2}} = -\frac{p^2}{(y^2 + p^2)^{3/2}}.$$

Note that the negative sign is correct because, as $s$ increases, $y$ increases and the slope decreases. Clearly, $|K|$ is greatest when $y = 0$.

**Example 3**  If the curve is given in rectangular parametric form one simply computes $K$ by

$$\frac{d\tau}{ds} = \frac{d\tau}{dt}\frac{dt}{ds}$$

if $t$ is the parameter. Thus, to find the curvature of the cycloid, we have, with $\theta$ the parameter:

$$x = a(\theta - \sin\theta), \qquad y = a(1 - \cos\theta);$$

$$\dot{x} = a(1 - \cos\theta), \qquad \dot{y} = a\sin\theta;$$

$$\left(\frac{ds}{d\theta}\right)^2 = a^2(1 - \cos\theta)^2 + a^2\sin^2\theta = 2a^2(1 - \cos\theta).$$

$$\tan\tau = \frac{\dot{y}}{\dot{x}} = \frac{\sin\theta}{1 - \cos\theta} = \cot\frac{\theta}{2} = \tan\left(\frac{\pi}{2} - \frac{\theta}{2}\right),$$

whence $\qquad\tau = \dfrac{\pi}{2} - \dfrac{\theta}{2}$ $\qquad$ if $\qquad 0 \leq \theta \leq 2\pi.$

Thus, if $s$ increases with $\theta$, $\tau$ decreases with $\theta$, and

$$K = \frac{d\tau}{ds} = \frac{d\tau}{d\theta}\frac{d\theta}{ds}$$

$$= \left(-\frac{1}{2}\right)\frac{1}{\sqrt{2a}\sqrt{1 - \cos\theta}}$$

$$= \frac{-1}{2\sqrt{2a}\sqrt{1 - \cos\theta}}.$$

Observe that $K$ is negative, and the curve bends to the right as $s$ increases.

**Example 4**   In polar coordinates one must remember that $\tau = \theta + \psi + n\pi$, and compute $K$ by

$$K = \frac{d\tau}{d\theta}\frac{d\theta}{ds}.$$

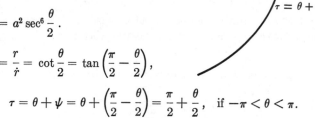

Thus, to find the curvature of the parabola $r = a\sec^2(\theta/2)$, we have:

$$\dot{r} = a\sec^2\frac{\theta}{2}\tan\frac{\theta}{2};$$

$$\left(\frac{ds}{d\theta}\right)^2 = a^2\sec^4\frac{\theta}{2}\tan^2\frac{\theta}{2} + a^2\sec^4\frac{\theta}{2}$$

$$= a^2\sec^6\frac{\theta}{2}.$$

$$\tan\psi = \frac{r}{\dot{r}} = \cot\frac{\theta}{2} = \tan\left(\frac{\pi}{2} - \frac{\theta}{2}\right),$$

whence $\qquad\tau = \theta + \psi = \theta + \left(\frac{\pi}{2} - \frac{\theta}{2}\right) = \frac{\pi}{2} + \frac{\theta}{2},$ $\quad$ if $-\pi < \theta < \pi.$

Thus, $\qquad\qquad K = \dfrac{d\tau}{d\theta}\dfrac{d\theta}{ds} = \dfrac{1}{2}\dfrac{1}{a\sec^3(\theta/2)} = \dfrac{1}{2a}\cos^3\dfrac{\theta}{2},$

if $\theta$ increases with $s$. Observe that then $K$ is positive, and the curve bends left.

### Problems

**In Problems 1 to 12 find the curvature and the radius of curvature directly from the definition, (1).**

1.   $y = \sqrt{x}$;   $s$ increases with $x$ $\qquad\qquad\qquad\qquad\qquad\qquad [K = -2(4x + 1)^{-3/2}]$

2.   $y = \log\sec x$;   $s$ increases with $x$ $\qquad\qquad\qquad\qquad\qquad\qquad\qquad [K = \cos x]$

3. $x^{1/2} + y^{1/2} = a^{1/2}$; $s$ increases with $x$ $\qquad [K = \frac{1}{2}a^{1/2}(x+y)^{-3/2}]$

4. $3y^2 = x^3$, $y < 0$; $s$ increases with $x$ $\qquad [K = -2\sqrt{3}x^{-1/2}(4+3x)^{-3/2}]$

5. $x = 3t^2$, $y = 3t - t^3$; $s$ increases with $t$ $\qquad [K = -\frac{2}{3}(1+t^2)^{-2}]$

6. $x = e^t \cos t$, $y = e^t \sin t$; $s$ increases with $t$ $\qquad [K = e^{-t}/\sqrt{2}]$

7. $x = a \cos^3 \theta$, $y = a \sin^3 \theta$; $s$ increases with $\theta$ $\qquad [K = -1/3a \sin \theta \cos \theta]$

8. $x = \cos t$, $y = \cos 2t$; $s$ increases with $t$ for $0 \le t \le \pi$
$$[K = -4(1 + 16 \cos^2 t)^{-3/2}]$$

9. $r = a \cos \theta$; $s$ increases with $\theta$ $\qquad [K = 2/a]$

10. $r = a(1 + \cos \theta)$; $s$ increases with $\theta$ $\qquad [K = 3/2\sqrt{2ar}]$

11. $r^2 = a^2 \cos 2\theta$; $s$ increases with $\theta$ $\qquad [K = 3r/a^2]$

12. $r = a\theta$; $s$ increases with $\theta$ $\qquad \left[ K = \dfrac{r^2 + 2a^2}{(r^2 + a^2)^{3/2}} \right]$

*13. If $\tau$ and $s$ are expressed in terms of the coordinate $x$, and if $s$ increases with $x$, then

$$K = \frac{y''}{(1 + y'^2)^{3/2}}.$$

Derive this formula.

*14. If the curve is given parametrically, and $s$ increases with $t$, then

$$K = \frac{\dot{x}\ddot{y} - \dot{y}\ddot{x}}{(\dot{x}^2 + \dot{y}^2)^{3/2}}.$$

Derive this formula.

*15. If the curve is given in polar coordinates and $s$ increases with $\theta$ then

$$K = \frac{r^2 + 2r'^2 - rr''}{(r^2 + r'^2)^{3/2}}.$$

Derive this formula.

(Hint: $d\tau/ds = (d\theta/ds) + (d\psi/ds)$ and first get a formula for $d\psi/ds$.)

16. Do Problem 1, using Problem 13.

17. Do Problem 2, using Problem 13.

18. Do Problem 6, using Problem 14.

19. Do Problem 7, using Problem 14.

20. Do Problem 10, using Problem 15.

21. Do Problem 11, using Problem 15.

In Problems 22 to 31 compute the radius of curvature at the point indicated by any method. Then draw the curve and the circle of curvature.

22.  $y = x^3 - 3x;\quad x = 1$ $\qquad\qquad\qquad\qquad\qquad\qquad\qquad\quad$ $[\frac{1}{6}]$

23.  $b^2x^2 + a^2y^2 = a^2b^2;\quad (a, 0),\quad a > 0$ $\qquad\qquad\qquad\qquad$ $[b^2/a]$

24.  $b^2x^2 - a^2y^2 = a^2b^2;\quad (a, 0),\quad a > 0$ $\qquad\qquad\qquad\qquad$ $[b^2/a]$

25.  $x^{2/3} + y^{2/3} = a^{2/3};\quad (2^{-3/2}a, 2^{-3/2}a)$ $\qquad\qquad\qquad$ $[3a/2]$

26.  $x = t - 1,\quad y = t^3;\quad t = 1$ $\qquad\qquad\qquad\qquad\qquad\qquad$ $[5\sqrt{10}/3]$

27.  $x = a(\cos\theta + \theta\sin\theta),\quad y = a(\sin\theta - \theta\cos\theta);\quad \theta = 0$ $\qquad$ $[0]$

28.  $x = -2 + 3\sin\theta,\quad y = 3 - 2\cos\theta;\quad \theta = 0$ $\qquad\qquad$ $[\frac{9}{2}]$

29.  $r = a(1 + 2\cos\theta);\quad \theta = \pi/2$ $\qquad\qquad\qquad\qquad\qquad$ $[5\sqrt{5}a/9]$

30.  $x = e^{-t},\quad y = 2e^t;\quad t = 0$ $\qquad\qquad\qquad\qquad\qquad\qquad$ $[5\sqrt{5}/4]$

31.  $r = a\sin\theta$ $\qquad\qquad\qquad\qquad\qquad\qquad\qquad\qquad\qquad\quad$ $[a/2]$

32.  What is the curvature of the graph of $y = f(x)$ at a flex point? $\qquad$ $[0]$

## *8   Acceleration; Normal and Tangential Components

An important application of curvature is to the resolution of the acceleration into its tangential and normal components. Up to now we have resolved velocity and acceleration into their components along either rectangular axes or the directions of increasing polar coordinates. Both of these relate to special ways of representing the curve. There is another way of decomposing the acceleration that is *intrinsic* in that it depends on the curve and not on the coordinate system.

Assume that arc length increases with time. The positive tangential direction is along the tangent line in the direction of increasing arc length. The positive normal direction is obtained by rotating the positive tangential direction $\pi/2$ radians counterclockwise.

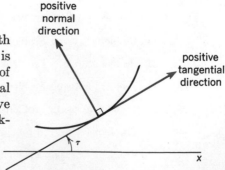

We shall compute the *normal*, $a_n$, and *tangential*, $a_t$, components of the acceleration. These are expressed in terms of the rectangular components of acceleration. From the figure (by the same method as we used on page 165) we obtain

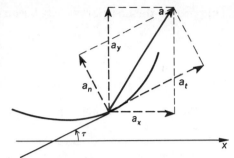

$$\begin{cases} a_t = a_x \cos \tau + a_y \sin \tau, \\ a_n = -a_x \sin \tau + a_y \cos \tau. \end{cases}$$

Now we use $a_x = \ddot{x}$, $a_y = \ddot{y}$ and $dx/ds = \cos \tau$, $dy/ds = \sin \tau$ to get

(1)
$$\begin{cases} a_t = \ddot{x}\dfrac{dx}{ds} + \ddot{y}\dfrac{dy}{ds}, \\[2mm] a_n = -\ddot{x}\dfrac{dy}{ds} + \ddot{y}\dfrac{dx}{ds} \end{cases}$$

Equations (1) are often useful for computation. However, there is no obvious geometric interpretation of them. Nevertheless they can be interpreted geometrically as the following theorem asserts:

**THEOREM**

(2)
$$\begin{cases} a_t = \dfrac{d^2 s}{dt^2}, \quad \{\text{just as one might guess} \\[3mm] a_n = \left(\dfrac{ds}{dt}\right)^2 K = v^2 K = \dfrac{\pm v^2}{R}. \quad \{\text{surprising!} \end{cases}$$

**PROOF**    Our proof will make use of equations (1) but with a specially selected rectangular coordinate system. We wish to compute $a_t$ and $a_n$ at a point $P$ on a plane curve. Let us choose the origin at $P$ and the positive $x$-axis and $y$-axis as shown.

Then, from (1), we have for the components at $P$,

(3)
$$\begin{cases} a_t = a_x = \ddot{x}, \\ a_n = a_y = \ddot{y}, \end{cases}$$

because $dx/ds = 1$ and $dy/ds = 0$ at $P$.

To complete the proof we compute $d^2s/dt^2$ and $v^2/R$ and compare them with (2):

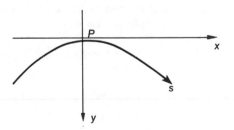

$$\frac{d^2 s}{dt^2} = \frac{d}{dt}\sqrt{\dot{x}^2 + \dot{y}^2} = \frac{\dot{x}\ddot{x} + \dot{y}\ddot{y}}{\sqrt{\dot{x}^2 + \dot{y}^2}} = \ddot{x} \text{ at } P.$$

This proves the first part of (2) because $\dot{x} = 1$ and $\dot{y} = 0$ at $P$.

$$K = \frac{\dot{x}\ddot{y} - \dot{y}\ddot{x}}{(\dot{x}^2 + \dot{y}^2)^{3/2}} = \frac{\ddot{y}}{\dot{x}^2} = \frac{\ddot{y}}{v^2} \quad \{\text{Cf. Problem 14, Section 7}$$

Thus, $a_n = \ddot{y} = v^2 K$. This completes the proof.

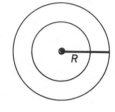

*Example*   For uniform speed on a circle of radius $R$, $a_t = 0$ and $a_n = v^2/R$.

Therefore, the larger the circle, the smaller the normal acceleration for a given speed.

In the case of a satellite in a circular orbit we can find its speed. Thus, if $R = 4000$ mi and the acceleration is $g = 32$ ft/sec$^2$ (an orbit not far from the earth), then

$$v^2 = gR = 32\,(4000)\,(5280),$$

$$v = 26{,}000 \text{ fps} = 4.92 \text{ mi/sec}.$$

### Problems

In Problems 1 to 5 the parameter is time. Draw the path. Find $a_x$, $a_y$, $a_t$, and $a_n$ at the point indicated. Draw the acceleration vector.

1.   $x = t$,  $y = \cos \pi t$;  $t = 0$                           $[a_t = 0,\ a_n = -\pi^2]$

2.   $x = a \sin \omega t$,  $y = a \cos \omega t$;  $t = t_1$                 $[a_t = 0,\ a_n = -a\omega^2]$

3.   $x = a \cos^3 t$,  $y = a \sin^3 t$;  $t = \pi/4$                 $[a_t = 0,\ a_n = -3a/2]$

4.   $x = t^2$,  $y = t^3$;  $t = 1$                           $[a_t = 22/\sqrt{13},\ a_n = 6/\sqrt{13}]$

5.   $x = \cos t$,  $y = \sin 2t$;  $t = \pi/4$                 $[a_t = 1/\sqrt{2},\ a_n = 4]$

6.   A point moves on the parabola $2y = x^2$ with $v_x$ constant at $\frac{1}{2}$. Find $a_t$ and $a_n$ at $x = 2$.                           $[a_t = 1/2\sqrt{5},\ a_n = 1/4\sqrt{5}]$

7.   A point moves on the parabola $2y = x^2$ with constant speed $v = \frac{1}{2}$. Find $a_t$ and $a_n$ at $x = 2$.                           $[a_t = 0,\ |a_n| = 1/20\sqrt{5}]$

8.   A point moves according to the equations $x = e^t$, $y = e^{-2t}$. Find $a_t$ and $a_n$ at $(1, 1)$.                           $[a_t = -7/\sqrt{5},\ a_n = 6/\sqrt{5}]$

9.   A point moves on the ellipse $b^2 x^2 + a^2 y^2 = a^2 b^2$, $a > b > 0$, with constant speed. Where is the magnitude of the acceleration greatest?                           $[\text{at } (\pm a, 0)]$

*10.   Show, using equations (2), that the acceleration always points to the concave side of the curve if $v$ and $K$ are not zero.

## *9 Evolutes

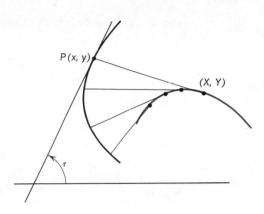

As a point $P(x, y)$ moves along a curve its center of curvature $(X, Y)$ moves and traces a curve called the *evolute*. Many interesting curves are obtained in this way.*

To fix ideas, suppose that we select a direction on the curve for $s$ to be measured positively. Then the tangent line is directed and has direction cosines $dx/ds$ and $dy/ds$. The normal, directed toward the concave side, will have direction cosines either

$$-\frac{dy}{ds}, \frac{dx}{ds} \quad \text{or} \quad \frac{dy}{ds}, -\frac{dx}{ds},$$

depending on whether $K > 0$ or $K < 0$.

Then the center of curvature $(X, Y)$ is given by

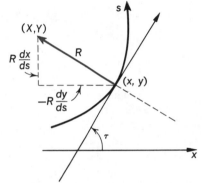

$$X = x - R\frac{dy}{ds}, \quad Y = y + R\frac{dx}{ds} \quad \text{if } K > 0,$$

or

$$X = x + R\frac{dy}{ds}, \quad Y = y - R\frac{dx}{ds} \quad \text{if } K < 0.$$

If we replace $R$ by $1/|K|$, then we get, in both cases, the center of curvature as

(1) $$X = x - \frac{1}{K}\frac{dy}{ds}, \qquad Y = y + \frac{1}{K}\frac{dx}{ds}.$$

Observe that if $s$ is measured positively in the opposite direction, $K$, $dx/ds$, and $dy/ds$ all change signs, and the formulas remain unchanged.

---

* Evolutes have an interesting property illustrated by the figure. Imagine a string wrapped around a given curve $C_1$ and then unwrapped. The end $P$ of the string will describe a curve $C_2$ called the involute of the given curve $C_1$. It should be observed that there are many involutes, corresponding to different lengths of string.

Then the evolute of the involute is the given curve:

*Evolute of $C_2$ = the evolute of the involute of $C_1$ = $C_1$.*

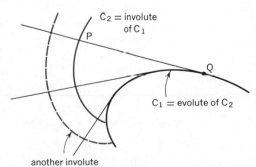

*Example*   Find the evolute of the parabola $y^2 = 2px$. We compute using $y$ as the independent variable:

$$\frac{dx}{dy} = \frac{y}{p}; \qquad \frac{d^2x}{dy^2} = \frac{1}{p}.$$

If $s$ increases with $y$, then

$$\frac{ds}{dy} = \sqrt{1 + \left(\frac{dx}{dy}\right)^2}$$

$$= \sqrt{1 + \frac{y^2}{p^2}} = \frac{\sqrt{p^2 + y^2}}{p},$$

so

$$\frac{dy}{ds} = \frac{p}{\sqrt{p^2 + y^2}},$$

$$\frac{dx}{ds} = \frac{dx}{dy}\frac{dy}{ds} = \frac{y}{\sqrt{p^2 + y^2}}.$$

$$K = \frac{d^2x/dy^2}{-[1 + (dx/dy)^2]^{3/2}} = \frac{-1/p}{[1 + (y^2/p^2)]^{3/2}} = \frac{-p^2}{(p^2 + y^2)^{3/2}}.$$

The minus sign is selected because the curve bends to the right as $s$ increases. $K < 0$. Therefore, the evolute is

(2)
$$\begin{cases} X = x + \dfrac{(p^2 + y^2)^{3/2}}{p^2} \cdot \dfrac{p}{\sqrt{p^2 + y^2}} = 3x + p = \dfrac{3y^2}{2p} + p, \\[4mm] Y = y - \dfrac{(p^2 + y^2)^{3/2}}{p^2} \cdot \dfrac{y}{\sqrt{p^2 + y^2}} = -\dfrac{y^3}{p^2}. \end{cases}$$

Equations (2) are parametric equations of the evolute with parameter $y$. Elimination of the parameter $y$ yields

$$Y^2 = \frac{8}{27p}(X - p)^3,$$

which is a semi-cubical parabola with vertex at $(p, 0)$.

### Problems

1.  Find the evolute of the parabola $y = x^2$. First obtain parametric equations and then a rectangular equation. Draw a figure.   $[X = -4x^3,\ Y = 3x^2 + \frac{1}{2}]$

2.  Find a rectangular equation for the evolute of the ellipse $2x^2 + 3y^2 = 12$. Sketch the ellipse and its evolute.   $[(\sqrt{6}x)^{2/3} + (2y)^{2/3} = 2^{2/3}]$

3.  Give the coordinates of center of curvature of the hypocycloid $x^{2/3} + y^{2/3} = a^{2/3}$ in terms of $x$ and $y$.   $[X = x + 3x^{1/3}y^{2/3},\ Y = y + 3x^{2/3}y^{1/3}]$

4.  Find, in terms of the parameter $\theta$, the evolute of the hypocycloid $x = a\cos^3\theta$, $y = a\sin^3\theta$.
$$\left[\begin{array}{l} X = a\cos^3\theta + 3a\cos\theta\sin^2\theta \\ Y = a\sin^3\theta + 3a\cos^2\theta\sin\theta \end{array}\right]$$

5.  Find the evolute of the curve $x = a\sin\theta$, $y = a\cos\theta$.    [the point $(0,0)$]

6.  Find the evolute of the hyperbola $xy = 1$.    $\left[X = \dfrac{3x}{2} + \dfrac{1}{2x^3},\ Y = \dfrac{3}{2x} + \dfrac{x^3}{2}\right]$

7.  Find the evolute of the hyperbola $x = a\sec t$, $y = b\tan t$, in terms of the parameter.
$$\left[X = \frac{a^2+b^2}{a}\sec^3 t,\ Y = -\frac{a^2+b^2}{b}\tan^3 t\right]$$

8.  Find parametric equations for the evolute of the curve $x = a(e^t + e^{-t})$, $y = a(e^t - e^{-t})$.    $[X = a(e^t + e^{-t})^3/2,\ Y = -a(e^t - e^{-t})^3/2]$

*9.  Show that the evolute of the cycloid $x = a(\theta - \sin\theta)$, $y = a(1 - \cos\theta)$, is also a cycloid.

## CHAPTER 7

# THE MEAN VALUE THEOREM, INDETERMINATE FORMS, TAYLOR'S FORMULA

## 1    The Mean Value Theorem (MVT) Again

In Chapter 4 we used the MVT as an aid to graphing. We now sketch a proof and then give Cauchy's* generalized MVT, which we need to evaluate indeterminate forms.

The proof of the MVT is based on a special case called Rolle's theorem (M. Rolle, 1652–1719) after a minor French mathematician who applied it to polynomials.

---

* **Historical Note.** The significance of the MVT for calculus theory was apparently first recognized by the great French analyst Augustin-Louis Cauchy (1787–1837). Cauchy was the founder of modern function theory and a pioneer in the study of substitution groups. He was an amazingly prolific writer of literally hundreds of papers (some of them several hundred pages in length) on all sorts of topics in physics as well as mathematics.

His lectures at l'École Polytechnique were published in 1821 at the urging of his friends and constitute the first satisfactory exposition of the theory of calculus. These lectures have had great influence on calculus texts, and even today are pleasurable reading.

The box at the right gives Rolle's theorem. The only hard part of the proof is to show that $f$ reaches its maximum, or minimum at some point. We shall assume this is the case here, but for a proof, see Appendix C.

**PROOF OF ROLLE'S THEOREM**   If $f$ is constant in $[a, b]$, then for every $c$ in $(a, b)$, $f'(c) = 0$. If $f$ is not constant, and $f(x) > f(a)$ somewhere, then $f$ has a maximum at a point $c$ in $(a, b)$. If $f$ is not constant, and $f(x) \leq f(a)$ for all $x$ in $[a, b]$, then $f$ has a minimum at a point $c$ in $(a, b)$. In either case (see Chapter 4, page 97), $f'(c) = 0$.

To refresh your memory, the MVT is reproduced in the box. A proof is obtained by means of a trick that reduces the theorem to Rolle's theorem.

An equation of the secant line is

$$y = f(a) + \frac{f(b) - f(a)}{b - a}(x - a).$$

Now subtract this linear function from $f$. One gets

$$\varphi(x) = f(x) - y$$

$$= f(x) - f(a) - \frac{f(b) - f(a)}{b - a}(x - a).$$

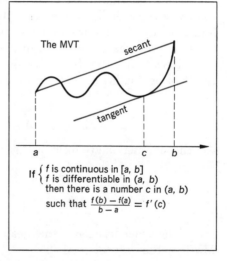

Rolle's Theorem

If $\begin{cases} f \text{ is continuous in } [a, b] \\ f \text{ is differentiable in } (a,b) \\ f(a) = f(b) \end{cases}$

then

$f'(c) = 0$
for some $c$ in $(a, b)$

The MVT

secant

tangent

If $\begin{cases} f \text{ is continuous in } [a, b] \\ f \text{ is differentiable in } (a, b) \end{cases}$
then there is a number $c$ in $(a, b)$
such that $\frac{f(b) - f(a)}{b - a} = f'(c)$

Observe that $\varphi(x)$ is the difference, at $x$, between the height of the curve and the height of the secant line. The function $\varphi$ satisfies the hypotheses of Rolle's theorem with $\varphi(a) = \varphi(b) = 0$. Therefore, $\varphi'(c) = 0$ for some $c$, $a < c < b$,

$$\varphi'(c) = 0 = f'(c) - \frac{f(b) - f(a)}{b - a},$$

and this implies the MVT.

An alternative form of the MVT that is sometimes met is as follows:

$$f(b) = f(a) + (b - a)f'[a + \theta(b - a)], \quad \text{where } 0 < \theta < 1.$$

This follows because any number $c$ between $a$ and $b$ can be expressed as $c = a + \theta(b - a)$, where $0 < \theta < 1$. We will use this form for the MVT in many applications.

Now that the MVT has been proved, let us recall how it has been used in the past. There were two main theorems whose proofs required the MVT.

(1) If the derivative of $f$ is positive on an interval, then $f$ is increasing on that interval (Theorem 1, page 112).

From this theorem we obtained tests for maxima, minima, and concavity.

(2) If two functions have the same derivative on an interval, $f' = g'$, then they differ by a constant, $f = g +$ constant on that interval (Theorem 2, page 112).

This theorem was used in problems involving antiderivatives in Chapter 1.

## 2 *The Indeterminate Form 0/0*

There is a generalization of the MVT that is used to evaluate certain limits that are called indeterminate forms. In this section we are concerned only with one of these forms that is described in the box at the right.

Recall that the derivative of a function is the result of evaluating an indeterminate form 0/0:

$$f'(x) = \lim_{\Delta x \to 0} \frac{f(x + \Delta x) - f(x)}{\Delta x}.$$

Another indeterminate form of this type is

$$\lim_{x \to 0} \frac{\sin x}{x}.$$

> One has the indeterminate form
> $$\frac{0}{0}$$
> if $\lim_{x \to a} f(x) = 0$
> $$\lim_{x \to a} g(x) = 0$$
> and one wants
> $$\lim_{x \to a} \frac{f(x)}{g(x)}$$

The generalization of the mean value theorem that we need is called Cauchy's MVT or the *parametric* MVT.

### THEOREM 1 (CAUCHY'S MVT)

If $\begin{bmatrix} f, g \text{ are continuous in } [a, b] \\ f, g \text{ are differentiable in } (a, b) \\ g'(x) \neq 0 \text{ for } a < x < b \end{bmatrix}$, then $\dfrac{f(b) - f(a)}{g(b) - g(a)} = \dfrac{f'(c)}{g'(c)}$

*for some number $c$ such that $a < c < b$.*

**PROOF**  First we observe that because $g'(x) \neq 0$, $g$ is either increasing or decreasing in $[a, b]$, and so $g(b) - g(a) \neq 0$. Thus, the statement makes sense.

The proof is similar to the proof of the ordinary MVT. Consider the function $\varphi$ given by

$$\varphi(x) = [f(b) - f(a)][g(x) - g(a)] - [f(x) - f(a)][g(b) - g(a)].$$

Then $\varphi$ satisfies the hypotheses of Rolle's theorem, and $\varphi'(c) = 0$ for some $c$ in $(a, b)$:

$$\varphi'(c) = 0 = [f(b) - f(a)]g'(c) - f'(c)[g(b) - g(a)],$$

and this implies the theorem.

The figure at the right shows why Theorem 1 is called the *parametric* MVT. If the parametric curve $x = g(t)$, $y = f(t)$ has a tangent at $(0, 0) = (g(a), f(a))$, with slope given by $f'(a)/g'(a)$, then

$$\lim_{t \to a} \frac{f(t)}{g(t)} = \frac{f'(a)}{g'(a)}.$$

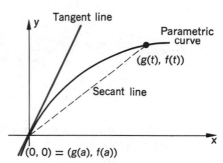

as $t \longrightarrow a$ the secant approaches the tangent

The technique for evaluating indeterminate forms is called *l'Hospital's Rule** and is an immediate consequence of Cauchy's MVT.

**THEOREM 2 (l'HOSPITAL'S RULE)**    *If $f$, $g$ are differentiable for $x \neq a$, and if*

$$\lim_{x \to a} f(x) = 0, \qquad \lim_{x \to a} g(x) = 0,$$

*and if $g'(x) \neq 0$ for $x \neq a$, and if*

$$\lim_{x \to a} \frac{f'(x)}{g'(x)} = L, \qquad then \qquad \lim_{x \to a} \frac{f(x)}{g(x)} = L.$$

**PROOF**    Define $f(a) = 0$, $g(a) = 0$. Then $f$ and $g$ are continuous at $x = a$ so that Theorem 1 applies. In Theorem 1, put $x = b$. Then

(1) $$\frac{f(x)}{g(x)} = \frac{f'(c)}{g'(c)}, \quad \text{where } c \text{ is between } a \text{ and } x.$$

---

* **Historical Note.** The first textbook on calculus, *Analyse des infiniments petits pour l'intelligence des lignes courbes*, appeared in 1696. It was written by the wealthy Marquis de l'Hospital, a pupil of Jakob Bernoulli, and was a significant factor in popularizing the Leibnizian calculus on the continent.

When young and struggling for recognition, Bernoulli made a bargain with the Marquis. In return for the receipt of a regular income, Bernoulli agreed to teach l'Hospital mathematics and granted the Marquis permission to use his discoveries as he wished. And so the rule due to Bernoulli is credited to l'Hospital.

The reader should bear in mind that at this time knowledge of calculus coupled with power to use it was essentially confined to Newton, Leibniz, and the Bernoulli brothers.

Thus, in (1), if $x$ approaches $a$, $c$ also approaches $a$, so *if*

$$\lim_{c \to a} \frac{f'(c)}{g'(c)} = L, \qquad then\ also \qquad \lim_{x \to a} \frac{f(x)}{g(x)} = L.$$

But this is the conclusion of the Theorem:

*If* $\qquad\qquad \lim_{x \to a} \frac{f'(x)}{g'(x)} = L, \qquad then \qquad \lim_{x \to a} \frac{f(x)}{g(x)} = L.$

**Example 1** $\qquad \lim_{x \to 2} \frac{x^2 - 4}{x^2 + 3x - 10} = \lim_{x \to 2} \frac{2x}{2x + 3} = \frac{4}{7}.$

$$\lim_{x \to 0} \frac{1 - \cos x}{x^2} = \lim_{x \to 0} \frac{\sin x}{2x}.$$

This last limit is also an indeterminate form for which we can again use l'Hospital's rule:

$$\lim_{x \to 0} \frac{\sin x}{2x} = \lim_{x \to 0} \frac{\cos x}{2} = \frac{1}{2}.$$

Thus, $\qquad\qquad\qquad \lim_{x \to 0} \frac{1 - \cos x}{x^2} = \frac{1}{2}.$

**Example 2** l'Hospital's rule can also be used when $\lim_{x \to a} [f'(x)/g'(x)]$ is infinite. Thus,

$$\lim_{\theta \to 0} \frac{\tan 2\theta}{\sin^2 \theta} = \lim_{\theta \to 0} \frac{2 \sec^2 2\theta}{2 \sin \theta \cos \theta}$$

$$= \begin{cases} +\infty & \text{if } \theta \to 0 \text{ with } \theta > 0, \\ -\infty & \text{if } \theta \to 0 \text{ with } \theta < 0. \end{cases}$$

Here the symbol $+\infty$ is used as a shorthand for the statement that the function is arbitrarily large when $\theta$ is positive and near 0. To indicate that $\theta$ approaches 0 through positive values we write $\lim_{\theta \to 0^+}$. Thus,

$$\lim_{\theta \to 0^+} \frac{\tan 2\theta}{\sin^2 \theta} = +\infty,$$

and similarly $\qquad\qquad \lim_{\theta \to 0^-} \frac{\tan 2\theta}{\sin^2 \theta} = -\infty.$

*Remark.* One should always be sure that one actually has an indeterminate form. Moreover, one should check that the hypotheses of l'Hospital's rule are satisfied. Usually this will be clear, and no comment will be necessary.

## Problems

Evaluate the following limits by any method.

1. $\lim\limits_{x \to -1} \dfrac{x^3 + x^2 - x - 1}{x^3 - 3x - 2}$ $\quad \left[\frac{2}{3}\right]$

15. $\lim\limits_{\theta \to \pi} \dfrac{\sin a\theta}{\theta}$ $\quad \left[\dfrac{\sin a\pi}{\pi}\right]$

2. $\lim\limits_{\theta \to \pi} \dfrac{\tan 2\theta}{\sin \theta}$ $\quad [-2]$

16. $\lim\limits_{x \to 1} \dfrac{x - 1}{\log x}$ $\quad [1]$

3. $\lim\limits_{x \to -2} \dfrac{x^3 - 2x^2 - 4x + 8}{x^4 - 3x^2 - 4}$ $\quad \left[-\frac{4}{5}\right]$

17. $\lim\limits_{t \to 0} \dfrac{e^t + t - 1}{\log (1 + t)}$ $\quad [2]$

4. $\lim\limits_{x \to -1} \dfrac{2x^3 + 3x^2 + 1}{x^2 + 2x + 1}$ $\quad [+\infty]$

18. $\lim\limits_{x \to 1} \dfrac{(\log ex)^2 - 1}{x^2 - 1}$ $\quad [1]$

5. $\lim\limits_{x \to 0} \dfrac{e^{2x} - 1}{\cos x - 1}$ $\quad [\pm\infty]$

19. $\lim\limits_{x \to 0} \dfrac{\sin^3 x - \sin x}{x^3 - x}$ $\quad [1]$

6. $\lim\limits_{x \to 0} \dfrac{\sin x}{x}$ $\quad [1]$

20. $\lim\limits_{s \to 0} \dfrac{\tan s - s}{\sin s - s}$ $\quad [-2]$

7. $\lim\limits_{x \to 0} \dfrac{\sin 2x}{x}$ $\quad [2]$

21. $\lim\limits_{x \to 0} \dfrac{x^3}{\cos^3 x - 1}$ $\quad [0]$

8. $\lim\limits_{x \to 0} \dfrac{\sin x}{2x}$ $\quad \left[\frac{1}{2}\right]$

22. $\lim\limits_{\varphi \to \pi/2} \dfrac{\log \sin \varphi}{\cos \varphi}$ $\quad [0]$

9. $\lim\limits_{x \to 0} \dfrac{\sin (x^2)}{x}$ $\quad [0]$

23. $\lim\limits_{t \to 0} \dfrac{\tan t - \sin t}{t^3}$ $\quad \left[\frac{1}{2}\right]$

10. $\lim\limits_{x \to 0} \dfrac{\sin x}{x^2}$ $\quad [\pm\infty]$

24. $\lim\limits_{x \to 0} \dfrac{ax - \sin ax}{x^3}$ $\quad \left[\dfrac{a^3}{6}\right]$

11. $\lim\limits_{x \to \pi/2} \dfrac{\sin x}{x}$ $\quad \left[\dfrac{2}{\pi}\right]$

25. $\lim\limits_{t \to 0} \dfrac{e^t + e^{-t} - 2}{t \sin t}$ $\quad [1]$

* 12. $\lim\limits_{x \to 0} \dfrac{e^x - 2^x}{\sqrt{x}}$ $\quad [0]$

26. $\lim\limits_{t \to 0} \dfrac{e^t - e^{-t} - 2 \sin t}{t - \sin t}$ $\quad [4]$

* 13. $\lim\limits_{x \to 0} \dfrac{a^x - b^x}{x}$ $\quad \left[\log \dfrac{a}{b}\right]$

27. $\lim\limits_{x \to 0} \dfrac{ax^2 + b\sqrt{x}}{cx^3 + e\sqrt{x}}$ $\quad \left[\dfrac{b}{e}\right]$

14. $\lim\limits_{\theta \to 0} \dfrac{\sin a\theta}{\theta}$ $\quad [a]$

28. $\lim\limits_{\theta \to 0} \dfrac{\text{Arc} \sin \theta}{\theta}$ $\quad [1]$

29. $\lim\limits_{t\to 0}\dfrac{\text{Arc}\sin t}{\text{Arc}\tan t}$     [1]     30. $\lim\limits_{x\to 1}\dfrac{\text{Arc}\cos x}{x-1}$     $[-\infty]$

In Problems 31 to 36 the limits can be found without direct use of l'Hospital's rule if use is made of algebra and what has been learned in the previous problems. Evaluate these limits.

31. $\lim\limits_{x\to\infty} x\sin\dfrac{1}{x}$     [1]     34. $\lim\limits_{x\to 1}\dfrac{x^{1/3}-1}{x^{1/2}-1}$     $\left[\frac{2}{3}\right]$

32. $\lim\limits_{x\to 0} x\cot x$     [1]     35. $\lim\limits_{x\to a}\dfrac{x^5-a^5}{x-a}$     $[5a^4]$

33. $\lim\limits_{\theta\to 0}\dfrac{1-\cos\theta}{\sin^2\theta}$     $\left[\frac{1}{2}\right]$     36. $\lim\limits_{t\to 0}\dfrac{\sin^3 t}{\sin t^3}$     [1]

37. Evaluate $\lim\limits_{h\to 0}\dfrac{f(x+h)-f(x)}{h}$ by l'Hospital's rule.

38. Show that (assuming $f''$ is continuous)
$$\lim_{h\to 0}\frac{f(x+h)-f(x)-f'(x)h}{h^2}=\tfrac{1}{2}f''(x).$$

39. Show that (assuming $f''$ is continuous)
$$\lim_{h\to 0}\frac{f(x+h)-2f(x)+f(x-h)}{h^2}=f''(x).$$

40. Show that (assuming $f'''$ is continuous)
$$\lim_{h\to 0}\frac{f(x+h)-f(x)-f'(x)h-f''(x)h^2/2}{h^3}=\frac{f'''(x)}{6}.$$

*41. Show that the ordinary MVT can be used to evaluate $\lim_{x\to a}(f(x)/g(x))$ providing the rule does not need to be applied twice.

## 3  Other Indeterminate Forms

The indeterminate form "0/0" arises as the limit of the quotient of two functions, both approaching zero. Other indeterminate forms that are obtained from two functions are these:

$$\frac{\infty}{\infty},\quad 0\cdot\infty,\quad \infty-\infty,\quad 0^0,\quad \infty^0,\quad 1^\infty.$$

These will be illustrated in the examples and problems. In each case one tries to change the form so that l'Hospital's rule applies. First we observe that there are three other forms for l'Hospital's rule:

**OTHER FORMS OF l'HOSPITAL'S RULE**    *In all cases f and g are differentiable and $g'(x) \neq 0$.*

(1)  *If*  $\qquad \lim_{x \to \infty} f(x) = 0 \qquad$ *and* $\qquad \lim_{x \to \infty} g(x) = 0,$

$\quad$ *and if* $\qquad \lim_{x \to \infty} \dfrac{f'(x)}{g'(x)} = L, \qquad$ *then* $\qquad \lim_{x \to \infty} \dfrac{f(x)}{g(x)} = L.$

(2)  *If*  $\qquad \lim_{x \to a} f(x) = \infty \qquad$ *and* $\qquad \lim_{x \to a} g(x) = \infty,$

$\quad$ *and if* $\qquad \lim_{x \to a} \dfrac{f'(x)}{g'(x)} = L, \qquad$ *then* $\qquad \lim_{x \to a} \dfrac{f(x)}{g(x)} = L.$

(3)  *If*  $\qquad \lim_{x \to \infty} f(x) = \infty \qquad$ *and* $\qquad \lim_{x \to \infty} g(x) = \infty,$

$\quad$ *and if* $\qquad \lim_{x \to \infty} \dfrac{f'(x)}{g'(x)} = L, \qquad$ *then* $\qquad \lim_{x \to \infty} \dfrac{f(x)}{g(x)} = L.$

Note that (1) simply states that the rule is valid over an infinite interval. Rule (2) states that the method for the indeterminate form $\infty/\infty$ is the same as for 0/0. Rule (2) is difficult to establish, and will not be proved here. Rules (1) and (3), however, are easy to establish. See Problems 38 and 39, page 188.

*Example 1*    The form $0 \cdot \infty$ can be converted either to 0/0 or $\infty/\infty$. Sometimes one way is better.
$\qquad$ To evaluate

$$\lim_{x \to 0} \frac{1}{x} e^{-1/x}$$

we must restrict $x$ to be positive because otherwise the limit certainly does not exist. To indicate that $x$ approaches zero through positive values we write $\lim_{x \to 0^+}$. As given, we have the indeterminate form $\infty \cdot 0$. We convert it to the form 0/0 and apply l'Hospital's rule. Then

$$\lim_{x \to 0^+} \frac{1}{x} e^{-1/x} = \lim_{x \to 0^+} \frac{e^{-1/x}}{x} = \lim_{x \to 0^+} \frac{(e^{-1/x})(1/x^2)}{1} = \lim_{x \to 0^+} \frac{e^{-1/x}}{x^2},$$

and things are worse instead of better. However we also have, on converting to the form $\infty/\infty$,

$$\lim_{x \to 0^+} \frac{1}{x} e^{-1/x} = \lim_{x \to 0^+} \frac{1/x}{e^{1/x}} = \lim_{x \to 0^+} \frac{-1/x^2}{-e^{1/x}/x^2} = \lim_{x \to 0^+} e^{-1/x} = 0.$$

*Example 2*    The form $\infty - \infty$ can often be converted to $0/0$ or $\infty/\infty$ with simple algebra:

$$\lim_{t \to 0} \left( \frac{1}{t^2} - \frac{1}{t \sin t} \right) = \lim_{t \to 0} \frac{\sin t - t}{t^2 \sin t} = \lim_{t \to 0} \frac{\cos t - 1}{t^2 \cos t + 2t \sin t}$$

$$= \lim_{t \to 0} \frac{-\sin t}{2 \sin t + 4t \cos t - t^2 \sin t}$$

$$= \lim_{t \to 0} \frac{-1}{2 + 4 \dfrac{t}{\sin t} \cos t - t^2} = -\frac{1}{6}.$$

*Example 3*    The indeterminate forms $0^0$, $1^\infty$, $\infty^0$ all arise from trying to evaluate $\lim_{x \to a} f(x)^{g(x)}$. If one lets

$$y = f(x)^{g(x)}, \qquad \text{then} \qquad \log y = g(x) \log f(x),$$

and the indeterminate form for $\log y$ becomes one of the form $0 \cdot \infty$.

Consider $\lim_{h \to 0} (1 + h)^{1/h}$, which we know to be equal to $e$. Let

$$y = (1 + h)^{1/h}; \qquad \text{then} \qquad \log y = \frac{1}{h} \log (1 + h).$$

$$\lim_{h \to 0} \log y = \lim_{h \to 0} \frac{\log (1 + h)}{h} = \lim_{h \to 0} \frac{1}{1 + h} = 1.$$

Therefore, since $\log y \to 1$, $\lim_{h \to 0} y = \lim_{h \to 0} e^{\log y} = e$.

## Problems

**Evaluate the following limits.**

1. $\lim_{x \to \infty} \dfrac{ax^2 + bx + c}{ex^2 + fx + g}$, $e \neq 0$ $\qquad \left[ \dfrac{a}{e} \right]$

6. $\lim_{x \to 0} \dfrac{1}{x^2} e^{-1/x^2}$ $\qquad [0]$

2. $\lim_{x \to \infty} \dfrac{(\pi/2) - \text{Arc} \tan x}{e^{-x}}$ $\qquad [\infty]$

7. $\lim_{x \to 0^+} \dfrac{\cot x}{\log x}$ $\qquad [-\infty]$

3. $\lim_{x \to 1/2^-} \dfrac{\tan \pi x}{\log \cos \pi x}$ $\qquad [-\infty]$

8. $\lim_{x \to 0^+} x^x$ $\qquad [1]$

4. $\lim_{\theta \to \pi/2} (\sec \theta - \tan \theta)$ $\qquad [0]$

9. $\lim_{x \to 0^+} x^{\sqrt{x}}$ $\qquad [1]$

5. $\lim_{x \to 0^+} \left( \dfrac{1}{x} + \log x \right)$ $\qquad [\infty]$

10. $\lim_{x \to \infty} \left( 1 + \dfrac{c}{x} \right)^x$ $\qquad [e^c]$

11. $\lim\limits_{x\to 0+} (\csc x)^{\sin^2 x}$ [1]

12. $\lim\limits_{\theta\to \pi/2} \dfrac{\tan\theta}{\tan 3\theta}$ [3]

13. $\lim\limits_{t\to 0} \left[\dfrac{1+t}{t} - \dfrac{1}{\log (1+t)}\right]$ $[\tfrac{1}{2}]$

14. $\lim\limits_{\theta\to 0} (1+\sin\theta)^{\csc\theta}$ [e]

15. $\lim\limits_{x\to 2} (2-x)\tan\dfrac{\pi x}{4}$ $\left[\dfrac{4}{\pi}\right]$

16. $\lim\limits_{x\to\infty} \dfrac{x}{\log\log x}$ $[\infty]$

17. $\lim\limits_{t\to 0} (\cos 2t)^{1/t^2}$ $[e^{-2}]$

18. $\lim\limits_{x\to\infty} x\left(\operatorname{Arc}\tan x - \dfrac{\pi}{2}\right)$ $[-1]$

19. $\lim\limits_{t\to 0}\left(\dfrac{1}{2t} - \dfrac{1}{te^{\pi t}}\right)$ $[\pm\infty]$

20. $\lim\limits_{t\to 0}\left(\dfrac{1}{t} - \dfrac{1}{te^{\pi t}}\right)$ $[\pi]$

21. $\lim\limits_{x\to 0+} x^n \log x, \quad n>0$ $[0]$

22. $\lim\limits_{x\to 0+} x \log^n x, \quad n>0$ $[0]$

23. $\lim\limits_{x\to\infty} x^n e^{-x}$ $[0]$

24. $\lim\limits_{x\to\infty} \dfrac{(1.01)^x}{x^{101}}$ $[\infty]$

25. $\lim\limits_{\theta\to 0} \dfrac{\operatorname{Arc}\sin\theta - \theta}{\sin^3\theta}$ $[\tfrac{1}{6}]$

26. $\lim\limits_{\theta\to 0} \left(\dfrac{\sin\theta}{\theta}\right)^{1/\theta^2}$ $[e^{-1/6}]$

27. $\lim\limits_{s\to 0} \dfrac{e^s + e^{-s} - 2\cos s}{s\sin s}$ $[2]$

28. $\lim\limits_{t\to 0+} t^{(t-c)};\quad$ (a) $c>0,\quad$ (b) $c<0$ $[$ (a) 0, (b) 1$]$

29. $\lim\limits_{\theta\to 0} \dfrac{a\sin\theta - \sin a\theta}{a\tan\theta - \tan a\theta}$ $[-\tfrac{1}{2}]$

30. $\lim\limits_{\theta\to\infty} \left(\theta\operatorname{Arc}\tan\theta - \dfrac{\pi}{2}\sqrt{1+\theta^2}\right)$ $[-1]$

*31. $\lim\limits_{x\to\infty} \left(\dfrac{1}{n}\sum\limits_{i=1}^{n} a_i^{1/x}\right)^{nx}, \quad a_i>0,\quad i=1,\cdots,n$ $\left[\prod\limits_{i=1}^{n} a_i\right]$

32. Sketch the graph of $y = x\log x$.

33. Sketch the graph of $y = e^{-1/x}$.

34. Sketch the graph of $y = e^{-1/x^2}$.

*35. If $f$ is defined by $f(x) = e^{-1/x^2}$ for $x \neq 0$ and $f(0) = 0$, compute $f'(0)$, and $f''(0)$.

36. Sketch the graph of $y = xe^{-1/x}$.

**37.** Show that even though $\lim\limits_{x\to\infty} \dfrac{\sin x - x}{x} = -1$ it cannot be evaluated

by using l'Hospital's rule.

**\*38.** Prove part (1) of l'Hospital's rule, page 185.

**\*39.** Prove part (3) of l'Hospital's rule, page 185, assuming part (2) is valid.

**40.** Show that $\dfrac{1}{x} - x + 2 \log x$ is positive if $0 < x < 1$.

## 4 *The Extended MVT (Taylor's Formula)*

The MVT can be considered as providing a constant approximation to $f$ in the neighborhood of $a$:

$$f(x) = f(a) + f'(c)(x - a) = f(a) + R.$$

The term $f'(c)(x - a) = R$ is the remainder, the error at $x$ in the approximation $f(x) \approx f(a)$.

The extended MVT provides an approximation to $f$ by a certain polynomial. What is vital is that one has a formula for the error, or remainder.

**THE EXTENDED MEAN VALUE THEOREM (TAYLOR'S\* FORMULA)** *If $f$ has $n + 1$ continuous derivatives in an interval, and if $a$ and $x$ are in that interval, and if*

$$(1) \quad f(x) = f(a) + \frac{f'(a)}{1!}(x - a) + \frac{f''(a)}{2!}(x - a)^2 + \cdots + \frac{f^{(n)}(a)}{n!}(x - a)^n + R_n,$$

*then there is a number $c$ between $a$ and $x$ such that*

$$(2) \qquad\qquad R_n = \frac{f^{(n+1)}(c)}{(n + 1)!}(x - a)^{n+1}.$$

**PROOF SKETCH** The proof involves a clever use of Rolle's theorem. First we consider $x$ as fixed in equation (1) and regard $a$ as the variable, which we replace by $t$. Then

$$(3) \quad f(x) = f(t) + f'(t)(x - t) + \frac{f''(t)}{2!}(x - t)^2 + \cdots + \frac{f^{(n)}(t)}{n!}(x - t)^n + R_n.$$

---

**\* Historical Note.** Though named after the Englishman Brook Taylor (1685–1731), the formula was actually the work of J. L. Lagrange, who recognized its importance in connection with Taylor's series. See the footnote on page 195.

In the early eighteenth century Taylor was one of the few Englishmen who could compete with the Bernoullis. Taylor is also known as the founder of the "calculus of finite differences."

Now we consider a function $F$, nearly equal to the difference of the two sides of equation (3):

$$(4) \quad F(t) = f(x) - f(t) - f'(t)(x - t) - \frac{f''(t)}{2!}(x - t)^2 - \cdots$$

$$- \frac{f^{(n)}(t)}{n!}(x - t)^n - R_n \frac{(x - t)^{n+1}}{(x - a)^{n+1}}.$$

Then $F(x) = 0$, and $F(a) = 0$ because of (1). Therefore, there is a number $c$ between $a$ and $x$ such that $F'(c) = 0$. But on differentiating (4) all the terms drop out (telescope) except two (the reader should check this), giving

$$(5) \qquad F'(t) = -\frac{f^{(n+1)}(t)}{n!}(x - t)^n + R_n \frac{(n + 1)(x - t)^n}{(x - a)^{n+1}}.$$

Then $F'(c) = 0$ implies that

$$R_n = \frac{f^{(n+1)}(c)}{(n + 1)!}(x - a)^{n+1}.$$

$R_n$ is called the *remainder* after $n + 1$ terms. It is *Lagrange's form* for the remainder. In Chapter 13 another form for the remainder will be found. Note that $R_n = f - P_n$ where $P_n$ is the *Taylor polynomial*:

$$P_n(x) = f(a) + \frac{f'(a)}{1!}(x - a) + \frac{f''(a)}{2!}(x - a)^2 + \cdots + \frac{f^{(n)}(a)}{n!}(x - a)^n.$$

The Taylor polynomial will be a good approximation to $f(x)$ if $R_n$ is small. Usually, when this is the case, one computes the value of the polynomial only for values $x$ close to $a$, since then $|x - a|$ is small, and fewer terms are needed.

When $n = 1$, the Taylor polynomial,

$$f(x) \approx f(a) + f'(a)(x - a),$$

gives the differential approximation. With $x - a = \Delta x = dx$ we have

$$f(x) - f(a) - f(a + \Delta x) - f(a) \approx f'(a) \, dx.$$

*Remark.* $P_n(x)$ is the *unique* polynomial of degree $\leq n$ that has the same value and the same derivatives as $f$ at $a$, up to the $n$th derivative:

$$P_n(a) = f(a), \quad P_n'(a) = f'(a), \quad \cdots, \quad P_n^{(n)}(a) = f^{(n)}(a).$$

**Example 1**   Any polynomial $F(x) = c_0 + c_1 x + c_2 x^2 + \cdots + c_n x^n$ can also be expressed in powers of $x - a$, so that

$$F(x) = a_0 + a_1(x - a) + a_2(x - a)^2 + \cdots + a_n(x - a)^n.$$

As is familiar from algebra, the coefficients $a_0, \cdots, a_n$ can be found by performing successive divisions. For example, $a_0 = F(a) =$ the remainder on dividing $F(x)$ by $x - a$; $a_1$ is the remainder on dividing that quotient by $x - a$; etc. The coefficients $a_0, \cdots, a_n$ can also be found from Taylor's formula:

$$a_k = \frac{F^{(k)}(a)}{k!}.$$

If $F(x) = x^4 - 4x^3 + 2x^2 + 3$, write $F(x)$ in powers of $x - 2$.

$$
\begin{array}{lll}
F(x) = x^4 - 4x^3 + 2x^2 + 3, & F(2) = -5, & a_0 = -5; \\
F'(x) = 4(x^3 - 3x^2 + x) \quad, & F'(2) = -8, & a_1 = -8; \\
F''(x) = 4(3x^2 - 6x + 1) \quad, & F''(2) = \phantom{-}4, & a_2 = 2; \\
F'''(x) = 24(x - 1) \quad, & F'''(2) = \phantom{-}24, & a_3 = 4; \\
F^{\mathrm{iv}}(x) = 24 \quad, & F^{\mathrm{iv}}(2) = \phantom{-}24, & a_4 = 1.
\end{array}
$$

Hence, $F(x) = -5 - 8(x - 2) + 2(x - 2)^2 + 4(x - 2)^3 + (x - 2)^4$. In this case $R_4 = 0$ because $F^{(\mathrm{v})}(x) = 0$.

**Example 2**   Apply Taylor's formula with $a = 0$ and $n = 1$ if $f(x) = \sqrt{1 + x}$. Compute $\sqrt{1.2}$ and estimate the error $R_1$.

$$
\begin{array}{ll}
f(x) = (1 + x)^{1/2} \quad, & f(0) = 1; \\
f'(x) = \tfrac{1}{2}(1 + x)^{-1/2}, & f'(0) = \tfrac{1}{2}; \\
f''(x) = -\tfrac{1}{4}(1 + x)^{-3/2}, & f''(c) = -\tfrac{1}{4}(1 + c)^{-3/2}.
\end{array}
$$

Then
$$f(x) = (1 + x)^{1/2} = 1 + \tfrac{1}{2}x + R_1,$$

where
$$R_1 = -\tfrac{1}{8}(1 + c)^{-3/2}x^2$$

and $c$ is between 0 and $x$. Then in case $x = 0.2$, we have

$$\sqrt{1.2} = 1 + \tfrac{1}{2}(0.2) - \frac{1}{8(1 + c)^{3/2}}(0.2)^2.$$

The first two terms give the approximation $\sqrt{1.2} = 1.1$. (This is the differential approximation.) The remainder $R_1$ can be estimated as follows: Because $0 < c < 0.2$, $1 + c > 1$, and so $|R_1| < \tfrac{1}{8}(0.2)^2 = 0.005$. Thus, 1.1 differs from $\sqrt{1.2}$ by less than 0.005. Because $R_1$ is negative, the true value is a bit less than 1.1, whence $1.095 < \sqrt{1.2} < 1.1$.

For this same function, if we were to take $n = 2$, then

$$f(x) = \sqrt{1 + x} = 1 + \tfrac{1}{2}x - \tfrac{1}{8}x^2 + R_2,$$

where
$$R_2 = \frac{1}{16}\frac{x^3}{(1 + c)^{5/2}}$$

and $c$ is between 0 and $x$. When $x = 0.2$ we have

$$\sqrt{1.2} = 1 + \frac{1}{2}(0.2) - \frac{1}{8}(0.04) + \frac{1}{16}\frac{(0.2)^3}{(1+c)^{5/2}}.$$

The first three terms give the approximation $\sqrt{1.2} = 1.095$. The remainder $R_2$ is positive, and $R_2 < 0.0005$, whence

$$1.095 < \sqrt{1.2} < 1.0955.$$

**Example 3** Obtain the Taylor polynomial of degree 3 for $\cos x$ near $x = \pi/3$. Here $a = \pi/3$ and $n = 3$.

$$f(x) = \cos x, \qquad f(\pi/3) = \tfrac{1}{2};$$
$$f'(x) = -\sin x, \qquad f'(\pi/3) = -\sqrt{3}/2;$$
$$f''(x) = -\cos x, \qquad f''(\pi/3) = -\tfrac{1}{2};$$
$$f'''(x) = \sin x, \qquad f'''(\pi/3) = \sqrt{3}/2;$$
$$f^{(iv)}(x) = \cos x, \qquad f^{(iv)}(c) = \cos c.$$

Then

$$\cos x = \frac{1}{2} - \frac{\sqrt{3}}{2}\left(x - \frac{\pi}{3}\right) - \frac{1}{4}\left(x - \frac{\pi}{3}\right)^2 + \frac{\sqrt{3}}{12}\left(x - \frac{\pi}{3}\right)^3 + \frac{\cos c}{24}\left(x - \frac{\pi}{3}\right)^4.$$

The Taylor polynomial approximation consists of the first four terms.

We use the polynomial to compute $\cos 61°$. Since an angle of $61°$ has measure $\pi/3 + \pi/180$ radians we have

$$(6) \quad \cos 61° \approx \frac{1}{2} - \frac{\sqrt{3}}{2}\frac{\pi}{180} - \frac{1}{4}\left(\frac{\pi}{180}\right)^2 + \frac{\sqrt{3}}{12}\left(\frac{\pi}{180}\right)^3$$

$$\approx 0.50000 - 0.01511 - 0.00008 + 0.00000 \ \{\text{rounding to 5 places*}$$

$$= 0.48481$$

which agrees with the five-place table value. The actual error in (6), *assuming we had not rounded off is*

$$R_3 = \frac{\cos c}{24}\left(\frac{\pi}{180}\right)^4 < \frac{1}{48}\left(\frac{\pi}{180}\right)^4 < (0.02)^5 = 0.0000000032,$$

where the last inequality is obtained by making a rough estimate:

$$1/48 \approx 1/50 \qquad \text{and} \qquad \pi/180 < 1/50.$$

---

* When $n$ is large, and the terms do not decrease rapidly, one has to be careful that round-off errors do not accumulate. This can be a serious problem with digital computers.

### Problems

In Problems 1 to 8 apply Taylor's formula for the indicated values of $n$ and $a$. Give $P_n$ and $R_n$.

1. $x^3 + 2x^2 - 3x + 1$;   $n = 4, a = -1$ (also obtain $P_4$ by long division)

$$[P_4 = 5 - 4(x+1) - (x+1)^2 + (x+1)^3, R_4 = 0]$$

2. $x^3 + 2x^2 - 3x + 1$;   $n = 2, a = 0$        $[P_2 = 1 - 3x + 2x^2, R_2 = x^3]$

3. $\cos x$;   $n = 2, a = 0$        $\left[ P_2 = 1 - \dfrac{1}{2}x^2, R_2 = \dfrac{\sin c}{6}x^3 \right]$

4. $\cos x$;   $n = 5, a = \dfrac{\pi}{2}$   $\left[ \begin{array}{l} P_5 = -\left(x - \dfrac{\pi}{2}\right) + \dfrac{1}{6}\left(x - \dfrac{\pi}{2}\right)^3 - \dfrac{1}{120}\left(x - \dfrac{\pi}{2}\right)^5, \\[2mm] R_5 = \dfrac{-\cos c}{720}\left(x - \dfrac{\pi}{2}\right)^6 \end{array} \right]$

5. $\log(1 + x)$;   $n = 4, a = 0$   $\left[ P_4 = x - \dfrac{x^2}{2} + \dfrac{x^3}{3} - \dfrac{x^4}{4}, R_4 = \dfrac{1}{5}\dfrac{x^5}{(1+c)^5} \right]$

6. $\text{Arc} \sin x$;   $n = 2, a = 0$        $\left[ P_2 = x, R_2 = \dfrac{x^3(1 + 2c^2)}{6(1 - c^2)^{5/2}} \right]$

7. $\text{Arc} \tan x$;   $n = 2, a = 1$   $\left[ \begin{array}{l} P_2 = \dfrac{\pi}{4} + \dfrac{1}{2}(x - 1) - \dfrac{1}{4}(x - 1)^2, \\[3mm] R_2 = \dfrac{3c^2 - 1}{3(1 + c^2)^3}(x - 1)^3 \end{array} \right]$

8. $e^x$;   $n = 4, a = 0$   $\left[ P_4 = 1 + x + \dfrac{x^2}{2!} + \dfrac{x^3}{3!} + \dfrac{x^4}{4!}, R_4 = \dfrac{e^c}{5!}x^5 \right]$

In Problems 9 to 12 compute the indicated Taylor polynomial and estimate the remainder for the indicated values of $x$, $n$, and $a$.

9. $(1 + x)^{1/3}$;   $a = 0, n = 2, x = 0.2$   $[P_2(0.2) = 1.0622, 0 < R_2 < 0.0005]$

10. $\dfrac{1}{1 - x}$;   $a = 0, n = 3, x = 0.1$   $[P_3(0.1) = 1.111, 0 < R_3 < 0.00017]$

11. $e^x$;   $a = 0, n = 4, x = 0.1$   $[P_4(0.1) = 1.105171, 0 < R_4 < 0.0000003]$

12.  $\sin x$;   $a = \pi/3$, $n = 2$, $x = 61\pi/180$
$$[\sin 61° \approx P_2(61°) = 0.874620,\ -0.0000008 < R_2 < 0]$$

13.  Compute $\sqrt[3]{26}$ to four decimal places. (Choose $a = 27$ and $f(x) = x^{1/3}$. Choose $n$ large enough and compute.)   $[-0.00002 < R_2 < 0]$

14.  Compute $\cos 5°$ to five decimal places.

15.  Compute $e^{-0.2}$ to four decimal places.

16.  Compute $(0.99)^{20}$ to four decimal places.   $[P_3 = 0.81786, 0 < R_3 < 0.00005]$

$\sqrt{\phantom{}}$17.  How good is the approximate formula $(1 + x)^{-3/2} = 1 - \frac{3}{2}x$ if $|x| \leq 0.1$?

18.  How good is the approximation $\cos x = 1 - \frac{1}{2}x^2$ for angles less than 5 deg?

19.  Find $P_3$ for $f(x) = \sin x$ and $a = 0$. Graph both $P_3(x)$ and $\sin x$ for $-\pi/2 \leq x \leq \pi/2$.

20.  Find $P_2$ for $f(x) = \cos x$ and $a = 0$. Graph both $P_2(x)$ and $\cos x$ for $0 \leq x \leq \pi/2$.

21.  Verify formula (5), page 189.

*22.  Show that, if $f(a) = f'(a) = \cdots = f^{(n)}(a) = 0$ and $g(a) = g'(a) = \cdots = g^{(n)}(a) = 0$ and $f$ and $g$ have continuous derivatives of order $n + 1$, then

$$\lim_{x \to a} \frac{f(x)}{g(x)} = L \qquad \text{if} \qquad \frac{f^{(n+1)}(a)}{g^{(n+1)}(a)} = L.$$

*23.  Use the definition of $R_2$ for $x = b$,

$$f(b) = f(a) + f'(a)(b - a) + \tfrac{1}{2}f''(a)(b - a)^2 + R_2$$

and the function $F$ given by

$$F(x) = f(x) - f(a) - (x - a)f'(a) - \frac{1}{2}(x - a)^2 f''(a) - \frac{(x - a)^3}{(b - a)^3} R_2$$

to conclude that $F(a) = F(b) = 0$, whence $F'(c_1) = 0$ for some $c_1$ between $a$ and $b$. But $F'(a) = 0$, whence $F''(c_2) = 0$ for some $c_2$ between $a$ and $c_1$. But $F''(a) = 0$, too, whence $F'''(c) = 0$ for some $c$ between $a$ and $c_2$. Then solve for $R_2$.

*24.   Consider Taylor's formula in the form

$$f(a + x) = f(a) + f'(a)x + \frac{f''(a)}{2!} x^2 + \cdots + \frac{f^{(n-1)}(a)}{(n-1)!} x^{n-1}$$

$$+ \frac{f^{(n)}(a + \theta x)}{n!} x^n, \quad \text{where } 0 < \theta < 1.$$

Show that as $x \to 0$, $\theta \to 1/(n+1)$ if $f^{(n+1)}(a) \neq 0$.

*Hint:* Show that

$$R_{n-1} = \frac{f^{(n)}(a)}{n!} x^n + \frac{f^{(n+1)}(a + \theta_1 \theta x)}{n!} \theta x^{n+1}, \quad 0 < \theta_1 < 1,$$

and consider the limit as $x \to 0$ of

$$1 = \frac{R_{n-1} - f^{(n)}(a)x^n/n!}{R_n}.$$

## CHAPTER 8

## TAYLOR'S SERIES

1  **Taylor's Series**

Taylor's formula provides a polynomial approximation to a function:

$$f(x) \approx P_n(x) = f(a) + f'(a)(x-a) + \cdots + \frac{f^{(n)}(a)}{n!}(x-a)^n.$$

Suppose now that $f$ has derivatives of all orders; then one can at least write an expression like this:

$$(1) \qquad f(x) \frown f(a) + f'(a)(x-a) + \cdots + \frac{f^{(n)}(a)}{n!}(x-a)^n + \cdots,$$

where the last three dots "$\cdots$" signify that the terms continue without end. Such an expression is called a *Taylor series*\* for the function $f$—or also, *Taylor's series at a*, for the function $f$.

The special case where $a = 0$ in (1) is called *Maclaurin's series*.†

--------

\* **Historical Note.** The series appeared for the first time in Taylor's book, *Methodus Incrementorum Directa et Inversa*, London, 1715. He had no *proof* that the series represented the function. All he did was show that *if* a series of powers of $(x-a)$ represented $f$, and *if* one could differentiate the series term by term, then the coefficients had to be given by equation (1).

† **Historical Note.** After the Scotsman Colin Maclaurin (1698–1746), whose *Treatise on Fluxions*, published in 1742 was the first systematic exposition of Newton's method of fluxions. The series occurs in that book, but the series had been given by James Stirling in 1730.

Naturally, the expression (1) cannot mean that one is to "add infinitely many numbers." This is impossible. Addition applies to finite sets of numbers. The precise meaning of (1) will be given in Section 3. In this section we shall be content with exhibiting Taylor's series for a few functions. The whole theory of infinite series will be examined in greater depth in Chapter 13.

Series (1) is a "sum of powers of $x - a$," and so is called a *power series*. The Taylor series (1) is called, therefore, "the expansion of $f$ in powers of $x - a$."

For convenience of reference, five particularly important Maclaurin series are given below. These series occur frequently, and although one need not memorize them, it does sometimes help to have them in mind. Values of $x$ for which the equality sign holds will be established in Section 3.

*The binomial series*

$$(1 + x)^m = 1 + \frac{m}{1}x + \frac{m(m-1)}{1 \cdot 2}x^2 + \cdots$$

$$+ \frac{m(m-1) \cdots (m-k+1)}{1 \cdot 2 \cdot 3 \cdots k} x^k + \cdots$$

*The geometric series*
A special case of the binomial series with $m = -1$ and $x$ replaced by $-x$:

$$(1 - x)^{-1} = 1 + x + x^2 + \cdots + x^n + \cdots$$

*The exponential series*

$$e^x = 1 + x + \frac{x^2}{2!} + \cdots + \frac{x^n}{n!} + \cdots$$

*The sine series*

$$\sin x = x - \frac{x^3}{3!} + \frac{x^5}{5!} - \cdots + \frac{(-1)^n x^{2n+1}}{(2n+1)!} + \cdots$$

*The cosine series*

$$\cos x = 1 - \frac{x^2}{2!} + \frac{x^4}{4!} - \cdots + \frac{(-1)^n x^{2n}}{(2n)!} + \cdots$$

Before proceeding with examples and problems we shall introduce a shorthand notation that will abbreviate the long sums that we have just encountered. The abbreviation is made using a summation sign, $\sum$, which is the Greek capital letter *sigma*. Examples will illustrate its use.

$$a_1 + a_2 + a_3 = \sum_{k=1}^{3} a_k.$$

$$2 + 4 + 6 + 8 = \sum_{n=1}^{4} (2n).$$

$$a_1 + a_2 + \cdots + a_n = \sum_{k=1}^{n} a_k.$$

$$1 + 2x + 3x^2 + 4x^3 + 5x^4 = \sum_{n=0}^{4} (n+1)x^n.$$

In each case the start of the sum is indicated at the bottom of the $\sum$. The last term of the sum is indicated at the top of the $\sum$. When the symbol "$\infty$" occurs at the top the terms continue without end.

$$a_1 + a_2 + \cdots + a_n + \cdots = \sum_{k=1}^{\infty} a_k$$

$$1 + \frac{1}{2} + \frac{1}{2^2} + \cdots + \frac{1}{2^n} + \cdots = \sum_{k=0}^{\infty} \frac{1}{2^k}.$$

The Taylor series (1) can be written as

$$f(c) + f'(a)(x-a) + \cdots + \frac{f^{(n)}(a)}{n!}(x-a)^n + \cdots = \sum_{n=0}^{\infty} \frac{f^{(n)}(a)}{n!}(x-a)^n.$$

The binomial, geometric, exponential, sine, and cosine series then can be written:

$$(1+x)^m = \sum_{k=0}^{\infty} \frac{m(m-1)\cdots(m-k+1)}{k!} x^k;$$

$$(1-x)^{-1} = \sum_{k=0}^{\infty} x^k;$$

$$e^x = \sum_{k=0}^{\infty} \frac{1}{k!} x^k;$$

$$\sin x = \sum_{k=0}^{\infty} \frac{(-1)^k x^{2k+1}}{(2k+1)!};$$

$$\cos x = \sum_{k=0}^{\infty} \frac{(-1)^k x^{2k}}{(2k)!}.$$

*Example*  Find the Taylor series for $\log x$ in powers of $x - a$.

$$f(x) = \log x \qquad\qquad f(a) = \log a$$
$$f'(x) = x^{-1} \qquad\qquad f'(a) = a^{-1}$$
$$\cdots \qquad\qquad\qquad \cdots$$
$$f^{(n)}(x) = (-1)^{n-1}(n-1)!x^{-n} \qquad f^{(n)}(a) = (-1)^{n-1}(n-1)!a^{-n}.$$
$$\cdots \qquad\qquad\qquad \cdots$$

The Taylor series is

$$\log x = \log a + \frac{x-a}{a} - \frac{1}{2}\frac{(x-a)^2}{a^2} + \frac{1}{3}\frac{(x-a)^3}{a^3} - \cdots$$

$$+ \frac{(-1)^{n-1}}{n}\frac{(x-a)^n}{a^n} + \cdots$$

$$= \log a + \sum_{n=1}^{\infty}\frac{(-1)^{n-1}}{n}\frac{(x-a)^n}{a^n}.$$

*Problems*

1.    Expand $x^3 + 2x^2 - x - 1$ in powers of $x - 1$.
$$[1 + 6(x-1) + 5(x-1)^2 + (x-1)^3]$$

2.    Find the Taylor series of $x^2 + 2x + 2$ at $x = -1$.    $[1 + (x+1)^2]$

3.    Find the Taylor expansion of $4y^4 + 3y^3 + 2y^2 + y + 2$ in powers of $y + 2$.
$$[48 - 99(y+2) + 80(y+2)^2 - 29(y+2)^3 + 4(y+2)^4]$$

4.    Find the Taylor expansion of $e^x$ at $x = -2$
$$\left[\sum_{n=0}^{\infty} e^{-2}\frac{(x+2)^n}{n!} = e^{-2}\left(1 + (x+2) + \frac{(x+2)^2}{2!} + \cdots + \frac{(x+2)^n}{n!} + \cdots\right)\right]$$

5.    Find the Maclaurin series for $e^x$.

6.    Expand $\sin x$ in powers of $x$.

7.    Find the Taylor series for $\cos x$ at $x = 0$.

8.    What is Taylor's series for $\log(1+x)$ at $x = 0$?
$$\left[\sum_{n=1}^{\infty}(-1)^{n-1}\frac{x^n}{n} = x - \frac{x^2}{2} + \frac{x^3}{3} - \cdots + (-1)^{n-1}\frac{x^n}{n} + \cdots\right]$$

9.    Expand $e^x$ in powers of $x - a$.
$$\left[\sum_{n=0}^{\infty} e^a\frac{(x-a)^n}{n!} = e^a\left(1 + (x-a) + \frac{(x-a)^2}{2!} + \cdots + \frac{(x-a)^n}{n!} + \cdots\right)\right]$$

10.    Expand $e^{a+x}$ in powers of $x$.
$$\left[e^a\left(1 + x + \frac{x^2}{2!} + \cdots + \frac{x^n}{n!} + \cdots\right)\right]$$

11.    Give the Maclaurin series for $(1-x)^{-1}$.

12.    Expand $(1-x)^m$ in powers of $x$.

13.    Give the Taylor series for $\sin x$ in powers of $x - a$.
$$\left[\sin a + \cos a(x-a) - \frac{\sin a}{2!}(x-a)^2 - \frac{\cos a}{3!}(x-a)^3 + \cdots\right]$$

14. Expand $\cos x$ in powers of $x - a$.

$$\left[\cos a - \sin a\,(x - a) - \frac{\cos a}{2!}\,(x - a)^2 + \frac{\sin a}{3!}\,(x - a)^3 + \cdots\right]$$

15. Give the Maclaurin series for $a^x$.

$$\left[\sum_{n=0}^{\infty}\frac{(x \log a)^n}{n!} = 1 + x \log a + \frac{(x \log a)^2}{2!} + \cdots + \frac{(x \log a)^n}{n!} + \cdots\right]$$

16. Expand $e^{kx}$ in a Taylor series at $x = 0$. $\left[1 + kx + \dfrac{k^2 x^2}{2!} + \cdots + \dfrac{k^n x^n}{n!} + \cdots\right]$

17. Give six non-zero terms of the Maclaurin series for $e^x \sin x$.

$$\left[x + \frac{2x^2}{2!} + \frac{2x^3}{3!} - \frac{2^2}{5!}x^5 - \frac{2^3}{6!}x^6 - \frac{2^3}{7!}x^7 + \cdots\right]$$

18. Expand $e^x \sin x$ in powers of $x - (\pi/2)$. Find four terms.

$$\left[e^{\pi/2}\left[1 + \left(x - \frac{\pi}{2}\right) - \frac{2}{3!}\left(x - \frac{\pi}{2}\right)^3 - \frac{2^2}{4!}\left(x - \frac{\pi}{2}\right)^4 - \cdots\right]\right]$$

19. Expand $\sin^2 x$ in powers of $x$.

$$\left[-\frac{1}{2}\sum_{n=1}^{\infty}\frac{(-1)^n(2x)^{2n}}{(2n)!} = \frac{2x^2}{2!} - \frac{2^3 x^4}{4!} + \cdots + \frac{(-1)^{n-1}(2x)^{2n}}{2(2n)!} + \cdots\right]$$

(Hint: $\sin^2 x = \frac{1}{2} - \frac{1}{2}\cos 2x$.)

20. Find the Maclaurin series for $1/(1 + x^2)$.

$$\left[\sum_{n=0}^{\infty}(-1)^n x^{2n} = 1 - x^2 + x^4 - \cdots + (-1)^n x^{2n} + \cdots\right]$$

21. Obtain the expansion

$$\frac{1}{\sqrt{1 - x}} = 1 + \tfrac{1}{2}x + \frac{1\cdot 3}{2\cdot 4}x^2 + \frac{1\cdot 3\cdot 5}{2\cdot 4\cdot 6}x^3 + \cdots.$$

22. Obtain the Maclaurin expansion of Arc tan $x$ through $x^5$. $\left[x - \dfrac{x^3}{3} + \dfrac{x^5}{5} - \cdots\right]$

23. Obtain the Maclaurin expansion of Arc sin $x$ through $x^5$.

$$\left[x + \frac{1}{2\cdot 3}x^3 + \frac{1\cdot 3}{2\cdot 4\cdot 5}x^5 + \cdots\right]$$

24. Expand $e^{\cos x}$ in powers of $x$ up to $x^5$.

$$\left[e\left(1 - \frac{x^2}{2} + \frac{x^4}{6} - \cdots\right)\right]$$

*25.* Find the Taylor expansion of sec $x$ in powers of $x$ through $x^6$.

$$\left[ 1 + \frac{x^2}{2} + \frac{5x^4}{24} + \frac{61x^6}{720} + \cdots \right]$$

*26.* Show that the Taylor expansion of $f$ at $a$ is the same as the Maclaurin expansion of $f(a + t)$.

## 2   Computation with Power Series

Although we have not discussed how the infinite sum in Taylor's series might be found, let us just start computing. We shall see that at least in some cases there is no question but that there is a "sum."

Surely it was in this intuitive computational sense that early users considered infinite series.* In the examples and problems of this section we suppose that the Taylor series does represent the function. So that our computations give the value of the function.

*Example 1*   A highly important series is the *geometric series*. From elementary algebra,

$$\frac{1}{1 - x} = 1 + x + x^2 + \cdots + x^n + \frac{x^{n+1}}{1 - x}.$$

Taylor's formula gives

$$\frac{1}{1 - x} = 1 + x + x^2 + \cdots + x^n + R_n,$$

so that we actually have a formula for the remainder $R_n$ without using calculus.

Then

$$R_n = \frac{x^{n+1}}{1 - x},$$

so

$$\lim_{n \to \infty} \frac{x^{n+1}}{1 - x} = 0 \qquad \text{if} \qquad |x| < 1.$$

---

**\* Historical Note.** Newton regarded the general applicability of his method as at least partly based on infinite series. Because differentiation of polynomials is extremely easy (and antidifferentiation too), it was "clear" that infinite series *should* behave the same way. Thus, if

$$\frac{1}{1 - x} = 1 + x + x^2 + \cdots + x^n + \cdots,$$

then

$$D\left(\frac{1}{1 - x}\right) = \frac{1}{(1 - x)^2} = 1 + 2x + 3x^2 + \cdots + nx^{n-1} + \cdots.$$

We shall see later that this is in fact the case.

Taylor never gave a definition of what the infinite sum might mean. He simply used the series.

In other words, if $|x| < 1$, the more terms of the series

$$1 + x + x^2 + x^3 + \cdots + x^n + \cdots$$

that one takes, the closer one gets to $1/(1-x)$.

The value $x = \frac{1}{2}$ is particularly instructive because the sum can be easily represented on the line, as is shown in the figure.

$$1 + \frac{1}{2} + \frac{1}{4} + \cdots + \frac{1}{2^n} = 2 - \frac{1}{2^n}$$

Repeating decimals can be regarded as geometric series. Thus

$$0.5323232\ldots = 0.5 + 0.032 + 0.00032 + 0.0000032 + \cdots$$

$$= 0.5 + 0.032 \left[ 1 + (0.01) + (0.01)^2 + \cdots \right]$$

$$= 0.5 + 0.032 \, \frac{1}{1 - 0.01}$$

$$= \frac{1}{2} + \frac{32}{990} = \frac{527}{990}.$$

Consequently $0.5323232\ldots$ is a rational number.

**Example 2**    The binomial series for $m = \frac{1}{2}$ is

$$(1 + x)^{1/2} = 1 + \tfrac{1}{2}x + \frac{\tfrac{1}{2}(\tfrac{1}{2} - 1)}{2!} x^2 + \cdots + \frac{\tfrac{1}{2}(\tfrac{1}{2} - 1) \cdots (\tfrac{1}{2} - n + 1)}{n!} x^n + \cdots$$

$$= 1 + \tfrac{1}{2}x - \frac{1}{2^2 \cdot 2!} x^2 + \cdots + (-1)^{n-1} \frac{1 \cdot 3 \cdot 5 \cdots (2n - 3)}{2^n n!} x^n + \cdots$$

Rather obviously, the terms will decrease rapidly if $|x|$ is small. Let us compute $(1 + 0.1)^{1/2} = \sqrt{1.1}$, keeping seven decimal places.

$$\sqrt{1.1} = 1.0000000 + 0.0500000 - 0.0012500$$

$$+ 0.0000625 - 0.0000039 + 0.0000002 - \cdots$$

$$= 1.0488088\ldots$$

How good is this result? The next term is obtained by multiplying the sixth term, $2 \times 10^{-7}$ by $[-9/2\,(6)](0.1) = -0.075$, and so cannot affect the seventh decimal place. In general the $(n + 1)$st term is obtained from the $n$th term by multiplying by $[(2n - 3)/2n](0.1)$. Thus, each succeeding term is less than $1/10$ the preceding, and since they alternate in sign, the sum at any stage is either a bit too large or a bit too small. Therefore, $\sqrt{1.1}$ is slightly less than $1.0488088$ but is certainly more than $1.0488087$.

**Example 3**    The Maclaurin series for $e^x$ is

$$e^x = 1 + x + \frac{x^2}{2!} + \cdots + \frac{x^n}{n!} + \cdots.$$

If $x = 0.1$ we get

$$e^{0.1} = 1 + 0.1 + 0.005 + 0.000\ 166\ 666\ 7$$
$$+\ 0.000\ 004\ 166\ 7 + 0.000\ 000\ 083\ 3$$
$$+\ 0.000\ 000\ 001\ 4 + \cdots$$
$$=\ 1.105\ 170\ 918\ 1,$$

Table of $1/n!$ rounded to ten decimal places

| $n$ | $1/n!$ |
|---|---|
| 3 | 0.166 666 666 7 |
| 4 | 0.041 666 666 7 |
| 5 | 0.008 333 333 3 |
| 6 | 0.001 388 888 9 |
| 7 | 0.000 198 412 7 |
| 8 | 0.000 024 801 6 |
| 9 | 0.000 002 755 7 |
| 10 | 0.000 000 275 6 |
| 11 | 0.000 000 025 1 |
| 12 | 0.000 000 002 1 |
| 13 | 0.000 000 000 2 |

accurate to eight decimals. The next term is $1.98 \times 10^{-11}$, and the following term is obtained from that by multiplying by $0.1/8 = 0.0125$. Later, terms are even smaller.

Thus, the "8" in the ninth decimal place is not far wrong.

When $x = 1$, the convergence is much less rapid, yet even here we can *see* how it goes. Suppose we wish to compute $e$ accurately to six decimal places. To avoid round-off errors we keep more places than we will need until it is clear how many are needed.

$$e = 1 + 1 + \frac{1}{2!} + \frac{1}{3!} + \cdots + \frac{1}{n!} + \cdots.$$

Using the factorial table through $n = 13$, we get

(1)    $$e \approx 2.718281829,$$

and more terms cannot increase the tenth digit by more than 1. But there are round-off errors in the tenth digit which might affect the ninth digit by 1 at most. Thus, eight decimal places in (1) are exact. To six decimals (rounded off), $e = 2.718282$.

**Example 4**    Maclaurin's series for $\log (1 + x)$ is

(2)    $$\log (1 + x) = x - \frac{x^2}{2} + \frac{x^3}{3} - \cdots + \frac{(-1)^{n-1}x^n}{n} + \cdots.$$

As will be proved later, the equality sign is valid for $|x| < 1$. But for $|x|$ near 1 the convergence is slow. For $x = \frac{1}{2}$ it is not too bad. Then

$$\log (1 + \tfrac{1}{2}) = \log \tfrac{3}{2} = 0.500\ 000 - 0.125\ 000 + 0.041\ 667 - 0.015\ 625$$
$$+\ 0.006\ 250 - 0.002\ 604 + 0.001\ 116 - 0.000\ 488$$
$$+\ 0.000\ 217 - 0.000\ 098 + 0.000\ 045 - 0.000\ 021 + \cdots$$

(3)    $$=\ 0.40546 \qquad \text{(rounded to five decimal places).}$$

When $x = -\frac{1}{2}$ we get all negative signs and

(4) $$\log (1 - \tfrac{1}{2}) = \log \tfrac{1}{2} = -\log 2 = -0.69313.$$

Thus, from (4) we get $\log 2 = 0.69313$, and from (3), using $\log 2$, $\log 3 = 1.09859$.

But series (2) is not rapidly enough convergent for easy computation because the terms decrease too slowly. The following device leads to better series. We have

$$\log (1 + x) = x - \frac{x^2}{2} + \frac{x^3}{3} - \cdots,$$

$$\log (1 - x) = -x - \frac{x^2}{2} - \frac{x^3}{3} - \cdots;$$

(5) $$\log \frac{1 + x}{1 - x} = 2\left( x + \frac{x^3}{3} + \frac{x^5}{5} + \cdots + \frac{x^{2n+1}}{2n + 1} + \cdots \right),$$

assuming that we can subtract the series term by term. This is justified in Chapter 13.

Series (5) is already better because the powers of $x$ are larger. Now set

$$x = \frac{M - N}{M + N} \qquad \text{so that} \qquad \frac{1 + x}{1 - x} = \frac{M}{N} \quad \text{and}$$

(6) $$\log \frac{M}{N} = 2\left[ \frac{M - N}{M + N} + \frac{1}{3}\left( \frac{M - N}{M + N} \right)^3 + \cdots \right].$$

If $M$ and $N$ are positive integers, one can use (6) to compute logarithms of integers. For example, if $M = 3$, $N = 2$ we get

$$\log \frac{3}{2} = 2\left( \frac{1}{5} + \frac{1}{3}\frac{1}{5^3} + \frac{1}{5}\frac{1}{5^5} + \cdots \right)$$

$$= 0.40546 \qquad \text{(rounded to five decimal places)}.$$

This is the same as (3) but obtained with much less labor.*

---

* **Historical Note.** In 1614 John Napier (1550–1617), the eighth Laird of Merchiston published a small booklet *Mirifici Logarithmorun Canonis Descriptio*. There was immediate interest because of the great saving of labor made possible in astronomical calculations. Attention soon centered on the construction of tables for which purpose many special and laborious methods were used. It was only after the invention of calculus that simple direct devices were at hand to construct tables.

Once tables have been constructed interest wanes. Yet as recently as 1941 new 16-place tables were made by the Work Projects Administration under the guidance of the National Bureau of Standards. These comprise four large volumes of the natural logarithms of the integers from 1 to 100,000 and decimal fractions from 0 to 10 at intervals of 0.0001.

As of this date, 1968, high-speed digital computers do not store tables of logarithms (or of other elementary functions) in their memories. They are equipped with subroutines that *compute* any tabular value when it is needed in just a few microseconds.

Equation (6) is especially convenient for $M = N + 1$, since then it becomes

(7) $\qquad \log (N + 1) = \log N + 2\left( \dfrac{1}{(2N + 1)} + \dfrac{1}{3}\dfrac{1}{(2N + 1)^3} + \cdots \right).$

This formula can be used to find logarithms of the integers one after the other.

*Example 5*   The Maclaurin series for Arc tan $x$ is*

(8) $\qquad\qquad\qquad \text{Arc tan } x = x - \dfrac{x^3}{3} + \dfrac{x^5}{5} - \cdots$

and is valid if $-1 \leq x \leq 1$, but for $|x|$ near 1 the convergence is *very* slow. The value $x = 1/\sqrt{3}$ yields a reasonable series and permits the calculation of $\pi$.

$$\text{Arc tan } \dfrac{1}{\sqrt{3}} = \dfrac{\pi}{6} = \dfrac{1}{\sqrt{3}}\left( 1 - \dfrac{1}{3}\dfrac{1}{3} + \dfrac{1}{5}\dfrac{1}{3^2} - \dfrac{1}{7}\dfrac{1}{3^3} + \cdots \right).$$

The series (8) has been the basis of most of the efforts to compute $\pi$. Perhaps the best use of (8) can be made by using Machin's relation†

(9) $\qquad\qquad\qquad \dfrac{\pi}{4} = 4 \text{ Arc tan } \dfrac{1}{5} - \text{Arc tan } \dfrac{1}{239},$

which can be established by considering the tangent of both sides. (To *discover* it is an altogether harder problem.) Then Gregory's series (8) permits the calculation of the right member of (9). See Problem 14. The first 15 places in $\pi$ are

$$\pi \approx 3.141592653589793.$$

*Problems*

1.  Use the geometric series to represent the number 0.037037... (the "037" repeats) as a quotient of two integers.                           [1/27]

2.  Use the binomial series to compute $\sqrt[3]{7.9}$ to four decimal places, rounded off.                           [1.9916]

---

* **Historical Note.** The series was discovered by James Gregory (1638–1675) about 1671. By 1667 he had found the expansions in powers of $x$ of sin $x$, cos $x$, Arc sin $x$, Arc cos $x$. It is worth noting that all this predates the invention of calculus.

† **Historical Note.** John Machin (1680–1751) found his formula in 1706 and computed $\pi$ to 100 decimal places.

In 1873 William Shanks gave $\pi$ to 707 decimal places using Machin's Formula. The calculation took him more than 15 years!

In 1946 D. F. Ferguson discovered errors in Shank's value beginning at the 528th place.

In 1961 John W. Wrench Jr. and Daniel Shanks (no relation) of Washington, D.C., using an IBM 7090, computed $\pi$ to 100,625 decimal places. Machine time was less than nine hours.

3. Use the binomial series to compute $(1.2)^{0.22}$ to four decimal places.      [1.0410]

4. Compute $\sqrt{e}$ to four decimal places.      [1.6487]

5. Compute $\sin 5°$ to five decimal places.

6. Compute $\sin 55°$ to five decimal places.

7. Compute $\log 2$ to five decimal places.      [0.693147]

8. Compute $\log 3$ to five decimal places.      [1.098608]

9. Compute $\log 4$ and $\log 6$ to five decimal places, using Problems 7 and 8.
      [1.386294, 1.791755]

10. Compute $\log 5$ to five decimal places, using Problem 9 and Formula (7).
      [1.609438]

11. Show that $\log_{10} x = \log x/\log 10$. Hence, obtain $\log_{10} x = 0.43429 \log x$.

12. Use Gregory's formula to compute $\pi$ to five decimal places.

13. Prove Machin's formula.

14. Use Machin's formula to compute $\pi$ to ten decimal places.

## 3    Taylor's Theorem

Taylor's *theorem** states that under suitable conditions the power series actually represents the function. That is, one has equality in

$$(1) \quad f(x) = f(a) + f'(a)(x - a) + \cdots + \frac{f^{(n)}(a)}{n!}(x - a)^n + \cdots$$

$$= \sum_{n=0}^{\infty} \frac{f^{(n)}(a)}{n!}(x - a)^n.$$

We now investigate what we are to mean by this and circumstances under which it is so.

The sum of the first $n + 1$ terms of the Taylor Series is the Taylor polynomial:

$$P_n(x) = f(a) + f'(a)(x - a) + \cdots + \frac{f^{(n)}(a)}{n!}(x - a)^n = \sum_{k=0}^{n} \frac{f^{(k)}(a)}{k!}(x - a)^k.$$

This is just what we compute when we use the first $n + 1$ terms of the series.

---

* **Historical Note.** Lagrange called Taylor's theorem the fundamental theorem of differential calculus. Even Lagrange did not concern himself with questions of *convergence* of the series, although he gave a formula for the remainder $R_n$. He assumed, as did others, that continuous functions could be represented by power series at all but a few exceptional points.

Therefore, this number will be close to $f(x)$ if the remainder, which is a function of $x$:

$$R_n(x) = f(x) - P_n(x)$$

is a small number (in absolute value). But Taylor's *formula* provides a formula for $R_n(x)$:

$$R_n(x) = \frac{f^{(n+1)}(c)}{(n+1)!} (x-a)^{n+1},$$

where $c$ is between $a$ and $x$. From the formula for $R_n(x)$ we *may* be able to estimate its magnitude and hence determine whether the Taylor polynomial is a good approximation.

The Taylor series is said to *converge* if

$$\lim_{n\to\infty} P_n(x)$$

exists. Taylor's theorem gives a condition under which the series converges, and converges to the given function.

**TAYLOR'S THEOREM**    *The Taylor series of $f$ will converge to $f(x)$ if*

$$\lim_{n\to\infty} R_n(x) = 0 = \lim_{n\to\infty} \frac{f^{(n+1)}(c)}{(n+1)!} (x-a)^{n+1}.$$

**PROOF**    If $\lim_{n\to\infty} R_n(x) = 0$, then

$$\lim_{n\to\infty} P_n(x) = f(x) - \lim_{n\to\infty} R_n(x)$$

$$= f(x) - 0 = f(x).$$

Taylor's theorem will be useful only in those cases where we can estimate $R_n$ and prove that it approaches 0. In this section we wish to show that this is, in fact, the case for some of the series we have encountered.

*Example 1*    In the case of the geometric series the remainder can be found from simple algebra.

$$\frac{1}{1-x} = 1 + x + \cdots + x^n + \frac{x^{n+1}}{1-x}$$

whence $R_n(x) = x^{n+1}/(1-x)$ and $R_n(x) \to 0$ if $|x| < 1$.

*Example 2*    The binomial series is surprisingly troublesome.

$$(1+x)^m = 1 + mx + \cdots + \frac{m(m-1)\cdots(m-n+1)}{n!} x^n + R_n(x)$$

with

$$R_n(x) = \frac{m(m-1)\cdots(m-n)}{(n+1)!} (1+c)^{m-n-1} x^{n+1}.$$

This formula for $R_n$ is not easy to handle. In Chapter 13 it will be shown that, if $|x| < 1$, then, just as with the geometric series, $\lim R_n(x) = 0$.

**Example 3**    Consider $f(x) = \sin x$. Taylor's formula gives

$$\sin x = x - \frac{x^3}{3!} + \cdots + \frac{(-1)^n x^{2n+1}}{(2n+1)!} + R_{2n+1}(x),$$

$$R_{2n+1}(x) = \frac{f^{(2n+2)}(c)}{(2n+2)!} x^{2n+2} = \frac{(-1)^n \sin c}{(2n+2)!} x^{2n+2}.$$

And now the question is whether, for a *fixed* $x$, $R_{2n+1}(x) \to 0$. Because $|\sin c| \leq 1$,

$$|R_{2n+1}(x)| \leq \frac{|x|^{2n+2}}{(2n+2)!}.$$

To see that this last expression approaches 0, choose $n_0$ so large that $|x|/2n_0 < \frac{1}{2}$. Then, for $n > n_0$,

$$|R_{2n+1}(x)| \leq \frac{x^{2n_0}}{(2n_0)!} \frac{x^{2(n-n_0)+2}}{(2n+2)(2n+1)\cdots(2n_0+1)} < \frac{x^{2n_0}}{(2n_0)!} \frac{x^{2(n-n_0)+2}}{(2n_0)^{2(n-n_0)+2}}$$

$$= \frac{x^{2n_0}}{(2n_0)!} \left(\frac{x}{2n_0}\right)^{2(n-n_0+1)} < \frac{x^{2n_0}}{(2n_0)!} \left(\frac{1}{2}\right)^{2(n-n_0+1)\cdot}$$

The first factor $x^{2n_0}/(2n_0)!$ is fixed, though perhaps large. The second factor approaches 0 as $n \to \infty$. Thus $R_{2n+1}(x) \to 0$ for any $x$. (However, the $n_0$ in the proof will naturally have to be chosen differently for different values of $x$.) We conclude from Taylor's theorem that the Maclaurin series for the sine function converges to $\sin x$ for all $x$.

*Remark*. The Maclaurin series for $\cos x$ can be discussed analogously. In fact, the convergence of the series for $\sin x$ and $\cos x$ in powers of $x - a$ are handled the same way.

**Example 4**    Consider $f(x) = e^x$. Taylor's formula gives

$$e^x = 1 + x + \frac{x^2}{2!} + \cdots + \frac{x^n}{n!} + R_n(x),$$

$$R_n(x) = \frac{e^c}{(n+1)!} x^{n+1}, \qquad \text{where } c \text{ is between 0 and } x.$$

To estimate the magnitude of $R_n(x)$ observe that, if $x > 0$, then $e^c < e^x$, whereas, if $x < 0$, then $e^c < 1$. In either case, for a *fixed* $x$, there is a constant $M$ such that $e^c \leq M$. Then

$$|R_n(x)| \leq M \frac{|x|^{n+1}}{(n+1)!}.$$

Now we can proceed as in Example 3. Choose $n_0$ so large that $|x|/n_0 < \frac{1}{2}$; then, if $n > n_0$

$$|R_n(x)| \leq M \frac{|x|^{n_0}}{n_0!} \frac{|x|^{n-n_0+1}}{(n+1)n\cdots(n_0+1)} < M \frac{|x|^{n_0}}{n_0!} \frac{|x|^{n-n_0+1}}{(n_0)^{n-n_0+1}}$$

$$= M \frac{|x|^{n_0}}{n_0!} \left(\frac{|x|}{n_0}\right)^{n-n_0+1} < M \frac{|x|^{n_0}}{n_0!} \left(\frac{1}{2}\right)^{n-n_0+1},$$

and $R_n(x) \to 0$ as $n \to \infty$.

We conclude from Taylor's theorem that the Maclaurin series for the exponential function converges to $e^x$ for *all* $x$.

*Remark.* Although the series converges to $e^x$ for all $x$, it is not computationally useful for large $x$. For example, even for $x = 10$ some 50 terms would be needed for modest accuracy.

### Problems

1.  Prove (using the method of Example 3) that the Taylor series for $\sin x$ in powers of $x - a$ converges to $\sin x$ for all $x$.

2.  Prove that the Maclaurin series for $\cos x$ converges to the function for all $x$.

3.  Use the identity $\sin x = \sin (a + (x - a))$ to obtain the Taylor expansion of $\sin x$ in powers of $x - a$. Assume that you can add power series as you can polynomials. Where is the expansion valid and why?

4.  Assuming that the exponential series converges to $e^x$ for all $x$, prove that the series for $e^x$ in powers of $x - a$ also converges to $e^x$ for all $x$.

5.  For what values of $x$ does the Maclaurin series for $1/(1 + x^2)$ converge to the function?

6.  Show that

$$\sum_{n=0}^{\infty} (2 \cos^2 \theta)^n = -\sec 2\theta \qquad \text{if } \frac{\pi}{4} < \theta < \frac{3\pi}{4}.$$

7.  Show that if $|x| < 1$, then

$$\sum_{n=0}^{\infty} (n + 2)x^n = \frac{-1}{x - 1} + \frac{1}{(x - 1)^2}.$$

8.  Compute $(\sin 0.1)/(0.1)$ to six decimal places.   [0.998334]

## 4   Analytic Functions

We have seen that many of the elementary functions can be represented by power series—their Taylor series. One might suppose that, for any function that is infinitely differentiable, its Taylor series would converge to it. The following example shows that this is not the case.

**Example 1**   Suppose $f$ is given by

$$f(x) = \begin{cases} e^{-1/x^2}, & \text{if } x \neq 0. \\ 0 & \text{if } x = 0. \end{cases}$$

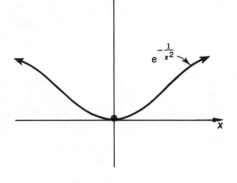

When $x \neq 0$, the function is an elementary function and infinitely differentiable. The hard part is to evaluate the derivatives at 0. If $x \neq 0$

$$f'(x) = \frac{2}{x^3} e^{-1/x^2} \quad \text{and} \quad \lim_{x \to 0} f'(x) = 0,$$

but that does not necessarily mean that $f'(0) = 0$ because we do not know that $f'$ is continuous. We must compute $f'(0)$ from the definition of derivative.

$$f'(0) = \lim_{\Delta x \to 0} \frac{e^{-1/\Delta x^2} - 0}{\Delta x} = 0 \quad \text{(using l'Hospital's rule).}$$

Moreover, it is possible to show that at 0 *all* derivatives are 0.

$$f(0) = f'(0) = f''(0) = \cdots = f^{(n)}(0) = \cdots = 0.$$

Therefore, the Maclaurin series for $f$ is identically zero. Taylor's formula then gives

$$f(x) = e^{-1/x^2} = 0 + 0 + \cdots + 0 + R_n(x) = R_n(x).$$

The entire function is the remainder! Clearly $R_n(x)$ does not approach 0 as $n \to \infty$.

This example raises the question: *Which differentiable functions are represented by their Taylor series?*

The following definition now is pertinent.

**DEFINITION**   A function $f$ is said to be *analytic* at $a$, if it is represented by its Taylor series in a neighborhood of $a$. That is, if there is a positive number $r$ such that

Taylor's series converges to $f(x)$

$$f(x) = \sum_{n=0}^{\infty} \frac{f^{(n)}(a)}{n!} (x - a)^n \qquad \text{if} \qquad |x - a| < r.$$

It is shown in Chapter 13 that sums, products, and quotients (except where the denominator is 0) of analytic functions are also analytic functions. Moreover, the composition of two analytic functions also is an analytic function. The theory of analytic functions constitutes one of the most elegant and satisfying parts of analysis. Within an interval in which the power series converges to the function one can treat the infinite power series almost like a polynomial.* In particular, series for derivatives and antiderivatives can be written at once—and these series converge to the derivative or antiderivative function.

In the examples and problems that follow the reader is to assume that all these theorems just mentioned have been established. The reader is expected to proceed freely, as did the analysts of the eighteenth century, and differentiate, find antiderivatives, add, subtract, multiply, and divide power series to obtain new power series. These examples and problems will demonstrate strength and flexibility of power series. In Chapter 13 we will return to this subject and justify the computations we now make.

**Example 2**    The Maclaurin series for $1/(1 + x^2)$ is

$$\frac{1}{1 + x^2} = 1 - x^2 + x^4 - \cdots + (-1)^n x^{2n} + \cdots,$$

convergent for $|x| < 1$. Since $D \operatorname{Arc} \tan x = 1/(1 + x^2)$, Arc tan $x$ is an antiderivative. Thus,

$$\operatorname{Arc} \tan x = C + x - \frac{x^3}{3} + \frac{x^5}{5} - \cdots + (-1)^n \frac{x^{2n+1}}{2n + 1} + \cdots.$$

All that remains is to calculate the constant $C$. But Arc tan $0 = 0$, whence $C = 0$, and we obtain Gregory's series.

$$\operatorname{Arc} \tan x = x - \frac{x^3}{3} + \frac{x^5}{5} - \cdots = \sum_{n=0}^{\infty} \frac{(-1)^n x^{2n+1}}{2n + 1}.$$

*Remark.* Even though $1/(1 + x^2)$ is infinitely differentiable at all real numbers $x$, the Maclaurin series converges only for the interval $x^2 < 1$. See Chapter 13 for a reason why.

**Example 3**    The binomial series converges to $(1 + x)^m$ if $|x| < 1$,

$$(1 + x)^m = 1 + mx + \frac{m(m - 1)}{2} x^2 + \cdots$$

$$+ \frac{m(m - 1) \cdots (m - n + 1)}{n!} x^n + \cdots.$$

---

**\* Historical Note.** Mathematicians of the late seventeenth and the eighteenth century were in large measure concerned with analytic functions. They sought power series expansions because (perhaps) they could use them just as they would polynomials.

Differentiating the series term by term, we have

$m(1+x)^{m-1}$

$$= m + m(m-1)x + \cdots + \frac{m(m-1)\cdots(m-n+1)}{(n-1)!}x^{n-1} + \cdots$$

$$= m\left[1 + (m-1)x + \cdots + \frac{(m-1)(m-2)\cdots(m-n+1)}{(n-1)!}x^{n-1} + \cdots\right]$$

$$= m(1+x)^{m-1},$$

as it should be.

*Example 4*    Because products of analytic functions are also analytic one should be able to "multiply their Taylor expansions at $a$" to obtain the expansion of the product.

$$\frac{1}{1-x} = 1 + x + x^2 + \cdots + x^n + \cdots,$$

$$\frac{1}{(1-x)^2} = \frac{1}{1-x}\frac{1}{1-x}$$

$$= (1 + x + x^2 + \cdots + x^n + \cdots)(1 + x + x^2 + \cdots + x^n + \cdots)$$

For each power $x^n$ of $x$ there are but finitely many terms in each factor, which, when multiplied together give $x^n$. We must use all of them. Then

$$\frac{1}{(1-x)^2} = 1 + (x+x) + (x^2 + x\cdot x + x^2) + (x^3 + x^2\cdot x + x\cdot x^2 + x^3) + \cdots$$

$$+ (x^n + x^{n-1}x + \cdots + x\cdot x^{n-1} + x^n) + \cdots$$

$$= 1 + 2x + 3x^2 + 4x^3 + \cdots + (n+1)x^n + \cdots = \sum_{n=0}^{\infty}(n+1)x^n.$$

Observe that for each power of $x$ there are but finitely many products that yield that power. Observe also that we have generated the series for the derivative of $1/(1-x)$.

*Example 5*    Because the quotient of two functions analytic at $a$ is a function analytic at $a$ (if the denominator does not vanish at $a$), one should be able to "divide their power series." Unlike division of polynomials where one divides by the highest power first, with power series we begin with the lowest power.

An extremely simple example is provided by the series for $1/(1-x)$.

$$\frac{1}{1-x} = 1 + x + x^2 + \cdots + x^n + \cdots$$

Then $$\frac{1}{1/(1-x)} = \frac{1}{1+x+x^2+\cdots+x^n+\cdots}$$

and "long division" appears below. Hence the quotient series is just $1 - x$, as it should be.

$$
\begin{array}{r}
1 - x \\
\hline
1 + x + x^2 + \cdots + x^n + \cdots \, \big) \, 1 \\
\end{array}
$$

$$
\begin{array}{l}
1 + x + x^2 + \cdots + x^n + \cdots \\
\underline{- x - x^2 - \cdots - x^n - \cdots} \\
- x - x^2 - \cdots - x^n - \cdots \\
\hline
0
\end{array}
$$

**Example 6**    The composite of two analytic functions is an analytic function. Simple substitution should yield the series for the composite function.

Thus, if $f(x) = \dfrac{1}{1 - x}$    and    $x = \sin t$,

then    $f(\sin t) = \dfrac{1}{1 - \sin t} = 1 + \sin t + \sin^2 t + \sin^3 t + \cdots$,

which is a series in powers of $\sin t$. To get a series in powers of $t$ we simply substitute the series for $\sin t$.

$$
\frac{1}{1 - \sin t} = 1 + \left( t - \frac{t^3}{3!} + \frac{t^5}{5!} - \cdots \right) + \left( t - \frac{t^3}{3!} + \frac{t^5}{5!} - \cdots \right)^2
$$

$$
+ \left( t - \frac{t^3}{3!} + \frac{t^5}{5!} - \cdots \right)^3 + \cdots
$$

$$
= 1 + t + t^2 + \frac{5t^3}{6} - \frac{t^4}{3} - \frac{13}{40} t^5 - \cdots.
$$

The coefficient of $t^n$ is troublesome to evaluate.

### Problems

*1.*    Find in two ways the Maclaurin expansion of $e^{2x}$.    $\left[ \displaystyle\sum_{n=0}^{\infty} \frac{(2x)^n}{n!} \right]$

*2.*    Differentiate the Maclaurin series for $\sin x$ and $\cos x$.

*3.*    Find an antiderivative of $\cos x$ from its Maclaurin series.

*4.*    Obtain the Maclaurin series for $1/(1 - x)$ by division.

*5.*    Expand $\sin x/(1 - x)$ in powers of $x$.    $[x + x^2 + \frac{5}{6}x^3 + \frac{5}{6}x^4 + \frac{101}{120}x^5 + \cdots]$

*6.*    Expand $\cos x/(1 + x)$ in powers of $x$.    $[1 - x + \frac{1}{2}x^2 - \frac{1}{2}x^3 + \frac{13}{24}x^4 + \cdots]$

7.  The function defined by $f(x) = (e^x - e^{-x})/2$ is called the *hyperbolic sine*:

$$\sinh x = \frac{e^x - e^{-x}}{2}$$

(See Chapter 13 for the reasons behind this nomenclature.) Obtain its Maclaurin expansion.

$$\left[ x + \frac{x^3}{3!} + \frac{x^5}{5!} + \cdots = \sum_{n=0}^{\infty} \frac{x^{2n+1}}{(2n+1)!} \right]$$

8.  Obtain the expansion in powers of $x$ of the *hyperbolic cosine* $= \cosh x = (e^x + e^{-x})/2$.

$$\left[ 1 + \frac{x^2}{2!} + \frac{x^4}{4!} + \cdots = \sum_{n=0}^{\infty} \frac{x^{2n}}{(2n)!} \right]$$

9.  What is the derivative of the hyperbolic sine? of the hyperbolic cosine?
$$[D \sinh = \cosh, \ D \cosh = \sinh]$$

10. Obtain, in two ways, three non-zero terms in the expansion of $\sin^2 x$ in powers of $x$.

(Hint: $\sin^2 x = (1 - \cos 2x)/2$.)
$$\left[ x^2 - \frac{2^3 x^4}{4!} + \frac{2^5 x^6}{6!} - \cdots \right]$$

11. Expand $e^{x^2}$ in powers of $x$. Differentiate and show that the differentiated series is obtained from the series for $e^{x^2}$ by multiplying by $2x$.

12. Expand $\sqrt{1 - 4x}$ in powers of $x$.
$$\left[ 1 - 2 \sum_{n=1}^{\infty} \frac{(2n-2)!}{n!(n-1)!} x^n \right]$$

13. Differentiate the Maclaurin series for $\sqrt{1 - x}$. Identify the series.
$$\left[ -\tfrac{1}{2}(1-x)^{-1/2} = -\tfrac{1}{2}\left( 1 + \tfrac{1}{2}x + \frac{1 \cdot 3}{2 \cdot 4} x^2 + \frac{1 \cdot 3 \cdot 5}{2 \cdot 4 \cdot 6} x^3 + \cdots \right) \right]$$

14. Obtain the expansion of Arc sin $x$ from the series for its derivative.
$$\left[ x + \frac{1}{2} \frac{x^3}{3} + \frac{1 \cdot 3}{2 \cdot 4} \frac{x^5}{5} + \frac{1 \cdot 3 \cdot 5}{2 \cdot 4 \cdot 6} \frac{x^7}{7} + \cdots \right]$$

*15. Verify the identity $\sin (a + x) = \sin a \cos x + \cos a \sin x$ from the Maclaurin series for $\sin a$, $\cos a$, $\sin x$, and $\cos x$.

16. Expand $e^{x+(x^2/2)}$ in powers of $x$. Obtain the series in two ways, one by substitution in the exponential series, the other by multiplying two series.

17. Use Taylor's theorem to obtain four non-zero terms in the expansion of $\sec x$ in powers of $x$. Verify your answer by division, using the cosine series.
$$[1 + \tfrac{1}{2}x^2 + \tfrac{5}{24}x^4 + \tfrac{61}{720}x^6 + \cdots]$$

18. Obtain in two ways three non-zero terms in the Maclaurin expansion of $\tan x$.
$$[x + \tfrac{1}{3}x^3 + \tfrac{2}{15}x^5 + \tfrac{17}{315}x^7 + \cdots]$$

19. From the answer to Problem 17 obtain four non-zero terms of the Maclaurin series for log (sec $x$ + tan $x$). $[x + \frac{1}{6}x^3 + \frac{1}{24}x^5 + \frac{61}{5040}x^7 + \cdots]$

20. From the answer to Problem 18 obtain three non-zero terms of the Maclaurin series for log sec $x$. $[\frac{1}{2}x^2 + \frac{1}{12}x^4 + \frac{1}{45}x^6 + \frac{17}{2520}x^8 + \cdots]$

21. Expand $1/(1 - x + x^2)$ in powers of $x$.
$$[1 + x - x^3 - x^4 + x^6 + x^7 - \cdots + (-1)^n x^{3n} + (-1)^n x^{3n+1} + \cdots]$$

(*Hint:* Besides division there is another method based on simple algebra.)

22. Expand $x/(1 - x - 2x^2)$ in powers of $x$ up to the fourth power.
$$[x + x^2 + 3x^3 + 5x^4 + \cdots]$$

23. Give six non-zero terms in the expansion of log $(1 - x + x^2)$ in powers of $x$.
$$[\log (1 + x^3) - \log (1 + x) = -x + \frac{1}{2}x^2 + \frac{2}{3}x^3 + \frac{1}{4}x^4 - \frac{1}{5}x^5 - \frac{1}{3}x^6 + \cdots]$$

24. Give the Maclaurin series for log $(x + \sqrt{1 + x^2})$.

(*Hint:* Consider its derivative.) $$\left[ x - \frac{1}{2}\frac{x^3}{3} + \frac{1 \cdot 3}{2 \cdot 4}\frac{x^5}{5} - \frac{1 \cdot 3 \cdot 5}{2 \cdot 4 \cdot 6}\frac{x^7}{7} + \cdots \right]$$

25. Obtain five terms of the Maclaurin series of $e^{\sin x}$. $[1 + x + \frac{1}{2}x^2 - \frac{1}{8}x^4 + \cdots]$

26. Obtain four non-zero terms in the expansion of $\sqrt{1 + \sin x}$ in powers of $x$.
$$[1 + \frac{1}{2}x - \frac{1}{8}x^2 - \frac{1}{48}x^3 - \cdots]$$

In Problems 27 to 31 identify functions with the given series expansions—not necessarily power series.

27. $(x - 1) - \frac{1}{2}(x - 1)^2 + \frac{1}{3}(x - 1)^3 - \cdots$ $[\log x]$

28. $1 + \dfrac{1}{1 - x} + \dfrac{1}{(1 - x)^2} + \dfrac{1}{(1 - x)^3} + \cdots$ $\left[ 1 - \dfrac{1}{x} \right]$

29. $1 + 2\cos^2 x + 3\cos^4 x + 4\cos^6 x + \cdots$ $[\csc^4 x]$

30. $1 + mx + \dfrac{m(m + 1)}{2!}x^2 + \dfrac{m(m + 1)(m + 2)}{3!}x^3 + \cdots$ $[(1 - x)^{-m}]$

31. $(e^{-x} - 1) - \frac{1}{2}(e^{-x} - 1)^2 + \frac{1}{3}(e^{-x} - 1)^3 - \cdots$ $[-x]$

PART II

*INTEGRAL CALCULUS*

Gottfried Wilhelm von Leibniz
(1646–1716)
Born in Leipzig July 1, 1646
Entered the University of Leipzig, 1661
Legal study, 1663–1666, Doctor of Laws, 1666
Entered service with the elector of Mainz 1667
Went to Paris 1672 and began study of
mathematics under the physicist C. Huygens.
Moved to Hanover 1676 to begin forty years of
diplomatic service for the family of the Duke of
Brunswick.
Had essentially completed the invention of calculus
1677. First printed account of the calculus (six
pages) in Acta Eruditorum, 1684.
Devotion to philosophy 1690–1716.

A man of universal genius, the variety of his intellectual effort is truly astounding. He left a huge collection of manuscripts and papers that even today are not fully explored. (However, the steps in his evolution of calculus symbolism are well documented.) In his essay of 1666, *De arte combinatoria,* he conceived of a "universal logical calculus" that would reduce reasoning to calculation with symbols. This work anticipates the modern development of symbolic logic by 200 years!

He invented a calculating machine in 1671, an improvement on Pascal's, and about this time invented the binary system, which represents numbers in base two instead of ten. He saw clearly its convenience (as is the case today) for computing machinery.

His brilliance in law may have been his scientific undoing for from his essays on law he came to the attention of, and entered the legal and diplomatic service of, a succession of petty royalty. He had a marvelous capacity for work in all circumstances, and he travelled much on various political missions. In 1672 Leibniz proposed (in order to distract Louis XIV from an attack on Germany) that the French invade Egypt—this more than a century before Napoleon's ill-fated invasion! In a somewhat different direction he proposed the unification of the Protestant and Catholic churches in 1686.

Leibniz is a central figure in the history of philosophy and would be in the history books on the basis of his philosophical writing alone. In fact, his philosophical work extends over a period of fifty years. As with Newton, his active mathematical and scientific life is confined to his younger years. The reader is referred to books on philosophy for an appreciation of his intricate view of the world.

## CHAPTER 9

# THE DEFINITE INTEGRAL

## 1   The Problem of Area

In the middle of the seventeenth century mathematicians were concerned with the "problem of area" as
well as with the "problem of tangents." They wished
to find the area under the graphs of functions because
they saw the wide applicability of a simple process for
finding such areas. The reader should recall the
numerous applications of the derivative, a concept
that came out of the problem of tangents. Analogous

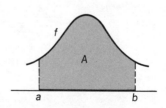

to this will be the applications of the definite integral, a concept that arises from the
problem of area.

In our development we shall assume that the reader knows "all about areas
in the plane" from his previous studies. And we shall use this intuition about area*
to introduce the concept of the definite integral. To illustrate the method we shall
find the area of the region under the graph of $y = x$, above the $x$-axis and between

---

* Area of arbitrary subsets of the plane is a sophisticated concept, much more complex than might
be expected.

$x = 0$ and $x = a$. The area is obvious from elementary geometry, but the method used here for computing it is new and is applicable to other areas.

*Example 1*    To find the area of the triangular region we first approximate the area by sums of areas of rectangles. For any natural number $n$ divide the interval $[0, a]$ into $n$ intervals of length $h = a/n$, as shown in the figure.

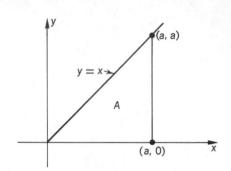

One set of rectangles on these intervals is "inscribed" in the triangular region, and one set "circumscribes" the region.

The area $s_n$ of the inscribed region is

(1)  $s_n = 0 + h \cdot h + (2h)h + (3h)h$

$\qquad + \cdots + [(n-1)h]h$

$\qquad = [1 + 2 + 3 + \cdots$

$\qquad\qquad + (n-1)]h^2$

$\qquad = \dfrac{a^2}{n^2} \displaystyle\sum_{i=1}^{n-1} i.$

The area $S_n$ of the circumscribed region is

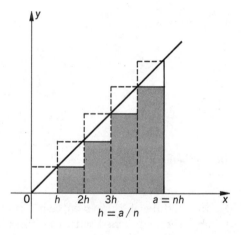

(2)  $S_n = h \cdot h + (2h)h + \cdots + (nh)h$

$\qquad = (1 + 2 + 3 + \cdots + n)h^2$

$\qquad = \dfrac{a^2}{n^2} \displaystyle\sum_{i=1}^{n} i.$

Then, if $A$ is the area of the triangular region, it is obvious that*

(3)  $$s_n < A < S_n.$$

It is intuitively clear that both $s_n$ and $S_n$ are good approximations to $A$ when $n$ is large. Our problem is that of evaluating the limits of $s_n$ and $S_n$ as $n \to \infty$.

Fortunately, there is a formula from algebra that is exactly what we require. The sum (an arithmetic series) of the first $k$ natural numbers is

$$\sum_{i=1}^{k} i = 1 + 2 + 3 + \cdots + k = \frac{k}{2}(k+1).$$

---

* Obvious on the basis of natural belief about area.

Then, using this in $s_n$ and $S_n$, we get

$$s_n = \frac{a^2}{n^2} \frac{(n-1)(n)}{2} = \frac{a^2}{2}\left(1 - \frac{1}{n}\right);$$

$$S_n = \frac{a^2}{n^2} \frac{n(n+1)}{2} = \frac{a^2}{2}\left(1 + \frac{1}{n}\right).$$

Now the limits of $s_n$ and $S_n$ as $n \to \infty$ are clear, and from (3) we have*

$$\lim_{n \to \infty} s_n = \frac{a^2}{2} \leq A \leq \lim_{n \to \infty} S_n = \frac{a^2}{2}.$$

The area must be equal to $a^2/2$.

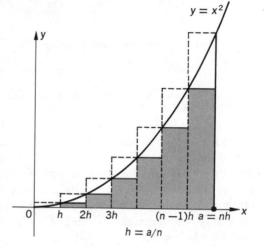

**Example 2**    We use the same method to find the area of the region under the parabola† $y = x^2$, above the $x$-axis and between $x = 0$ and $x = a$, where $a > 0$.

Divide, as in Example 1, the interval $[0, a]$ into $n$ intervals of length $h = a/n$, as shown in the figure.

One set of rectangles on these intervals is inscribed in the parabolic region, and one set circumscribes the region.

The area of the inscribed region is

$$\begin{aligned}
(4) \quad s_n &= 0 + (h^2)h + (2h)^2h + (3h)^2h + \cdots + [(n-1)h]^2 h \\
&= [1 + 2^2 + 3^3 + \cdots + (n-1)^2]h^3 \\
&= \frac{a^3}{n^3} \sum_{i=1}^{n-1} i^2.
\end{aligned}$$

The area of the circumscribed region is

$$\begin{aligned}
(5) \quad S_n &= (h^2)h + (2h)^2h + \cdots + (nh)^2h \\
&= (1 + 2^2 + 3^2 + \cdots + n^2)h^3 \\
&= \frac{a^3}{n^3} \sum_{i=1}^{n} i^2.
\end{aligned}$$

---

* The number we want, namely the area $A$, is "squeezed" between $\lim s_n$ and $\lim S_n$.

† The problem for the parabola was solved by Archimedes (who died 212 BC) by a different method which is, nevertheless, similar in spirit!

If $A$ is the desired area, it is obvious that

$$s_n < A < S_n.$$

To obtain $s_n$ and $S_n$ in more convenient form we use the following formula*

$$\sum_{i=1}^{k} i^2 = \tfrac{1}{6} k(k+1)(2k+1).$$

Then   $$s_n = \frac{a^3}{n^3} \frac{(n-1)n(2n-1)}{6} = \frac{a^3}{3}\left(1 - \frac{3}{2n} + \frac{1}{2n^2}\right),$$

$$S_n = \frac{a^3}{n^3} \frac{n(n+1)(2n+1)}{6} = \frac{a^3}{3}\left(1 + \frac{3}{2n} + \frac{1}{2n^2}\right).$$

The area is squeezed between $\lim s_n$ and $\lim S_n$ and is

$$A = \tfrac{1}{3}a^3.$$

In these examples it is not necessary to use both $s_n$ and $S_n$ because they have the same limit.

The areas of Examples 1 and 2, and also areas under graphs of cubic, quartic, and higher powers, were known before Newton. In the seventeenth century the problem for other functions was the real challenge. The simplicity of Examples 1 and 2, and the problems below suggested that there must be a shortcut.

Here is another example where the necessary algebra is pleasantly easy.

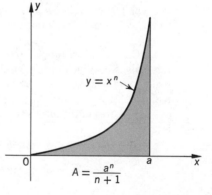

$y = x^n$

$$A = \frac{a^n}{n+1}$$

---

* A *proof* of this formula can be given by induction, but induction does not help one to *discover* the formula. The following trick gives the formula and generalizes to sums of higher powers. Observe that

$$(\checkmark) \quad \sum_{i=1}^{k} [(i+1)^3 - i^3] = (1+1)^3 - 1^3 + (2+1)^3 - 2^3 + (3+1)^3 - 3^3 + \cdots + (k+1)^3 - k^3$$
$$= (k+1)^3 - 1^3 = k^3 + 3k^2 + 3k$$

because of "telescoping." But also one has

$$(\checkmark\checkmark) \qquad \sum_{i=1}^{k} [(i+1)^3 - i^3] = \sum_{i=1}^{k} [3i^2 + 3i + 1]$$

$$= 3\sum_{i=1}^{k} i^2 + 3\frac{k(k+1)}{2} + k.$$

Solving $(\checkmark)$ and $(\checkmark\checkmark)$ for $\sum i^2$ will yield the desired formula.

*Example 3*   Find the area under the graph of $y = e^x$ between $x = 0$ and $x = a,\ a > 0$.

We shall use only the inscribed rectangles. Then with $h = \Delta x = a/n$ we have

$$s_n = e^0\,\Delta x + e^{\Delta x}\,\Delta x + e^{2\Delta x}\,\Delta x + \cdots + e^{(n-1)\Delta x}\,\Delta x$$

$$= (1 + e^{\Delta x} + e^{2\Delta x} + \cdots + e^{(n-1)\Delta x})\,\Delta x.$$

Fortunately, the expression in parentheses is a geometric series with ratio $e^{\Delta x}$ (see Appendix A, page 492). Then

$$s_n = \frac{1 - e^{n\Delta x}}{1 - e^{\Delta x}}\,\Delta x = (1 - e^a)\,\frac{\Delta x}{1 - e^{\Delta x}}.$$

The limit of $s_n$ as $n \to \infty$ is

$$\lim_{\Delta x \to 0} (1 - e^a)\,\frac{\Delta x}{1 - e^{\Delta x}}.$$

The indeterminate form

$$\lim_{\Delta x \to 0} \left(\frac{\Delta x}{1 - e^{\Delta x}}\right)$$

is easily evaluated using l'Hospital's rule:

$$\lim_{\Delta x \to 0} \frac{\Delta x}{1 - e^{\Delta x}} = \lim_{\Delta x \to 0} \frac{1}{-e^{\Delta x}} = -1.$$

Therefore,          $\text{Area} = \lim_{n \to \infty} s_n = e^a - 1.$

## Problems

In Problems 1 to 10 find the area of the region under the graph of the function, above the x-axis and between the given values of x. Use either inscribed or circumscribed rectangles and the method of the examples.

1.  $y = 3x$          ;   $x = 0$ and $x = 2$          $[6]$

2.  $y = \frac{1}{3}x^2$          ;   $x = 0$ and $x = 2$          $\left[\frac{8}{9}\right]$

3.  $y = 2x$          ;   $x = 0$ and $x = a$          $[a^2]$

✓4.  $y = cx$          ;   $x = 0$ and $x = a$          $\left[\frac{1}{2}ca^2\right]$

5.  $y = 3x^2$          ;   $x = 0$ and $x = x_0$          $[x_0^3]$

6.  $y = 1 - x^2$          ;   $x = 0$ and $x = 1$          $\left[\frac{2}{3}\right]$

7.  $y = x^2 + x$          ;   $x = 0$ and $x = x_0$          $\left[\frac{1}{3}x_0^3 + \frac{1}{2}x_0^2\right]$

✓8.  $y = x^2 + x + 1$;   $x = a$ and $x = b$          $\left[\frac{1}{3}(b^3 - a^3) + \frac{1}{2}(b^2 - a^2) + (b - a)\right]$

9.  $y = e^x$      ;   $x = -a$ and $x = a$          $[e^a - e^{-a}]$

10.  $y = e^{2x}$      ;   $x = 0$ and $x = a$          $[\frac{1}{2}e^{2a} - \frac{1}{2}]$

11.  Find, in any way you choose, the area under the graph of $y = 2x^2 - x + 1$, above the $x$-axis, and between $x_1$ and $x_2$.
$$[\tfrac{2}{3}(x_2^3 - x_1^3) - \tfrac{1}{2}(x_2^2 - x_1^2) + (x_2 - x_1)]$$

12.  Use answers to Examples 1 and 2 to find the area underneath the parabola $y = x - x^2$ and above the $x$-axis.      $[\frac{1}{6}]$

13.  Find the area of the parabolic sector bounded by $4y = 3x^2$ and $3x - 4y + 6 = 0$.
$$[\tfrac{27}{8}]$$

*14.  Find the area under the graph of $y = \sqrt{2x}$ between $x = 0$ and $x = 2$.
$$[\tfrac{8}{3}]$$

*15.  Use the device of the footnote for Example 2 to show that
$$\sum_{i=1}^{k} i^3 = \frac{k^2(k+1)^2}{4}$$

**Hint: Begin with**      $\displaystyle\sum_{i=1}^{k} [(i+1)^4 - i^4]$.

*16.  By analogy with Problem 15 obtain a formula for $\displaystyle\sum_{i=1}^{k} i^4$.
$$\left[\frac{k(k+1)(2k+1)(3k^2 + 3k - 1)}{30}\right]$$

17.  Obtain the area under $y = x^3$ between $x = 0$ and $x = x_0$. Use the method of the examples and the formula of Problem 15.      $[\frac{1}{4}x_0^4]$

18.  Find the area under the graph of $y = cx^3\ (c > 0)$ and between $x = x_1$ and $x = x_2$.      $[\frac{1}{4}c(x_2^4 - x_1^4)]$

19.  Use Problem 16 to obtain the area under the graph of $y = x^4$ between $x = 0$ and $x = x_0$.      $[\frac{1}{5}x_0^5]$

*20.  In Example 2 how large should $n$ be in order that $A - s_n$ be less than 0.001? less than $10^{-5}$?      $[500a^3,\ 10^5a^3/2]$

*21.  Set up a sum of areas of rectangles that approximates the area in the first quadrant inside the ellipse $b^2x^2 + a^2y^2 = a^2b^2$.

       Then compare this sum for the sum for a corresponding set of rectangles inscribed in the part of the circle $x^2 + y^2 = a^2$ that lies in the first quadrant. From this comparison determine the area of the ellipse.      $[\pi ab]$

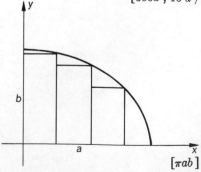

## 2  *The Definite Integral*

The method we used in Section 1 for finding areas involved finding the limits of certain sums. We now generalize and *forget about the area interpretation.*

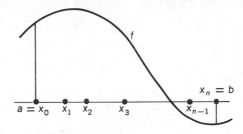

(i) Suppose $f$ is a continuous function on an interval $[a, b]$.

(ii) If $n$ is a positive integer, choose $n + 1$ points

$$x_0 = a < x_1 < \cdots < x_n = b.$$

This is a *subdivision* of $[a, b]$ that subdivides $[a, b]$ into $n$ subintervals of lengths

$$\Delta x_i = x_i - x_{i-1}, \qquad i = 1, \cdots, n.$$

The *mesh* of the subdivision is the greatest length of the subintervals

$$\text{mesh} = \max \Delta x_i, \qquad i = 1, \cdots, n.$$

(iii) In each of the subintervals choose any point $\xi_i$

$$x_{i-1} \leqq \xi_i \leqq x_i, \qquad i = 1, \cdots, n,$$

and form the sum

(1) $$\sum_{i=1}^{n} f(\xi_i)\, \Delta x_i = f(\xi_1)\, \Delta x_1 + \cdots + f(\xi_n)\, \Delta x_n, \quad \{\text{a Riemann sum.}$$

Such a sum is called a *Riemann\* sum* for the function $f$ and the interval $[a, b]$.

We now investigate what happens to the Riemann sum

$$\sum_{i=1}^{n} f(\xi_i)\, \Delta x_i$$

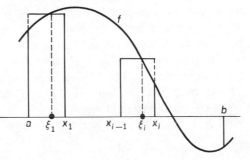

as $n$ increases and the mesh approaches zero. In the case where $f$ is positive the considerations of Section 1 lead us to believe that the Riemann sums approach a limit, which is the area under the graph of $f$.

---

\* After G. F. B. Riemann (1826–1866), one of the most original of mathematicians and the founder of what is today called Riemannian geometry.

Before proceeding to discuss this limit we generalize a bit more. We have assumed that $a < b$. But if $a > b$ we can still subdivide the interval with

$$a = x_0 > x_1 > x_2 > \cdots > x_n = b.$$

Then all the increments are negative:

$$\Delta x_i = x_i - x_{i-1} < 0, \qquad i = 1, \cdots, n.$$

Thus, the terms $f(\xi_i)\,\Delta x_i$ in the Riemann sum will have signs depending on the sign of $f(\xi_i)$ and $\Delta x_i$. In this case the mesh is the maximum of $|\Delta x_i|$:

$$\text{mesh} = \max|\Delta x_i|, \qquad i = 1, \cdots, n.$$

Finally, for completeness, we permit the case where $a = b$, and define the Riemann sum to be 0.

**DEFINITION**   Suppose there is a number $I$ such that the Riemann sums approach $I$, regardless of how the points $\xi_i$ are chosen, as the mesh approaches* zero. This number $I$ is the *definite integral* of $f$ from $a$ to $b$, and we denote this limit by

$$(2) \qquad I = \int_a^b f(x)\,dx = \lim_{\text{mesh}\to 0} \sum_{i=1}^n f(\xi_i)\,\Delta x_i.$$

The symbol "$\int$" is called an *integral sign*.† The numbers $a$ and $b$ are the *lower* and *upper limits* of the integral, respectively. The whole symbol is read as

$$(3) \qquad \int_a^b f(x)\,dx = \text{"the integral from } a \text{ to } b \text{ of } f \text{ of } x\,dx."$$

The function $f$ is called the *integrand function*.

The symbols $x$ and $dx$ in (3) are "dummy variables" because any letter would do as well. The integral clearly depends only on the function $f$ and the interval. It does not depend on the symbol used to represent the independent variable. This being the case, the reader may wonder why the $x$ and $dx$ are retained when we could simply write

$$\int_a^b f.$$

The integral depends only on $f$, and $a$ and $b$.

$$\int_a^b f = \int_a^b f(x)\,dx$$

$$= \int_a^b f(t)\,dt$$

---

* If the mesh $\to 0$, then necessarily $n \to \infty$, but not conversely.

† **Historical Note.** In the seventeenth century Leibniz and others regarded the area under curves as "an infinite sum of infinitely thin rectangles." The integral sign was first used by Leibniz on October 25, 1676. He regarded it as an elongated S, standing for "sum."

In 1690 Jakob Bernoulli first used the word "integral." In 1696 Leibniz and Bernoulli agreed upon the term "integral calculus."

Obviously the $dx$ is a reminder of the increment $\Delta x_i$, and this will be useful in applications. Also, the $dx$ will be important when we seek antiderivatives in Section 5.

The simple examples of Section 1 indicate that for at least some functions the definite integral exists—that is, the Riemann sums have a limit as the mesh goes to zero. The same is true for any continuous function.

**THEOREM**   *If $f$ is continuous in the interval $[a, b]$, then*

$$\int_a^b f(x)\ dx \quad exists.$$

*Remarks about a proof.* In finding area, we approximated the area by sums of areas of rectangles, obtaining areas of inscribed and circumscribed regions. The same idea applies to the integral of any continuous function.

If the points $\xi_i'$ are chosen so that, for each $i$, $f(\xi_i')$ is the maximum of $f$ on the interval $[x_{i-1}, x_i]$, then the sum

$$S = \sum_{i=1}^n f(\xi_i')\ \Delta x_i$$

is called the *upper Riemann sum for this subdivision.*

If the points $\xi_i''$ are chosen so that $f(\xi_i'')$, for each $i$, is the minimum of $f$ on the interval $[x_{i-1}, x_i]$, then the sum

$$s = \sum_{i=1}^n f(\xi_i'')\ \Delta x_i$$

is called the *lower Riemann sum for this subdivision.*

The main idea of the proof of the theorem consists in showing that as the mesh of the subdivision approaches zero the upper and lower Riemann sums have a common limit. The details are troublesome. However, if the function $f$ is increasing (or decreasing) on $[a, b]$, a proof is much easier, and we give below a proof for that case.

**PROOF FOR f INCREASING**   Choose a subdivision of the interval. Then, because $f$ is increasing, $\xi_i' = x_i$ and $\xi_i'' = x_{i-1}$. Therefore,

$$s_n = \sum_{i=1}^n f(x_{i-1})\ \Delta x_i$$

is a *lower* Riemann sum; and

$$S_n = \sum_{i=1}^{n} f(x_i)\,\Delta x_i$$

is an *upper* Riemann sum. Let

$$\sum_{i=1}^{n} f(\xi_i)\,\Delta x_i$$

be *any* Riemann sum. Because $f$ is increasing we have

$$s_n \leqq \sum_{i=1}^{n} f(\xi_i)\,\Delta x_i \leqq S_n.$$

Now consider the difference

$$S_n - s_n = \sum_{i=1}^{n} \left[\,f(x_i) - f(x_{i-1})\,\right]\Delta x_i.$$

Because $f$ is increasing $f(x_i) - f(x_{i-1}) \geqq 0$ and because $a < b$, $\Delta x_i > 0$. Therefore, if we replace each $\Delta x_i$ by the mesh $\Delta$, then

(4) $$S_n - s_n \leqq \sum_{i=1}^{n} \left[\,f(x_i) - f(x_{i-1})\,\right]\Delta = \left[\,f(b) - f(a)\,\right]\Delta$$

because of the "telescoping" of terms. The situation is shown in the figure. The contributions to the difference $S - s$ from each subinterval, when "slid over to the right" are contained in a slender rectangle of height $f(b) - f(a)$. Upper and lower Riemann sums differ by very little when the mesh $\Delta$ is small.

Now we observe that if a new point of subdivision is added, say $x_i^*$ to the interval $[x_{i-1}, x_i]$, then the lower sum is increased and the upper sum is decreased. To see that this is so, examine the figures at the right.

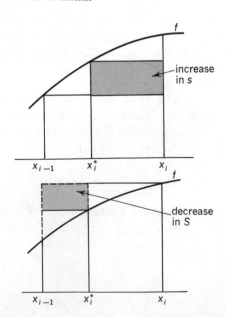

Thus, as more points are added to a subdivision the lower sums increase and the upper sums decrease. This fact, coupled with the inequality (4) can be used to show that the Riemann sums do have a limit as the mesh $\Delta \to 0$.

*Remark.* The theorem assumes that $a < b$. Clearly, a similar proof could be made with $a > b$. The case $a = b$ is trivial, since all Riemann sums are zero.

*Example* Compute upper and lower Riemann sums for

$$\int_0^1 \frac{dx}{x^2 + 1}$$

with $n = 4$ and equal subintervals.

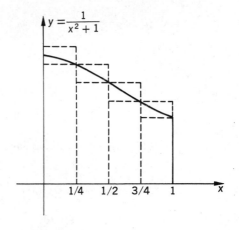

$y = \dfrac{1}{x^2 + 1}$

$$S_4 = \frac{1}{0 + 1}\frac{1}{4} + \frac{1}{\frac{1}{16} + 1}\frac{1}{4}$$

$$+ \frac{1}{\frac{1}{4} + 1}\frac{1}{4} + \frac{1}{\frac{9}{16} + 1}\frac{1}{4}$$

$$= \frac{1437}{1700}.$$

$$s_4 = \frac{1}{\frac{1}{16} + 1}\frac{1}{4} + \frac{1}{\frac{1}{4} + 1}\frac{1}{4} + \frac{1}{\frac{9}{16} + 1}\frac{1}{4} + \frac{1}{1 + 1}\frac{1}{4} = \frac{2449}{3400}.$$

### Problems

In Problems 1 to 8 compute the lower and upper Riemann sums for the integral for the indicated subdivision.

1. $\displaystyle\int_0^1 (x^2 + 1)\, dx;$  $n = 2$ equal subintervals.  $\left[s = \frac{9}{8},\ S = \frac{13}{8}\right]$

2. $\displaystyle\int_0^1 (x^2 + 1)\, dx;$  $n = 4$ equal subintervals.  $\left[s = \frac{39}{32},\ S = \frac{21}{16}\right]$

3. $\displaystyle\int_0^1 (x^2 + 1)\, dx;$  $n = 1.$  $\left[s = 1,\ S = 2\right]$

4. $\displaystyle\int_{-1}^0 e^x\, dx$  ;  $n = 5$ equal subintervals.  $\left[s = 0.571,\ S = 0.697\right]$

5. $\displaystyle\int_{-1}^1 \frac{t\, dt}{t^2 + 1}$  ;  $n = 8$ equal subintervals.  $\left[s = -\frac{1}{8},\ S = \frac{1}{8}\right]$

6. $\displaystyle\int_0^\pi \sin x\, dx$  ;  $n = 4$ equal subintervals.  $\left[\begin{array}{l}s = \pi/2\sqrt{2}, \\ S = \pi(\sqrt{2} + 1)/2\sqrt{2}\end{array}\right]$

7. $\displaystyle\int_1^2 x^2\, dx$  ;  $x_0 = 1,\ x_1 = 1.2,\ x_2 = 1.4,\ x_3 = 1.5,\ x_4 = 1.6,$
$x_5 = 1.7,\ x_6 = 1.8,\ x_7 = 1.9,\ x_8 = 2$

$\left[s = 2.139,\ S = 2.339\right]$

8.  $\displaystyle\int_0^2 \frac{dx}{1+x}$    ;    $x_0 = 0$, $x_1 = 0.1$, $x_2 = 0.2$, $x_3 = 0.4$, $x_4 = 0.6$,

$x_5 = 0.8$, $x_6 = 1$, $x_7 = 1.3$, $x_8 = 1.6$, $x_9 = 2$

$[s = 1.0323, \; S = 1.1709]$

9.  In Problem 1 bisect the first subinterval and see what happens to $s$ and $S$.

10.  Compare $s$ and $S$ for Problems 1 and 3. See the proof of the theorem.

11.  Without working hard, what is the value of $\displaystyle\int_{-1}^1 \frac{t}{t^2+1}\,dt$?

*12.  Does the integral $\displaystyle\int_{-1}^1 \frac{dx}{x}$ exist? Show that the Riemann sums can be arbitrarily

large. The integrand function is not defined at $x = 0$, but one can still form Riemann sums.

13.  Draw a region whose area is the integral $\displaystyle\int_{-1}^2 \sqrt{x+1}\,dx$.

In Problems 14, 15, and 16 make an estimate of the integral $\displaystyle\int_a^b f\,(x)\,dx$.

14.                          15.                          16.

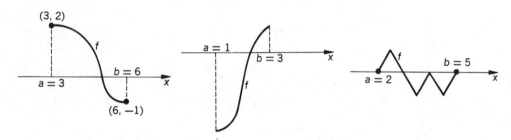

Some discontinuous functions have integrals. This can happen if the functions are not too

badly discontinuous. In Problems 17, 18, and 19 estimate $\displaystyle\int_a^b f(x)\,dx$.

17.                          18.                          19.

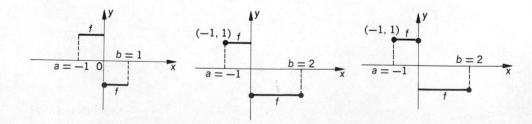

*20.   Prove that $\displaystyle\int_a^b f(x)\,dx = -\int_b^a f(x)\,dx.$

## 3  Properties of the Definite Integral

In this section we examine general properties of the integral. The first theorem gives four simple properties that are valid for any integrable functions, where by integrable on an interval we mean only that the definite integral of the function over that interval exists.

**THEOREM 1**    *If f and g are integrable from a to b, then*

(i)   $\displaystyle\int_a^b cf(x)\,dx = c\int_a^b f(x)\,dx,\quad c = constant;$

(ii)   $\displaystyle\int_a^b f(x)\,dx = -\int_b^a f(x)\,dx;$

(iii)   $\displaystyle\int_a^b [f(x) + g(x)]\,dx = \int_a^b f(x)\,dx + \int_a^b g(x)\,dx;$

(iv)   *If c is a number between a and b, then**

$$\int_a^b f(x)\,dx = \int_a^c f(x)\,dx + \int_c^b f(x)\,dx.$$

The proofs of all four assertions are rather easy consequences of the definition of the definite integral. Proofs of (i), (ii), and (iii) are left as exercises for the reader. A proof of (iv) (restricted to the case of a continuous $f$) is as follows.

**PROOF**    In the first place, we may assume that the point $c$ is one of the points of subdivision, say $c = x_j$, because given any subdivision we can add, if necessary, this additional point. Then a Riemann sum is

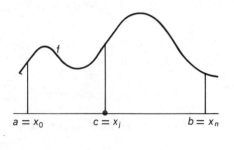

(1)   $\displaystyle\sum_{i=1}^{n} f(\xi_i)\,\Delta x_i$

$$= \sum_{i=1}^{j} f(\xi_i)\,\Delta x_i + \sum_{i=j+1}^{n} f(\xi_i)\,\Delta x_i.$$

$a = x_0$     $c = x_j$     $b = x_n$

---

* Property (iv) is also true if $c$ is not between $a$ and $b$. If, for example, $a < b < c$, then by (iv)

$$\int_a^b f(x)\,dx = \int_a^b f(x)\,dx + \int_b^c f(x)\,dx = \int_a^c f(x)\,dx - \int_c^b f(x)\,dx.$$

Then, transposing the last integral, we obtain (iv) for this case.

As the mesh $\Delta \to 0$ each of the above sums approaches a definite integral, because $f$ is continuous and we get the desired equality (iv).

**COROLLARY 1**     *If on an interval* $[-a,\, a]$

$$f(-x) = -f(x) \qquad \text{for all } x,$$

then     $$\int_{-a}^{a} f(x)\, dx = 0.$$

(Such a function is called an *odd* function.)

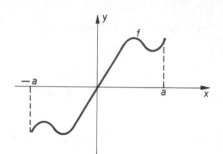

an odd function

**COROLLARY 2**     *If on an interval* $[-a,\, a]$

$$f(-x) = f(x) \qquad \text{for all } x,$$

then     $$\int_{-a}^{a} f(x)\, dx = 2 \int_{0}^{a} f(x)\, dx.$$

(Such a function is called an *even* function.)

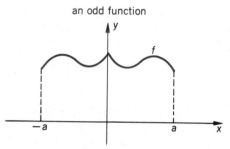

an even function

Proofs are left as exercises.

Highly important for the next section is the following theorem, the *Mean Value Theorem of the Integral Calculus*, which we state for the case where $a < b$, although it is also valid for $a > b$.

**THEOREM 2**     (MVT of the integral calculus). *If $f$ is continuous on $[a,\, b]$, then there is a number $c$ such that*

$$\int_{a}^{b} f(x)\, dx = (b - a)\, f(c), \qquad \text{where} \quad a \leqq c \leqq b.$$

**PROOF**     Let $M$ and $m$ be the maximum and minimum values of $f$ on $[a,\, b]$. Then, for each Riemann sum,

$$m(b - a) \leqq \sum_{i=1}^{n} f(\xi_i)\, \Delta x_i \leqq M(b - a).$$

Therefore, as the mesh goes to 0, we approach the integral and get

$$m(b - a) \leqq \int_{a}^{b} f(x)\, dx \leqq M(b - a).$$

Therefore, the integral is equal to $K(b - a)$, where $m \leqq K \leqq M$:

$$\int_{a}^{b} f(x)\, dx = K(b - a), \qquad \text{where} \quad m \leqq K \leqq M.$$

Because $f$ is continuous it assumes all values between its minimum and maximum.* Therefore, there is a number $c$, $a \leq c \leq b$ such that $K = f(c)$. This proves the theorem.

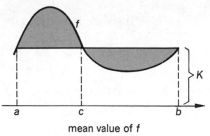

mean value of $f$

**DEFINITION**    The number $K$ is the "average height" of $f$ on the interval. It is called the *mean value of $f$ over* $[a, b]$.

The mean value of $f$ on $[a, b] = \dfrac{1}{b - a} \displaystyle\int_a^b f(x) \, dx$.

### Problems

1.    Give a geometric argument to prove that $\displaystyle\int_a^b 2x \, dx = 2 \int_a^b x \, dx$.

2.    Set up Riemann sums to show that $\displaystyle\int_0^a x \, dx = -\int_a^0 x \, dx$.

3.    Prove part (i) of Theorem 1.

4.    Prove part (ii) of Theorem 1.

5.    Prove part (iii) of Theorem 1.

6.    Use the definition of the integral to prove that $\displaystyle\int_a^b dx = b - a$.

      Then use the MVT of the integral calculus.

7.    Prove Corollary 1.

8.    Prove Corollary 2.

9.    Determine which of the following functions are even, or odd, or neither.

      (a)    $f(x) = x/(x^2 + 3)$

      (b)    $f(x) = 1/(x^2 + 1)$

      (c)    $g(x) = 1/(x^3 + 1)$

      (d)    $f(x) = x^3/(x^2 + 1)(x^2 + 2)$

      (e)    $F(\theta) = \sin^3 \theta \cos^3 \theta$

      (f)    $H(\theta) = \tan^2 \theta \sec \theta + \tan \theta \sec^2 \theta$

10.    Interpret the MVT of the integral calculus in terms of area.

---

* This is a consequence of the *intermediate value theorem*. See Appendix C.

*11.   The arithmetic mean of $n$ numbers $a_1, \cdots, a_n$ is $(\sum_{i=1}^{n} a_i)/n$.

Show that the mean of $f$ on the interval $[a, b]$ is the limit of arithmetic means of the values of $f$ at equally spaced points of the interval.

12.   What is the mean of $f(x) = x$ over $[a, b]$?                $[(a + b)/2]$

13.   What is the mean of $f(x) = x^2$ over $[a, b]$?          $[\frac{1}{3}(a^2 + ab + b^2)]$

14.   What is the mean of $f(x) = \sqrt{a^2 - x^2}$ over the interval $[-a, a]$?      $[\pi a/4]$

15.   What is the mean of $f(x) = x^2 + 2x + 1$ over the interval $[0, 1]$? Draw a picture.                $[\frac{7}{3}]$

16.   Suppose $f(x) = 0$ for $0 \leq x \leq 1$ and $f(x) = x$ for $1 \leq x \leq 2$. What is the mean of $f$ over $[0, 2]$? Draw a picture.                $[\frac{3}{4}]$

17.   Prove, from the definition of the integral, that $\int_1^2 x^2 \, dx < \int_1^2 x^3 \, dx$.

*18.   How does $\int_a^x f(t) \, dt$ behave as $x \to \infty$, if $f(x) \geq 1$ for all $x$.

## 4   *The Fundamental Theorem of Calculus*

The two basic concepts of the calculus are the derivative and the definite integral. Both are defined as limits. Each concept had its origin in a geometric problem and each can be given a purely arithmetic, non-geometric, formulation. The Fundamental Theorem of calculus asserts that the two concepts are related.

In Part I of this text algorithms (shortcuts) were developed which circumvented the laborious calculation of derivatives directly from the definition. In Part II we have, until now, computed definite integrals by brute force. Riemann sums have been calculated and their limits have been obtained by various ingenious devices. The Fundamental Theorem provides an algorithm, a shortcut, to evaluate certain definite integrals.

Consider the function $F$ defined by

$$F(x) = \int_a^x f(t) \, dt.$$

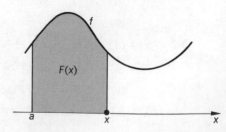

Note that $F(a) = 0$ and that the upper limit on the integral is the independent variable. Clearly (by the Theorem on page 225) $F$ is defined for all $x$ in an interval that contains $a$ and in which $f$ is continuous.

**THEOREM (THE FUNDAMENTAL THEOREM OF CALCULUS)**    *If f is continuous,* *then*

$$\frac{d}{dx} F(x) = \frac{d}{dx} \int_a^x f(t)\ dt = f(x).$$

**PROOF**    We simply compute the derivative of $F$ where

$$F(x) = \int_a^x f(t)\ dt.$$

$$\Delta F = F(x + \Delta x) - F(x)$$

$$= \int_a^{x+\Delta x} f(t)\ dt - \int_a^x f(t)\ dt$$

$$= \int_x^{x+\Delta x} f(t)\ dt \qquad \text{(why?)}$$

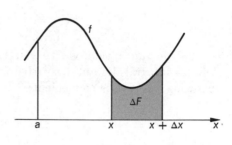

$$= \Delta x f(c), \ c = x + \theta\ \Delta x, \ 0 \leq \theta \leq 1, \text{ by the MVT of the integral calculus.}$$
Therefore,

$$\frac{\Delta F}{\Delta x} = f(x + \theta\ \Delta x)$$

and

$$\frac{dF}{dx} = \lim_{\Delta x \to 0} \frac{\Delta F}{\Delta x} = \lim_{\Delta x \to 0} f(x + \theta\ \Delta x) = f(x).$$

The last equality sign is valid because $f$ is assumed to be continuous.

The theorem asserts that $F$ is an antiderivative of $f$. This is the connection between differentiation and integration* that provides the shortcut for calculating definite integrals. The following corollary is the shortcut.

**COROLLARY**    *If f is continuous on* $[a, b]$ *and* $\varphi$ *is any antiderivative of f, then*

$$\int_a^b f(x)\ dx = \varphi(b) - \varphi(a).$$

**PROOF**    As in the proof of the theorem, define

$$F(x) = \int_a^x f(t)\ dt.$$

---

**\* Historical Note.** That the two fundamental operations, differentiation and integration, are inverses of one another was recognized by Fermat and Barrow. The general concept of a *calculus*, or an algorithmic procedure, was not recognized. Moreover, the class of functions with which they could deal was restricted. These men were almost inventors of the calculus.

Then, because $F$ and $\varphi$ are antiderivatives of $f$, we have

$$F(x) = \varphi(x) + C, \quad \text{for some constant } C. \quad \text{(Why?)}$$

Because $F(a) = 0$, we have

$$0 = \varphi(a) + C,$$

whence $C = -\varphi(a)$. Therefore, $F(x) = \varphi(x) - \varphi(a)$. In particular,

$$F(b) = \int_a^b f(t) \, dt = \varphi(b) - \varphi(a).$$

Often an antiderivative can be obtained by inspection. Then to get the definite integral the antiderivative is evaluated at the limits of the integral. The following notation is convenient:

$$\int_a^b f(x) \, dx = \varphi(x) \Big|_a^b = \varphi(b) - \varphi(a).$$

**Example 1**
$$\int_0^a x^3 \, dx = \tfrac{1}{4}x^4 \Big|_0^a = \tfrac{1}{4}a^4.$$

This is the result of Section 1.

**Example 2**
$$\int_{-1}^3 (x^2 - x - 1) \, dx = (\tfrac{1}{3}x^3 - \tfrac{1}{2}x^2 - x) \Big|_{-1}^3$$

$$= (9 - \tfrac{9}{2} - 3) - (-\tfrac{1}{3} - \tfrac{1}{2} + 1) = \tfrac{4}{3}.$$

**Example 3**
$$\int_0^{\pi/2} \sin \theta \, d\theta = -\cos \theta \Big|_0^{\pi/2} = 0 - (-1) = 1.$$

**Example 4**   Often antiderivatives can be far less obvious, and we shall spend much time on devices for finding them. But one can always *test by differentiation*. Because

$$\frac{d}{dx} \sqrt{x^2 + 4} = \frac{x}{\sqrt{x^2 + 4}},$$

$$\int_0^2 \frac{x \, dx}{\sqrt{x^2 + 4}} = \sqrt{x^2 + 4} \Big|_0^2 = 2\sqrt{2} - 2.$$

**Problems**

In the Problems evaluate the definite integral. Use any method.

1. $\displaystyle\int_{-1}^1 u^2 \, du$ $\qquad\qquad [\tfrac{2}{3}]$ $\qquad$ 2. $\displaystyle\int_{-1}^1 u^3 \, du$

3. $\displaystyle\int_a^b cu^2\, du$   $\left[\dfrac{c}{3}\,(b^3 - a^3)\right]$

4. $\displaystyle\int_0^a (a^2 - x^2)\, dx$   $\left[\dfrac{2a^3}{3}\right]$

5. $\displaystyle\int_0^a \sqrt{a^2 - x^2}\, dx$   $\left[\dfrac{\pi a^2}{4}\right]$

6. $\displaystyle\int_0^{1/2} \dfrac{dx}{\sqrt{1 - x^2}}$   $\left[\dfrac{\pi}{6}\right]$

7. $\displaystyle\int_0^1 \dfrac{du}{1 + u^2}$   $\left[\dfrac{\pi}{4}\right]$

8. $\displaystyle\int_0^{\pi/4} \sec^2\theta\, d\theta$   $[1]$

9. $\displaystyle\int_{-\pi/3}^{\pi/3} \cos t\, dt$   $[\sqrt{3}]$

10. $\displaystyle\int_{-\pi/3}^{\pi/3} \cos 2t\, dt$   $\left[\dfrac{\sqrt{3}}{2}\right]$

11. $\displaystyle\int_1^x \dfrac{dt}{t},\quad x > 0$   $[\log x]$

12. $\displaystyle\int_1^x \dfrac{dt}{2t},\quad x > 0$   $[\log \sqrt{x}]$

13. $\displaystyle\int_2^3 \dfrac{dx}{(x - 1)^2}$   $\left[\tfrac{1}{2}\right]$

14. $\displaystyle\int_0^1 (1 + x)^{10}\, dx$   $\left[\dfrac{2047}{11}\right]$

15. $\displaystyle\int_0^{1/2} 2\,(1 + x)^{10}\, dx$   $\left[\tfrac{2}{11}\,((\tfrac{3}{2})^{11} - 1)\right]$

16. $\displaystyle\int_0^2 \dfrac{dx}{x + 1}$   $[\log 3]$

17. $\displaystyle\int_1^2 \dfrac{x^2 + 1}{x}\, dx$   $\left[\tfrac{3}{2} + \log 2\right]$

18. $\displaystyle\int_{-1}^2 (2x + 1)\,(x + 2)\, dx$   $\left[19\tfrac{1}{2}\right]$

19. $\displaystyle\int_1^4 (1 + t^{1/2} + t - t^{-3/2})\, dt$   $\left[14\tfrac{1}{6}\right]$

20. $\displaystyle\int_1^9 \left(\sqrt{x} - \dfrac{1}{\sqrt{x}}\right) dx$   $\left[13\tfrac{1}{3}\right]$

21. $\displaystyle\int_1^9 \left(\sqrt{x} - \dfrac{1}{\sqrt{x}}\right)^2 dx$   $[24 + \log 9]$

22. $\displaystyle\int_{-1}^1 \dfrac{x\, dx}{(x^2 + 1)^2}$   $[0]$

23. $\displaystyle\int_{-1}^1 \dfrac{x\, dx}{(x^4 + 1)^2}$

## 5   The Indefinite Integral

Antiderivatives are also called *indefinite integrals*. The indefinite integral of $f$ is denoted by

$$\int f(x)\, dx = \text{an indefinite integral of } f = \text{an antiderivative of } f.$$

In many (but not all) problems $f$ will be an elementary function and will have an antiderivative that is also an elementary function.

*Example 1*
$$\int x^2 \, dx = \tfrac{1}{3}x^3 + C,$$

$$\int \sin x \, dx = -\cos x + C,$$

where the arbitrary constant $C$ is called the *constant of integration*. Adding the arbitrary constant assures us that we have obtained *all* antiderivatives.

Because of the arbitrary constant of integration, an antiderivative is not unique, and so when we write the indefinite integral we should always append the arbitrary constant.

In finding indefinite integrals the differential $dx$ plays an important role. The *differential of the indefinite integral is the integrand*:

$$F(x) = \int f(x) \, dx \qquad \textit{if and only if} \qquad dF(x) = f(x) \, dx.$$

*Example 2*
$$\int (2t)^3 \, (2 \, dt) = \tfrac{1}{4}(2t)^4 + C,$$

because the integrand is of the form $x^3 \, dx$ with $x = 2t$.

$$\int \sin 2t \, (2 \, dt) = -\cos 2t + C,$$

because the integrand is of the form $\sin x \, dx$ with $x = 2t$.

*Example 3*
$$\int x\sqrt{1 + 3x^2} \, dx = \int (1 + 3x^2)^{1/2}x \, dx,$$

and the integrand is not quite of the form $u^{1/2} \, du$. The form of the integrand makes the following substitution reasonable:

$$u = 1 + 3x^2 \qquad \text{and} \qquad du = 6x \, dx.$$

Then the integral becomes

$$\int (1 + 3x^2)^{1/2}x \, dx = \int u^{1/2}\tfrac{1}{6} \, du = \tfrac{1}{6} \cdot \tfrac{2}{3}u^{3/2} + C$$

$$= \tfrac{1}{9}(1 + 3x^2)^{3/2} + C.$$

Observe that if the integral had been $\int (1 + 3x^2)^{1/2} \, dx$, we would have been unable to proceed in the same way because we could not get the differential $du$ by simply multiplying by a constant.

These examples illustrate the use of simple substitutions in finding anti-derivatives. The use of substitutions is a common and powerful method for obtaining antiderivatives and will be discussed again in Chapter 10. At this stage we simply try to find a substitution that will render the integrand recognizable.

When a substitution is made in order to find a *definite* integral, it is usually simplest to change the limits on the integral. The following theorem justifies the method.

**THEOREM**    *Suppose that in the integral,*

$$\int_a^b f(x) \, dx,$$

*the substitution* $u = u(x)$, *where u is a differentiable function, results in*

$$\varphi(u) \, du = f(x) \, dx.$$

*Suppose further that as x changes from a to b, u changes from $\alpha$ to $\beta$ and that the function $\varphi$ is continuous in* $[\alpha, \beta]$. *Then*

$$\int_a^b f(x) \, dx = \int_\alpha^\beta \varphi(u) \, du.$$

**PROOF**    Suppose that $\Phi$ is an indefinite integral of $\varphi$,

$$\Phi(u) = \int \varphi(u) \, du.$$

Then
$$F(x) = \Phi[u(x)]$$

is an indefinite integral of $f$ because

$$dF(x) = d\Phi[u(x)] = \Phi'(u) \, du = \varphi(u) \, du = f(x) \, dx.$$

Therefore,

$$\Phi(u) \Big|_\alpha^\beta = \Phi[u(x)] \Big|_{x=a}^{x=b} = F(x) \Big|_{x=a}^{x=b} = \int_a^b f(x) \, dx.$$

*Remark.* Observe that this substitution theorem is simply an application of the chain rule.

*Example 4*    To evaluate    $\int_{-1}^{2} x^2 \sqrt{1 + x^3} \, dx,$

let $u = 1 + x^3$. Then $du = 3x^2\,dx$, and as $x$ increases from $-1$ to $2$, $u$ changes continuously from $0$ to $9$. Thus,

$$\int_{-1}^{2} x^2\sqrt{1+x^3}\,dx = \int_{-1}^{2} (1+x^3)^{1/2}x^2\,dx$$

$$= \int_{0}^{9} u^{1/2}\tfrac{1}{3}\,du = \tfrac{1}{3}\cdot\tfrac{2}{3}u^{3/2}\Big|_{0}^{9} = 6.$$

*Example 5*    To evaluate

$$\int_{\pi/8}^{\pi/6} \sin^2 2\theta \cos 2\theta\,d\theta,$$

let $u = \sin 2\theta$. Then $du = 2\cos 2\theta\,d\theta$, and as $\theta$ increases from $\pi/8$ to $\pi/6$, $u$ increases from $\sqrt{2}/2$ to $\sqrt{3}/2$. Thus,

$$\int_{\pi/8}^{\pi/6} \sin^2 2\theta \cos 2\theta\,d\theta = \int_{\sqrt{2}/2}^{\sqrt{3}/2} u^2\tfrac{1}{2}\,du = \tfrac{1}{2}\cdot\tfrac{1}{3}u^3\Big|_{\sqrt{2}/2}^{\sqrt{3}/2}$$

$$= \tfrac{1}{48}(3\sqrt{3} - 2\sqrt{2}).$$

*Problems*

**In Problems 1 to 13 find the indefinite integral. Check by differentiation.**

1.  $\displaystyle\int (x+5)^3\,dx$    $[\tfrac{1}{4}(x+5)^4 + C]$

2.  $\displaystyle\int (x+5)^{3/2}\,dx$    $[\tfrac{2}{5}(x+5)^{5/2} + C]$

3.  $\displaystyle\int 7(2x+5)^{3/2}\,dx$    $[\tfrac{7}{5}(2x+5)^{5/2} + C]$

4.  $\displaystyle\int (x+1)^2 x\,dx$    $[\tfrac{1}{4}x^4 + \tfrac{2}{3}x^3 + \tfrac{1}{2}x^2 + C]$

5.  $\displaystyle\int (x^2+1)^2 x\,dx$    $[\tfrac{1}{6}(x^2+1)^3 + C]$

6.  $\displaystyle\int (2x^2+1)^{3/2} x\,dx$    $[\tfrac{1}{10}(2x^2+1)^{5/2} + C]$

7.  $\displaystyle\int \frac{dx}{\sqrt{1+5x}}$    $[\tfrac{2}{5}\sqrt{1+5x} + C]$

8. $\displaystyle\int \frac{dx}{x}$ $\qquad\qquad\qquad$ $[\log |x| + C]$

(Consider $x > 0$ and $x < 0$ separately.)

9. $\displaystyle\int \frac{dx}{x+1}$ $\qquad\qquad\qquad$ $[\log |x+1| + C]$

(Consider $x + 1 > 0$ and $x + 1 < 0$ separately.)

10. $\displaystyle\int \left(1 + \frac{1}{x}\right)^{1/2} \frac{dx}{x^2}$ $\qquad\qquad$ $\left[-\frac{2}{3}\left(1 + \frac{1}{x}\right)^{3/2} + C\right]$

11. $\displaystyle\int \sin x \cos x \, dx$ $\qquad\qquad$ $[\tfrac{1}{2}\sin^2 x + C]$

12. $\displaystyle\int \frac{x \, dx}{x^2 + 1}$ $\qquad\qquad$ $[\log \sqrt{x^2 + 1} + C]$

13. $\displaystyle\int \frac{dx}{x^2 + 1}$ $\qquad\qquad$ $[\text{Arc}\tan x + C]$

In Problems 14 to 21 evaluate the definite integrals.

14. $\displaystyle\int_{1/2}^{2} (2x)^{5/2} \, dx$ $\qquad\qquad$ $[\tfrac{127}{2}]$

15. $\displaystyle\int_{-1}^{1} (2x + 1)^2 \, dx$ $\qquad\qquad$ $[\tfrac{14}{3}]$

16. $\displaystyle\int_{-1}^{1} (2x + 1)^2 x \, dx$ $\qquad\qquad$ $[\tfrac{8}{3}]$

17. $\displaystyle\int_{0}^{a} x\sqrt{x^2 + a^2} \, dx, \quad a > 0$ $\qquad$ $[\tfrac{1}{3}a^3 (2\sqrt{2} - 1)]$

18. $\displaystyle\int_{-a}^{a} x\sqrt{x^2 + a^2} \, dx$ $\qquad\qquad$ $[0]$

19. $\displaystyle\int_{0}^{a} x\sqrt{a^2 - x^2} \, dx$ $\qquad\qquad$ $[\tfrac{1}{3}a^3]$

20. $\displaystyle\int_{-1}^{0} \sqrt{53 - x} \, dx$ $\qquad\qquad$ $[\tfrac{2}{3}(54^{3/2} - 53^{3/2})]$

21. $\displaystyle\int_{-\pi/2}^{\pi/2} \sin^3 x \, dx$ $\qquad\qquad$ $[0]$

22. Find the area under the graph of $y = \frac{1}{2}x^2$ between $x = 0$ and $x = x_0$.

$$[\tfrac{1}{6}x_0^3]$$

23. Find, by integration the area of the region in the first quadrant bounded by $2x + 3y = 6$ and the axes.

$$[3]$$

24. Find the area under the graph of $y = \sqrt{2px}$, $p > 0$, between $x = 0$ and $x = a$.

$$[2a\sqrt{2pa}/3]$$

In the following problems interpret the limit of the sum as a definite integral and so evaluate the limit.

*25. $\lim\limits_{n\to\infty} \dfrac{1}{n}\left(\dfrac{1}{1} + \dfrac{1}{1+\dfrac{1}{n}} + \dfrac{1}{1+\dfrac{2}{n}} + \cdots + \dfrac{1}{1+\dfrac{n-1}{n}}\right)$ $\qquad [\log 2]$

*26. $\lim\limits_{n\to\infty} \dfrac{1}{n}\left(\dfrac{1}{1^2} + \dfrac{1}{\left(1+\dfrac{1}{n}\right)^2} + \cdots + \dfrac{1}{\left(1+\dfrac{n-1}{n}\right)^2}\right)$ $\qquad [\tfrac{1}{2}]$

*27. $\lim\limits_{n\to\infty} \left(\dfrac{1}{2n} + \dfrac{1}{2n+1} + \dfrac{1}{2n+2} + \cdots + \dfrac{1}{2n+(n-1)}\right)$ $\qquad [\log \tfrac{3}{2}]$

*28. $\lim\limits_{n\to\infty} \left(\dfrac{\log\left(1+\dfrac{e-1}{n}\right)}{n+(e-1)} + \dfrac{\log\left(1+\dfrac{2(e-1)}{n}\right)}{n+2(e-1)} + \cdots + \dfrac{\log\left(1+\dfrac{n(e-1)}{n}\right)}{n+n(e-1)}\right)$

$$\left[\dfrac{1}{2}\right] \qquad [\tfrac{1}{2}(c-1)]$$

## 6 Area

The definite integral arose out of the problem of finding areas of regions under the graphs of functions. Now we shall see that other areas can be expressed as definite integrals. Moreover, we shall see that Leibniz' notation is flexible and suggestive.

The area of the region (see the figure) between the graphs of $f$ and $g$, and between $x = a$ and $x = b$, is approximated by the Riemann sums

$$\sum_{i=1}^{n} [f(\xi_i) - g(\xi_i)]\, \Delta x_i$$

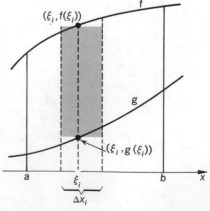

The limit of these sums as the mesh approaches 0 is the area*

(1)
$$\text{area} = \int_a^b [f(x) - g(x)]\,dx$$

Observe that we can obtain formula (1) by writing the *differential element of area*:

$$l(x)\,dx = [f(x) - g(x)]\,dx$$

which is the area of the "differential rectangle."

Also note that we get the same result in this case if we find the area of the region under $f$ and subtract the area of the region under $g$:

$$\text{area} = \int_a^b f(x)\,dx - \int_a^b g(x)\,dx$$

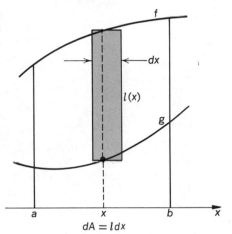

$dA = l\,dx$

**Example 1**  Find the area of the region bounded between the curves $y = 2x - \frac{1}{2}x^2$ and $2y + 3x = 6$.

The curves intersect at $(1, \frac{3}{2})$ and $(6, -6)$ as shown.

The differential element of area is

$$l\,dx = (y_{\text{parabola}} - y_{\text{line}})\,dx$$
$$= [(2x - \tfrac{1}{2}x^2) - (3 - \tfrac{3}{2}x)]\,dx$$
$$= (-3 + \tfrac{7}{2}x - \tfrac{1}{2}x^2)\,dx,$$

whence, the

$$\text{area} = \int_1^6 (-3 + \tfrac{7}{2}x - \tfrac{1}{2}x^2)\,dx = \tfrac{125}{12}.$$

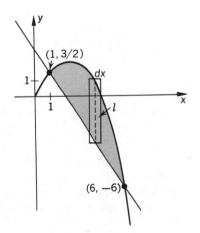

It is often wise to examine one's answer for gross errors by making a rough estimate of the answer. In this problem one can count squares, and parts of squares, in the region. A rough count will verify that the answer is of the correct order of magnitude.

---

* To establish beyond doubt that the limit is precisely the area one could consider rectangles of width $\Delta x_i$ that are "inscribed" and "circumscribed." Then the area would be squeezed between the sums for these two kinds of rectangles, and each sum would approach the definite integral.

*Example 2*    In some problems it is more ex-
peditious to view the area as a "sum" of differential rectangles parallel to the $x$-axis.

To find the area of the region bounded by
the parabola $y^2 - 2y + x = 0$, and the line
$x + y = 0$, we use the differential element of area,

$$l(y)\, dy = [(2y - y^2) - (-y)]\, dy$$

$$= (3y - y^2)\, dy.$$

Then the

$$\text{area} = \int_0^3 (3y - y^2)\, dy$$

$$= \tfrac{9}{2}.$$

A rough count of squares gives a satisfactory
check.

Note that the area *could* have
been obtained as the sum of the two
areas shown:

$$\text{area} = \int_{-3}^0 l_1\, dx + \int_0^1 l_2\, dx$$

$$= \int_{-3}^0 (1 + \sqrt{1 - x} + x)\, dx$$

$$+ \int_0^1 2\sqrt{1 - x}\, dx,$$

but this way involves more work.

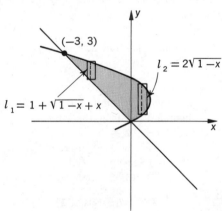

### Problems

In each of Problems 1 to 14 the equations form part of the boundary of a bounded
region. Find its area. Sketch each region.

1.   $y = x, y = 3x, x = 1, x = 3$                           $[8]$

2.   $y = x, y = x^2$                                       $[\tfrac{1}{6}]$

3.   $y = \sqrt{x}, y = x$                                   $[\tfrac{1}{6}]$

4.   $y = |x|, y = x^3$                                 $[\tfrac{1}{4}]$

5.   $x = \sqrt{y}, y = x^4$                               $[\tfrac{2}{15}]$

6.   $y^2 = -x, 2x - y = -1$                        $[\tfrac{9}{16}]$

7.   $y = 1/x, x = 0, y = 1, y = 3$           $[\log 3]$

8.  $x = 2 - y^2, x = \frac{1}{2}(2 - y^2)$ $[\frac{4}{3}\sqrt{2}]$

9.  $y = \frac{1}{2}x^3, x = 0, y = 4$ $[6]$

10. $y = x, x + 1 = (y - 1)^2$ $[\frac{9}{2}]$

11. $x^{1/2} + y^{1/2} = a^{1/2}, x = 0, y = 0$ $[\frac{1}{6}a^2]$

12. $3y^2 = x^3, y^2 = 3x, y \geq 0$ $[\frac{4.2}{5}]$

13. $x - y = 7, x = 2y^2 - y + 3$ $[9]$

14. $y = \tan x, y = 0, x = \pi/4$ $[\log \sqrt{2}]$

15. Find the area, using calculus, of the triangle whose vertices are $(-1, 1)$, $(3, 5)$ and $(8, -1)$. $[22]$

16. Find, using calculus, the area of the triangle bounded by the lines $2y = x + 4$, $2y = 3x - 4, 2y = -x - 4$. $[16]$

17. A parabola with vertex at the origin and axis along the $x$-axis passes through $(a, b)$ where $a, b > 0$. Find the area bounded by the parabola, the line $y = a$, and the line $x = 0$. $[a^4/3b^2]$

18. Find the area of the region between the parabola $y = \frac{1}{2}x^2$ and the witch $y = 1/(x^2 + 1)$. $[(\pi/2) - \frac{1}{3}]$

19. Find the area under one arch of the sine curve. $[2]$

20. Find the area inside the circle $x^2 + y^2 = a^2$ and outside the ellipse $b^2x^2 + a^2y^2 = a^2b^2$, where $a > b$. $[\pi a(a - b)]$

## 7  Volumes of Solids of Revolution

The power of the definite integral rests with the fact that so many quantities that we wish to find can be expressed as a limit of Riemann sums. Area in the plane is one example, and now we regard volume the same way.

Suppose that a solid is generated by revolving about the $x$-axis the plane region under the graph of $f$ and between $x = a$ and $x = b$.

The interval $[a, b]$ is subdivided and the thin rectangles are revolved about the axis. With the notation of the figure, the volume $\Delta V_i$ of the thin slab between $x_{i-1}$ and $x_i$ will be such that

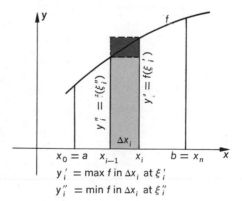

$y_i' = \max f$ in $\Delta x_i$ at $\xi_i'$
$y_i'' = \min f$ in $\Delta x_i$ at $\xi_i''$

$$\pi f^2(\xi_i'') \; \Delta x_i \leqq \Delta V_i \leqq \pi f^2(\xi_i') \; \Delta x_i,$$

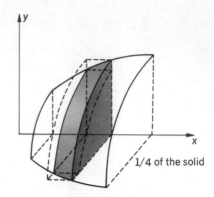

because the thin slab is circumscribed by one thin cylinder and inscribed by another. Therefore the volume $V$ of the solid is caught between two Riemann sums:

$$\pi \sum_{i=1}^{n} f^2(\xi_i'') \; \Delta x_i \leqq V \leqq \pi \sum_{i=1}^{n} f^2(\xi_i') \; \Delta x_i.$$

1/4 of the solid

As the mesh of the subdivision approaches 0, the volume is caught between two equal integrals,

$$\pi \int_a^b f^2(x) \; dx \leqq V \leqq \pi \int_a^b f^2(x) \; dx.$$

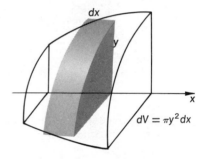

Observe that the differential element of area $f(x) \; dx = y \; dx$, when rotated about the $x$-axis, generates the differential element of volume

$$dV = \pi y^2 \; dx.$$

$dV = \pi y^2 dx$

**Example 1**  To find the volume of a sphere of radius $a$ we regard it as generated by revolving about the $x$-axis the upper half of the disc bounded by $x^2 + y^2 = a^2$.

The element of volume is

$$dV = \pi y^2 \; dx,$$

where $y^2 = a^2 - x^2$. Then

$$V = \pi \int_{-a}^{a} y^2 \; dx$$

$$= \pi \int_{-a}^{a} (a^2 - x^2) \; dx = \tfrac{4}{3}\pi a^3.$$

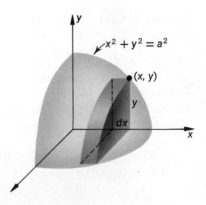

**Example 2**  If a solid is obtained by revolving a plane region about some line other than $y = 0$, then the radius and thickness of the differential element of volume must be expressed appropriately.

Find the volume of the solid generated by revolving about the line $x = -1$, the region bounded by $y^2 = x$, $y = 0$, $y = 1$, and $x = -1$.

The differential element of volume is $dV = \pi r^2 \, dy$ where $r = x - (-1) = y^2 - (-1)$. Therefore,

$$V = \pi \int_0^1 (y^2 + 1)^2 \, dy = \frac{28\pi}{15}.$$

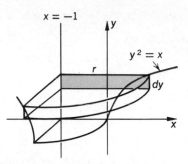

As a rough check, the volume should be less than the volume of a cylinder of radius 2 and height 1, namely $4\pi$, and more than that of a cylinder of radius 1 and height 1, namely $\pi$.

## Problems

In Problems 1 to 22 plane regions bounded by the given curves are revolved about the specified line. Draw a figure of the plane *region* and find the volume generated.

1.   $y = ax, x = b > 0, x = c > b, y = 0$;   about $y = 0$    $[\pi a^2 (c^3 - b^3)/3]$

2.   $y^2 = 2px, x = a$    ; . about $y = 0$    $[\pi p a^2]$

3.   $y^2 = 2px, x = a > 0$    ; about $x = 0$    $[\frac{8}{5}\pi a^2 \sqrt{2pa}]$

4.   $\dfrac{x}{a} + \dfrac{y}{b} = 1, x = 0, y = 0, a, b > 0$   ; about $y = 0$    $[\frac{1}{3}\pi ab^2]$

5.   $b^2 x^2 + a^2 y^2 = a^2 b^2$    ; about $y = 0$    $[\frac{4}{3}\pi ab^2]$

6.   $x^2 - 4x = y, y = 0$    ; about $y = 0$    $[512\pi/15]$

7.   $x^2 - 4x + 2 = y, y = 2$    ; about $y = 2$    $[512\pi/15]$

8.   $x^{1/2} + y^{1/2} = a^{1/2}, x = a, y = a$    ; about $x = 0$    $[\frac{14}{15}\pi a^3]$

9.   $x^{2/3} + y^{2/3} = a^{2/3}$    ; about $y = 0$    $[32\pi a^3/105]$

10.   $y^2 = x^3, x = a$    ; about $y = 0$    $[\pi a^4/4]$

11.   $y^2 = x^3, x = a$    ; about $x = a$    $[16\pi a^{7/2}/35]$

12.   $y = x^3, y = 0, x = a$    ; about $y = 0$    $[\pi a^7/7]$

13.   $y = x^3, y = 0, x = a$    ; about $x = 0$    $[2\pi a^5/5]$

14.   $y = x^3, y = 0, x = a$    ; about $y = a^3$    $[5\pi a^7/14]$

15.   $y = x^3, y = 0, x = a$    ; about $x = a$    $[\pi a^5/10]$

16.   $a^2 y^2 = x^2 (a^2 - x^2), a > 0$    ; about $y = 0$    $[4\pi a^3/15]$

17.   $y^2 = (x + 3)^3, x = -2$    ; about $y = 0$    $[\pi/4]$

18.   $y^2 = (x + 3)^2, x = -2$    ; about $x = -2$    $[2\pi/3]$

19. $y = e^x, x = 0, x = a, y = 0$ ; about $y = 0$ $\left[\frac{1}{2}\pi\,(e^{2a} - 1)\right]$

20. $y = \sqrt{x}e^{x^2}, y = 0, x = 1$ ; about $y = 0$ $\left[\frac{1}{4}\pi\,(e^2 - 1)\right]$

21. $(x - a)^2 + y^2 = b^2, a > b > 0$ ; about $x = 0$ $\left[2\pi^2ab^2\right]$

*22. $y = x^2, x = y$ ; about $x = y$ $\left[\sqrt{2}\pi/60\right]$

23. Find, using calculus, the volume of a right circular cone of radius $r$ and altitude $h$.

24. A solid has a flat circular base of radius $a$, and all cross sections perpendicular to a fixed diameter of the base are squares, as shown. Find its volume.
$\qquad\qquad\qquad[16a^3/3]$

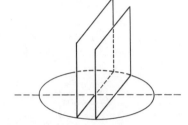

25. A solid has a flat circular base of radius $a$, and all cross sections perpendicular to a diameter of the base are equilateral triangles. Find its volume. $\quad[4a^3/\sqrt{3}]$

26. A solid has a flat circular base of radius $a$, and all cross sections perpendicular to a diameter of the base are isosceles triangles of height $h$. Find the volume.
$\qquad\qquad\qquad\qquad\qquad\qquad[\pi a^2h/2]$

## CHAPTER 10

## THE ELEMENTS OF FORMAL INTEGRATION

### 1 The Problem

The Fundamental Theorem of calculus assures us that continuous functions have antiderivatives, and the Corollary shows how to make use of an antiderivative if we can find one. But unless we can express the antiderivative in terms of known (elementary or computable) functions, the Fundamental Theorem does not provide any shortcut for evaluating definite integrals.

The problem therefore is to find antiderivatives. Unlike the situation with derivatives (the derivative of an elementary function is also an elementary function) *the indefinite integral of an elementary function need not be an elementary function.**

The first such example was given by Joseph Liouville in 1835. In this chapter and the next we examine an extensive variety of indefinite integrals that *can* be expressed in terms of elementary functions. The reader should be aware that he is to develop a *skill* that depends partly on good guessing. One tries to arrange the integrand in

$$\int f(x)\, dx$$

> The indefinite integral
>
> $$F(x) = \int e^{-x^2}\, dx$$
>
> is not an elementary function

---

* There is an elaborate and deep theory (beyond the scope of this text) concerning conditions under which indefinite integrals are elementary functions. See J. F. Ritt, *Integration in Finite Terms*, Columbia University Press, 1948. We are interested only in being able to recognize simple cases.

so that he can say, "Aha! I know a function $F$ such that

$$dF(x) = f(x) \, dx."$$

To begin with, the basic differentiation formulas need to be recalled. Because differentials are used in integration, we reproduce the formulas here in differential form, paired with the associated indefinite integral.

---

(I)    $\displaystyle \int (du + dv) = \int du + \int dv$        $d(u + v) = du + dv$

(II)    $\displaystyle \int c \, du = c \int du$        $d(cu) = c \, du$

(III)    $\displaystyle \int u^n \, du = \frac{u^{n+1}}{n+1} + C, \quad n \neq -1$        $du^n = nu^{n-1} \, du$

(IV)    $\displaystyle \int \frac{du}{u} = \log |u| + C$        $d \log |u| = \frac{1}{u} \, du$

(V)    $\displaystyle \int e^u \, du = e^u + C$        $de^u = e^u \, du$

    $\displaystyle \int a^u \, du = \frac{a^u}{\log a} + C$        $da^u = a^u \log a \, du$

(VI)    $\displaystyle \int \sin u \, du = -\cos u + C$        $d \cos u = -\sin u \, du$

(VII)    $\displaystyle \int \cos u \, du = \sin u + C$        $d \sin u = \cos u \, du$

(VIII)    $\displaystyle \int \sec^2 u \, du = \tan u + C$        $d \tan u = \sec^2 u \, du$

(IX)    $\displaystyle \int \csc^2 u \, du = -\cot u + C$        $d \cot u = -\csc^2 u \, du$

(X)    $\displaystyle \int \sec u \tan u \, du = \sec u + C$        $d \sec u = \sec u \tan u \, du$

(XI)    $\displaystyle \int \csc u \cot u \, du = -\csc u + C$        $d \csc u = -\csc u \cot u \, du$

(XII)    $\displaystyle \int \frac{du}{\sqrt{a^2 - u^2}} = \text{Arc} \sin \frac{u}{a} + C, \quad a > 0$        $d \, \text{Arc} \sin \frac{u}{a} = \frac{du}{\sqrt{a^2 - u^2}}, \quad a > 0$

(XIII)    $\displaystyle \int \frac{du}{a^2 + u^2} = \frac{1}{a} \text{Arc} \tan \frac{u}{a} + C$        $d \, \text{Arc} \tan \frac{u}{a} = \frac{a \, du}{a^2 + u^2}$

---

All these integral formulas can be readily verified by the reader. However, a few comments may be in order.

The absolute value $|u|$ is needed in (IV) because $\log u$ is defined only for $u > 0$. If $u$ is negative, say $u = -t$, $t > 0$, then $du/u = (-dt)/(-t) = dt/t$ and

$$\int \frac{du}{u} = \log t + C = \log |u| + C.$$

The integral $\int a^u \, du$ is seldom needed and can always be avoided by

$$\int a^u \, du = \int (e^{\log a})^u \, du.$$

Formulas (XII) and (XIII) have alternate forms

$$\int \frac{du}{\sqrt{a^2 - u^2}} = -\text{Arc cos} \frac{u}{a} + C \qquad \text{and} \qquad \int \frac{du}{a^2 + u^2} = -\frac{1}{a} \text{Arc cot} \frac{u}{a} + C,$$

which come from the differentiation formulas for Arc cos and Arc cot. These integrals differ from those of the formulas by a constant. They will not be needed.

To begin with, Formulas I to XIII are all we have. We try to arrange an integrand so one of these formulas applies. Later we will have other formulas.

*Example*    In order to apply the formulas when $u$ is a function, say $u = u(x)$, the *form* of the integrand must be exactly right. For example,

$$\int \cos 2x \, dx \neq \sin 2x + C$$

because $dx$ is not the differential of $2x$. We put the integrand in the correct form to use (VII) by writing

$$\int \cos 2x \, dx = \frac{1}{2} \int \cos 2x \, d(2x) = \frac{1}{2} \int \cos 2x \, (2 \, dx) = \frac{1}{2} \sin 2x + C.$$

Observe that the middle integrand is now of the form $\cos u \, du$ (where $u = 2x$).

### Problems

In Problems 1 to 20 write the differentials. Do not simplify.

1.  $d\sqrt{a^2 - x^2}$

2.  $d(a^2 - x^2)^{-1/2}$

3.  $d(a^2 + x^2)^n$

4.  $d(x^2 + 2x - 1)^{3/2}$

5.  $d \log (1 + x^2)$

6.  $de^{x^2}$

7.  $de^{x^2+1}$

8.  $d \log e^{x^2}$

9.  $d(\log x)^2$

10.  $d[\log (x^2 + 1)]^2$

11.  $d \log \sin \theta$

12.  $d \sin^3 2x$

13.  $d \tan (x + 1)$

14.  $d \operatorname{Arc} \tan \dfrac{x}{5}$

15.  $d \operatorname{Arc} \sin \dfrac{x}{2}$

16.  $d \sec 3x$

17.  $d \csc x^2$

18.  $d \operatorname{Arc} \sin \sqrt{x}$

19.  $d \operatorname{Arc} \sin \dfrac{x - a}{a}$

20.  $d \operatorname{Arc} \tan \sqrt{x}$

In Problems 21 to 36 the integrand is either exactly right to use one of Formulas I to XIII or else it differs only by a constant factor. Write the indefinite integral as an elementary function and check by differentiation.

21.  $\displaystyle \int e^{2t} 2 \, dt$

22.  $\displaystyle \int (4 - x)^3 \, dx$

23.  $\displaystyle \int (4 - x^2)^{3/2} \, (2x \, dx)$

24.  $\displaystyle \int \frac{\cos \theta}{1 + \sin \theta} \, d\theta$

25.  $\displaystyle \int \sin^3 \theta \cos \theta \, d\theta$

26.  $\displaystyle \int \frac{2 \, dx}{9 + 4x^2}$

27.  $\displaystyle \int \frac{2x \, dx}{9 + 4x^2}$

28.  $\displaystyle \int \sec^2 3\theta \, d\theta$

29.  $\displaystyle \int (1 + \tan \theta)^{-2} \sec^2 \theta \, d\theta$

30.  $\displaystyle \int \frac{dt}{\sqrt{1 - 4t^2}}$

31.  $\displaystyle \int (1 - 4t^2)^{-1/2} \, (t \, dt)$

32.  $\displaystyle \int e^{\sqrt{x}} \frac{dx}{\sqrt{x}}$

33.  $\displaystyle \int \sec e^t \tan e^t \, (e^t \, dt)$

34.  $\displaystyle \int \sin \sqrt{x} \, (x^{-1/2} \, dx)$

35.  $\displaystyle \int \frac{(2x + 2) \, dx}{4 + (x + 1)^2}$

36.  $\displaystyle \int \frac{x^{1/2} \, dx}{9 + x^3}$

## 2    Substitution

The basic technique for finding indefinite integrals is that of substitution. The method has already been given in Section 5 of Chapter 9. We restate it here for convenience:

*If $u = u(x)$ is a differentiable function, and if $\varphi$ and $f$ are continuous, and if*

$$\varphi[u(x)] \, du(x) = f(x) \, dx \qquad and \qquad \Phi(u) = \int \varphi(u) \, du,$$

*then*

$$\int f(x) \ dx = \int \varphi(u) \ du = \Phi(u(x))$$

*because* $d\Phi(u) = \Phi'(u) \ du = \varphi[u(x)] \ du(x) = f(x) \ dx.$

Observe that the chain rule has been used.

The trick is to recognize a good substitution. Experience is the best guide.

**Example 1**    To find    $\int \sec^2 2x \ dx,$

set $u = 2x$ so that $du = 2 \ dx.$

$$\int \sec^2 2x \ dx = \int \sec^2 u \ (\tfrac{1}{2} \ du) = \frac{1}{2} \int \sec^2 u \ du = \tfrac{1}{2} \tan u + C$$

$$= \tfrac{1}{2} \tan 2x + C.$$

Simple substitutions such as this are usually done mentally.

$$\int \sec^2 2x \ dx = \frac{1}{2} \int \sec^2 2x \ (2 \ dx) \quad \{\text{In order to get } du = d(2x) = 2 \ dx$$

$$= \tfrac{1}{2} \tan 2x + C.$$

**Example 2**

$$\int \frac{\sin \sqrt{x}}{\sqrt{x}} \ dx = \int (\sin x^{1/2}) (x^{-1/2} \ dx), \quad \{\text{Note that parentheses avoid ambiguity}$$

whence, if $u = x^{1/2}$, then $du = \tfrac{1}{2} x^{-1/2} \ dx$, and

$$\int (\sin x^{1/2}) (x^{-1/2} \ dx) = \int \sin u \ (2 \ du) = -2 \cos u + C$$

$$= -2 \cos \sqrt{x} + C.$$

But we could have avoided the explicit substitution.

$$\int (\sin x^{1/2}) (x^{-1/2} \ dx) = 2 \int (\sin x^{1/2}) (\tfrac{1}{2} x^{-1/2} \ dx)$$

$$= -2 \cos x^{1/2} + C.$$

**Example 3**    $\int \dfrac{dt}{9 + 4t^2} = \dfrac{1}{2} \int \dfrac{2 \ dt}{3^2 + (2t)^2} \quad \{\text{So the numerator} = d(2t)$

$$= (\tfrac{1}{2}) \tfrac{1}{3} \text{ Arc tan } \frac{2t}{3} + C = \tfrac{1}{6} \text{ Arc tan } \frac{2t}{3} + C.$$

**Example 4**    Sometimes a bit of algebra helps one to find the substitution. Completing the square is one of the algebraic devices.

$$ax^2 + bx + c = a\left(x^2 + \frac{b}{a}x\right) + c$$

$$= a\left(x^2 + \frac{b}{a}x + \frac{b^2}{4a^2}\right) + c - \frac{b^2}{4a}$$

$$= a\left(x + \frac{b}{2a}\right)^2 + \frac{4ac - b^2}{4a}.$$

$$\int \frac{x\,dx}{\sqrt{4x - x^2}} = \int \frac{x\,dx}{\sqrt{4 - (x - 2)^2}} \quad \{\text{Completing the square}$$

This suggests letting $u = x - 2$, $du = dx$; then

$$\int \frac{x\,dx}{\sqrt{4x - x^2}} = \int \frac{(u + 2)\,du}{\sqrt{4 - u^2}} = \int \frac{u\,du}{\sqrt{4 - u^2}} + \int \frac{2\,du}{\sqrt{4 - u^2}}$$

$$= -(4 - u^2)^{1/2} + 2\operatorname{Arc\,sin}\frac{u}{2} + C$$

$$= -(4x - x^2)^{1/2} + 2\operatorname{Arc\,sin}\frac{x - 2}{2} + C.$$

### Problems

1.    Verify by differentiation Formulas I to XIII.

**In Problems 2 to 97 find the indefinite integrals.**

2.  $\displaystyle\int t^{5/2}\,dt$                    $\left[\frac{2}{7}t^{7/2} + C\right]$

3.  $\displaystyle\int \sqrt{2t}\,dt$                    $\left[\frac{1}{3}(2t)^{3/2} + C\right]$

4.  $\displaystyle\int \frac{du}{u^2}$                    $\left[-\frac{1}{u} + C\right]$

5.  $\displaystyle\int \frac{d\theta}{2\theta}$

6.  $\displaystyle\int \sqrt{2px}\,dx$                    $\left[\frac{2\sqrt{2p}}{3}x^{3/2} + C\right]$

7.  $\displaystyle\int (1 + x^2)^{1/2}x\,dx$                    $\left[\frac{1}{3}(1 + x^2)^{3/2} + C\right]$

8. $\displaystyle\int \frac{du}{1+\sqrt{2u}}$ $\qquad\qquad$ $[\sqrt{2u} - \log(1+\sqrt{2u}) + C]$

9. $\displaystyle\int (2x^3 - 7x^2 + \sqrt{3}x - \pi)\, dx$ $\qquad\qquad$ $\left[\dfrac{1}{2}x^4 - \dfrac{7}{3}x^3 + \dfrac{\sqrt{3}}{2}x^2 - \pi x + C\right]$

10. $\displaystyle\int (\sqrt{2} - \sqrt{x})^3\, dx$

11. $\displaystyle\int (a + bx^2)^m x\, dx$ $\qquad\qquad$ $\left[\dfrac{1}{2b\,(m+1)}\,(a+bx^2)^{m+1} + C\right]$

12. $\displaystyle\int \frac{dx}{\sqrt{1-2x}}$

13. $\displaystyle\int 2at\sqrt{\frac{t^2}{b^2}+1}\, dt, \quad b<0$ $\qquad\qquad$ $\left[\dfrac{-2a}{3b}\,(t^2+b^2)^{3/2} + C\right]$

14. $\displaystyle\int e^{2x}\, dx$ $\qquad\qquad$ $[\frac{1}{2}e^{2x} + C]$

15. $\displaystyle\int e^{2x^2}x\, dx$

16. $\displaystyle\int \left(\sqrt[3]{t^2} - \sqrt[3]{t} + \frac{1}{\sqrt[3]{t}}\right) dt$ $\qquad\qquad$ $[\frac{3}{5}t^{5/3} - \frac{3}{4}t^{4/3} + \frac{3}{2}t^{2/3} + C]$

17. $\displaystyle\int (a+bx)^{10}\, dx$

18. $\displaystyle\int \left(3a\sqrt[3]{x} + \frac{b}{x^2} - \frac{c}{\sqrt{2x}}\right) dx$ $\qquad\qquad$ $\left[\dfrac{9a}{4}x^{4/3} - \dfrac{b}{x} - c\sqrt{2x} + C\right]$

19. $\displaystyle\int (2-t^2)^2\sqrt{t}\, dt$ $\qquad\qquad$ $[\frac{8}{3}t^{3/2} - \frac{8}{7}t^{7/2} + \frac{2}{11}t^{11/2} + C]$

20. $\displaystyle\int (x^2 - a^2)^2 x^3\, dx$ $\qquad\qquad$ $[\frac{1}{8}x^8 - \frac{1}{3}a^2x^6 + \frac{1}{4}a^4x^4 + C]$

21. $\displaystyle\int \frac{x\, dx}{x^2 \pm a^2}$

22. $\displaystyle\int \frac{dx}{2x+a}$ $\qquad\qquad$ $[\frac{1}{2}\log|2x+a| + C]$

23. $\displaystyle\int \sin^2\theta \cos\theta\, d\theta$

24. $\displaystyle\int \frac{\cos\theta}{\sin\theta}\, d\theta$ $\qquad\qquad [\log|\sin\theta| + C]$

25. $\displaystyle\int \frac{ax\, dx}{b + cx^2}$ $\qquad\qquad \left[\dfrac{a}{2c}\log|b + cx^2| + C\right]$

26. $\displaystyle\int e^{u+u^2}(1 + 2u)\, du$

27. $\displaystyle\int e^{\sin\theta}\cos\theta\, d\theta$ $\qquad\qquad [e^{\sin\theta} + C]$

28. $\displaystyle\int 2^x\, dx$ $\qquad\qquad [2^x/\log 2 + C]$

29. $\displaystyle\int (\sin\theta^2)\theta\, d\theta$ $\qquad\qquad [-\tfrac{1}{2}\cos\theta^2 + C]$

30. $\displaystyle\int \sqrt{\cos\theta}\,\sin\theta\, d\theta$

31. $\displaystyle\int \frac{(a^{2/3} - x^{2/3})^2}{\sqrt{x}}\, dx$ $\qquad [2a^{4/3}x^{1/2} - \tfrac{12}{7}a^{2/3}x^{7/6} + \tfrac{6}{11}x^{11/6} + C]$

32. $\displaystyle\int \csc^2 at\, dt$ $\qquad\qquad \left[-\dfrac{1}{a}\cot at + C\right]$

33. $\displaystyle\int \frac{\sec^2 2t\, dt}{1 + \tan 2t}$ $\qquad\qquad [\tfrac{1}{2}\log|1 + \tan 2t| + C]$

34. $\displaystyle\int \frac{\log z}{z}\, dz$

35. $\displaystyle\int \frac{e^{-x}\, dx}{1 + e^{-x}}$ $\qquad\qquad [-\log(1 + e^{-x}) + C]$

36. $\displaystyle\int \frac{dx}{e^x + 1}$

37. $\displaystyle\int \frac{e^x + 1}{e^x - 1}\, dx$ $\qquad\qquad [\log(e^x - 1)^2 - x + C]$

38. $\displaystyle\int \frac{x^2 + x + 1}{x + 1}\, dx$ $\qquad\qquad [\tfrac{1}{2}x^2 + \log|x + 1| + C]$

39. $\displaystyle\int (\log y)^{3/2}\frac{dy}{y}$

40. $\displaystyle\int \frac{x^a}{b + cx^{a+1}}\, dx$    $\left[\dfrac{1}{c\,(a+1)}\,\log|\,b + cx^{a+1}\,| + C\right]$

41. $\displaystyle\int \frac{\sin t}{a + b\cos t}\, dt$

42. $\displaystyle\int \frac{dx}{1 + \sqrt{x+1}}$    $[\,2\sqrt{x+1} - 2\log\,(1 + \sqrt{x+1}) + C\,]$

43. $\displaystyle\int \frac{dx}{a + \sqrt{bx + c}}$

44. $\displaystyle\int \frac{y\, dy}{\sqrt[3]{a^2 - y^2}}$    $[\,-\tfrac{3}{4}(a^2 - y^2)^{2/3} + C\,]$

45. $\displaystyle\int \cos^3 k\theta \, \sin k\theta \, d\theta$

46. $\displaystyle\int \frac{\sin \log t}{t}\, dt$

47. $\displaystyle\int \frac{4\, du}{e^{2u}}$    $[\,-2e^{-2u} + C\,]$

48. $\displaystyle\int \frac{a\, dx}{\sqrt{e^{bx}}}$    $\left[\dfrac{-2a}{b\sqrt{e^{bx}}} + C\right]$

49. $\displaystyle\int \frac{5\, dt}{a^3 t}$    $\left[\dfrac{5}{a^3}\,\log|\,t\,| + C\right]$

50. $\displaystyle\int 2^y 3^y \, dy$

51. $\displaystyle\int (e^{ax} + e^{-ax})^2 \, dx$

52. $\displaystyle\int \frac{e^{\sin \sqrt{\theta}} \cos \sqrt{\theta}}{\sqrt{\theta}}\, d\theta$    $[\,2e^{\sin \sqrt{\theta}} + C\,]$

53. $\displaystyle\int e^{\sin^2 \theta} \sin 2\theta \, d\theta$

54. $\displaystyle\int \sec \theta^2 \tan \theta^2 \, (\theta \, d\theta)$    $[\,\tfrac{1}{2}\sec \theta^2 + C\,]$

55. $\displaystyle\int \frac{dv}{\cos^2 3v}$

$\qquad$ $[\frac{1}{3}\tan 3v + C\,]$

56. $\displaystyle\int \frac{ds}{\sqrt{s}\,\sin^2 \sqrt{s}}$

57. $\displaystyle\int \frac{d\theta}{1+\sin\theta} = \int \frac{1-\sin\theta}{\cos^2\theta}\,d\theta$

$\qquad$ $[\tan\theta - \sec\theta + C\,]$

58. $\displaystyle\int \frac{dt}{1-\cos t}$

$\qquad$ $[-\cot t - \csc t + C\,]$

59. $\displaystyle\int \frac{dv}{\sqrt{4-3v^2}}$

$\qquad$ $\left[\dfrac{1}{\sqrt{3}}\,\text{Arc}\sin\dfrac{\sqrt{3}v}{2} + C\right]$

60. $\displaystyle\int \frac{5x^2\,dx}{a^3+x^3}$

61. $\displaystyle\int \frac{dr}{r^2+4r+13}$

$\qquad$ $\left[\dfrac{1}{3}\,\text{Arc}\tan\dfrac{r+2}{3} + C\right]$

62. $\displaystyle\int \frac{dy}{9y^2-12y+13}$

63. $\displaystyle\int \frac{dx}{\sqrt{12+4x-x^2}}$

$\qquad$ $\left[\text{Arc}\sin\dfrac{x-2}{4} + C\right]$

64. $\displaystyle\int \frac{dx}{ax^2+bx+c}$,

$\qquad$ $\left[\dfrac{2}{\sqrt{4ac-b^2}}\,\text{Arc}\tan\dfrac{2ax+b}{\sqrt{4ac-b^2}} + C\right]$

where $4ac - b^2 > 0$, $a > 0$

65. $\displaystyle\int \frac{dx}{2x^2-3x+2}$

$\qquad$ $\left[\dfrac{2}{\sqrt{7}}\,\text{Arc}\tan\dfrac{4x-3}{\sqrt{7}} + C\right]$

66. $\displaystyle\int \frac{\cos\theta\,d\theta}{1+\sin^2\theta}$

67. $\displaystyle\int \frac{ds}{s\log s}$

68. $\displaystyle\int \frac{y\,dy}{\sqrt{y-y^2}}$

$\qquad$ $[-\sqrt{y-y^2} + \frac{1}{2}\,\text{Arc}\sin\,(2y-1) + C\,]$

69. $\displaystyle\int \frac{dv}{\sqrt{4v-9v^2}}$

70. $\displaystyle\int \frac{\text{Arc sin } t}{\sqrt{1-t^2}}\, dt$ $\qquad\qquad$ $\left[\frac{1}{2}(\text{Arc sin } t)^2 + C\right]$

71. $\displaystyle\int \frac{\text{Arc sin } t^2}{\sqrt{1-t^4}}\, t\, dt$

72. $\displaystyle\int \frac{\text{Arc sin } \sqrt{t}}{\sqrt{t-t^2}}\, dt$ $\qquad\qquad$ $\left[(\text{Arc sin } \sqrt{t})^2 + C\right]$

73. $\displaystyle\int \tan^3 a\theta \sec^2 a\theta\, d\theta$ $\qquad\qquad$ $\left[\dfrac{1}{4a}\tan^4 a\theta + C\right]$

74. $\displaystyle\int \frac{x^2 - x - 2}{x+1}\, dx$

75. $\displaystyle\int \frac{x^2 - 2x + 3}{x+1}\, dx$ $\qquad$ $\left[\frac{1}{2}x^2 - 3x + 3\log (x+1)^2 + C\right]$

76. $\displaystyle\int \cot^2 \sqrt{\theta}\, \csc^2 \sqrt{\theta}\, \frac{d\theta}{\sqrt{\theta}}$ $\qquad\qquad$ $\left[-\frac{2}{3}\cot^3 \sqrt{\theta} + C\right]$

77. $\displaystyle\int e^{\tan 3t} \sec^2 3t\, dt$ $\qquad\qquad$ $\left[\frac{1}{3}e^{\tan 3t} + C\right]$

78. $\displaystyle\int \frac{\text{Arc tan } \sqrt{\theta}}{\sqrt{\theta}\,(1+\theta)}\, d\theta$ $\qquad\qquad$ $\left[(\text{Arc tan } \sqrt{\theta})^2 + C\right]$

79. $\displaystyle\int \frac{dw}{w\sqrt{\log w}}$ $\quad \log w = u$ $\qquad\qquad$ $\left[2\sqrt{\log w} + C\right]$

80. $\displaystyle\int \sqrt{e^{2x} + e^{3x}}\, dx$

81. $\displaystyle\int \frac{ax+b}{cx+d}\, dx$ $\qquad\qquad$ $\left[\dfrac{a}{c}x + \dfrac{bc-ad}{c^2}\log |\, cx+d\,| + C\right]$

82. $\displaystyle\int \theta^{-1/2} \csc \sqrt{\theta}\, \cot \sqrt{\theta}\, d\theta$ $\qquad\qquad$ $\left[-2\csc \sqrt{\theta} + C\right]$

83. $\displaystyle\int \frac{t-5}{\sqrt{25-9t^2}}\, dt$ $\qquad\qquad$ $\left[-\dfrac{1}{9}\sqrt{25-9t^2} - \dfrac{5}{3}\text{Arc sin } \dfrac{3t}{5} + C\right]$

84. $\displaystyle\int \frac{dt}{\sqrt{1-t-2t^2}}$ $\qquad\qquad$ $\left[\dfrac{1}{\sqrt{2}}\text{Arc sin } \dfrac{4t+1}{3} + C\right]$

85.  $\displaystyle\int (a^{1/2} - y^{1/2})^3 \, dy$ $\qquad\qquad [a^{3/2}y - 2ay^{3/2} + \frac{3}{2}a^{1/2}y^2 - \frac{2}{5}y^{5/2} + C]$

86.  $\displaystyle\int \sqrt[3]{1 + e^{\theta}} \, e^{\theta} \, d\theta$ $\qquad\qquad [\frac{3}{4}(1 + e^{\theta})^{4/3} + C]$

87.  $\displaystyle\int \frac{e^{2\theta}}{e^{\theta} - 1} \, d\theta$

88.  $\displaystyle\int a^{2x-b} \, dx$ $\qquad\qquad \left[\dfrac{a^{2x}}{2a^b \log a} + C\right]$

89.  $\displaystyle\int \csc^2 \frac{y}{b} \, dy$ $\qquad\qquad \left[-b \cot \dfrac{y}{b} + C\right]$

90.  $\displaystyle\int \sec^2 (a + by) \, dy$ $\qquad\qquad \left[\dfrac{1}{b} \tan (a + by) + C\right]$

91.  $\displaystyle\int \frac{du}{u^2 + 2u + 5}$ $\qquad\qquad \left[\dfrac{1}{2} \, \text{Arc tan} \, \dfrac{u+1}{2} + C\right]$

92.  $\displaystyle\int \frac{d\theta}{2\theta^2 - 2\theta + 1}$ $\qquad\qquad [\text{Arc tan} \, (2\theta - 1) + C]$

93.  $\displaystyle\int \frac{dz}{\sqrt{3z - 2 - z^2}}$ $\qquad\qquad [\text{Arc sin} \, (2z - 3) + C]$

94.  $\displaystyle\int \frac{ay + b}{\sqrt{1 - b^2 y^2}} \, dy$

95.  $\displaystyle\int \frac{e^{-\theta}}{1 + e^{-2\theta}} \, d\theta$ $\qquad e^{-\theta} = u \qquad\qquad [-\text{Arc tan} \, e^{-\theta} + C]$

96.  $\displaystyle\int \frac{dt}{t\sqrt{a^2 - (\log t)^2}}, \quad a < 0$

97.  $\displaystyle\int \frac{\cot^6 2\theta}{\sin^2 2\theta} \, d\theta$ $\qquad\qquad [-\frac{1}{14} \cot^7 2\theta + C]$

In Problems 98 to 111 evaluate the definite integrals.

98.  $\displaystyle\int_{-1}^{1} \frac{t \, dt}{\sqrt{4 - t^2}}$ $\qquad\qquad$ 99.  $\displaystyle\int_{-1}^{0} \frac{t \, dt}{\sqrt{4 - t^2}}$ $\qquad\qquad [\sqrt{3} - 2]$

100. $\displaystyle\int_0^{-1} \frac{dt}{\sqrt{4-t^2}}$  $\left[-\dfrac{\pi}{6}\right]$   106. $\displaystyle\int_{-5\pi/4}^{-\pi} \frac{\sin\theta\,d\theta}{\cos^3\theta}$

101. $\displaystyle\int_0^a \frac{d\theta}{a^2+\theta^2}$   107. $\displaystyle\int_0^{\pi/2} \frac{\cos\theta}{1+\sin^2\theta}\,d\theta$  $\left[\dfrac{\pi}{4}\right]$

102. $\displaystyle\int_{-a}^a \frac{\theta\,d\theta}{a^4+\theta^4}$  $[0]$   108. $\displaystyle\int_{1/2}^{3/2} \frac{dx}{\sqrt{2x-x^2}}$

103. $\displaystyle\int_{-3}^3 \frac{2x-5}{x^2+9}\,dx$  $\left[-\dfrac{5\pi}{6}\right]$   109. $\displaystyle\int_{-e}^e te^{-t^2}\,dt$  $[0]$

104. $\displaystyle\int_{-1}^{-2} \frac{dx}{x}$   110. $\displaystyle\int_0^{\log e} te^{-t^2}\,dt$  $\left[\dfrac{e-1}{2e}\right]$

105. $\displaystyle\int_{-5\pi/4}^{-\pi} \frac{\sin\theta\,d\theta}{\cos\theta}$  $[-\log\sqrt{2}]$   111. $\displaystyle\int_{1/2}^{7/8} \frac{(2x-1)\,dx}{\sqrt{2-x^2+x}}$  $[3-\tfrac{3}{4}\sqrt{15}]$

112. Find the area of the region under one arch of the sine curve.

113. The region under the graph of $y=e^{-x}$, and between $x=-\log a$ and $x=\log a$, $(a>1)$ is revolved about the $x$-axis. Find the volume of the solid generated.
$[\pi(a^4-1)/2a^2]$

114. Find the area of the region bounded by $x=0$, $y=e^2$, and $y=e^{2x}$.
$[(e^2+1)/2]$

115. Find the area of the region under the curve $y=(\log x)/x$ and between $x=1$ and $x=a$.
$[\tfrac{1}{2}\log^2 a]$

116. The region under the catenary, $y=(a/2)(e^{x/a}+e^{-x/a})$ and between $x=0$ and $x=a$ is revolved about the $x$-axis. Find the volume of the solid generated.
$[(\pi a^3/8)(e^2-e^{-2}+4)]$

## 3  Some Additional Formulas

The basic list of integration formulas appears to have certain gaps. For example, $\int \sec x\,dx$ does not appear and one would expect it to be simple enough. Moreover, while one can integrate $\int dx/(a^2+x^2)$ and $\int dx/\sqrt{a^2-x^2}$, there are no formulas for $\int dx/(a^2-x^2)$ and $\int dx/\sqrt{x^2\pm a^2}$. These gaps in the basic list are filled by the following additional formulas.

(**XIV**)  $\displaystyle\int \tan u\, du = \log |\sec u| + C$

(**XV**)  $\displaystyle\int \cot u\, du = \log |\sin u| + C$

(**XVI**)  $\displaystyle\int \sec u\, du = \log |\sec u + \tan u| + C$

(**XVII**)  $\displaystyle\int \csc u\, du = \log |\csc u - \cot u| + C$

(**XVIII**)  $\displaystyle\int \frac{du}{a^2 - u^2} = \frac{1}{2a} \log \left| \frac{a+u}{a-u} \right| + C; \int \frac{du}{u^2 - a^2} = \frac{1}{2a} \log \left| \frac{u-a}{u+a} \right| + C$

(**XIX**)  $\displaystyle\int \frac{du}{\sqrt{u^2 \pm a^2}} = \log |u + \sqrt{u^2 \pm a^2}| + C$

All these can be established by differentiation.
The following device generates XVI.

$$\int \sec u\, du = \int \frac{\sec u\, (\sec u + \tan u)\, du}{\sec u + \tan u}$$

$$= \int \frac{d(\sec u + \tan u)}{\sec u + \tan u} = \log |\sec u + \tan u| + C.$$

Formula XVIII will be derived again in the next chapter as a special case of a general method. It is valid because

$$\frac{1}{a^2 - u^2} = \frac{1}{2a}\left( \frac{1}{a+u} + \frac{1}{a-u} \right).$$

Formula XIX will also be derived in the next chapter. At present the only proof is by differentiation.

**Example**    Sometimes the proper formula is disguised.

$$\int \frac{dt}{\sqrt{(t-2)^2 + 1}} = \log |t - 2 + \sqrt{(t-2)^2 + 1}| + C.$$

$$\int \frac{dx}{\sqrt{x^2 - x - 1}} = \int \frac{dx}{\sqrt{(x - \frac{1}{2})^2 - \frac{5}{4}}} = \log |x - \tfrac{1}{2} + \sqrt{x^2 - x - 1}| + C.$$

$$\int \frac{dx}{x^2 - x} = \int \frac{dx}{(x - \frac{1}{2})^2 - \frac{1}{4}} = \frac{1}{2(\frac{1}{2})} \log \left| \frac{(x - \frac{1}{2}) - \frac{1}{2}}{(x - \frac{1}{2}) + \frac{1}{2}} \right| + C = \log \left| \frac{x-1}{x} \right| + C$$

## Problems

1. Verify **XIV** to **XIX** by differentiation.

2. Derive $\int \csc u \, du$ by a trick similar to the one used to derive $\int \sec u \, du$.

**In Problems 3 to 36 give the indefinite integrals.**

3. $\displaystyle\int \tan{(ax - b)}\, dx$

4. $\displaystyle\int \cot{(b - ax)}\, dx$

5. $\displaystyle\int \theta^{-1/2} \sec{\theta^{1/2}}\, d\theta$ $\qquad\qquad [\tfrac{1}{2}\log|\sec\sqrt{\theta} + \tan\sqrt{\theta}| + C\,]$

6. $\displaystyle\int \frac{dy}{2 - 3y^2}$

7. $\displaystyle\int \frac{du}{a^2u^2 - b^2}$ $\qquad\qquad \left[\dfrac{1}{2ab}\log\left|\dfrac{b - au}{b + au}\right| + C\right]$

8. $\displaystyle\int \frac{dr}{\sqrt{4r^2 - 9}}, \quad r > 0$ $\qquad [\log\sqrt{2r} + \sqrt{4r^2 - 9} + C\,]$

9. $\displaystyle\int \frac{dt}{\sqrt{2t^2 + 3}}$

10. $\displaystyle\int \frac{\tan\log\theta\, d\theta}{\theta}$ $\qquad\qquad [\log|\sec{(\log\theta)}| + C\,]$

11. $\displaystyle\int \frac{ax^2\, dx}{a^2 - b^2x^6}$

12. $\displaystyle\int \sec 2\theta\, d\theta$

13. $\displaystyle\int \frac{dx}{x\sqrt{1 - \log^2 x}}$ $\qquad\qquad [\text{Arc}\sin{(\log x)} + C\,]$

14. $\displaystyle\int \sqrt{\theta}\tan{(\theta\sqrt{\theta})}\, d\theta$

15. $\displaystyle\int \csc a\theta\, d\theta$

16. $\displaystyle\int \frac{e^x\, dx}{\sqrt{e^{2x}+e^{2a}}}$      $[\log\,(e^x + \sqrt{e^{2x}+e^{2a}}) + C]$

17. $\displaystyle\int \frac{\cos a\theta\, d\theta}{a^2 - \sin^2 a\theta}$

18. $\displaystyle\int \frac{2\, dz}{8z^2 - 8z + 1}$      $\left[\dfrac{1}{2\sqrt{2}}\log\left|\dfrac{1+\sqrt{2}-2\sqrt{2}z}{1-\sqrt{2}+2\sqrt{2}z}\right| + C\right]$

19. $\displaystyle\int \frac{dt}{3 - 4t - 4t^2}$

20. $\displaystyle\int \frac{\sqrt{3}y - 3}{\sqrt{3y^2 - 9}}\, dy$      $\left[\dfrac{1}{\sqrt{3}}\sqrt{3y^2 - 9} - \sqrt{3}\log|\sqrt{3}y + \sqrt{3y^2-9}| + C\right]$

21. $\displaystyle\int \frac{v + 2}{\sqrt{v^2 + 4}}\, dv$      $[\sqrt{v^2 + 4} + \log\,(v + \sqrt{v^2+4})^2 + C]$

22. $\displaystyle\int \frac{y - 3}{3y^2 - 4}\, dy$

23. $\displaystyle\int \frac{e^{ax}\, dx}{e^{2ax} - 1}$      $\left[\dfrac{1}{2a}\log\left|\dfrac{e^{ax}-1}{e^{ax}+1}\right| + C\right]$

24. $\displaystyle\int \frac{3\, ds}{\sqrt{9 - 2s^2}}$      $\left[\dfrac{3}{\sqrt{2}}\,\text{Arc sin}\,\dfrac{\sqrt{2}s}{3} + C\right]$

25. $\displaystyle\int \frac{\sin\theta\, d\theta}{\sin^2\theta - 1}$

26. $\displaystyle\int \frac{\sin 2\theta\, d\theta}{\sin^2\theta - 1}$      $[\log\,(\cos^2\theta) + C]$

27. $\displaystyle\int \frac{\cos\theta\, d\theta}{\sin^2\theta - 1}$      $\left[\log\sqrt{\dfrac{1 - \sin\theta}{1 + \sin\theta}} + C\right]$

28. $\displaystyle\int \frac{a\, dx}{x^2 + 2ax + a^2}$

29. $\displaystyle\int \frac{dy}{3y^2 - 5y - 2}$

30. $\displaystyle\int \frac{dy}{3y^2 - 2y - 1}$      $\left[\dfrac{1}{4}\log\left|\dfrac{3 - 3y}{1 + 3y}\right| + C\right]$

31. $\displaystyle\int \frac{dy}{\sqrt{3y^2 - 2y - 1}}$ $\qquad\left[\dfrac{1}{\sqrt{3}} \log \left| \dfrac{3y - 1}{\sqrt{3}} + \sqrt{3y^2 - 2y - 1} \right| + C\right]$

32. $\displaystyle\int \frac{dy}{\sqrt{1 + 2y - 3y^2}}$ $\qquad\left[\dfrac{1}{\sqrt{3}} \text{ Arc sin} \left(\dfrac{3y - 1}{2}\right) + C\right]$

33. $\displaystyle\int \frac{dy}{3y^2 - 2y + 4}$

34. $\displaystyle\int \frac{dy}{\sqrt{3y^2 - 2y + 4}}$

35. $\displaystyle\int \frac{dx}{\sqrt{8 - 4x - 4x^2}}$ $\qquad\left[\dfrac{1}{2} \text{ Arc sin} \dfrac{2x + 1}{3} + C\right]$

36. $\displaystyle\int \frac{dx}{\sqrt{4x^2 + 4x - 8}}$ $\qquad[\tfrac{1}{2} \log | 2x + 1 + \sqrt{4x^2 + 4x - 8}| + C]$

**In Problems 37 to 42 evaluate the definite integrals.**

37. $\displaystyle\int_1^2 \frac{dx}{\sqrt{2x^2 - 1}}$ $\qquad\left[\dfrac{1}{\sqrt{2}} \log (4 + \sqrt{14} - 2\sqrt{2} - \sqrt{7})\right]$

38. $\displaystyle\int_0^{1/2} \frac{dx}{\sqrt{1 - 2x^2}}$ $\qquad[\pi/4\sqrt{2}]$

39. $\displaystyle\int_{-1/2}^{1/2} \frac{\text{Arc sin } \sqrt{2}x}{\sqrt{1 - 2x^2}} \, dx$

40. $\displaystyle\int_{-a}^a \frac{x \, dx}{\sqrt{x^2 + a^2}}$ $\qquad[0]$

41. $\displaystyle\int_{-a}^{\sqrt{3}a} \frac{dx}{\sqrt{x^2 + a^2}}, \quad a > 0$ $\qquad[\log (2\sqrt{2} + \sqrt{6} + \sqrt{3} + 2)]$

42. $\displaystyle\int_{-2}^2 \frac{3\theta - 2}{\sqrt{2\theta^2 + 1}} \, d\theta$

43. Try to integrate $\displaystyle\int \frac{dx}{\sqrt{1 - x^2}}$ by letting $u = \sqrt{1 - x^2}$.

44. Try to integrate $\displaystyle\int \frac{dx}{\sqrt{x^2 \pm a^2}}$ by letting $u = \sqrt{x^2 \pm a^2}$.

## 4  *Some Trigonometric Integrals*

Integrands involving powers of the trigonometric functions occur frequently enough to merit attention. The formulas are not to be memorized—rather, one should remember the methods of attack. Knowledge of trigonometric identities is vital.

Integrals of the form

$$\int \sin^m u \cos^n u \, du$$

where at least one of $m$ or $n$ is an odd positive integer, can be found by reducing either $m$ or $n$ to 1 by using the identity $\sin^2 u + \cos^2 u = 1$.

**Example 1**
$$\int \sin^{-5/2} u \cos^3 u \, du = \int \sin^{-5/2} u \, (1 - \sin^2 u) \cos u \, du$$

$$= \int \sin^{-5/2} u \cos u \, du - \int \sin^{-1/2} u \cos u \, du$$

$$= -\tfrac{2}{3} \sin^{-3/2} u - 2 \sin^{1/2} u + C.$$

Integrals of the form

$$\int \sin^m u \cos^n u \, du$$

with both $m$ and $n$ even positive integers can be found by use of the half-angle formulas.

$$\sin^2 u = \frac{1 - \cos 2u}{2}, \quad \cos^2 u = \frac{1 + \cos 2u}{2}, \quad \sin u \cos u = \tfrac{1}{2} \sin 2u.$$

These formulas, possibly by repeated use, reduce the integral to the case of Example 1.

**Example 2**
$$\int \sin^4 2x \, dx = \int \left( \frac{1 - \cos 4x}{2} \right)^2 dx$$

$$= \frac{1}{4} \int (1 - 2 \cos 4x + \cos^2 4x) \, dx$$

$$= \frac{1}{4} \int \left( 1 - 2 \cos 4x + \frac{1 + \cos 8x}{2} \right) dx$$

$$= \frac{3x}{8} - \frac{1}{8} \sin 4x + \frac{1}{64} \sin 8x + C.$$

When powers of $\tan u$ or $\cot u$ occur, the identities $1 + \tan^2 u = \sec^2 u$ and $1 + \cot^2 u = \csc^2 u$ *may* prove useful. These substitutions will handle the integrals

$$\int \tan^m u \sec^n u \, du \qquad \text{and} \qquad \int \cot^m u \csc^n u \, du$$

when $n$ is a positive even integer.

**Example 3**
$$\int \tan^3 u \, du = \int \tan u \, (\sec^2 u - 1) \, du$$
$$= \tfrac{1}{2} \tan^2 u + \log |\cos u| + C.$$

The same identities as in Example 3 can be used on the integrals

$$\int \tan^m u \sec^n u \, du \qquad \text{and} \qquad \int \cot^m u \csc^n u \, du,$$

when $m$ is a positive odd integer, to reduce to integrals of the form

$$\int \sec^k u \, (\sec u \tan u) \, du \qquad \text{or} \qquad \int \csc^k u \, (\csc u \cot u) \, du,$$

to which the power formula applies.

**Example 4** $\displaystyle \int \tan^5 u \sec^{3/2} u \, du = \int \tan u \, (\sec^2 u - 1)^2 \sec^{3/2} u \, du$

$$= \int (\sec^{9/2} u - 2 \sec^{5/2} u + \sec^{1/2} u) \sec u \tan u \, du$$

$$= \tfrac{2}{11} \sec^{11/2} u - \tfrac{4}{7} \sec^{7/2} u + \tfrac{2}{3} \sec^{3/2} u + C.$$

Integrals of the form

$$\int \sin mu \cos nu \, du, \quad \int \sin mu \sin nu \, du, \quad \int \cos mu \cos nu \, du,$$

can be reduced to simple integrals by the identities:

$$\sin mx \cos nx = \tfrac{1}{2} \sin (m+n)x + \tfrac{1}{2} \sin (m-n)x$$
$$\sin mx \sin nx = -\tfrac{1}{2} \cos (m+n)x + \tfrac{1}{2} \cos (m-n)x$$
$$\cos mx \cos nx = \tfrac{1}{2} \cos (m+n)x + \tfrac{1}{2} \cos (m-n)x.$$

**Example 5** $\displaystyle \int \sin 4\theta \cos 6\theta \, d\theta = \int \left[ \tfrac{1}{2} \sin 10\theta + \tfrac{1}{2} \sin (-2\theta) \right] d\theta$

$$= -\tfrac{1}{20} \cos 10\theta + \tfrac{1}{4} \cos 2\theta + C.$$

### Problems

In Problems 1 to 37 find the indefinite integral, or evaluate the definite integral.

1. $\displaystyle\int \sin^{1/2}\theta \cos^3\theta \, d\theta$ $\qquad\qquad$ $\left[\frac{2}{3}\sin^{3/2}\theta - \frac{2}{7}\sin^{7/2}\theta + C\right]$

2. $\displaystyle\int \cos^{-3/2}\theta \sin\theta \, d\theta$

3. $\displaystyle\int \cos^m 2\theta \sin^3 2\theta \, d\theta$ $\qquad$ $\left[-\dfrac{1}{2(m+1)}\cos^{m+1}2\theta + \dfrac{1}{2(m+3)}\cos^{m+3}2\theta + C\right]$

4. $\displaystyle\int_{-\pi/2}^{\pi/2} \cos^3\theta \sin^3\theta \, d\theta$ $\qquad\qquad$ $[0]$

5. $\displaystyle\int \sin\theta \cos\theta \, d\theta$ $\qquad$ $\left[\frac{1}{2}\sin^2\theta + C_1 = -\frac{1}{2}\cos^2\theta + C_2 = -\frac{1}{4}\cos 2\theta + C\right]$

6. $\displaystyle\int \cos^3 3\theta \, d\theta$

7. $\displaystyle\int \frac{\sin^3 x}{\cos^2 x}\, dx$ $\qquad\qquad$ $[\sec x + \cos x + C]$

8. $\displaystyle\int \cos^3\theta \csc^3\theta \, d\theta$ $\qquad\qquad$ $\left[-\frac{1}{2}\cot^2\theta - \log|\sin\theta| + C\right]$

9. $\displaystyle\int \sin^3\frac{\theta}{2}\, d\theta$

10. $\displaystyle\int \cot^3 2\theta \, d\theta$

11. $\displaystyle\int \tan^5 x \, dx$ $\qquad\qquad$ $\left[\frac{1}{4}\tan^4 x - \displaystyle\int \tan^3 x \, dx\right]$

12. $\displaystyle\int \tan^2\theta \sec^4\theta \, d\theta$

13. $\displaystyle\int \tan^3\theta \sec^4\theta \, d\theta$ $\qquad\qquad$ $\left[\frac{1}{6}\sec^6\theta - \frac{1}{4}\sec^4\theta + C\right]$

14. $\displaystyle\int \tan^3\theta \sec^{3/2}\theta \, d\theta$

15. $\int \cot^{3/2} \theta \csc^4 \theta \, d\theta$

16. $\int (\tan^2 \theta + \tan^4 \theta) \, d\theta$     $[\frac{1}{3} \tan^3 \theta + C]$

17. $\int \cot^2 \dfrac{x}{3} \, dx$     $\left[ -3 \cot \dfrac{x}{3} - x + C \right]$

18. $\int \csc^4 \alpha x \, dx$     $\left[ -\dfrac{1}{\alpha} \cot \alpha x - \dfrac{1}{3\alpha} \cot^3 \alpha x + C \right]$

19. $\int \dfrac{a \, d\theta}{\tan^4 b\theta}$     $\left[ -\dfrac{a}{3b} \cot^3 b\theta - a \int \cot^2 b\theta \, d\theta \right]$

20. $\int \cot \theta \csc^2 \theta \, d\theta$

21. $\int \csc^4 \theta \tan^2 \theta \, d\theta$

22. $\int_{-\pi/4}^{\pi/4} \tan \theta \sec^4 \theta \, d\theta$     $[0]$

23. $\int (\tan \theta + \cot \theta)^2 \, d\theta$

24. $\int (\tan \theta + \cot \theta)^3 \, d\theta$     $[\frac{1}{2} \tan^2 \theta - \frac{1}{2} \cot^2 \theta + \log \tan^2 \theta + C]$

25. $\int \cos^2 \theta \, d\theta$     $[\frac{1}{2}\theta + \frac{1}{4} \sin 2\theta + C]$

26. $\int_{-a}^{a} \sin \theta \cos \theta \, d\theta$

27. $\int_{-\pi/2}^{\pi/2} \sin^2 \theta \, d\theta$     $[\pi/2]$

28. $\int \cos^4 \dfrac{t}{2} \, dt$

29. $\int \sin^6 \theta \, d\theta$

30. $\displaystyle\int \sin^4\theta\,\cos^2\theta\,d\theta$ $\qquad\qquad\qquad\left[\displaystyle\int \sin^4\theta\,d\theta - \int \sin^6\theta\,d\theta\right]$

31. $\displaystyle\int \tan^2\theta\,\cos^4\theta\,d\theta$

32. $\displaystyle\int \sin^4\theta\,\cos^4\theta\,d\theta$

33. $\displaystyle\int \cos^6\theta\,\sin^2\theta\,d\theta$

$\qquad\qquad\qquad\qquad\left[\tfrac{1}{128}\left(35\theta + 32\sin 2\theta - \tfrac{16}{3}\sin^3 2\theta + 7\sin 4\theta + \tfrac{1}{8}\sin 8\theta + C\right)\right]$

34. $\displaystyle\int_{-\pi/2}^{\pi/2} \cos 2x\,\cos 3x\,dx$ $\qquad\qquad\qquad\qquad\qquad\left[\tfrac{2}{5}\right]$

35. $\displaystyle\int \cos 2x\,\sin 3x\,dx$ $\qquad\qquad\qquad\left[\tfrac{1}{2}\sin 5x + \tfrac{1}{2}\sin x + C\right]$

36. $\displaystyle\int_{-\pi/2}^{\pi/2} \sin 2x\,\cos 3x\,dx$ $\qquad\qquad\qquad\qquad\qquad\left[0\right]$

37. $\displaystyle\int \cos 5x\,\cos 7x\,dx - \int \sin 5x\,\sin 7x\,dx$ $\qquad\left[\tfrac{1}{12}\sin 12x + C\right]$

*38. Find the area of the circle using the parametric equations $x = a\cos\theta$, $y = a\sin\theta$.

*39. Find the volume of the sphere of radius $a$. Use the parametric equations $x = a\cos\theta$, $y = a\sin\theta$, and revolve about the $x$-axis.

*40. Find the area of the region under one arch of the cycloid $x = a(\theta - \sin\theta)$, $y = a(1 - \cos\theta)$.

*41. Find the volume of the solid generated by revolving the region under one arch of the cycloid of Problem 40 about the $x$-axis.

42. The root-mean-square (rms) voltage of an alternating voltage source is the square root of the mean of the square of the voltage, where the mean is taken over one cycle. If the voltage is given by $E = E_0 \sin 2\pi kt$, where $k$ is the frequency and $t$ is in seconds, show that the rms voltage is $E_0/\sqrt{2}$.

## 5  Review Problems

So far we have found integrals which were easily obtained by reference to the simplest formulas, or by a substitution. For the most part, each problem set was devoted to a limited variety of types so that the choice of what to do was not too burdensome. The real test of skill in formal integration is whether one can tell quickly what technique to use on a given integral. Once a correct decision as to method has been made, working out the details is simply a matter of time.

The following list of problems will test the reader's skill in discerning a suitable method. (There may be several that work.) He should first go through the list mentally and note possible approaches to solutions. Certainly any problem for which there is some doubt in the reader's mind should be worked out. Some are very easy, a few are not.

### Problems

1. $\displaystyle\int \frac{\sin x}{1 + \cos^2 x}\, dx$

2. $\displaystyle\int \frac{\sin x}{1 + \cos x}\, dx$

3. $\displaystyle\int \frac{d\theta}{\sin \theta \cos \theta}$

4. $\displaystyle\int \frac{dt}{1 + \sqrt{t}}$

5. $\displaystyle\int \frac{t\, dt}{1 + \sqrt{t}}$

6. $\displaystyle\int \frac{\sin \theta}{\cos 2\theta}\, d\theta$

7. $\displaystyle\int \cot \theta \sqrt{\log \sin \theta}\, d\theta$

8. $\displaystyle\int \frac{\sin \sqrt{2t}}{\sqrt{t}}\, dt$

9. $\displaystyle\int \frac{\sec \sqrt{t}\, \tan \sqrt{t}}{\sqrt{t}}\, dt$

10. $\displaystyle\int \frac{d\theta}{1 - \sin \theta}$

11. $\displaystyle\int \frac{\cos \theta\, d\theta}{1 - 2 \sin \theta}$

12. $\displaystyle\int \frac{d\theta}{\sqrt{4\theta + \theta^{3/2}}}$

13. $\displaystyle\int \frac{e^x}{1 + e^{2x}}\, dx$

14. $\displaystyle\int \frac{dx}{e^{-x} + e^x}$

15. $\displaystyle\int e^{\log x^{1/3}}\, dx$

16. $\displaystyle\int \frac{\text{Arc cos } (x/a)}{\sqrt{a^2 - x^2}}\, dx$

17. $\displaystyle\int \frac{dx}{\sqrt{x^2 - 4x + 5}}$

18. $\displaystyle\int \frac{dx}{\sqrt{4x - x^2 - 3}}$

19. $\displaystyle\int \frac{\sec^2 2\theta}{\tan 2\theta}\, d\theta$

20. $\displaystyle\int \frac{\cos \theta\, d\theta}{(1 + \sin \theta)^m}$

21. $\displaystyle\int \sqrt{1 + \sin \theta}\, d\theta$

22. $\displaystyle\int \sqrt{1 + e^t}\, e^{2t}\, dt$

23. $\displaystyle\int \frac{dx}{x^{1/3}\sqrt{1+x^{2/3}}}$

24. $\displaystyle\int \frac{x\,dx}{x+1}$

25. $\displaystyle\int \frac{x\,dx}{(x+1)^2}$

26. $\displaystyle\int \frac{x\,dx}{x^2+2x+2}$

27. $\displaystyle\int \frac{t\,dt}{\sqrt{1-t}}$

28. $\displaystyle\int x\sqrt{x+1}\,dx$

29. $\displaystyle\int \frac{\sin 3\theta\,d\theta}{\cos^3 3\theta}$

30. $\displaystyle\int \frac{\tan\theta\,d\theta}{\sec\theta+\tan\theta}$

31. $\displaystyle\int \frac{e^{-t}\,dt}{1+e^{2t}}$

32. $\displaystyle\int \frac{1-\cos\theta}{1+\cos\theta}\,d\theta$

33. $\displaystyle\int \frac{e^{2t}\,dt}{(1+e^t)^{1/3}}$

34. $\displaystyle\int \frac{d\theta}{\sin 3\theta}$

35. $\displaystyle\int \frac{d\theta}{\cot^2 2\theta}$

36. $\displaystyle\int \frac{\alpha\theta\,d\theta}{\beta+\gamma\theta^2}$

37. $\displaystyle\int \frac{t\,dt}{(1+2t^2)^2}$

38. $\displaystyle\int \frac{du}{u^2+6u+5}$

39. $\displaystyle\int_{-\pi/2}^{\pi/6} \sin^2 2t\,dt$

40. $\displaystyle\int \frac{\log^3(x+1)}{x+1}\,dx$

41. $\displaystyle\int \tan^2 z\,\sec^4 z\,dz$

42. $\displaystyle\int \frac{\alpha x+\beta}{\beta x+\gamma}\,dx$

43. $\displaystyle\int (e^{-ax}-e^{ax})^2\,dx$

44. $\displaystyle\int x^3\sqrt{1+x^2}\,dx$

45. $\displaystyle\int \frac{t^3}{2-t}\,dt$

46. $\displaystyle\int \frac{dx}{\sqrt{3-x^2+2x}}$

47. $\displaystyle\int \frac{dx}{\sqrt{x^2-2x-3}}$

48. $\displaystyle\int \frac{d\theta}{\sec^2 2\theta}$

49. $\displaystyle\int \frac{dt}{t^2-4}$

50. $\displaystyle\int \frac{dt}{t^2-2t}$

51. $\displaystyle\int \frac{a+bx}{c^2x^2-d^2}\,dx$

52. $\displaystyle\int \sin ax\,\cos ax\,dx$

53. $\displaystyle\int \sin 3x \cos 4x \, dx$

54. $\displaystyle\int \sin^2 x \cos^2 x \, dx$

55. $\displaystyle\int \frac{dy}{y\sqrt{y^2 - 1}}$

56. $\displaystyle\int \frac{3 - 4u}{9 - 4u^2} \, du$

57. $\displaystyle\int \frac{ax + b}{\sqrt{x^2 + a^2}} \, dx$

58. $\displaystyle\int \frac{(2 - 3t)^3}{t^{1/2}} \, dt$

59. $\displaystyle\int \csc^4 \sqrt{x} \, \frac{dx}{\sqrt{x}}$

60. $\displaystyle\int e^{-x^2} x \, dx$

61. $\displaystyle\int_{-a}^{a} (a + bx^2)^{-n} x \, dx$

62. $\displaystyle\int \cot 2t \, dt$

63. $\displaystyle\int \frac{d\theta}{\sqrt{9 - 4\theta^2}}$

64. $\displaystyle\int \frac{du}{\sqrt{9 + 4u^2}}$

65. $\displaystyle\int \frac{ax^2 \, dx}{b + cx^6}$

66. $\displaystyle\int \frac{a \, dx}{b\,(x^2 - c^2)}$

67. $\displaystyle\int \frac{a \, dx}{b^2 x^2 + c^2}$

68. $\displaystyle\int \frac{a \, dx}{\sqrt{b^2 x^2 \pm c^2}}$

69. $\displaystyle\int \frac{a \, dx}{\sqrt{c^2 - b^2 x^2}}$

70. $\displaystyle\int \csc 2t \, dt$

71. $\displaystyle\int \theta \sec \theta^2 \, d\theta$

72. $\displaystyle\int_{-\beta/\alpha}^{(\beta/\alpha)+(2/\alpha)} A \sin\,(\alpha t + \beta) \, dt$

73. $\displaystyle\int_{e-1}^{e} \sin\,(\log t) \, \frac{dt}{t}$

74. $\displaystyle\int \frac{dy}{\sqrt{2 + 2y - y^2}}$

75. $\displaystyle\int \frac{dy}{\sqrt{y^2 - 2y - 2}}$

76. $\displaystyle\int_{-1}^{1} \frac{2x - 3}{\sqrt{4 + x^2}} \, dx$

## CHAPTER 11

## FURTHER INTEGRATION TECHNIQUES

*Integration by Parts*

Thus far, indefinite integrals have been found by changing the form of the integrand by a substitution. Another powerful method, of considerable theoretical importance, is provided by *integration by parts*:

If $u$ and $v$ are functions, then

$$d(uv) = u\,dv + v\,du \quad \leftrightarrow \quad uv = \int u\,dv + \int v\,du$$

or

$$\boxed{\int u\,dv = uv - \int v\,du.}$$  {Integration by parts.

For definite integrals this becomes

$$\int_a^b u(x)\,dv(x) = u(x)v(x)\Big|_a^b - \int_a^b v(x)\,du(x).$$

The formula for integration by parts changes the integrand to a form that *may* be recognizable. There are no certain rules of procedure, but this much can be said:

272

(1) The factor $dv$ should be selected so that one can get $v$ at once, by inspection. A constant of integration is not needed for $v$.

(2) Often one selects for $u$ a factor that would tend to become simpler when differentiated.

*Example 1*
$$\int x \sin x \, dx = \int x \, (\sin x \, dx).$$

If
$$u = x, \qquad dv = \sin x \, dx,$$

then
$$du = dx, \qquad v = -\cos x$$

and
$$\int x \sin x \, dx = -x \cos x - \int (-\cos x) \, dx$$
$$= -x \cos x + \sin x + C.$$

Observe that we could easily integrate $dv$, and that $du$ became simply $dx$.

*Example 2*    One need not write out the substitution for $u$ and $v$. It can be clear from context.

$$\int \underbrace{\text{Arc sin } x}_{u} \underbrace{dx}_{dv} = \underbrace{x}_{v} \underbrace{\text{Arc sin } x}_{u} - \int \underbrace{x}_{v} \underbrace{\frac{dx}{\sqrt{1 - x^2}}}_{du}$$

$$= x \text{ Arc sin } x + \sqrt{1 - x^2} + C.$$

*Example 3*    Sometimes one must apply integration by parts more than once.

$$\int x^5 e^{x^2} \, dx = \int \underbrace{x^4}_{u} \underbrace{x e^{x^2} \, dx}_{dv} = \underbrace{x^4}_{u} \underbrace{(\tfrac{1}{2} e^{x^2})}_{v} - \int \underbrace{\tfrac{1}{2} e^{x^2}}_{v} \underbrace{(4x^3 \, dx)}_{du}.$$

We now apply integration by parts to the last integral.

$$\int \tfrac{1}{2} e^{x^2} (4x^3 \, dx) = \int \underbrace{x^2}_{u} \underbrace{(2x e^{x^2} \, dx)}_{dv} = \underbrace{x^2}_{u} \underbrace{e^{x^2}}_{v} - \int \underbrace{e^{x^2}}_{v} \underbrace{(2x \, dx)}_{du}$$

$$= x^2 e^{x^2} - e^{x^2} + C.$$

Thus, after collecting all terms,

$$\int x^5 e^{x^2} \, dx = \tfrac{1}{2} x^4 e^{x^2} - x^2 e^{x^2} + e^{x^2} + C = e^{x^2} (\tfrac{1}{2} x^4 - x^2 + 1) + C.$$

*Example 4*    Occasionally the original integral will appear on the right after integrating by parts twice.

$$\int e^{ax} \sin bx \, dx = e^{ax} \left( -\frac{1}{b} \cos bx \right) - \int \left( -\frac{1}{b} \cos bx \right) (ae^{ax} \, dx).$$

$$\underbrace{\qquad}_{u} \underbrace{\qquad}_{dv} \quad \underbrace{\qquad}_{u} \underbrace{\qquad}_{v} \quad \underbrace{\qquad}_{v} \underbrace{\qquad}_{du}$$

We now apply integration by parts to the last integral.

$$\frac{a}{b} \int e^{ax} \cos bx \, dx = \frac{a}{b} e^{ax} \left( \frac{1}{b} \sin bx \right) - \frac{a}{b} \int \frac{1}{b} \sin bx \, (ae^{ax} \, dx).$$

$$\underbrace{\qquad}_{u} \underbrace{\qquad}_{dv} \quad \underbrace{\qquad}_{u} \underbrace{\qquad}_{v} \quad \underbrace{\qquad}_{v} \underbrace{\qquad}_{du}$$

If we call

$$I = \int e^{ax} \sin bx \, dx$$

(the original integral) and collect all terms, we obtain

$$I = -\frac{1}{b} e^{ax} \cos bx + \frac{a}{b^2} e^{ax} \sin bx - \frac{a^2}{b^2} I.$$

The desired integral reappears on the right! Consequently we simply solve this algebraic equation for the integral:

$$I = \int e^{ax} \sin bx \, dx = \frac{e^{ax}}{a^2 + b^2} (-b \cos bx + a \sin bx) + C.$$

### Problems

**In Problems 1 to 42 find the integral.**

1.  $\displaystyle \int x \cos x \, dx$                                 $[x \sin x + \cos x + C]$

2.  $\displaystyle \int x \sin 2x \, dx$                                $[-\tfrac{1}{2} x \cos 2x + \tfrac{1}{4} \sin 2x + C]$

3.  $\displaystyle \int x^2 \sin 2x \, dx$

4.  $\displaystyle \int \text{Arc} \tan x \, dx$

5.  $\displaystyle \int x \, \text{Arc} \tan x \, dx$                   $[\tfrac{1}{2} x^2 \, \text{Arc} \tan x + \tfrac{1}{2} \, \text{Arc} \tan x - \tfrac{1}{2} x + C]$

6. $\displaystyle\int \text{Arc sin } \frac{x}{a} \, dx, \quad a > 0$

$$\left[ x \text{ Arc sin } \frac{x}{a} + \sqrt{a^2 - x^2} + C \right]$$

7. $\displaystyle\int \sqrt{1 - x^2} \, dx$

$[\frac{1}{2}x\sqrt{1 - x^2} + \frac{1}{2} \text{ Arc sin } x + C]$

8. $\displaystyle\int \log x \, dx$

9. $\displaystyle\int x \log x \, dx$

$[\frac{1}{2}x^2 \log x - \frac{1}{4}x^2 + C]$

10. $\displaystyle\int x^k \log x \, dx \quad (k \neq -1)$

$$\left[ \frac{1}{k + 1} x^{k+1} \log x - \frac{1}{(k + 1)^2} x^{k+1} + C \right]$$

11. $\displaystyle\int x^k \log^2 x \, dx \quad (k \neq -1)$

$$\left[ x^{k+1} \left( \frac{1}{k + 1} \log^2 x - \frac{2}{(k + 1)^2} \log x + \frac{2}{(k + 1)^3} \right) + C \right]$$

12. $\displaystyle\int x^{-1} \log x \, dx$

13. $\displaystyle\int \sec^3 \theta \, d\theta$

$[\frac{1}{2} \sec \theta \tan \theta + \frac{1}{2} \log | \sec \theta + \tan \theta | + C]$

14. $\displaystyle\int \cos \sqrt{y} \, dy$

$[2\sqrt{y} \sin \sqrt{y} + 2 \cos \sqrt{y} + C]$

(Let $\sqrt{y} = u$.)

15. $\displaystyle\int \sqrt{y} \cos \sqrt{y} \, dy$

$[2y \sin \sqrt{y} + 4\sqrt{y} \cos \sqrt{y} - 4 \sin \sqrt{y} + C]$

16. $\displaystyle\int \log^2 x \, dx$

17. $\displaystyle\int e^x \sin x \, dx$

$[\frac{1}{2}e^x (\sin x - \cos x) + C]$

18. $\displaystyle\int e^{-2x} \cos x \, dx$

19. $\displaystyle\int e^{ax} \cos bx \, dx$

$$\left[ \frac{1}{a^2 + b^2} e^{ax} (b \sin bx + a \cos bx) + C \right]$$

20. $\displaystyle\int \frac{\text{Arc sin } \sqrt{x}\, dx}{\sqrt{x}}$

21. $\displaystyle\int \text{Arc tan } \sqrt{x}\, dx$    $[(x+1) \text{ Arc tan } \sqrt{x} - \sqrt{x} + C]$

22. $\displaystyle\int x^3 e^{x^2}\, dx$    $[\frac{1}{2}x^2 e^{x^2} - \frac{1}{2}e^{x^2} + C]$

23. $\displaystyle\int x^3 e^{2x}\, dx$

24. $\displaystyle\int \theta \sec^2 a\theta\, d\theta$    $\left[\dfrac{1}{a}\theta \tan a\theta + \dfrac{1}{a^2}\log |\cos a\theta| + C\right]$

25. $\displaystyle\int x^2 \text{ Arc sin } x\, dx$    $[\frac{1}{3}x^3 \text{ Arc sin } x + \frac{1}{9}(2+x^2)\sqrt{1-x^2} + C]$

26. $\displaystyle\int \sin^3 x\, dx$

27. $\displaystyle\int \cos 2x \sin x\, dx$    $[-\frac{2}{3}\cos^3 x + \cos x + C]$

28. $\displaystyle\int x^2 e^{\sqrt{x}}\, dx$    $[e^{\sqrt{x}}(2x^{5/2} - 10x^2 + 40x^{3/2} - 120x + 240x^{1/2} - 240) + C]$

29. $\displaystyle\int x^2 e^{-x}\, dx$

30. $\displaystyle\int \sin \theta \log \cos \theta\, d\theta$    $[\cos \theta (1 - \log \cos \theta) + C]$

31. $\displaystyle\int xa^x\, dx$    $\left[\dfrac{1}{\log a}xa^x - \dfrac{1}{\log^2 a}a^x + C\right]$

32. $\displaystyle\int \theta \sin \theta \cos \theta\, d\theta$

33. $\displaystyle\int \text{Arc cot } z\, dz$    $[z \text{ Arc cot } z + \frac{1}{2}\log (1+z^2) + C]$

34. $\displaystyle\int \frac{\log \log x}{x}\, dx$    $[\log x \log \log x - \log x + C]$

35. $\displaystyle\int x^3 \text{ Arc tan } 2x^2\, dx$    $[\frac{1}{16}(4x^4 + 1) \text{ Arc tan } (2x^2) - \frac{1}{8}x^2 + C]$

36. $\displaystyle\int \frac{\log \sqrt{x+a}}{x+a}\,dx$ $\qquad \left[\dfrac{1}{4}\log^2 (x+a) + C\right]$

37.✓ $\displaystyle\int x^3 \sqrt{a^2 - x^2}\,dx$ $\qquad \left[-\tfrac{1}{3}x^2 (a^2 - x^2)^{3/2} - \tfrac{2}{15}(a^2 - x^2)^{5/2} + C\right]$

38. $\displaystyle\int x^3 (x^2 - a^2)^{3/2}\,dx$ $\qquad \left[\tfrac{1}{5}x^2 (x^2 - a^2)^{5/2} - \tfrac{2}{35}(x^2 - a^2)^{7/2} + C\right]$

39.✓ $\displaystyle\int \sqrt{a^2 - x^2}\,dx, \quad a > 0$ $\qquad \left[\tfrac{1}{2}x\sqrt{a^2 - x^2} + \dfrac{a^2}{2} \operatorname{Arc\,sin} \dfrac{x}{a} + C\right]$

40. $\displaystyle\int \frac{x^2\,dx}{\sqrt{a^2 - x^2}}$ $\qquad \left[-\dfrac{x}{2}\sqrt{a^2 - x^2} + \dfrac{a^2}{2} \operatorname{Arc\,sin} \dfrac{x}{a} + C\right]$

41.✓ $\displaystyle\int \frac{x^3\,dx}{\sqrt{a^2 - x^2}}$ $\qquad \left[-x^2\sqrt{a^2 - x^2} - \tfrac{2}{3}(a^2 - x^2)^{3/2} + C\right]$

42. $\displaystyle\int \sqrt{a^2 + x^2}\,dx$ $\qquad \left[\dfrac{x}{2}\sqrt{a^2 + x^2} + \dfrac{a^2}{2}\log (x + \sqrt{a^2 + x^2}) + C\right]$

*43.  If $u = dw/dx$ and $v$ is a function of $x$, obtain $\int u\,dv$ in terms of an integral of $v''$ and $w$.

$$[wv' - \textstyle\int wv''\,dx]$$

*44.  Show that, if $f(x) = a_0 x^n + \cdots + a_n$ is a polynomial of degree $n$, then

$$\int f(x)e^{ax}\,dx = e^{ax}\left(\frac{1}{a}f(x) - \frac{1}{a^2}f'(x) + \cdots + \frac{(-1)^n}{a^{n+1}}f^{(n)}(x)\right) + C.$$

45.✓  Use integration by parts on $\int (\log x/x)\,dx$ and so compute the integral.

*46.  Show that $\int x^n e^{x^2}\,dx$ is an elementary function if $n$ is an odd positive integer.

*47.  By induction show that

$$\int_0^{\pi/2} \sin^n x\,dx = \begin{cases} \dfrac{1 \cdot 3 \cdots (n-1)}{2 \cdot 4 \cdots (n)}\,\dfrac{\pi}{2} & \text{if } n \text{ is even;} \\[3ex] \dfrac{2 \cdot 4 \cdots (n-1)}{1 \cdot 3 \cdots (n)} & \text{if } n \text{ is odd, and } n \geq 3. \end{cases}$$

*48.  Evaluate $\qquad\qquad \displaystyle\lim_{x \to 0} \int_1^x t \log t\,dt.$

Hence find the area of the region bounded by $y = x \log x$ and the $x$-axis. $\quad [\tfrac{1}{4}]$

## 2   *Rational Functions*

The principal theoretical result of this section and the next is that the indefinite integral of any rational function is an elementary function. The validity of this theorem will be evident from the examples considered.

Therefore, we are interested in integrals of the form

$$\int R(x)\, dx \qquad \text{where } R(x) = \frac{f(x)}{g(x)} \quad \text{and } f \text{ and } g \text{ are polynomials with real coefficients.}$$

First we observe that, for the discussion here, we may restrict attention to the case where the degree of $f$ is less than the degree of $g$, since by division of polynomials one can reduce the degree of the numerator. For example:

$$\frac{x^3 + 1}{x^2 + x + 1} = x - 1 + \frac{2}{x^2 + x + 1},$$

and the quotient, in this case $x - 1$, is a polynomial and, hence, easily integrated.

The method we use depends on the following theorem from algebra, which we accept without proof.

**THEOREM**     *Any real polynomial can be factored into real linear and real irreducible quadratic factors.\**

To illustrate this, consider

$$g(x) = x^4 - x^3 - x^2 - x - 2;$$

then

$$g(x) = (x - 2)(x + 1)(x^2 + 1).$$

There are two linear factors, and the quadratic factor is *irreducible* because it cannot be factored into real factors [though it does factor into complex factors: $x^2 + 1 = (x - i)(x + i)$]. The irreducible factors may be repeated, for example

$$x^8 + 7x^6 - 10x^5 + 8x^4 - 30x^3 + 9x^2 + 40x - 25$$

$$= (x + 1)(x - 1)^3(x^2 + x + 5)^2.$$

In all problems success depends on being able to factor the denominator $g(x)$ in $f(x)/g(x)$. This is a non-trivial problem. The theory proceeds on the assumption that it has been done. The problems are specially selected so this algebraic problem is easy of solution.

There are two cases to be considered in the integration of $f(x)/g(x)$. This section treats only Case I.

---

\* See Appendix A, page 495.

**CASE I** *The linear and quadratic factors of $g(x)$ are distinct.*

In advanced algebra it is proved that in Case I if

$$g(x) = (x - a_1) \cdots (x - a_m)(x^2 + b_1 x + c_1) \cdots (x^2 + b_n x + c_n),$$

then there are unique constants $A_1, \cdots, A_m, B_1, \cdots, B_n,$ and $C_1, \cdots, C_n$ such that

(1) $$\frac{f(x)}{g(x)} = \frac{A_1}{x - a_1} + \cdots + \frac{A_m}{x - a_m} + \frac{B_1 x + C_1}{x^2 + b_1 x + c_1} + \cdots + \frac{B_n x + C_n}{x^2 + b_n x + c_n}.$$

Notice that for each $i$, $1 \leq i \leq n$, $b_i^2 - 4c_i < 0$ because the quadratic factors are irreducible (see Appendix A). The right member of (1) is called the *partial fraction expansion of $f(x)/g(x)$*. The coefficients $A_1, \cdots, C_n$ are readily found by simple algebra. Then each term in (1) has an elementary integral.

**Example 1** $$\int \frac{2x^2 + 5x + 5}{x^3 + 2x^2 - x - 2} \, dx$$

(2) $$= \int \frac{2x^2 + 5x + 5}{(x + 2)(x + 1)(x - 1)} \, dx$$

$$= \int \left( \frac{A}{x + 2} + \frac{B}{x + 1} + \frac{C}{x - 1} \right) dx$$

(3) $$= \int \frac{A(x^2 - 1) + B(x^2 + x - 2) + C(x^2 + 3x + 2)}{(x + 2)(x + 1)(x - 1)} \, dx.$$

Comparing numerators in (2) and (3) we see that

(4) $$2x^2 + 5x + 5 = A(x^2 - 1) + B(x^2 + x - 2) + C(x^2 + 3x + 2)$$

$$= (A + B + C)x^2 + (B + 3C)x + (-A - 2B + 2C).$$

This identity leads to the simultaneous equations, obtained by equating coefficients* of the two polynomials in (4):

$$A + B + C = 2,$$

$$B + 3C = 5,$$

$$-A - 2B + 2C = 5.$$

This system has the solution†

$$A = 1, \quad B = -1, \quad C = 2.$$

---

* Two polynomials are identical if and only if they have the same coefficients.

† When the factors are linear and distinct the coefficients $A_1, \cdots, A_m$ are most easily obtained by substitution of $a_1, a_2, \cdots, a_n$ in the equation analogous to (4). In this example we substitute $x = 1$, $-1$, $-2$ to obtain: $12 = 6C$, $2 = 2B$, $3 = 3A$.

Thus,    $$\frac{2x^2 + 5x + 5}{x^3 + 2x^2 - x - 2} = \frac{1}{x+2} + \frac{-1}{x+1} + \frac{2}{x-1},$$

and    $$\int \frac{2x^2 + 5x + 5}{x^3 + 2x^2 - x - 2}\,dx = \int \frac{dx}{x+2} - \int \frac{dx}{x+1} + 2\int \frac{dx}{x-1}$$

$$= \log\left|\frac{(x-1)^2(x+2)}{x+1}\right| + C.$$

**Example 2** $$\int \frac{x\,dx}{x^3 - 1} = \int \frac{x\,dx}{(x-1)(x^2+x+1)} = \int\left(\frac{A}{x-1} + \frac{Bx+C}{x^2+x+1}\right)dx,$$

whence, adding fractions and equating numerators,

$$x = A(x^2 + x + 1) + (Bx + C)(x - 1)$$

$$= (A + B)x^2 + (A - B + C)x + (A - C).$$

On equating coefficients we obtain the three equations

$$A + B \qquad = 0,$$
$$A - B + C = 1,$$
$$A \qquad - C = 0,$$

which have the solution

$$A = \tfrac{1}{3}, \quad B = -\tfrac{1}{3}, \quad C = \tfrac{1}{3}.$$

Then

$$\int \frac{x\,dx}{x^3 - 1} = \frac{1}{3}\int \frac{dx}{x-1} - \frac{1}{3}\int \frac{x-1}{x^2+x+1}\,dx$$

$$= \frac{1}{3}\log|x-1| - \frac{1}{3}\int \frac{(x+\tfrac{1}{2}) - \tfrac{3}{2}}{(x+\tfrac{1}{2})^2 + \tfrac{3}{4}}\,dx$$

$$= \frac{1}{3}\log|x-1| - \frac{1}{6}\log(x^2+x+1) + \frac{1}{\sqrt{3}}\,\text{Arc tan}\,\frac{2x+1}{\sqrt{3}} + C.$$

**Problems**

1. $$\int \frac{dx}{a^2 - x^2} \qquad\qquad \left[\frac{1}{2a}\log\left|\frac{a+x}{a-x}\right| + C\right]$$

2. $$\int \frac{x\,dx}{x^2 + 2x - 3} \qquad\qquad \left[\tfrac{1}{4}\log|(x+3)^3(x-1)| + C\right]$$

3. $\displaystyle\int \frac{dx}{x^3 + x}$

$$\left[ \log \frac{|x|}{\sqrt{x^2 + 1}} + C \right]$$

4. $\displaystyle\int \frac{\cos\theta \, d\theta}{\sin^2\theta - \sin\theta - 2}$

$$\left[ \tfrac{1}{3} \log | \, (\sin\theta - 2)^2 (\sin\theta + 1) \, | + C \, \right]$$

5. $\displaystyle\int \frac{x^3 - 1}{x^3 + x} \, dx$

$$\left[ x - \log \frac{|x|}{\sqrt{x^2 + 1}} - \operatorname{Arc\ tan} x + C \right]$$

6. $\displaystyle\int \frac{dx}{e^{-x} + 3 + 2e^x}$

$$\left[ \log \frac{1 + 2e^x}{1 + e^x} + C \right]$$

7. $\checkmark$ $\displaystyle\int \frac{\csc^2\theta \, d\theta}{2 - 4\cot\theta + \csc^2\theta}$

$$\left[ \frac{1}{2} \log \left| \frac{\cot\theta - 1}{\cot\theta - 3} \right| + C \right]$$

8. $\displaystyle\int \frac{x \, dx}{x^4 - a^4}$

$$\left[ \frac{1}{4a^2} \log \left| \frac{x^2 - a^2}{x^2 + a^2} \right| + C \right]$$

9. $\displaystyle\int \frac{x^2 + 4x + 6}{x^3 + x^2 - 2x} \, dx$

10. $\displaystyle\int \frac{x^2 + 4x - 2}{x^3 + x^2 - 2x} \, dx$

$$\left[ \log \left| \frac{x^2 - x}{x + 2} \right| + C \right]$$

11. $\checkmark$ $\displaystyle\int \frac{2x^2 + 5x - 1}{(x + 1)(x^2 + 2x + 5)} \, dx$

$$\left[ \log \frac{(x^2 + 2x + 5)^{3/2}}{|x + 1|} + \frac{1}{2} \operatorname{Arc\ tan} \frac{x + 1}{2} + C \right]$$

12. $\displaystyle\int \frac{x^3 - 4}{x^4 + 5x^2 + 6} \, dx$

$$\left[ \log \frac{(x^2 + 3)^{3/2}}{x^2 + 2} + \frac{4}{\sqrt{3}} \operatorname{Arc\ tan} \frac{x}{\sqrt{3}} - \frac{4}{\sqrt{2}} \operatorname{Arc\ tan} \frac{x}{\sqrt{2}} + C \right]$$

13. $\displaystyle\int \frac{dx}{x^4 + x^2 + 1}$

$$\left[ \frac{1}{4} \log \frac{x^2 + x + 1}{x^2 - x + 1} + \frac{1}{2\sqrt{3}} \operatorname{Arc\ tan} \frac{2x - 1}{\sqrt{3}} + \frac{1}{2\sqrt{3}} \operatorname{Arc\ tan} \frac{2x + 1}{\sqrt{3}} + C \right]$$

14. $\displaystyle\int \frac{2\,dx}{x^4+4}$

$$\left[\frac{1}{8}\log\frac{x^2+2x+2}{x^2-2x+2}+\frac{1}{4}\operatorname{Arc\,tan}\,(x+1)+\frac{1}{4}\operatorname{Arc\,tan}\,(x-1)+C\right]$$

15. $\displaystyle\int \frac{x^2+4x+2}{(x-2)(x^2+5x+7)}\,dx$

$$\left[\frac{1}{6}\log\,(x-2)^4(x^2+5x+7)+\frac{1}{\sqrt{3}}\operatorname{Arc\,tan}\frac{2x+5}{\sqrt{3}}+C\right]$$

16. $\displaystyle\int \frac{6x^2-16x+2}{x^4-5x^3+5x^2+5x-6}\,dx$ $\qquad\left[\log\left|\dfrac{(x-2)^2(x-3)}{(x-1)^2(x+1)}\right|+C\right]$

17. $\displaystyle\int_0^2 \frac{x\,dx}{(x^2+1)(x^2+3)}$ $\qquad\qquad\qquad\qquad[\tfrac{1}{4}\log\tfrac{15}{7}]$

18. $\displaystyle\int_0^2 \frac{x^3-3x^2+3x-1}{(x^2-2x+2)(x^2-2x+3)}\,dx$ $\qquad\qquad[0]$

(There is a shortcut.)

19. $\displaystyle\int_0^{-2} \frac{dx}{2x^2-3x+1}$ $\qquad\qquad\qquad\qquad[\log\tfrac{3}{5}]$

20. $\displaystyle\int_{-1}^0 \frac{2x^2-x+3}{(x^2+1)(x-1)}\,dx$ $\qquad\qquad\left[-\dfrac{\pi}{4}-\log 4\right]$

21. $\displaystyle\int_{-1}^1 \frac{x^3+x+1}{x^4+3x^2+2}\,dx$ $\qquad\qquad\left[\dfrac{\pi}{2}-\sqrt{2}\operatorname{Arc\,tan}\dfrac{1}{\sqrt{2}}\right]$

22. $\displaystyle\int_{-1/2}^{1/2} \frac{(\theta^2+\theta)\,d\theta}{\theta^4+\theta^2-2}$ $\qquad\qquad\left[-\dfrac{1}{3}\log 3+\dfrac{2\sqrt{2}}{3}\operatorname{Arc\,tan}\dfrac{\sqrt{2}}{4}\right]$

## 3   *Rational Functions Concluded*

We now turn to Case II for $\int R(x)\,dx$.

**CASE II**    *Some of the linear or quadratic factors, of the denominator of the rational function, are repeated.*

In this case the partial fraction expansion, analogous to (1) of Section 2, is more complicated. We will content ourselves with a description of the procedure.

A proof that such partial fraction expansions always exist can be found in advanced algebra books.

If the linear factor $(x - a)$ occurs to the $k$th power in $g(x)$, then one must consider partial fractions of the form:

(1)
$$\frac{A_1}{(x - a)} + \frac{A_2}{(x - a)^2} + \cdots + \frac{A_k}{(x - a)^k}.$$

**Example 1** $\displaystyle\int \frac{x\, dx}{(x + 1)^2(x + 2)} = \int \left(\frac{A}{(x + 1)^2} + \frac{B}{x + 1} + \frac{C}{x + 2}\right) dx.$

For the two integrands to be identical we must have

$$x = A(x + 2) + B(x + 1)(x + 2) + C(x + 1)^2$$
$$= (B + C)x^2 + (A + 3B + 2C)x + (2A + 2B + C).$$

Equating coefficients leads to

$$B + C = 0,$$
$$A + 3B + 2C = 1,$$
$$2A + 2B + C = 0,$$

whence $\qquad\qquad A = -1, \quad B = 2, \quad C = -2,$

and $\qquad\displaystyle\int \frac{x\, dx}{(x + 1)^2(x + 2)} = \int \frac{-dx}{(x + 1)^2} + \int \frac{2\, dx}{x + 1} - \int \frac{2\, dx}{x + 2}$

$$= (x + 1)^{-1} + \log \frac{(x + 1)^2}{(x + 2)^2} + C.$$

We now consider repeated quadratic factors in the denominator. If $x^2 + bx + c$ occurs to the $k$th power in $g(x)$, then one must consider partial fractions of the form:

(2)
$$\frac{B_1 x + C_1}{x^2 + bx + c} + \frac{B_2 x + C_2}{(x^2 + bx + c)^2} + \cdots + \frac{B_k x + C_k}{(x^2 + bx + c)^k}.$$

We are thus led to the integration of terms such as

(3) $\qquad\displaystyle\int \frac{Bx + C}{(x^2 + bx + c)^n}\, dx \qquad$ where $n$ is a positive integer $> 1$,

and unless these terms can be integrated we are at an end. Fortunately integrals such as (3) can be dealt with as follows. First we complete the square on the quadratic factor

$$x^2 + bx + c = \left(x + \frac{b}{2}\right)^2 + \frac{4c - b^2}{4}.$$

Then if we make the substitution

$$u = x + \frac{b}{2}, \qquad du = dx,$$

the integral (3) assumes the form

(4) $$\int \frac{\alpha u + \beta}{(u^2 + \gamma^2)^n} \, du = \int \frac{\alpha u \, du}{(u^2 + \gamma^2)^n} + \int \frac{\beta \, du}{(u^2 + \gamma^2)^n}.$$

The first integral in the right member integrates by the power formula. The second integral is handled by the following *reduction formula* (for a proof sketch see Problem 14):

(5) $$\int \frac{du}{(u^2 + \gamma^2)^n} = \frac{u}{(2n - 2)\gamma^2(u^2 + \gamma^2)^{n-1}}$$

$$+ \frac{2n - 3}{(2n - 2)\gamma^2} \int \frac{du}{(u^2 + \gamma^2)^{n-1}} \left.\begin{matrix} \\ \\ \\ \\ \\ \end{matrix}\right\} \begin{array}{l}\text{Reduction}\\\text{formula}\end{array}$$

Successive use of (5) will reduce the final integral in (4) to a constant times Arc tan $u/\gamma$.

*Example 2*   Find the integral

$$I = \int \frac{(x - 1)\, dx}{(x - 2)(x^2 - 2x + 2)^2}$$

$$= \int \left( \frac{A}{x - 2} + \frac{B_1 x + C_1}{x^2 - 2x + 2} + \frac{B_2 x + C_2}{(x^2 - 2x + 2)^2} \right) dx.$$

For identical integrands we must have

$$x - 1 = A\,(x^2 - 2x + 2)^2 + (B_1 x + C_1)(x - 2)(x^2 - 2x + 2)$$

$$+ (B_2 x + C_2)(x - 2)$$

$$= (A + B_1)x^4 + (-4A - 4B_1 + C_1)x^3 + (8A + 6B_1 - 4C_1 + B_2)x^2$$

$$+ (-8A - 4B_1 + 6C_1 - 2B_2 + C_2)x + (4A - 4C_1 - 2C_2).$$

Equating coefficients and solving for $A, \cdots, C_2$, we get, after some effort,

$$A = \tfrac{1}{4}, \quad B_1 = -\tfrac{1}{4}, \quad C_1 = 0, \quad B_2 = -\tfrac{1}{2}, \quad C_2 = 1.$$

$$I = \int \frac{(x - 1)\, dx}{(x - 2)(x^2 - 2x + 2)^2} = \frac{1}{4} \int \frac{dx}{x - 2} - \frac{1}{4} \int \frac{x\, dx}{x^2 - 2x + 2}$$

$$- \frac{1}{2} \int \frac{(x - 2)\, dx}{(x^2 - 2x + 2)^2},$$

and we may as well complete the square and make the substitution at this point.

$$x^2 - 2x + 2 = (x - 1)^2 + 1, \qquad \text{so let} \qquad u = x - 1; \quad \text{then}$$

$$I = \frac{1}{4}\int \frac{du}{u-1} - \frac{1}{4}\int \frac{(u+1)\,du}{u^2+1} - \frac{1}{2}\int \frac{(u-1)\,du}{(u^2+1)^2}$$

(6)
$$= \frac{1}{4}\log|u-1| - \frac{1}{8}\log(u^2+1) - \frac{1}{4}\operatorname{Arc\,tan} u + \frac{1}{4}(u^2+1)^{-1}$$

$$+ \frac{1}{2}\int \frac{du}{(u^2+1)^2}.$$

To the last integral in (6) we apply the reduction formula (5).

(7)
$$\int \frac{du}{(u^2+1)^2} = \frac{u}{2(u^2+1)} + \frac{1}{2}\int \frac{du}{u^2+1}$$

$$= \frac{u}{2(u^2+1)} + \frac{1}{2}\operatorname{Arc\,tan} u + C.$$

Finally, from (6) and (7),

$$I = \frac{1}{8}\log \frac{(u-1)^2}{u^2+1} + \frac{1}{4}\frac{u+1}{u^2+1} + C,$$

and in terms of $x$

$$I = \frac{1}{8}\log \frac{(x-2)^2}{x^2-2x+2} + \frac{1}{4}\frac{x}{x^2-2x+2} + C.$$

### Problems

**Find the integrals.**

1. $\displaystyle \int \frac{dx}{x^3 - x^2}$  $\left[\dfrac{1}{x} + \log\left|\dfrac{x-1}{x}\right| + C\right]$

2. $\displaystyle \int \frac{(u^2 - 2u + 1)\,du}{(u+1)^2(u+3)^2}$  $\left[-\dfrac{5u+7}{u^2+4u+3} + \log\left(\dfrac{u+3}{u+1}\right)^2 + C\right]$

3. $\displaystyle \int \frac{(x-1)^2\,dx}{x^3 + 2x^2 + x}$  $\left[\log|x| + \dfrac{4}{x+1} + C\right]$

4. $\displaystyle \int \frac{x^2\,dx}{(x+1)^3}$  $\left[\log|x+1| + \dfrac{2}{x+1} - \dfrac{1}{2(x+1)^2} + C\right]$

5. $\displaystyle \int \frac{x\,dx}{(x^2 - x + 1)^2}$  $\left[\dfrac{x-2}{3(x^2-x+1)} + \dfrac{2}{3\sqrt{3}}\operatorname{Arc\,tan}\dfrac{2x-1}{\sqrt{3}} + C\right]$

6. $\displaystyle\int \frac{4y\,dy}{(y^2+1)^2(y+1)}$     $\left[\log \dfrac{\sqrt{y^2+1}}{|\,y+1\,|} + \dfrac{y-1}{y^2+1} + C\right]$

7. $\displaystyle\int \frac{x^4+1}{(x^2+1)^3}\,dx$     $\left[\dfrac{3}{4}\,\mathrm{Arc\,tan}\,x - \dfrac{x}{4\,(x^2+1)} + \dfrac{x}{2\,(x^2+1)^2} + C\right]$

8. $\displaystyle\int_{-1}^{1} \frac{x^3-1}{(x^2+1)^3}\,dx$     $\left[-\dfrac{1}{2} - \dfrac{3\pi}{16}\right]$

9. $\displaystyle\int \frac{x^3+2x^2-3}{(x^2+9)^2}\,dx$     $\left[\log\sqrt{x^2+9} + \dfrac{5}{18}\,\mathrm{Arc\,tan}\,\dfrac{x}{3} + \dfrac{27-7x}{6\,(x^2+9)} + C\right]$

10. $\displaystyle\int_{-1}^{1} \frac{2x^3-x-6}{(x^2+1)^4}\,dx$     $\left[-\dfrac{11}{4} - \dfrac{15\pi}{16}\right]$

11. $\displaystyle\int \frac{x^2\,dx}{(x^2+x+2)^2}$     $\left[\dfrac{-3x+2}{7\,(x^2+x+2)} + \dfrac{8}{7\sqrt{7}}\,\mathrm{Arc\,tan}\,\dfrac{2x+1}{\sqrt{7}} + C\right]$

12. $\displaystyle\int \frac{(3u+5)\,du}{(u^2-u+1)^2}$     $\left[\dfrac{13u-11}{3\,(u^2-u+1)} + \dfrac{26}{3\sqrt{3}}\,\mathrm{Arc\,tan}\,\dfrac{2u-1}{\sqrt{3}} + C\right]$

13. $\displaystyle\int \frac{4x\,dx}{(2x+1)^2(x^2+x+1)}$

$\left[\dfrac{2}{3}\log\dfrac{(2x+1)^2}{x^2+x+1} + \dfrac{4}{3\,(2x+1)} + \dfrac{4}{3\sqrt{3}}\,\mathrm{Arc\,tan}\,\dfrac{2x+1}{\sqrt{3}} + C\right]$

*14.   Prove the Reduction Formula (5) as follows:

$$\int \frac{du}{(u^2+\gamma^2)^n} = \frac{1}{\gamma^2}\int \frac{\gamma^2+u^2-u^2}{(u^2+\gamma^2)^n}\,du$$

$$= \frac{1}{\gamma^2}\int \frac{du}{(u^2+\gamma^2)^{n-1}} - \frac{1}{\gamma^2}\int \frac{u^2\,du}{(u^2+\gamma^2)^n}.$$

Now use integration by parts on the last integral, with $dv = u\,du/(u^2+\gamma^2)^n$.

## *4   *Other Rationalizing Substitutions*

The technique of Sections 2 and 3 shows that indefinite integrals of rational functions are elementary functions. It may, therefore, be advantageous in other integrals to make a substitution that leads to a rational integrand, providing that one is possible.

There are many possibilities. The examples provide suggestions.

A result of this section is that the following integrals are elementary:

$$\int R(x^{1/n}) \, dx, \quad \int R(x, \sqrt{x^2 + ax + b}) \, dx, \quad \int R(x, \sqrt{b + ax - x^2}) \, dx,$$

where $R$ is a rational function of one or two variables, and $n$ is an integer.

In practice, however, one uses tables of integrals whenever possible, only carrying out a substitution far enough to use the tables. Some remarks on the use of tables are made in Section 7.

***Example 1***   Integrands of the form $R(x, (ax + b)^{1/n})$ can be made rational by the substitution $u = (ax + b)^{1/n}$. Then

$$x = \frac{1}{a}(u^n - b) \quad \text{and} \quad dx = \frac{n}{a}u^{n-1} \, du.$$

Thus, for the integral

$$I = \int \frac{x^{1/3} - 2x^{1/2}}{1 + x^{1/3}} \, dx = \int R(x^{1/6}) \, dx,$$

let $x^{1/6} = u$. Then $x = u^6$ and $dx = 6u^5 \, du$, and the integral becomes

$$I = \int \frac{u^2 - 2u^3}{1 + u^2} (6)u^5 \, du = \int \frac{6u^7 - 12u^8}{1 + u^2} \, du.$$

This rational integrand is easily integrated.

For the integral

$$I = \int \frac{x + 2}{x\sqrt{x - 2}} \, dx = \int R[(x - 2)^{1/2}] \, dx,$$

let $u = (x - 2)^{1/2}$. Then $x = u^2 + 2$ and $dx = 2u \, du$, and the integral becomes

$$I = \int \frac{(u^2 + 4)2 \, du}{(u^2 + 2)},$$

which is elementary.

***Example 2***   For integrands of the form

$$R(x, \sqrt{a^2 - x^2}) \quad \text{or} \quad R(x, \sqrt{x^2 \perp a^2})$$

the obvious substitutions

$$u = \sqrt{a^2 - x^2} \quad \text{or} \quad u = \sqrt{x^2 \pm a^2}$$

will *often, but not always,* lead to rational integrands. Thus, if $u = \sqrt{a^2 - x^2}$, then

$$\int \frac{\sqrt{a^2 - x^2}}{x} \, dx = \int \frac{u(-u \, du)}{(a^2 - u^2)} = \int \left(1 - \frac{a^2}{a^2 - u^2}\right) du,$$

which is elementary.

On the other hand, if we try the same substitution in the integral $\int dx/\sqrt{a^2 - x^2}$, we get (depending on the sign of $x$)

$$\int \frac{dx}{\sqrt{a^2 - x^2}} = \int \frac{-u\,du}{\pm u\sqrt{a^2 - u^2}} = \mp \int \frac{du}{\sqrt{a^2 - u^2}}.$$

We are right back where we began.

**Example 3**   When the obvious substitutions suggested in Example 2 fail to rationalize the integrand, the following tricky ones *always* will work. It is assumed that $a > 0$.

For   $\displaystyle\int R(x, \sqrt{a^2 - x^2})\,dx$, use

$$\begin{cases} u^2 = \dfrac{a - x}{a + x}, \\[2mm] \text{then} \\[2mm] x = \dfrac{a(1 - u^2)}{1 + u^2}, \\[2mm] dx = \dfrac{-4au\,du}{(1 + u^2)^2}, \\[2mm] \sqrt{a^2 - x^2} = \dfrac{2au}{1 + u^2}. \end{cases}$$

For   $\displaystyle\int R(x, \sqrt{x^2 \pm a^2})\,dx$, use

$$\begin{cases} u = x + \sqrt{x^2 \pm a^2}, \\[2mm] \text{then} \\[2mm] x = \dfrac{u^2 \mp a^2}{2u}, \\[2mm] dx = \dfrac{u^2 \pm a^2}{2u^2}\,du, \\[2mm] \sqrt{x^2 \pm a^2} = \dfrac{u^2 \pm a^2}{2u}. \end{cases}$$

Thus, for example

$$\int \frac{dx}{x\sqrt{a^2 - x^2}} = \int \frac{-4au\,du}{\dfrac{a(1 - u^2)}{1 + u^2}\,\dfrac{2au}{1 + u^2}\,(1 + u^2)^2} = \int \frac{-2\,du}{a(1 - u^2)},$$

and the integral is elementary.

**Example 4**   Integrands of the form

$$R(x, \sqrt{x^2 + ax + b}) \qquad \text{or} \qquad R(x, \sqrt{b + ax - x^2})$$

are easily reduced to those of Example 3 by completing the square and a substitution. Thus,

$$I = \int \frac{dx}{x\sqrt{2 + x - x^2}} = \int \frac{dx}{x\sqrt{\frac{9}{4} - (x - \frac{1}{2})^2}}$$

$$= \int \frac{du}{(u + \frac{1}{2})\sqrt{\frac{9}{4} - u^2}}, \quad \text{if } u = x - \frac{1}{2}.$$

Then, using the first substitution of Example 3,

$$u = \frac{3}{2}\frac{1 - v^2}{1 + v^2},$$

the integral becomes

$$I = \int \frac{-6v\,dv}{(1 + v^2)^2 \left(\frac{4 - 2v^2}{2(1 + v^2)}\right) \frac{3v}{(1 + v^2)}} = \int \frac{2\,dv}{v^2 - 2},$$

which is elementary.

## Problems

**In Problems 1 to 6 find the indefinite integrals.**

1. $\int \dfrac{x^2\,dx}{(2x + 1)^{3/2}}$
$$\left[ \frac{(2x + 1)^{3/2}}{12} - \frac{(2x + 1)^{1/2}}{2} - \frac{1}{4(2x + 1)^{1/2}} + C \right]$$

2. $\int \dfrac{x\,dx}{(ax + b)^{2/3}}$
$$\left[ \frac{3}{4a^2}(ax + b)^{4/3} - \frac{3b}{a^2}(ax + b)^{1/3} + C \right]$$

3. $\int \dfrac{dx}{\sqrt{a^2 + x^2}}$
$$[\log(x + \sqrt{a^2 + x^2}) + C]$$

(Use Example 3.)

4. $\int \dfrac{dx}{(a^2 - x^2)^{3/2}}$
$$\left[ \frac{x}{a^2\sqrt{a^2 - x^2}} + C \right]$$

(Use Example 3.)

5. $\int \dfrac{\sqrt{x^2 - a^2}}{x}\,dx$
$$\left[ \sqrt{x^2 - a^2} - a\,\text{Arc tan}\,\frac{\sqrt{x^2 - a^2}}{a} + C \right]$$

(Use Example 2.)

6. $\int \dfrac{\sqrt{x + 2} - 2}{\sqrt{x + 2} + 2}\,dx$
$$[x - 8\sqrt{x + 2} + 16\log(\sqrt{x + 2} + 2) + C]$$

In Problems 7 to 10 transform, by a suitable substitution, to obtain an integral with a rational integrand. Do not integrate.

7. $\displaystyle\int \frac{\sqrt{x^2 - a^2}}{x}\,dx$    $\left[\displaystyle\int \frac{(u^2 - a^2)^2\,du}{2u^2(u^2 + a^2)}\right]$

(Use Example 3.)

8. $\displaystyle\int \frac{dx}{x\sqrt{x^2 + 2x - 1}}$    $\left[\displaystyle\int \frac{2\,du}{(u^2 - 2u + 2)},\quad u = x + 1 + \sqrt{x^2 + 2x - 1}\right]$

9. $\displaystyle\int \frac{x\,dx}{(3 + 4x - 4x^2)^{3/2}}$    $\left[\dfrac{1}{32}\displaystyle\int \frac{u^2 - 3}{u^2}\,du,\quad u^2 = \dfrac{3 - 2x}{1 + 2x}\right]$

10. $\displaystyle\int \frac{\sqrt{x^2 + a^2}}{x^2}\,dx$    $\left[\displaystyle\int \frac{(u^2 + a^2)^2\,du}{u(u^2 - a^2)^2}\right]$

11.   Verify the formulas for $\sqrt{a^2 - x^2}$, $x$, and $dx$ in Example 3.

12.   Verify the formulas for $\sqrt{x^2 \pm a^2}$, $x$, and $dx$ in Example 3.

## *5   Rational Functions of Sine and Cosine

Because rational functions have elementary integrals, we can also prove that integrands that are rational functions of sine and cosine can be integrated in terms of elementary functions. If $R(\sin\theta,\ \cos\theta)$ is a rational function of $\sin\theta$ and $\cos\theta$, then

$$\int R(\sin\theta,\ \cos\theta)\,d\theta = \text{elementary function of } \theta.$$

The device is to substitute

(1)   $\left[\begin{array}{l} u = \tan\dfrac{\theta}{2} \quad \text{or} \quad \theta = 2\,\text{Arc}\tan u, \quad -\pi < \theta < \pi; \\[2ex] d\theta = \dfrac{2\,du}{1 + u^2}. \end{array}\right.$

Then $-\pi/2 < \theta/2 < \pi/2$, and

$$\sin\frac{\theta}{2} = \frac{u}{\sqrt{1+u^2}}, \qquad \cos\frac{\theta}{2} = \frac{1}{\sqrt{1+u^2}},$$

whence

(2)
$$
\begin{bmatrix}
\sin\theta = 2\sin\frac{\theta}{2}\cos\frac{\theta}{2} = \frac{2u}{1+u^2}, \\[2mm]
\cos\theta = \cos^2\frac{\theta}{2} - \sin^2\frac{\theta}{2} = \frac{1-u^2}{1+u^2}.
\end{bmatrix}
$$

$u > 0$

$u < 0$

From (1) and (2),

$$\int R(\sin\theta,\cos\theta)\,d\theta = \int R\left(\frac{2u}{1+u^2}, \frac{1-u^2}{1+u^2}\right)\frac{2\,du}{1+u^2},$$

and the integrand is a rational function of $u$. Therefore, the integral is an elementary function of $u$ and, hence, an elementary function of $\theta$.

*Example*

$$\int \frac{d\theta}{3 - 2\sin\theta} = \int \frac{2\,du}{(1+u^2)\left(3 - \dfrac{4u}{1+u^2}\right)} = \int \frac{2\,du}{3u^2 - 4u + 3}$$

$$= \frac{2}{\sqrt{5}}\,\text{Arc tan}\,\frac{3u - 2}{\sqrt{5}} + C$$

$$= \frac{2}{\sqrt{5}}\,\text{Arc tan}\,\frac{3\tan\dfrac{\theta}{2} - 2}{\sqrt{5}} + C.$$

*Problems*

1. $\displaystyle\int_{-\pi/2}^{0} \frac{d\theta}{1 - \sin\theta}$  [1]

2. $\displaystyle\int \frac{d\theta}{\cos\theta}$

$$\left[\log\left|\frac{1 + \tan(\theta/2)}{1 - \tan(\theta/2)}\right| + C\right]$$

3. $\displaystyle\int \frac{d\theta}{a + \cos\theta}, \quad 0 < a < 1$

$$\left[\frac{1}{\sqrt{1 - a^2}}\log\left|\frac{\sqrt{\dfrac{a+1}{1-a}} + \tan\dfrac{\theta}{2}}{\sqrt{\dfrac{a+1}{1-a}} - \tan\dfrac{\theta}{2}}\right| + C\right]$$

4.  $\displaystyle\int_{-\pi/4}^{\pi/4} \frac{\sin\theta + \cos\theta}{1 + \sin^2\theta}\,d\theta$ $\qquad\qquad \left[2\,\text{Arc}\,\tan\dfrac{1}{\sqrt{2}}\right]$

5.  $\displaystyle\int \frac{\sin\theta\,d\theta}{1 + \sin\theta}$ $\qquad\qquad \left[\theta + \dfrac{2}{1 + \tan\,(\theta/2)} + C\right]$

6.  $\displaystyle\int_{\pi/3}^{\pi/2} \frac{d\theta}{1 - \cos\theta}$ $\qquad\qquad [\sqrt{3} - 1]$

## 6   *Trigonometric Substitutions*

Use of the trigonometric substitutions in the following table is by far the most common device to find integrals of the form

(1) $\qquad\qquad \displaystyle\int R(x,\,\sqrt{a^2 - x^2})\,dx, \qquad \int R(x,\,\sqrt{x^2 + a^2})\,dx,$

where $R$ is a rational function of two variables.

### THE TRIGONOMETRIC SUBSTITUTIONS

| with $\sqrt{a^2 - x^2}$, use | with $\sqrt{x^2 + a^2}$, use | with $\sqrt{x^2 - a^2}$, use |
|---|---|---|
| $x = a\sin\theta,$ | $x = a\tan\theta,$ | $x = a\sec\theta,$ |
| $-a \leqq x \leqq a,$ | $-\infty < x < \infty,$ | $a \leqq x < \infty,$ |
| $-\pi/2 \leqq \theta \leqq \pi/2;$ | $-\pi/2 < \theta < \pi/2;$ | $0 \leqq \theta < \pi/2;$ |
| $dx = a\cos\theta\,d\theta;$ | $dx = a\sec^2\theta\,d\theta;$ | $dx = a\sec\theta\tan\theta\,d\theta;$ |
| $\theta = \text{Arc}\,\sin\dfrac{x}{a}\,;$ | $\theta = \text{Arc}\,\tan\dfrac{x}{a}\,;$ | $\theta = \text{Arc}\,\cos\dfrac{a}{x}\,;$ |
| $\sqrt{a^2 - x^2} = a\cos\theta$ | $\sqrt{x^2 + a^2} = a\sec\theta$ | $\sqrt{x^2 - a^2} = a\tan\theta$ |

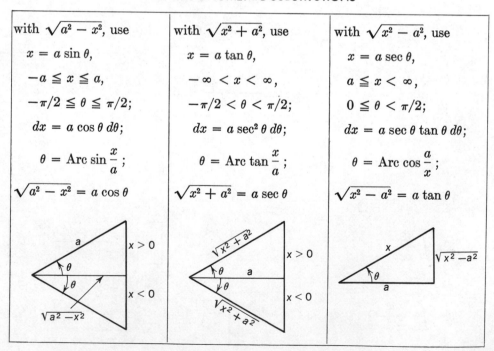

In all cases $a$ is assumed to be positive. $x$ and $\theta$ can be positive or negative in the first two cases, but not in the third.

Evidently, from the trigonometric substitutions, the integrals (1) transform into integrals of rational functions of $\sin \theta$ and $\cos \theta$. By the method of Section 5 these integrals are elementary functions. Frequently the problems work out very simply.

*Example 1*    To evaluate

$$\int_0^a \sqrt{a^2 - x^2} \, dx,$$

let $x = a \sin \theta$; then $dx = a \cos \theta \, d\theta$ and $\sqrt{a^2 - x^2} = a \cos \theta$.

$$\int_0^a \sqrt{a^2 - x^2} \, dx = \int_0^{\pi/2} a \cos \theta \, a \cos \theta \, d\theta$$

$$= a^2 \int_0^{\pi/2} \frac{1 + \cos 2\theta}{2} \, d\theta = \frac{\pi a^2}{4} \,.$$

*Example 2*    To find

$$\int \frac{\sqrt{x^2 - a^2}}{x} \, dx,$$

let $x = a \sec \theta$. Then

$$\int \frac{\sqrt{x^2 - a^2}}{x} \, dx = \int \frac{a \tan \theta \, a \sec \theta \tan \theta \, d\theta}{a \sec \theta}$$

$$= a \int \tan^2 \theta \, d\theta = a \tan \theta - a\theta + C$$

$$= \sqrt{x^2 - a^2} - a \operatorname{Arc} \cos \frac{a}{x} + C.$$

If a definite integral is over an interval where $x$ is negative, then the preliminary substitution $t = -x$ will reduce it to the case in the text. For example,

$$\int_{-2a}^{-a} \frac{\sqrt{x^2 - a^2}}{x} \, dx = \int_{2a}^{a} \frac{\sqrt{t^2 - a^2}}{-t} \, (-dt)$$

$$= \left( \sqrt{t^2 - a^2} - a \operatorname{Arc} \cos \frac{a}{t} \right) \Big|_{2a}^{a}$$

$$= \left( \frac{\pi}{3} - \sqrt{3} \right) a.$$

*Example 3*    The same trigonometric substitutions can sometimes be success-
fully used with rational integrands. Although

$$\int \frac{du}{(u^2 + a^2)^2}$$

was integrated in Section 3 by a reduction formula, the substitution $u = a \tan \theta$
leads to the following.

$$\int \frac{du}{(u^2 + a^2)^2} = \int \frac{a \sec^2 \theta \, d\theta}{a^4 \sec^4 \theta} = \frac{1}{a^3} \int \cos^2 \theta \, d\theta$$

$$= \frac{1}{a^3} \int \frac{1 + \cos 2\theta}{2} \, d\theta = \frac{\theta}{2a^3} + \frac{\sin 2\theta}{4a^3} + C$$

$$= \frac{1}{2a^3} \text{Arc} \tan \frac{u}{a} + \frac{1}{2a^2} \frac{u}{u^2 + a^2} + C.$$

### Problems

**Find the integrals in Problems 1 to 22.**

1. $\displaystyle\int \frac{dx}{\sqrt{x^2 - a^2}}$    $[\log | x + \sqrt{x^2 - a^2} | + C]$

2. $\displaystyle\int \frac{dx}{\sqrt{x^2 + a^2}}$    $[\log | x + \sqrt{x^2 + a^2} | + C]$

3. $\displaystyle\int \frac{dx}{(x^2 - a^2)^{3/2}}$    $\left[ \dfrac{-x}{a^2 \sqrt{x^2 - a^2}} + C \right]$

4. $\displaystyle\int \frac{dx}{(x^2 + a^2)^{3/2}}$    $\left[ \dfrac{x}{a^2 \sqrt{x^2 + a^2}} + C \right]$

5. $\displaystyle\int \frac{dx}{(a^2 - x^2)^{3/2}}$    $\left[ \dfrac{x}{a^2 \sqrt{a^2 - x^2}} + C \right]$

6. $\displaystyle\int_0^3 \frac{x^3 \, dx}{\sqrt{25 - x^2}}$    $\left[ \dfrac{14}{3} \right]$

7. $\displaystyle\int_{-2}^2 \frac{(x^2 - x) \, dx}{\sqrt{x^2 + 4}}$    $[4\sqrt{2} - 4 \log (1 + \sqrt{2})]$

8. $\displaystyle\int \frac{\sqrt{x^2 + a^2}}{x} \, dx$    $\left[ \sqrt{x^2 + a^2} - a \log \dfrac{a + \sqrt{x^2 + a^2}}{| x |} + C \right]$

9. $\displaystyle\int \frac{x^2\, dx}{\sqrt{a^2 - x^2}}$

$$\left[ -\frac{x}{2}\sqrt{a^2 - x^2} + \frac{a^2}{2}\, \text{Arc sin}\, \frac{x}{a} + C \right]$$

10. $\displaystyle\int_{-2}^{-4} \frac{\sqrt{x^2 - 4}}{x^2}\, dx$

$$\left[ \frac{\sqrt{3}}{2} + \log\, (2 - \sqrt{3}) \right]$$

11. $\displaystyle\int \frac{\sqrt{x^2 - a^2}}{x^4}\, dx$

$$\left[ \frac{(x^2 - a^2)^{3/2}}{3a^2 x^3} + C \right]$$

12. $\displaystyle\int x^3\sqrt{a^2 - x^2}\, dx$

$$\left[ -\frac{a^2}{3}\, (a^2 - x^2)^{3/2} + \frac{1}{5}\, (a^2 - x^2)^{5/2} + C \right]$$

13. $\displaystyle\int x^3\sqrt{a^2 + x^2}\, dx$

$$\left[ \frac{1}{5}\, (a^2 + x^2)^{5/2} - \frac{a^2}{3}\, (a^2 + x^2)^{3/2} + C \right]$$

14. $\displaystyle\int_0^a \frac{x^2\, dx}{(a^2 + x^2)^{3/2}}$

$$\left[ \log\, (\sqrt{2} + 1) - \frac{1}{\sqrt{2}} \right]$$

15. $\displaystyle\int \frac{\sqrt{a^2 - x^2}}{x^3}\, dx$

$$\left[ -\frac{\sqrt{a^2 - x^2}}{2x^2} + \frac{1}{2a}\, \log\, \frac{a + \sqrt{a^2 - x^2}}{|x|} + C \right]$$

16. $\displaystyle\int \frac{x^2\, dx}{(a^2 + x^2)^2}$

$$\left[ \frac{1}{2a}\, \text{Arc tan}\, \frac{x}{a} - \frac{x}{2\, (a^2 + x^2)} + C \right]$$

17. $\displaystyle\int \frac{dx}{a^2 - x^2}$

$$\left[ \frac{1}{2a}\, \log\, \left| \frac{a + x}{a - x} \right| + C \right]$$

18. $\displaystyle\int \sqrt{x^2 - a^2}\, dx$

$$\left[ \frac{1}{2}\, x\sqrt{x^2 - a^2} - \frac{a^2}{2}\, \log |\, x + \sqrt{x^2 - a^2}\, | + C \right]$$

19. $\displaystyle\int \frac{x\, dx}{\sqrt{2ax - x^2}}$

$$\left[ a\, \text{Arc sin}\, \frac{x - a}{a} - \sqrt{2ax - x^2} + C \right]$$

20. $\displaystyle\int \frac{\sqrt{x - x^2}}{x^2}\, dx$

$$\left[ \frac{2\, (x - 1)}{\sqrt{x - x^2}} - \text{Arc sin}\, (2x - 1) + C \right]$$

21. $\displaystyle\int \sqrt{x^2 + 2x}\, x\, dx$

$$\left[ -\frac{x + 1}{2}\, (x^2 + 2x)^{1/2} + \frac{1}{3}\, (x^2 + 2x)^{3/2} + \frac{1}{2}\, \log |\, x + 1 + \sqrt{x^2 + 2x}\, | + C \right]$$

22.  $\displaystyle\int x\sqrt{a^2 - x^2}\,\log x\,dx$    $\Bigg[\,-\dfrac{1}{3}\sqrt{a^2 - x^2}\,\log x + \dfrac{a^3}{3}\log\dfrac{(a - \sqrt{a^2 - x^2})}{|\,x\,|}$

$$+\,\frac{a^2}{3}\,(a^2 - x^2)^{1/2} + \frac{1}{9}\,(a^2 - x^2)^{3/2} + C\,\Bigg]$$

23.  $\displaystyle\int \frac{x^3\,dx}{(x^2 + 1)^3}:$    (a) by letting $u = x^2 + 1$; (b) by letting $x = \tan\theta$. Reconcile your answers.

24.  Find the area of the region inside the ellipse $16x^2 + 9y^2 = 144$ and between $x = -2$ and $x = 2$.

$$\left[\frac{16\sqrt{5}}{3} + 24\,\text{Arc}\sin\frac{2}{3}\right]$$

25.  Find the area of the region bounded by the hyperbola $y^2 + 6y - x^2 + 8 = 0$ and the line $y = 0$.    $[\,6\sqrt{2} - \log(3 + 2\sqrt{2})\,]$

## 7   *Tables of Integrals*

As was remarked on p. 247, $\int e^{-x^2}\,dx$ is not an elementary function. Thus, it happens that, although indefinite integrals of elementary functions exist,* they are not all elementary functions. So the problem of finding (representing) antiderivatives of elementary functions is essentially more complicated than the problem of finding derivatives, for which definite algorithms exist. The action of the derivative operator $D$ can be portrayed by the above diagram.

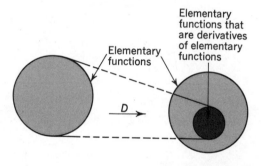

Nor is it possible to adjoin a few functions to the class of elementary functions and relieve the difficulty. As a consequence, no table of integrals of elementary functions can be fully adequate. There will always be integrals that are not listed.† The student should have in his library a good table of integrals, but, for the convenience of the reader, a very short table is to be found in Appendix B. The

---

* $\int_a^x f(t)\,dt = F(x)$ exists, and $F'(x) = f(x)$ when $f$ is continuous, by the Existence Theorem of Chapter 10.

† Today, this problem is less acute in applications. Numerical methods using large digital computers can tabulate indefinite integrals, to a high degree of approximation, rather quickly. See Chapter 12 for some discussion of numerical integration.

reader should consult this table and observe how the integrands are classified. A little practice will suffice to render the table a useful tool.

**Example 1**    In some cases a substitution is a necessary preliminary to using the table. For example,

$$\int \frac{x\, dx}{(2ax - x^2)^{3/2}} = \int \frac{x\, dx}{(a^2 - (x - a)^2)^{3/2}}$$

$$= \int \frac{(u + a)\, du}{(a^2 - u^2)^{3/2}} \qquad \text{(where } u = x - a)$$

$$= \int \frac{u\, du}{(a^2 - u^2)^{3/2}} + a \int \frac{du}{(a^2 - u^2)^{3/2}}.$$

The first integral is handled by a power formula, the second by No. 29 of Appendix B.

**Example 2**    In some cases the particular integral does not occur, but a reduction formula can be used. (Formulas in the Appendix numbered 21 to 24, 36, 37, 53, and 65 to 71 are all reduction formulas.) Thus,

$$\int \frac{x^4}{(a^2 - x^2)^{1/2}}\, dx = -\frac{x^3}{4} \sqrt{a^2 - x^2} + \frac{3a^2}{4} \int \frac{x^2\, dx}{(a^2 - x^2)^{1/2}},$$

by Formula 37, etc.

### Problems

Use the table of integrals as well as the standard methods to find the integrals.

1.  $\displaystyle\int \frac{du}{(u^2 - a^2)^2}$

2.  $\displaystyle\int \cos^8 \theta\, d\theta$

3.  $\displaystyle\int \frac{\sqrt{a^2 + x^2}}{x}\, dx$

4.  $\displaystyle\int \frac{dx}{(a^2 + x^2)^{3/2}}$

5.  $\displaystyle\int \sin^{7/2} \theta \cos^5 \theta\, d\theta$

6.  $\displaystyle\int \csc \theta \cos^5 \theta\, d\theta$

7.  $\displaystyle\int \sec^3 2\theta\, d\theta$

8.  $\displaystyle\int e^{3x} \sin \sqrt{2x}\, dx$

9.  $\displaystyle\int \frac{d\theta}{2 + 3 \cos \theta}$

10. $\displaystyle\int \frac{d\theta}{2 + \sqrt{2} \cos \theta}$

11. $\displaystyle\int \cos 2x \cos 3x \, dx$

12. $\displaystyle\int \sqrt{\frac{2-x}{2+x}} \, dx$

13. $\displaystyle\int \frac{\sqrt{2x^2+7}}{x^2} \, dx$

14. $\displaystyle\int x^3 \sqrt{a^2-x^2} \, dx$

15. $\displaystyle\int \frac{x^3}{\sqrt{a^2-x^2}} \, dx$

16. $\displaystyle\int \frac{\sqrt{a^2-x^2}}{x^3} \, dx$

## 8   Review Problems

Because extensive tables of integrals are readily available, one might think (incorrectly) that dexterity in formal integration can be dispensed with entirely. This is not the case because this dexterity sharpens the insight for algebraic patterns. The recognition of the *type* of the integral is important, and in addition considerable effort may be required to get an integral into a form where the tables are useable.

The problems below should be done *without* recourse to the tables. Only the list of basic formulas, I to XVIII, substitution, parts, rationalization, etc., are needed. The types are mixed, so judgment is necessary. Before working out the integrals in detail it is a worthwhile exercise to examine each problem in the list to see whether a plan of attack suggests itself.

### Problems

Find the integrals.

1. $\displaystyle\int \csc^{-1/2} \theta \cos \theta \, d\theta$

2. $\displaystyle\int \cos^4 \frac{\theta}{2} \, d\theta$

3. $\displaystyle\int \theta \sin \theta \cos \theta \, d\theta$

4. $\displaystyle\int \theta \sec \theta^2 \tan \theta^2 \, d\theta$

5. $\displaystyle\int \frac{\sqrt{x+2}-3}{\sqrt{x+2}+3} \, dx$

6. $\displaystyle\int x^2 \sqrt{4-x^2} \, dx$

7. $\displaystyle\int \frac{du}{u^3+2u^2-u-2}$

8. $\displaystyle\int \log(4+x^2) \, dx$

9. $\displaystyle\int_{-\pi/2}^{\pi/2} \frac{\sin x + \cos x}{2-\sin^2 x} \, dx$

10. $\displaystyle\int \sec^3 \theta \, d\theta$

11. $\displaystyle\int \frac{d\theta}{\sin^4 a\theta}$

12. $\displaystyle\int (\text{Arc} \sin x)^2 \, dx$

13. $\displaystyle\int \frac{\sqrt{2x - x^2}}{x^2}\, dx$

14. $\displaystyle\int \frac{\sqrt{x^2 - x}}{x}\, dx$

15. $\displaystyle\int \frac{\log \sqrt{2x}}{x}\, dx$

16. $\displaystyle\int \frac{t\, dt}{(a + bt^2)^{3/2}}$

17. $\displaystyle\int_{-\pi/4}^{\pi/4} |x|\, \tan x \sec x\, dx$

18. $\displaystyle\int \frac{\sec \sqrt{x}}{\sqrt{x}}\, dx$

19. $\displaystyle\int \frac{dt}{t^2 + 4t + 5}$

20. $\displaystyle\int \frac{dt}{t^2 + 4t + 3}$

21. $\displaystyle\int \frac{\cos\theta\, d\theta}{\sin^2\theta - \sin\theta - 2}$

22. $\displaystyle\int \frac{x^2\, dx}{(a^2 + x^2)^2}$

23. $\displaystyle\int e^{2x} \sin 3x\, dx$

24. $\displaystyle\int \frac{d\theta}{1 + \cos\theta}$

25. $\displaystyle\int \frac{x^{1/3}\, dx}{1 + x^{1/2}}$

26. $\displaystyle\int \frac{1 + x^{1/3}}{x^{1/2}}\, dx$

27. $\displaystyle\int \sqrt{\frac{1 + x}{1 - x}}\, dx$

28. $\displaystyle\int \frac{dx}{x + \sqrt{x^2 + 4}}$

29. $\displaystyle\int \frac{du}{1 + e^u}$

30. $\displaystyle\int x \log \sqrt[3]{2x}\, dx$

31. $\displaystyle\int \frac{\cos\theta\, d\theta}{\sqrt{2 - \cos^2\theta}}$

32. $\displaystyle\int \csc 2\theta\, d\theta$

33. $\displaystyle\int \frac{dx}{x^{1/3} \sqrt{a^{2/3} + x^{2/3}}}$

34. $\displaystyle\int \sqrt{1 - \cos y}\, dy, \quad 0 \le y \le \pi$

35. $\displaystyle\int \frac{du}{\sqrt{1 + e^u}}$

36. $\displaystyle\int \log (x + \sqrt{x^2 - a^2})\, dx$

37. $\displaystyle\int \sin x^{1/3}\, dx$

38. $\displaystyle\int \frac{(x^3 - x^2)\, dx}{x^2 - x - 6}$

39. $\displaystyle\int \frac{(x^2 - 1)\, dx}{(x^2 + 2)(x^2 - 2x + 3)}$

40. $\displaystyle\int \frac{dt}{t^4 - 1}$

41. $\displaystyle\int_2^4 \frac{dt}{1 - \sqrt{t}}$

42. $\displaystyle\int \cos^5\theta\, d\theta$

43. $\displaystyle\int (\tan\theta + \cot\theta)^2\, d\theta$

44. $\displaystyle\int \frac{\sec^4\varphi\, d\varphi}{\tan^4\varphi}$

45. $\displaystyle\int \frac{du}{\cot^3 2u}$

46. $\displaystyle\int \sin 2\theta \cos 3\theta\, d\theta$

47. $\displaystyle\int \frac{\sin^3 t}{\sqrt{\cos t}}\, dt$

48. $\displaystyle\int \frac{dx}{\sqrt{25 - 9x^2}}$

49. $\displaystyle\int_{-1}^{1} \frac{ax + b}{\sqrt{a^2x^2 + b^2}}\, dx$

50. $\displaystyle\int \frac{2x - 3}{3x^2 - 4}\, dx$

51. $\displaystyle\int_{-1/2}^{3/2} \frac{2t - 1}{\sqrt{15 - 4t^2 + 4t}}\, dt$

52. $\displaystyle\int \frac{x\, dx}{\sqrt{3 + 2x}}$

53. $\displaystyle\int \frac{x\, dx}{(3 + 2x)^{3/2}}$

54. $\displaystyle\int (x - 2)e^{2x}\, dx$

55. $\displaystyle\int e^{\sqrt{x}}\, dx$

56. $\displaystyle\int \frac{dx}{(2x - 1)\sqrt{x^2 - x}}, \quad x > 1$

## CHAPTER 12

# APPLICATIONS OF INTEGRATION

In Part I we saw that the concept of derivative had numerous applications: to slope, rate of change, velocity, acceleration, curvature, etc. Although the concept of derivative had its origin in the Problem of Tangents, many of the applications seem but remotely related to tangent lines. In an analogous way we will see in this chapter that the concepts of definite and indefinite integrals have applications not directly related to the origins of these concepts in the Problem of Area.

## 1 Differential Equations

We are familiar with equations in which the unknown is a *number*, such as polynomial equations or trigonometric equations. In many applications, equations arise in which the unknown is a *function*. An equation involving an unknown function and its derivative is called a *first-order differential equation*.

We have already been solving very simple differential equations by integration: If

$$\frac{dy}{dx} = f(x),$$

where $f$ is a given function, then

$$y = \int f(x)\, dx.$$

If $F$ is an antiderivative of $f$, then

$$y = F(x) + C.$$

The family of solutions appears as in the figure.

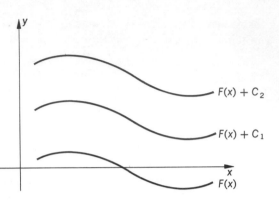

A general theory of differential equations belongs in a separate course, but a large class of equations, in which the *variables are separable*, is easily accessible to us now. These have the form,

(1)  $\dfrac{dy}{dx} = \dfrac{f(x)}{g(y)}$   or   $g(y)\, dy = f(x)\, dx.$

A solution of (1) is a function $y = y(x)$ such that

$$\frac{d}{dx} y(x) = \frac{f(x)}{g[y(x)]}$$

or

$$\frac{dy}{dx} = f(x, y),$$

a first order differential equation.

If   $\dfrac{dy}{dx} = \dfrac{f(x)}{g(y)}$

or   $g(y)\, dy = f(x)\, dx,$

the variables are separable, or separated.

(2)                 $g[y(x)]\, dy(x) = f(x)\, dx,$

identically in $x$.

We shall suppose there is a solution* and then calculate it. Because (2) is an identity, we integrate to obtain

(3)                 $\displaystyle \int g(y)\, dy = \int f(x)\, dx.$

Then if $G$ and $F$ are antiderivatives of $g$ and $f$,

(4)                 $G(y) = F(x) + C.$

---

* *Existence* of solutions is the most troublesome aspect of elementary differential equations, and existence proofs are not given here. In Section 2, however, we shall give geometric reasons for believing that solutions exist.

Equation (4) defines $y$ implicitly as a function of $x$ and gives a solution because implicit differentiation of (4) gives the differential equation (2).

The constant of integration will be determined if $y$ is known at some value of $x$; that is, by initial conditions:

$$y_0 = y(x_0) \quad \{\text{Initial conditions}$$

**Example 1**  Find $y$ if $y' = \dfrac{2x}{y}(1 + y^2)$ and $y(0) = 1$. We have

$$\frac{dy}{dx} = \frac{2x(1 + y^2)}{y}, \qquad \frac{y\,dy}{1 + y^2} = 2x\,dx,$$

and

$$\tfrac{1}{2}\log|1 + y^2| = x^2 + C,$$

or

$$1 + y^2 = e^{2x^2 + 2C} = A e^{2x^2}$$

where $A = e^{2C}$. Because $y(0) = 1$, we have $1 + 1 = A e^0$; thus, $A = 2$, and

$$1 + y^2 = 2e^{2x^2},$$

which gives the solution implicitly. If we solve for $y$, we get

$$y = \sqrt{2e^{2x^2} - 1}.$$

The positive root is used because of the initial conditions.

**Example 2**  Find an equation of the family of curves such that the slope at $(x, y)$ is $-y/x$. Find that one which passes through $(-3, 2)$.
We have $y' = -y/x$, or

$$\frac{dy}{y} = -\frac{dx}{x};$$

thus,  $\log|y| = -\log|x| + C,$

or  $|xy| = e^C.$

Thus, $xy = K$, where $|K| = e^C.$

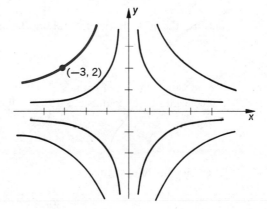

There are infinitely many solutions $y = K/x$, one for each value of $K$. Their graphs constitute a one-parameter family of curves. The one through $(-3, 2)$ is given by $xy = -6.$

*Example 3*    In numerous problems, for example, radioactive decay, the rate of change of a quantity $x$ is proportional to $x$:

$$\frac{dx}{dt} = kx, \qquad k = \text{constant}.$$

In the case of decay the constant $k$ is negative.

To solve the differential equation we have

$$\frac{dx}{x} = k\, dt, \qquad \text{so} \qquad \log |x| = kt + C.$$

Thus, $|x| = Ae^{kt}$, where the constant of integration $A = e^C$ is the value of $|x|$ at $t = 0$.

In the case of decay the constant $k$ is related to the "half-life" $t_0$, that is, the time required for $x$ to decrease from $A$ to $\frac{1}{2}A$. Then

$$\tfrac{1}{2}A = Ae^{kt_0}, \qquad \text{whence} \qquad kt_0 = -\log 2.$$

Thus, for example, if the half-life $t_0$ is ten years,

$$k = (-\log 2)/10 = -0.0693.$$

*Example 4*    A tank contains 500 gal of brine containing 200 lb of dissolved salt. If brine containing 0.1 lb of salt per gal runs into the tank at the rate of 10 gal/min and brine (thoroughly mixed) flows out at the same rate, how much salt remains at the end of one hour.

In problems like this one always has

rate of change = (rate of input) — (rate of output).

Because the brine flowing in contains 0.1 lb of salt per gallon, the

$$\text{rate of input} = (0.1)\,(10).$$

If $x(t)$ is the number of pounds of salt in the tank at the end of $t$ minutes, then the concentration (pounds of salt per gallon) is $x/500$, and the

$$\text{rate of output} = \frac{x}{500}\,(10).$$

Then the rate of change of $x$ is

$$\frac{dx}{dt} = (0.1)\,(10) - \frac{x}{500}\,(10).$$

So

$$\frac{dx}{dt} = \frac{50 - x}{50} \qquad \text{or} \qquad \frac{dx}{50 - x} = \frac{dt}{50}.$$

The solution of this differential equation is $x(t) = 50 + Ae^{-t/50}$. The constant of integration is determined from the initial conditions: $x = 200$ at $t = 0$. Therefore, $A = 150$ and $x(t) = 50 + 150e^{-t/50}$. At the end of one hour, $t = 60$ and

$$x = 50 + 150e^{-1.2} = 50 + 150(0.3012) = 95\text{ lb}$$

of salt, approximately.


### Problems

In Problems 1 to 12 solve the differential equation subject to the given initial conditions.

1.  $\dfrac{du}{dt} = t^2 - t - 1; \quad u = 0$ at $t = 1$ $\qquad\qquad [u = \frac{1}{6}(2t^3 - 3t^2 - 6t + 7)]$

2.  $\dfrac{dy}{dx} = \dfrac{1}{2x - 1}; \quad y = 1$ at $x = 1$ $\qquad\qquad [y = \log \sqrt{2x - 1} + 1]$

3.  $\dfrac{dy}{dx} = \dfrac{1}{2x - 1}; \quad y = 1$ at $x = 0$ $\qquad\qquad [y = \log \sqrt{1 - 2x} + 1]$

4.  $\dfrac{dz}{d\theta} = \tan\theta - \sec^2\theta; \quad z = 2$ at $\theta = \pi$ $\qquad\qquad [z = \log|\sec\theta| - \tan\theta + 2]$

5.  $\dfrac{dy}{dx} = \dfrac{\cos x}{\cos y}; \quad y\left(\dfrac{\pi}{2}\right) = \pi$ $\qquad\qquad [\sin y = \sin x - 1]$

6.  $y' = \dfrac{x^2}{y}; \quad y(2) = 2$ $\qquad\qquad [3y^2 = 2x^3 - 4,\, y > 0]$

7.  $\dfrac{ds}{dt} = Ate^{-t^2}; \quad s = A$ at $t = 0$ $\qquad\qquad [s = \frac{3}{2}A - \frac{1}{2}Ae^{-t^2}]$

8.  $y' = xy; \quad y(0) = 2$ $\qquad\qquad [y = 2e^{x^2/2}]$

9.  $y' = \dfrac{1 - x}{1 + y}; \quad y(0) = 0$ $\qquad\qquad [x^2 + y^2 - 2x + 2y = 0]$

10. $\dfrac{dy}{dx} = \dfrac{y + 1}{xy}; \quad y = 0$ at $x = 1$ $\qquad\qquad [x(y + 1) = e^y]$

11. $\dfrac{du}{dt} = e^{t+u}; \quad u = 0$ at $t = 0$ $\qquad\qquad [e^{-u} = 2 - e^t]$

12. $\dfrac{dx}{dt} = 1 - x^2; \quad x = 0$ at $t = 0$ $\qquad\qquad [1 + x = (1 - x)e^{2t}]$

In Problems 13 to 29 find an equation for the family of curves that solves the differential equation. Sketch a few curves of each family. In the problems indicated find the particular solution satisfying the initial conditions, IC.

13. $\dfrac{dy}{dx} = x - 1$ $\qquad\qquad\qquad\qquad\qquad\qquad [y = \tfrac{1}{2}x^2 - x + C]$

14. $\dfrac{dy}{dx} = y - 1$ $\qquad\qquad\qquad\qquad\qquad\qquad [y = 1 + Ae^x]$

15. $y' = 1/x$ $\qquad\qquad\qquad\qquad\qquad\qquad\qquad [y = \log|x| + C]$

16. $y' = 1/y;$ IC: $y = 2$ at $x = -1$ $\qquad\qquad\qquad [y^2 = 2x + 6]$

17. $\dfrac{dy}{dx} = \dfrac{b^2 x}{a^2 y}$; IC: through the point $(a, 0)$ $\qquad [b^2x^2 - a^2y^2 = a^2b^2]$

18. $\dfrac{dy}{dx} = -\dfrac{b^2 x}{a^2 y}$ $\qquad\qquad\qquad\qquad\qquad [a^2y^2 + b^2x^2 = C]$

19. $y' = 1/xy$ $\qquad\qquad\qquad\qquad\qquad\qquad [y^2 = \log x^2 + C]$

20. $y' = x/y,$ IC: $y = 1$ at $x = 1$ $\qquad\qquad\qquad [y = x]$

21. $y' = -x/y,$ IC: $(x_0, y_0)$ $\qquad\qquad\qquad [x^2 + y^2 = x_0^2 + y_0^2]$

22. $y' = y/x$ $\qquad\qquad\qquad\qquad\qquad\qquad\qquad [y = cx]$

23. $y' = m = $ constant $\qquad\qquad\qquad\qquad\qquad [y = mx + C]$

24. $y' = x^2/y$ $\qquad\qquad\qquad\qquad\qquad\qquad [3y^2 = 2x^3 + C]$

25. $y' = y/x^2$ $\qquad\qquad\qquad\qquad\qquad\qquad [y = Ae^{-1/x}]$

26. $y' = x/y^2$ $\qquad\qquad\qquad\qquad\qquad\qquad [2y^3 = 3x^2 + C]$

27. $xy' = \sqrt{y} + y\sqrt{y}$ $\qquad\qquad\qquad\qquad [y = \tan^2(\log\sqrt{x} + C)]$

28. $(x^2 - 2x + 2)\,dy = (y^2 - 2y + 2)\,dx$ $\qquad\left[y = \dfrac{2c + (1 - c)x}{1 + c - cx}\right]$

29. $(xy - x)y' = y^2$ $\qquad\qquad\qquad\qquad\qquad [x = Aye^{1/y}]$

30. The half-life of a certain radioactive isotope is 100 yr. How long does it take to decay to 10 per cent of its initial amount? $\qquad\qquad\qquad$ [332 yr]

31. A tank contains 100 gal of pure water. Brine, containing 0.2 lb of salt per gallon, runs into the tank at the rate of 2 gal/min, and the mixture runs out at the same rate. What is the amount of salt in the tank at the end of $t$ min? 50 min? $\qquad\qquad\qquad\qquad\qquad\qquad\qquad\qquad\qquad\qquad$ [12.6 lb]

32. A tank contains 100 gal of brine containing 0.05 lb of salt per gallon. Pure water begins to run in at the rate of 1 gal/min from one tap, brine containing 0.25 lb of

salt per gallon runs in from another tap at the rate of 2 gal/min, and the mixture runs out at the rate of 3 gal/min. What is the amount of salt in the tank at the end of $t$ min?    $[50/3 - (35/3)e^{-0.03t}]$

33.  If brakes on a car apply a constant deceleration of $a$ ft/sec$^2$, find a formula for velocity and distance if initially the velocity is $v_0$ and the distance $x_0$.
$$[v = v_0 - at, \ x = x_0 + v_0 t - \tfrac{1}{2}at^2]$$

34.  A projectile is fired at an angle of elevation of $\alpha$ and an initial velocity of $v_0$ ft/sec. Assuming that the only force on the projectile is that of gravity which imparts a downward acceleration of 32 ft/sec$^2$, find the coordinates of the projectile at the end of $t$ sec.
$$[x = v_0 t \cos \alpha, \ y = v_0 t \sin \alpha - 16t^2]$$

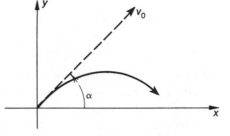

35. ✓  The force $F$ of wind resistance is proportional to the square of the velocity ($F = -kv^2$). Use Newton's second law ($F = $ mass $\times$ acceleration) to show that a body of mass $m$, initially at velocity $v_0$ ft/sec and acted on only by wind resistance, will have after $t$ sec a velocity $v = mv_0/(m + v_0 kt)$.

36. ✓  The resistance force $F$ on a body moving slowly in a viscous fluid is proportional to the velocity ($F = -ky$). Find a formula for the velocity after $t$ sec, of a body of mass $m$ that has initially a velocity $v_0$.    $[v = v_0 e^{-kt/m}]$

37.  Newton's law of cooling states that the time rate of change of the temperature $T$ of a body is proportional to the difference, $T - T_0$, between its temperature and that of its environment, $T_0$. If a body initially at 212° F (boiling water) cools to 152° F in 20 min when the air is at 32° F, find the temperature after $t$ min of a body initially at 100° F.    $[T = 32 + 68e^{-0.0202t}]$

38.  In a simple series circuit with resistance $R$ ohms, inductance $L$ henrys and electromotive force $E$ volts, the time rate of change of current $i$ (in amperes) is

$$L \frac{di}{dt} + Ri = E,$$

where the time $t$ is in seconds. Find the current at time $t$ if no current is flowing when the switch is closed at $t = 0$.

$$\left[ i = \frac{E}{R} \left(1 - e^{-Rt/L}\right) \right]$$

39. ✓  A tank contains 50 gal of brine containing 25 lb of dissolved salt. If water runs in at the rate of 2 gal/min and the mixture runs out at the same rate, find the salt concentration after $t$ minutes.    $[\tfrac{1}{2}e^{-0.04t} \text{ lb/gal}]$

## 2  *Geometric Aspects of Differential Equations*

In Section 1 the solutions of equations with the variables separable were families of curves. In this section we examine briefly the geometric problem posed by any first-order differential equation*

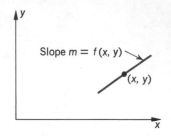

Slope $m = f(x, y)$
$(x, y)$

$$(1) \qquad \frac{dy}{dx} = f(x, y),$$

where, for our purposes, we suppose $f$ to be an elementary function. At each point $(x, y)$ of the plane, the value $f(x, y)$ gives the slope of a solution curve at that point.

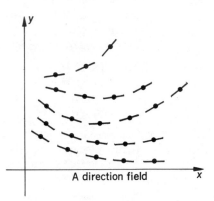

Thus, at each point the differential equation "tells which way to go"; it is a "road map."

If at many points $(x, y)$ of the plane we draw short segments (line elements) with slope $f(x, y)$, then we get a picture of a *direction field*.

A direction field

**Example 1**  The direction field of the differential equation

$$\frac{dy}{dx} = x^2 + y$$

can be sketched by tabulating the slope $x^2 + y$ at several points.

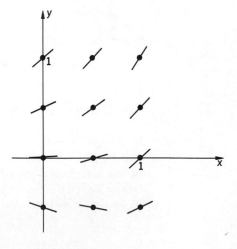

---

* The notation $f(x, y)$ represents the value of a function $f$ of two variables $x$ and $y$. For each point $(x, y)$ in some region of the plane, a number $f(x, y)$ is determined. For example: $f(x, y) = x \log y - x^2 - e^{x+y}$.

| $x$ | 0 | 0 | 0 | 0 | $\frac{1}{2}$ | $\frac{1}{2}$ | $\frac{1}{2}$ | $\frac{1}{2}$ | 1 | 1 | 1 | 1 |
|---|---|---|---|---|---|---|---|---|---|---|---|---|
| $y$ | 1 | $\frac{1}{2}$ | 0 | $-\frac{1}{2}$ | 1 | $\frac{1}{2}$ | 0 | $-\frac{1}{2}$ | 1 | $\frac{1}{2}$ | 0 | $-\frac{1}{2}$ |
| $x^2+y$ | 1 | $\frac{1}{2}$ | 0 | $-\frac{1}{2}$ | $\frac{5}{4}$ | $\frac{3}{4}$ | $\frac{1}{4}$ | $-\frac{1}{4}$ | 2 | $\frac{3}{2}$ | 1 | $\frac{1}{2}$ |

To get a clearer picture of the direction field and the solution curves many more line elements need to be drawn. A more systematic way of sketching the direction field is illustrated in Examples 3 and 4.

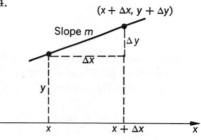

A graphical (approximate) solution of the differential equation would be obtained by drawing a curve that follows the line elements as closely as possible. In fact, numerical solutions of first order differential equations are simply arithmetic versions of this graphical idea. If one is at $(x, y)$ on the solution curve, then a nearby point on the curve is approximately $(x + \Delta x, y + \Delta y)$ where

$$\Delta y = m\,\Delta x = f(x, y)\,\Delta x.$$

***Example 2*** The direction field of the equation $dy/dx = y/(x + 1)$ is shown in the figure.

The solution of the differential equation is easy to obtain because the variables are separable:

$$\frac{dy}{y} = \frac{dx}{x + 1},$$

$$\log|y| = \log|x + 1| + C,$$

$$y = A(x + 1), \qquad \text{where } |A| = e^C.$$

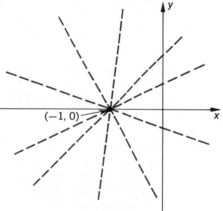

This is an equation of the family of (non-vertical) straight lines through $(-1, 0)$.

***Example 3*** The direction field of the equation

$$\frac{dy}{dx} = -\frac{2x}{y}$$

is shown in the figure. Observe that on lines through the origin the slope is constant:

$$-2x/y = m.$$

This observation facilitates sketching the direction field. We draw the

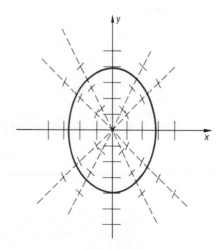

curves where the slope is constant $= m$: $-2x/y = m$ or $-2x = my$. These are shown dotted in the figure. The line elements along each of these lines are parallel.

The family of solutions is the family of ellipses $2x^2 + y^2 = C$.

*Example 4*   The equation

$$dy/dx = x - y$$

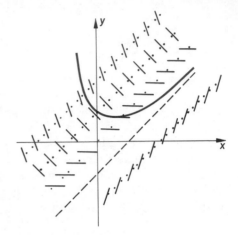

does not have the variables separable. The direction field is sketched in the figure. Note that the slope is constant along the lines $x - y = m$.

The solution through $(0, 1)$ has been sketched, free hand, to follow the direction field.

*Example 5*   An *orthogonal trajectory* to a family of curves is a curve that meets each member of the family at right angles. To find orthogonal trajectories of the family $y = cx^3$ we first find a differential equation satisfied by the given family.

From $y = cx^3$ we get $y' = 3cx^2$. However, this is not a differential equation giving $y'$ at an arbitrary point $(x, y)$. We must eliminate the parameter $c$ between $y = cx^3$ and $y' = 3cx^2$. Then we obtain

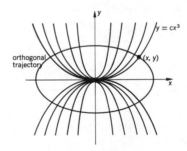

$$y' = 3\left(\frac{y}{x^3}\right)x^2 = \frac{3y}{x}.$$

Therefore, at the point $(x, y)$ the curve of the given family has slope $m = 3y/x$. An orthogonal trajectory must have the negative reciprocal slope, $-x/3y$. Therefore, an orthogonal trajectory will satisfy the differential equation

$$\frac{dy}{dx} = -\frac{x}{3y} \qquad \text{or} \qquad 3y\,dy = -x\,dx.$$

Integration gives the family of orthogonal trajectories. They are the ellipses $3y^2 + x^2 = k = $ constant.

***Example 6***   Orthogonal trajectories in polar coordinates are handled similarly. The family of circles

$$r = c \cos \theta$$

satisfies the differential equation

$$\frac{dr}{d\theta} = -r \tan \theta,$$

whence, on the original family of circles, $\tan \psi = r \, d\theta/dr = -\cot \theta$. Therefore, on an orthogonal trajectory (see the figures)

$$\tan \psi' = \tan\left(\psi \pm \frac{\pi}{2}\right) = -\cot \psi = -\frac{1}{\tan \psi}.$$

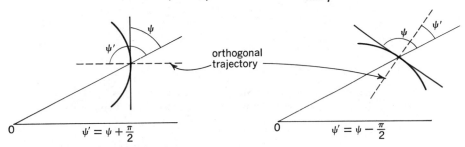

orthogonal
trajectory

$$\psi' = \psi + \frac{\pi}{2} \qquad\qquad\qquad \psi' = \psi - \frac{\pi}{2}$$

The orthogonal trajectory thus satisfies the equation

$$r \frac{d\theta}{dr} = \tan \theta.$$

The solution of this equation is

$$r = k \sin \theta \qquad (k = \text{constant}),$$

which represents a family of circles with centers on the $y$-axis.

### Problems

**In Problems 1 to 7 sketch the direction field and draw a solution through the point indicated.**

1.  $y' = -y/x$;   $(1, 1)$

2.  $y' = y/x$;   $(1, 1)$

3.  $y' = x/y$;   $(2, 1)$

4.  $y' = 2x + y$;   $(0, 1)$

5.  $y' = y^2$;   $(0, 1)$

6.  $y' = y^2 + x$;   $(0, 1)$

7.  $y' = xy$;   $(0, 1)$

In Problems 8 to 13 find the family of orthogonal trajectories to the given family. Sketch the given family and one orthogonal trajectory.

8. $y^2 = cx^3$ $\qquad\qquad\qquad\qquad\qquad\qquad\qquad\qquad$ $[3y^2 + 2x^2 = k]$

9. $xy = c$ $\qquad\qquad\qquad\qquad\qquad\qquad\qquad\qquad\qquad$ $[y^2 - x^2 = k]$

10. $x^2 + y^2 = c$ $\qquad\qquad\qquad\qquad\qquad\qquad\qquad\qquad$ $[y = kx]$

11. $c^2x^2 + y^2 = c^2$ $\qquad\qquad\qquad\qquad\qquad\qquad$ $[x^2 + y^2 = \log x^2 + k]$

12. $y^2 = cx$ $\qquad\qquad\qquad\qquad\qquad\qquad\qquad\qquad$ $[2x^2 + y^2 = k]$

13. $x^2 - 2y^2 = c$ $\qquad\qquad\qquad\qquad\qquad\qquad\qquad\qquad$ $[yx^2 = k]$

In Problems 14 to 17 find the orthogonal trajectories of the families given by the polar equations. Sketch the family and one orthogonal trajectory.

14. $r = c \sin \theta$ $\qquad\qquad\qquad\qquad\qquad\qquad\qquad$ $[r = k \cos \theta]$

15. $r = c \cos 2\theta$ $\qquad\qquad\qquad\qquad\qquad\qquad$ $[r^4 = k \sin 2\theta]$

16. $r^2 = c \sin 2\theta$ $\qquad\qquad\qquad\qquad\qquad\qquad$ $[r^2 = k \cos 2\theta]$

17. $r = c(1 - \cos \theta)$ $\qquad\qquad\qquad\qquad\qquad$ $[r = k(1 + \cos \theta)]$

*18. Find a family of curves such that the tangent line at each point $P$ meets the $x$-axis in a point with the $x$-coordinate twice the $x$-coordinate of $P$. $\qquad$ $[xy = c]$

*19. At each point $P$ on a curve the normal at $P$ meets the $x$-axis at a point $Q$, possibly depending on $P$, such that $|PQ|$ is a constant equal to $a$. Find the curve if it passes through $(0, a)$. $\qquad\qquad\qquad\qquad$ $[x^2 + y^2 = a^2]$

20. Find, in polar coordinates, the curves for which the angle between the radius vector and the tangent is a constant equal to Arc cot $a$.

$\qquad\qquad\qquad\qquad\qquad\qquad$ [equiangular spiral: $r = ce^{a\theta}$]

21. Find, in polar coordinates, curves for which the tangent of the angle between the radius vector and the tangent is proportional to $r$. [Use $\tan \psi = (1/a)r$.]

$\qquad\qquad\qquad\qquad\qquad\qquad$ [Archimedes spiral: $r = a\theta + C$]

## 3  *Approximate Integration*

The Fundamental Theorem of calculus provides a simple technique for evaluating definite integrals *if* an antiderivative can be found. However, not all antiderivatives are elementary functions, and so one must find methods to attack arbitrary integrals. The problem is of the highest importance so that methods to obtain approximations to integrals are vital.

In this section we give two such approximations. But before beginning, observe that the Riemann sums,

$$\sum_{i=1}^{n} f(\xi_i) \, \Delta x_i,$$

already give an approximation to the definite integral—obtained by inscribing rectangles under the graph of $f$.

Often, a somewhat better approximation is provided by inscribing trapezoids,

$$\int_a^b f(x)\ dx \approx \text{the sum of the areas of the trapezoids.}$$

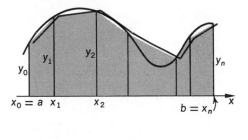

The area of a single trapezoid is

$$\tfrac{1}{2}(y_{i-1} + y_i)\ \Delta x_i.$$

Then, if we use $n$ equal subdivisions,

$$\Delta x = \frac{b-a}{n},$$

and we have the *Trapezoidal Rule*:

$$\int_a^b f(x)\ dx \approx \frac{\Delta x}{2}\ (y_0 + 2y_1 + 2y_2 + \cdots + 2y_{n-1} + y_n).$$

**Example 1**   We use $n = 5$ and the trapezoidal rule to approximate $\int_0^1 \sqrt{1 - x^2}\ dx$.

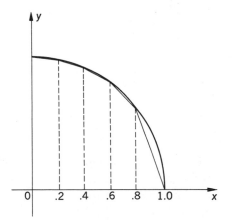

$$\int_0^1 \sqrt{1 - x^2}\ dx$$

$$\approx \frac{0.2}{2}[1 + 2\sqrt{1 - (0.2)^2}$$

$$+ 2\sqrt{1 - (0.4)^2} + 2\sqrt{1 - (0.6)^2}$$

$$+ 2\sqrt{1 - (0.8)^2} + 0]$$

$$\approx 0.759 \qquad (\text{correct value} \approx 0.785).$$

With $n = 10$ one gets 0.776.

In many cases an even better approximation is provided by *Simpson's rule*. We subdivide the interval $[a, b]$ into an *even* number of subintervals of equal length. Then we fit parabolic arcs through pairs of three adjacent points.

Suppose for convenience the two adjacent intervals are at the origin, as

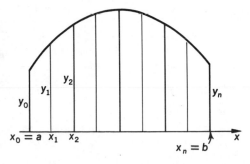

shown. And suppose the parabola

$$y = \alpha + \beta x + \gamma x^2$$

coincides with the graph of $f$ at $-\Delta x$, 0, $\Delta x$. The area under the parabola is

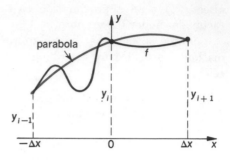

parabola

$f$

(1) $\displaystyle\int_{-\Delta x}^{\Delta x} (\alpha + \beta x + \gamma x^2)\, dx$

$$= 2\alpha\, \Delta x + \frac{2\gamma}{3}\, \Delta x^3$$

and is an approximation to $\int_{-\Delta x}^{\Delta x} f(x)\, dx$. To make use of (1) we compute $\alpha$ and $\gamma$ in terms of the values of $y$ at $-\Delta x$, 0, and $\Delta x$. We have

$$y_{i-1} = \alpha - \beta\, \Delta x + \gamma\, \Delta x^2,$$

$$y_i = \alpha,$$

$$y_{i+1} = \alpha + \beta\, \Delta x + \gamma\, \Delta x^2,$$

from which we conclude that

$$\alpha = y_i \qquad \text{and} \qquad \gamma = \frac{y_{i-1} + y_{i+1} - 2y_i}{2\, \Delta x^2}.$$

With these values substituted in (1) we get

(2) $\displaystyle\int_{-\Delta x}^{\Delta x} (\alpha + \beta x + \gamma x^2)\, dx = \frac{\Delta x}{3}\, (y_{i-1} + 4y_i + y_{i+1}).$

The value of $\beta$ could also be found if needed. That we have been able to solve for $\alpha$, $\beta$, $\gamma$ shows that one can fit a parabola through the three points. If we apply equation (2) to successive pairs of the $2n$ subintervals of $[a, b]$, we get as an approximation *Simpson's rule*:

$$\int_a^b f(x)\, dx \approx \frac{\Delta x}{3}\, (y_0 + 4y_1 + 2y_2 + 4y_3 + \cdots + 4y_{2n-1} + y_{2n}).$$

*Example 2*    The same integral as Example 1, when approximated by Simpson's rule with $n = 5$, comes out as

$$\int_0^1 \sqrt{1 - x^2}\, dx \approx \frac{0.1}{3}\, [1 + 4\sqrt{0.99} + 2\sqrt{0.96} + \cdots + 4\sqrt{0.19} + 0]$$

$$\approx 0.782.$$

To use these approximate formulas with complete assurance one needs estimates of the possible error. There are such estimates, for which the reader should consult a book on numerical analysis. But in practice such estimates are seldom used because (a) they are rather crude and give much too conservative estimates of error, and (b) one can get (usually) suitably accurate results by choosing $n$ large enough. However, one must be careful. As $n$ increases, the number of terms increases, and the round-off errors can accumulate. The whole subject of accuracy in numerical work is a subtle one and beyond the scope of this book.

### Problems

In Problems 1 to 8 use either the Trapezoidal Rule (T) or Simpson's Rule (S), as indicated. Use the given value of $n$ and keep the number of decimal places suggested. (With Simpson's Rule the number of intervals is $2n$.) Compare with the exact value in 1 to 7.

1. $\displaystyle\int_0^1 \frac{dx}{1+x^2}$;   T, $n = 5$, four decimals          [0.7837]

2. $\displaystyle\int_1^2 \frac{dx}{x}$;   T, $n = 10$, four decimals          [0.6938]

3. $\displaystyle\int_1^2 \frac{dx}{x}$;   S, $n = 2$, four decimals          [0.6932]

4. $\displaystyle\int_1^2 \frac{dx}{x}$;   S, $n = 4$, five decimals          [0.69315]

5. $\displaystyle\int_1^2 \log x \, dx$;   S, $n = 4$, five decimals

6. $\displaystyle\int_0^2 x^3 \, dx$;   S, $n = 4$, five decimals

7. $\displaystyle\int_0^1 \frac{x \, dx}{\sqrt{1+x^2}}$;   S, $n = 2$, four decimals

8. $\displaystyle\int_0^{\pi/2} \frac{dx}{\sqrt{1 - \frac{1}{2}\sin^2 x}}$;   S, $n = 2$, four decimals          [1.8537]

*9.   Prove that, if $f$ is a cubic polynomial, then Simpson's rule is exact.

## 4 Further Area Problems

In this section we calculate areas of regions bounded by curves whose equations are given either in rectangular parametric form, or in polar coordinates.

We saw in Chapter 9 that the area under the graph of $y = f(x)$ and between $a$ and $b$ is

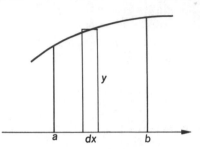

$$\int_a^b y\, dx = \int_a^b f(x)\, dx.$$

If the curve is given parametrically, say by

(1) $\quad x = x(t), \quad y = y(t), \quad \alpha \leq t \leq \beta,$

then the area integral is obtained by using equations (1) as a change of variable, or a substitution. Then

$$\text{Area} = \int_a^b y\, dx = \int_\alpha^\beta y(t)\, dx(t) = \int_\alpha^\beta y(t)x'(t)\, dt.$$

**Example 1**    Find the area of $\frac{1}{4}$ of the circle

$$x = a \cos\theta, \quad y = a \sin\theta, \quad 0 \leq \theta \leq \pi/2.$$

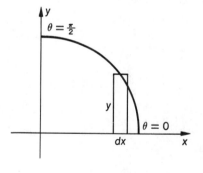

$$\text{Area} = \int_0^a y\, dx = \int_{\pi/2}^0 (a\sin\theta)(-a\sin\theta)\, d\theta$$

$$= a^2 \int_0^{\pi/2} \sin^2\theta\, d\theta = \frac{\pi a^2}{4}.$$

Observe that

$$\int_0^a y\, dx = \int_0^a \sqrt{a^2 - x^2}\, dx$$

and that use of the parametric equations is equivalent to a trigonometric substitution.

**Example 2**    Find the area of the region bounded by

$$x = \sin^2 t, \quad y = \sin t \cos^2 t,$$

$$-\pi/2 \leq t \leq \pi/2.$$

The parameterized curve is shown in red. It is part of the cubic curve whose equation is

$$y^2 = x(x-1)^2,$$

as may be seen by eliminating $t$.

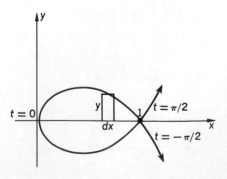

$$\text{Area} = 2 \int_0^1 y\,dx = 2 \int_0^{\pi/2} \sin t \cos^2 t\,(2 \sin t \cos t)\,dt$$

$$= 4 \int_0^{\pi/2} \sin^2 t\,(1 - \sin^2 t) \cos t\,dt = \frac{8}{15}.$$

We now turn to regions bounded by lines through $O$ and curves given in polar coordinates by

$$r = f(\theta), \qquad \alpha \leqq \theta \leqq \beta.$$

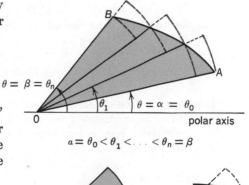

The area of the "polar sector $OAB$" is the limit of a sum of areas of circular sectors. From elementary geometry, the area of the circular sector shown in the figure is

$$\Delta A = \tfrac{1}{2} r^2\,\Delta\theta.$$

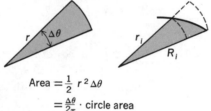

$$\text{Area} = \frac{1}{2} r^2 \Delta\theta$$
$$= \frac{\Delta\theta}{2\pi} \cdot \text{circle area}$$

Suppose the interval $[\alpha, \beta]$ is subdivided into $n$ subintervals:

$$\alpha = \theta_0 < \theta_1 < \cdots < \theta_n = \beta.$$

Let $r_i$ and $R_i$ be the minimum and maximum of $r = f(\theta)$ in the $i$th subinterval. Then the area $A$ of the polar sector lies between the areas of the inscribed and circumscribed unions of the circular sectors;

(2) $$\sum_{i=1}^n \tfrac{1}{2} r_i^2\,\Delta\theta_i \leqq A \leqq \sum_{i=1}^n \tfrac{1}{2} R_i^2\,\Delta\theta_i, \qquad \Delta\theta_i = \theta_i - \theta_{i-1}.$$

The two sums in (2) are lower and upper Riemann sums for the integral

$$\int_\alpha^\beta \tfrac{1}{2} r^2\,d\theta = \int_\alpha^\beta \tfrac{1}{2} f^2(\theta)\,d\theta.$$

As the mesh of the subdivision approaches 0, the two Riemann sums in (2) approach the area. Thus, we have

$$A = \tfrac{1}{2} \int_\alpha^\beta r^2\,d\theta.$$

**Example 3**    Find the area of the region enclosed by the rose $r = a \cos 3\theta$.

The full curve is obtained for $0 \leq \theta \leq \pi$; therefore, the area is

$$\text{Area} = \frac{1}{2} \int_0^\pi a^2 \cos^2 3\theta \, d\theta = \frac{\pi a^2}{4}.$$

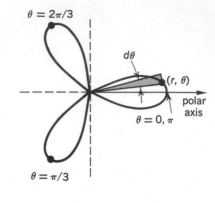

From symmetry, we could also calculate the area by finding the area in the first quadrant and multiplying by 6.

$$\text{Area} = 6 \left(\tfrac{1}{2}\right) \int_0^{\pi/6} a^2 \cos 3\theta \, d\theta.$$

**Example 4**    Find the area of the region enclosed by the lemniscate $r^2 = a^2 \cos 2\theta$.

The total area is four times the first quadrant area.

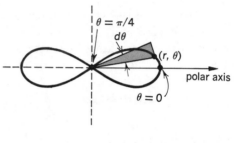

$$A = 4\left(\tfrac{1}{2}\right) \int_0^{\pi/4} r^2 \, d\theta$$

$$= 2 \int_0^{\pi/4} a^2 \cos 2\theta \, d\theta = a^2.$$

One must be careful with the lemniscate. It is tempting to write

$$A \overset{?}{=} \frac{1}{2} \int_0^{2\pi} r^2 \, d\theta = \frac{a^2}{2} \int_0^{2\pi} \cos 2\theta \, d\theta = 0?$$

The dilemma is resolved if we observe that, when $\pi/4 < \theta < 3\pi/4$ and $5\pi/4 < \theta < 7\pi/4$, $\cos 2\theta$ is negative, whence $r^2$ is negative, and there are no points on the lemniscate for these values of $\theta$.

**Example 5**    Areas of regions bounded by lines through the origin and curves given in rectangular parametric form are often conveniently found as "sums" of thin triangles.

The area $\Delta A$ of a triangle (see Appendix A) with vertices $(0,0)$, $(x, y)$, and $(x + \Delta x, y + \Delta y)$ is

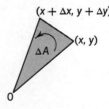

$$\Delta A = \tfrac{1}{2}(x \, \Delta y - y \, \Delta x),$$

if, as one traverses the triangle in a counterclockwise direction, the vertices occur in the order $(0,0)$, $(x, y)$, $(x + \Delta x, y + \Delta y)$. With this direction around the triangle, the region always lies on the left.

Suppose that the wedge-shaped region is bounded by lines through the origin and the curve given by

$$x = x(t), \quad y = y(t), \quad \alpha \leq t \leq \beta,$$

and the region lies on the left as $t$ increases.

If we subdivide the parameter interval, form Riemann sums (of areas of thin triangles), and pass to the limit, we find that the area of the region $OAB$ is given by

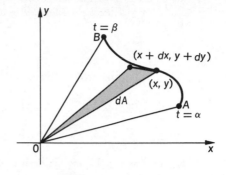

$$(3) \quad \text{Area} = \frac{1}{2} \int_{t=\alpha}^{t=\beta} (x \, dy - y \, dx)$$

$$= \frac{1}{2} \int_{\alpha}^{\beta} (xy' - yx') \, dt.$$

Thus, for the quarter circle of Example 1,

$$x = a \cos \theta, \qquad\qquad y = a \sin \theta$$

$$dx = -a \sin \theta \, d\theta, \qquad dy = a \cos \theta \, d\theta.$$

$$\text{Area} = \frac{1}{2} \int_{0}^{\pi/2} \big[ (a \cos \theta)(a \cos \theta \, d\theta) - (a \sin \theta)(-a \sin \theta \, d\theta) \big]$$

$$= \frac{1}{2} \int_{0}^{\pi/2} a^2 \, d\theta = \frac{\pi a^2}{4}.$$

Formula (3) can also be used with the nonparametric rectangular equation.

## Problems

In Problems 1 to 7 find the areas of the regions bounded by the given curves. Use either the method of Example 1 or Example 5. Sketch a figure.

1. The ellipse $x = a \cos \theta$, $y = b \sin \theta$. $[\pi ab]$

2. The hypocycloid $x = a \cos^3 \theta$, $y = a \sin^3 \theta$. $[\frac{3}{8}\pi a^2]$

3. The axes and the parabola $x = a \cos^4 \theta$, $y = a \sin^4 \theta$. $[a^2/6]$

4. The cycloid (one arch) $x = a(\theta - \sin \theta)$, $y = a(1 - \cos \theta)$ and the $x$-axis.
   $[3\pi a^2]$

5. The hyperbola $x = a \sec \theta$, $y = a \tan \theta$, the $x$-axis and the radius vector to $(x, y)$ on the hyperbola.
   $\left[ \dfrac{a^2}{2} \log \left| \dfrac{x+y}{a} \right| \right]$

6. The circle $x = h + a \cos \theta$, $y = k + a \sin \theta$. $[\pi a^2]$

7. The cardioid $x = a(2 \cos \theta - \cos 2\theta)$, $y = a(2 \sin \theta - \sin 2\theta)$. $[6\pi a^2]$

*8. The conchoid of Nicomedes (see problem 6, page 147) is parameterized by $x = b \tan \theta + a \sin \theta$, $y = a \cos \theta$. The loop is obtained for

$$\pi - \text{Arc cos} \frac{b}{a} \leq \theta \leq \pi + \text{Arc cos} \frac{b}{a}.$$

Find the area of the loop.

$$\left[ a^2 \, \text{Arc cos} \frac{b}{a} - 2ab \log \frac{a + \sqrt{a^2 - b^2}}{b} + b\sqrt{a^2 - b^2} \right]$$

**In Problems 9 to 28 find the area of the region bounded by the curves given in polar coordinates.**

9. $r = a \cos \theta$ $\quad\quad\quad [\pi a^2/4]$

10. $r = a(1 + \cos \theta)$ $\quad\quad\quad [\frac{3}{2}\pi a^2]$

11. $r = a\theta,\ \theta = 0,\ \theta = 2\pi$ $\quad\quad\quad [\frac{4}{3}\pi^3 a^2]$

12. $r = e^{a\theta},\ \theta = 0,\ \theta = \alpha$ $\quad\quad\quad [(e^{2a\alpha} - 1)/4a]$

13. $r = a + b \cos \theta,\ \ a > b$ $\quad\quad\quad [\pi a^2 + (\pi b^2/2)]$

*14. The small loop of the limaçon $r = a + b \cos \theta,\ \ b > a$

$$\left[ \left( a^2 + \frac{b^2}{2} \right) \text{Arc cos} \frac{a}{b} - \frac{3a}{2} \sqrt{b^2 - a^2} \right]$$

*15. The loop of the conchoid $r = a \csc \theta + b,\ \ b > a$

$$\left[ b^2 \left( \frac{\pi}{2} - \text{Arc sin} \frac{a}{b} \right) - 2ab \log \frac{b + \sqrt{b^2 - a^2}}{a} + a\sqrt{b^2 - a^2} \right]$$

16. The four-leaved rose $r = a \cos 2\theta$ $\quad\quad\quad [\pi a^2/2]$

17. $r = \dfrac{a}{\theta},\ \theta = \alpha,\ \theta = \beta$ $\quad\quad\quad [a^2 (\beta - \alpha)/2\alpha\beta]$

18. $r^2 = a^2 \sin 2\theta$ $\quad\quad\quad [a^2]$

19. Inside the limaçon $r = 3 + \cos \theta$ and outside the circle $r = 3 \cos \theta$

$\quad\quad\quad [29\pi/4]$

20. The region common to the circles $r = 2a \cos \theta$ and $r = a$ $\quad\quad\quad \left[ a^2 \left( \dfrac{2\pi}{3} - \dfrac{\sqrt{3}}{2} \right) \right]$

21. The rose-like curve $r^2 = a^2 \cos 4\theta$ $\quad\quad\quad [a^2]$

22. $r = \dfrac{a}{1 + \cos \theta},\ \theta = 0,\ \theta = \dfrac{\pi}{2}$ $\quad\quad\quad \left[ \dfrac{a^2}{3} \right]$

23. The region inside the circle $r = a(\cos\theta - \sin\theta)$ and inside the lemniscate $r^2 = a^2 \sin 2\theta$

$$\left[ \frac{\pi a^2}{6} + \frac{a^2}{2}(1 - \sqrt{3}) \right]$$

24. One of the small loops of the curve $r^2 = a^2(1 + \cos\theta)$ $\qquad$ $[a^2(\pi/2 - 1)]$

25. One loop of $r^2 = a^2 \cos\theta$ $\qquad$ $[a^2]$

26. Inside $r = 3\sin\theta$ and the limaçon $r = 2 - \sin\theta$ $\qquad$ $[(9\pi/4) - 3\sqrt{3}]$

27. Inside the circle $r = \sin\theta$ and outside the cardioid $r = 1 - \sin\theta$

$$[\sqrt{3} - (\pi/3)]$$

28. $r = a\sin\theta, \theta = 0, \theta = \alpha$

$$\left[ \frac{a^2}{8}(2\alpha - \sin 2\alpha) \right]$$

*29. Show, using the method of Example 5, that the area $A$ pictured is given by

$$A = \frac{1}{2}\int_a^b [xg'(x) - xf'(x) - g(x)$$
$$+ f(x)]\,dx + \tfrac{1}{2}\{b[f(b) - g(b)]$$
$$- a[f(a) - g(a)]\}.$$

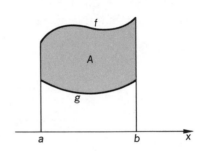

## 5 Improper Integrals

Up to now definite integrals have been integrals of continuous functions over finite intervals. In many applications either the interval of definition of the integral is infinite, or the integrand function becomes infinite at some point of the interval. In these cases the Riemann integral is inadequate, and we must *define* what we are to mean by the integral. Consider first two examples.

**Example 1** Find the area under the curve $y = 1/x^2$ from $x = 1$ to $x = \infty$.

We must first decide how to define the area. We choose to define

$$\text{Area} = A = \lim_{a\to\infty} \int_1^a \frac{1}{x^2}\,dx.$$

Thus, the area is defined to be the limit of the areas of expanding regions. Then

$$A = \lim_{a\to\infty}\left(1 - \frac{1}{a}\right) = 1.$$

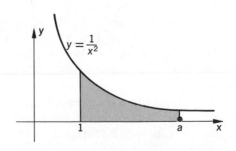

*Example 2* Find the area under the graph of $y = 1/\sqrt{a^2 - x^2}$ between $x = 0$ and $x = a$.

This time the interval $[0, a)$ is finite, but the function becomes infinite as $x \to a$. We define the area to be the limit of the areas of expanding regions obtained by stopping short of $a$.

$$\text{Area} = A = \lim_{b \to a} \int_0^b \frac{dx}{\sqrt{a^2 - x^2}}$$

$$= \lim_{b \to a} \text{Arc sin} \frac{b}{a} = \frac{\pi}{2}.$$

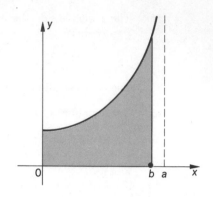

These examples illustrate the method. When either the integrand or the interval of integration becomes infinite, one restricts the interval of integration, and then lets it expand. We are thus led to the following definitions.

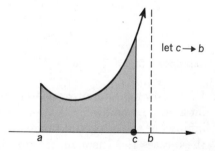

let $b \to \infty$

$$(1) \qquad \int_a^\infty f(x) \, dx = \lim_{b \to \infty} \int_a^b f(x),$$

$f$ continuous in $[a, \infty)$

$$(2) \qquad \int_a^b f(x) \, dx = \lim_{c \to b} \int_a^c f(x) \, dx,$$

$f$ continuous in $[a, b)$

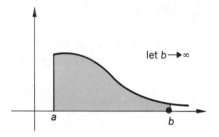

let $c \to b$

$$(3) \qquad \int_a^b f(x) \, dx = \lim_{c_1 \to c} \int_a^{c_1} f(x) \, dx$$

$$+ \lim_{c_2 \to c} \int_{c_2}^b f(x) \, dx,$$

if $f$ is continuous in $[a, b]$ except at $c$ and $a < c_1 < c < c_2 < b$.

The integrals (1), (2), and (3) are called *improper integrals*. If the limit exists, the integral is said to be *convergent*. If the limit does not exist, the integral is said to be *divergent*.

It should be observed that if $f$ is continuous in the whole interval $[a, b]$, then the limits in (2) and (3) exist and equal the ordinary definite integral.

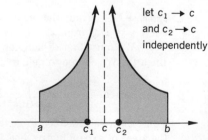

let $c_1 \to c$
and $c_2 \to c$
independently

**Example 3**   The integral

$$\int_1^\infty \frac{dx}{x} = \lim_{a\to\infty} \int_1^a \frac{dx}{x} = \lim_{a\to\infty} \log a = \infty.$$

Thus, the integral is divergent.

**Example 4**   Sometimes an improper integral becomes proper after a change of variable. If we let $x = a \sin\theta$, then

$$\int_0^a \frac{dx}{\sqrt{a^2 - x^2}} = \lim_{b\to a} \int_0^b \frac{dx}{\sqrt{a^2 - x^2}}$$

$$= \lim_{b\to a} \int_0^{\text{Arc sin } b/a} d\theta = \int_0^{\pi/2} d\theta = \frac{\pi}{2}.$$

**Example 5**   In some cases it may be difficult to determine whether an improper integral converges or diverges because an antiderivative cannot be found. In such cases one can try to compare the given integral with a known integral. Consider

$$\int_1^\infty e^{-x^2}\, dx = \lim_{b\to\infty} \int_1^b e^{-x^2}\, dx.$$

The integrand is positive, and, therefore, the integral

$$\int_1^b e^{-x^2}\, dx$$

increases with $b$. Thus, the limit of this integral is either infinite, or the integral is bounded as $b\to\infty$ (see Appendix C), and the limit exists. However, because $e^{-x^2} < e^{-x}$ for $x > 1$, we have

$$\int_1^b e^{-x^2}\, dx < \int_1^b e^{-x}\, dx = e^{-1} - e^{-b} < e^{-1}.$$

Therefore,

$$\lim_{b\to\infty} \int_1^b e^{-x^2}\, dx$$

exists, and is some number $< e^{-1}$. The original improper integral is convergent.

**Example 6**   One can also prove divergence in some problems by comparison. The integral

$$\int_2^\infty \frac{dx}{\sqrt{x}\, \log x}$$

can be compared with

$$\int_2^\infty \frac{dx}{x \log x}$$

because            $$\frac{1}{\sqrt{x}\, \log x} > \frac{1}{x \log x} \qquad \text{for} \qquad x \geq 2.$$

Then
$$\int_2^\infty \frac{dx}{\sqrt{x}\,\log x} \geqq \int_2^\infty \frac{dx}{x\,\log x} = \lim_{b \to \infty} \int_2^b \frac{dx}{x\,\log x}$$

$$= \lim_{b \to \infty} \left[\log\log b - \log\log 2\right] = \infty.$$

Therefore, the original integral is divergent.

### Problems

In Problems 1 to 21 evaluate the improper integrals if they are convergent. State if they are divergent.

1. $\displaystyle\int_0^1 \frac{dx}{\sqrt{x}}$  [2]

2. $\displaystyle\int_0^1 \frac{dx}{\sqrt{x}}$

   (Use substitution $u = \sqrt{x}$.)

3. $\displaystyle\int_{-1}^1 \frac{dx}{x}$  [divergent]

4. $\displaystyle\int_{-1}^1 \frac{dx}{x^{1/3}}$  [0]

5. $\displaystyle\int_0^1 \frac{dx}{x^n}, \quad 0 < n < 1$

6. $\displaystyle\int_0^1 \frac{dx}{x^n}, \quad n \geqq 1$

7. $\displaystyle\int_0^\infty e^{-x}\,dx$

8. $\displaystyle\int_0^1 -\log x\,dx$  [1]

   (Interpret as area and compare with 7.)

9. $\displaystyle\int_0^{\pi/2} \tan\theta\,d\theta$

10. $\displaystyle\int_0^\infty e^{-x}\sin x\,dx$  [$\frac{1}{2}$]

    (Sketch the graph of the integrand and compare with 7.)

11. $\displaystyle\int_0^1 \frac{dx}{\sqrt{x(1-x)}}$          $[\pi]$

12. $\displaystyle\int_0^a \frac{dx}{(x-a)^2}$          [divergent]

13. $\displaystyle\int_e^\infty \frac{dx}{x \log x^2}$

14. $\displaystyle\int_e^\infty \frac{dx}{x \log^2 x}$          [1]

15. $\displaystyle\int_{2a}^\infty \frac{dx}{x^2 - a^2}, \quad a > 0$

16. $\displaystyle\int_0^\infty t^2 e^{-t}\, dt$          [2]

17. $\displaystyle\int_0^2 \frac{x\, dx}{\sqrt{4 - x^2}}$

18. $\displaystyle\int_2^3 \frac{dx}{\sqrt{x-2}}$          [2]

19. $\displaystyle\int_0^{\pi/2} \sec\theta\, d\theta$

20. $\displaystyle\int_0^{2a} \frac{(x-a)\, dx}{\sqrt{2ax - x^2}}$          [0]

21. $\displaystyle\int_0^{\pi/2} \frac{\cos\theta\, d\theta}{\sqrt{1 - \sin\theta}}$          [2]

22. For what values of $n$ is $\int_0^\infty x^n\, dx$ convergent?      [none]

23. Find the area under the curve $y = 1/\sqrt{x+1}$ and between $x = -1$ and $x = 1$.
         $[2\sqrt{2}]$

24. The region under the graph of $y = e^{-x}$ between $x = 0$ and $x = \infty$ is revolved about the $x$-axis. Find the volume of the solid generated.      $[\pi/2]$

25. The region of Problem 24 is revolved about the $y$-axis. Find the volume of the solid generated.      $[2\pi]$

26. Is the area under the curve $y = 1/x$ and above the $x$-axis, where $1 \leq x < \infty$ finite? Is the volume of the solid obtained by revolving this region about the $x$-axis finite?

In Problems 27 to 32 decide whether the integrals are convergent or divergent by comparing them with known integrals. See Examples 5 and 6.

27. $\displaystyle\int_{e}^{\infty} \frac{dx}{x + e^{x^2}}$

*30. $\displaystyle\int_{0}^{\infty} \frac{dx}{1 + x^{10}}$ [convergent]

28. $\displaystyle\int_{1}^{\infty} \frac{|\sin x|}{x^2}\, dx$

31. $\displaystyle\int_{1}^{\infty} \frac{dx}{1 + \sqrt{x}}$

29. $\displaystyle\int_{1}^{\infty} \frac{dx}{\sqrt{x + x^3}}$

32. $\displaystyle\int_{0}^{\infty} \frac{dx}{2 + \sin x + x^{1/2}}$ [divergent]

## 6  More Volume Problems

In Chapter 10 the differential elements of volume were thin cylinders.

$$dV = \pi R^2\, dx \qquad \text{or} \qquad \pi R^2\, dy.$$

Another way of regarding solids of revolution considers them as sums of thin cylindrical *shells*. The shell is generated by revolving thin strips parallel to the axis of rotation.

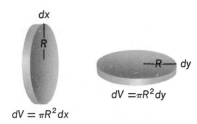

$dV = \pi R^2\, dx$

$dV = \pi R^2\, dy$

The volume of the thin shell of height $h$ and radii $R$, $R + \Delta R$ is

(1) $\qquad \Delta V = \pi (R + \Delta R)^2 h - \pi R^2 h$

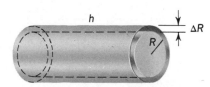

$\Delta V = 2\pi h(R + \frac{\Delta R}{2})\Delta R$

$$= 2\pi h \left( R + \frac{\Delta R}{2} \right) \Delta R$$

$$= 2\pi h R'\, \Delta R,$$

where $R'$ is the mean radius $= R + \Delta R/2$.

Formula (1) has the appearance of a typical summand in a Riemann sum for an integral for the volume. We will omit the development of the volume integral by means of inequalities on Riemann sums and simply state that the differential element of volume has the form

$$dV = 2\pi h R\, dR.$$

An example will illustrate the method.

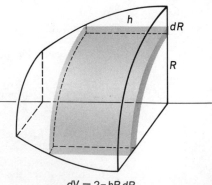

$dV = 2\pi h R\, dR$

**Example 1**    Find the volume of the solid obtained by revolving about the line $x = a$, the region under $y = x^2$, above the $x$-axis, and between $x = 0$ and $x = a$.

The thin strip shown in the figure will, when revolved about the line $x = a$, sweep out a thin shell of radius $R = a - x$. The height of the shell is $h = y = x^2$. Thus,

$$dV = 2\pi x^2 (a - x)\, dx,$$

and    $V = \displaystyle\int_0^a 2\pi x^2 (a - x)\, dx = \dfrac{\pi a^4}{6}\,.$

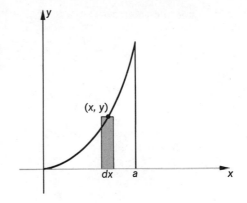

In addition to solids of revolution, for which cross sections are circular discs, volumes of other solids can be obtained as integrals, providing that a family of cross sections is simple enough.

Suppose that the cross sections of the solid pictured, by planes perpendicular to the $x$-axis have area $A(x)$. Then the solid is the sum of thin slices and the volume is*

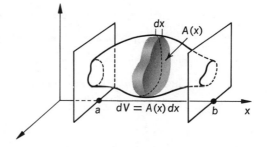

(2)        $V = \displaystyle\int_a^b A(x)\, dx.$

**Example 2**    A tent has a circular base of radius $a$ and height $h$, with the ridge pole above a diameter of the base. Find the volume.

The cross section perpendicular to the ridge pole is an isosceles triangle. With the notation of the figure

$$A(y) = hx = h\sqrt{a^2 - y^2}.$$

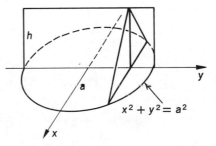

---

* Equation (2) implies that two solids with the same parallel cross-sectional areas have the same volume, though their shapes be different. This is Cavalieri's Theorem (Bonaventura Cavalieri, 1598–1647). In the seventeenth century volumes were often imagined as "infinite sums of infinitely thin elements." The modern point of view gives volume as a *finite* sum of volumes of slabs and then a passage to the limit.

Then the volume is

$$V = \int_{-a}^{a} h\sqrt{a^2 - y^2}\, dy = 2 \int_{0}^{a} h\sqrt{a^2 - y^2}\, dy$$

$$= 2h\left(\frac{y}{2}\sqrt{a^2 - y^2} + \frac{a^2}{2}\text{Arc sin }\frac{y}{a}\right)\Bigg|_{0}^{a} = \frac{\pi a^2 h}{2}.$$

This is a reasonable answer, as it is half the volume of a cylinder of radius $a$ and height $h$.

### Problems

In Problems 1 to 15 the plane region described is revolved about a line. Find the volume of the solid generated.

1. $x^2 - y^2 = a^2$, $x = 2a$;  about the $y$-axis $\qquad\qquad [4\pi\sqrt{3}a^3]$

2. $x^2 - y^2 = a^2$, $x = 2a$;  about $x = 2a$ $\qquad [4\pi\sqrt{3}a^3 + 4\pi a^3 \log(2 - \sqrt{3})]$

3. $y = e^{-x^2}$, $y = 0$, $x = 0$,  about the $y$-axis $\qquad\qquad [\pi]$

4. $xy = 2$, $x + y = 3$;  about the $x$-axis $\qquad\qquad [\pi/3]$

5. The region under one arch of $y = \sin x$;  about $y = 0$ $\qquad [\pi^2/2]$

6. The same region as in Problem 5;  about the $y$-axis $\qquad [2\pi^2]$

7. $y = x^3$, $x = 2$, $y = 0$;  about $x = 2$ $\qquad\qquad [16\pi/5]$

8. $2y^2 = x$, $y^2 = 3 - x$;  about $y = 0$ $\qquad\qquad [3\pi/2]$

9. $2y^2 = x$, $8y = x^3$;  about $x = 0$ $\qquad\qquad [8\pi/5]$

10. $y = \dfrac{x}{\sqrt{3 - x}}$, $y = 0$, $x = 0$, $x = 3$;  about $x = 3$ $\qquad \left[\dfrac{24\pi\sqrt{3}}{5}\right]$

11. $x = \sqrt{a^2 + y^2}$, $x = 0$, $y = 0$, $y = a$;  about the $x$-axis $\qquad \left[\dfrac{2\pi a^3}{3}(2\sqrt{2} - 1)\right]$

12. $y = x^3 - 8$, $y = 0$, $x = 1$;  about $x = 1$ $\qquad\qquad [31\pi/10]$

13. $x^2 + (y - b)^2 = a^2$, $b > a$;  about $y = 0$ $\qquad\qquad [2\pi^2 a^2 b]$

14. $x = a(\theta - \sin\theta)$, $y = a(1 - \cos\theta)$, one arch, and $y = 0$;  about $x = 0$ $\qquad\qquad [6\pi^3 a^3]$

15. Above $4y = x^2$ and inside $x^2 + y^2 = 32$;  about the $y$-axis $\qquad \left[\dfrac{32\pi}{3}(8\sqrt{2} - 7)\right]$

16. A hole of radius $a$ is bored through the center of a sphere of radius $b > a$. Find the volume cut out.
$$\left[ \frac{4\pi}{3} \left( b^3 - (b^2 - a^2)^{3/2} \right) \right]$$

17. Two circular cylinders of radius $a$ intersect at right angles. Find the volume of the common region. $[16a^3/3]$

18. A circular hole of radius $a$ is drilled at right angles through a cylinder of radius $b > a$. *Set up* an integral for the volume cut out.
$$\left[ 8 \int_0^a \sqrt{a^2 - x^2} \sqrt{b^2 - x^2}\, dx \right]$$

19. Find the volume of the region in the first octant bounded by the cylinders $x^2 = 1 - y$ and $z^2 = 1 - y$. $[\frac{1}{2}]$

20. ✓ A solid has a circular base of radius $a$. Plane sections perpendicular to a fixed diameter of the base are equilateral triangles. Find its volume. $[4a^3/\sqrt{3}]$

21. Establish the formula for the volume of a pyramid of height $h$ and area of base $B$.

22. Find the volume of the ellipsoid
$$\frac{x^2}{a^2} + \frac{y^2}{b^2} + \frac{z^2}{c^2} = 1.$$ $[\frac{4}{3}\pi abc]$

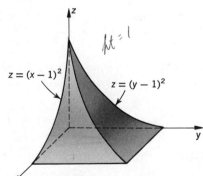

23. Find the volume of the "steeple-like" solid pictured. $[\frac{1}{6}]$

24. A cornucopia has circular parallel cross sections, with the diameter of the circle in inches equal to $\frac{1}{24}$ of the square of the distance, in inches, from the plane of the circle to the vertex. Find the volume of the cornucopia if the largest circle has a diameter of 1 ft.
$[432\sqrt{2}\pi/5 \text{ cu in.}]$

25. A circular hole is drilled through the center of a sphere leaving a ring shaped solid. If the width of the ring is 1, find the volume of the ring.
$[\pi/6]$

## 7   *Arc Length*

In Chapter 6 we used formulas for the differential of arc length. Now we establish these formulas. A definition of arc length is needed. We shall proceed as in plane geometry for the length of a circle. We define arc length as a limit of lengths of inscribed polygons.

**DEFINITION**   Consider a curve parameterized by

$$x = \varphi(t), \quad y = \psi(t), \quad a \leq t \leq b.$$

Let $a = t_0 < t_1 < \cdots < t_n = b$ be a subdivision of $[a, b]$, and let

$$x_i = \varphi(t_i), \quad y_i = \psi(t_i), \quad i = 0, \cdots, n$$

$$\Delta x_i = x_i - x_{i-1}, \quad \Delta y_i = y_i - y_{i-1}, \quad i = 1, \cdots, n.$$

(1)         $|P_{i-1}P_i| = \sqrt{\Delta x_i^2 + \Delta y_i^2}.$

Then the length of the curve is the limit (if it exists)*:

(2)         $$s = \lim_{n \to \infty} \sum_{i=1}^{n} |P_{i-1}P_i|,$$

the limit being taken as the mesh of the subdivision approaches zero.

In other words arc length is the limit of lengths of inscribed polygons.

**THEOREM**   *If the curve is the graph of a continuously differentiable function* $y = y(x)$, $a \leq x \leq b$, *then the limit* (2) *exists and is*

$$s = \int_a^b \sqrt{1 + y'^2}\, dx.$$

**PROOF**   From formula (1)

$$|P_{i-1}P_i| = \sqrt{(\Delta x_i)^2 + (\Delta y_i)^2} = \sqrt{1 + (\Delta y_i / \Delta x_i)^2}\, \Delta x_i$$

$$= \sqrt{1 + [y'(\xi_i)]^2}\, \Delta x_i,$$

where $x_{i-1} < \xi_i < x_i$ by the MVT of differential calculus. Therefore,

$$s = \lim \sum_{i=1}^{n} |P_{i-1}P_i| = \lim \sum_{i=1}^{n} \sqrt{1 + [y'(\xi_i)]^2}\, \Delta x_i.$$

---

* Curves for which the limit exists, that is, curves with finite length, are called *rectifiable* curves. In the past the expression "to rectify a curve" was used meaning "to find its length," that is, "to straighten it out." There are curves with infinite length that do not have infinite extent.

The limit exists because it is a limit of Riemann sums for the function $f(x) = \sqrt{1 + [y'(x)]^2}$. Therefore,

$$s = \int_a^b \sqrt{1 + [y'(x)]^2}\, dx.$$

If we integrate to a variable point $(x, y)$, then

$$s = s(x) = \int_a^x \sqrt{1 + y'^2}\, dx.$$

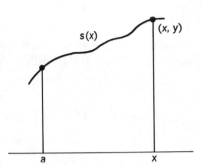

Thus,

$$(3)\begin{cases} ds = \sqrt{1 + y'^2}\, dx = \sqrt{1 + (dy/dx)^2}\, dx, \\ ds^2 = dx^2 + dy^2. \end{cases}$$

These are the formulas of Chapter 6.*

If the curve is given parametrically by $x = \varphi(t)$, $y = \psi(t)$, where $\varphi$ and $\psi$ are differentiable, then, from equations (3),

$$ds^2 = [\varphi'(t)\, dt]^2 + [\psi'(t)\, dt]^2$$
$$= (\varphi'^2 + \psi'^2)\, dt^2.$$

Thus, the length of the curve $x = \varphi(t)$, $y = \psi(t)$ from $t = t_0$ to $t = t_1$ is

$$s = \int_{t_0}^{t_1} \sqrt{\varphi'^2 + \psi'^2}\, dt = \int_{t_0}^{t_1} \sqrt{\dot{x}^2 + \dot{y}^2}\, dt.$$

If the curve is given in terms of polar coordinates by $r = r(\theta)$, where $r$ is a differentiable function of $\theta$, then

$$x = r \cos \theta, \qquad y = r \sin \theta.$$

Now we can compute $ds$ from equations (3). We have

(4)
$$\left.\begin{array}{l} dx = -r \sin \theta\, d\theta + \dot{r} \cos \theta\, d\theta \\ \\ dy = \phantom{-}r \cos \theta\, d\theta + \dot{r} \sin \theta\, d\theta \end{array}\right\} \quad \text{where } \dot{r} = \dfrac{dr}{d\theta};$$

$$ds^2 = dx^2 + dy^2 = (r^2 + \dot{r}^2)\, d\theta^2$$
$$= r^2\, d\theta^2 + dr^2.$$

_____

* Leibniz wrote $ds = \sqrt{dx\, dx + dy\, dy}$.

Thus, the length of the curve $r = r(\theta)$ from $\theta = \theta_0$ to $\theta = \theta_1$ is given by

$$s = \int_{\theta_0}^{\theta_1} \sqrt{r^2 + \dot{r}^2}\, d\theta.$$

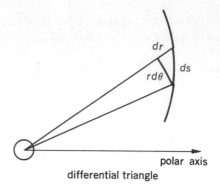

Equation (4) is easy to remember by the mnemonic device of the "differential triangle."

polar axis

differential triangle

*Example 1*   Find the length of the curve $y^2 = x^3$ from $(0,0)$ to $(2, 2\sqrt{2})$.

$\quad y = x^{3/2}$,     positive because we are on the upper half;

$\quad y' = \frac{3}{2}x^{1/2}$.

$$s = \int_0^2 \sqrt{1 + \frac{9x}{4}}\, dx = \frac{22\sqrt{22} - 8}{27} \approx 3.53.$$

*Example 2*   Find the length of the circle $r = a\cos\theta$.

$\quad ds^2 = r^2\, d\theta^2 + dr^2$

$\quad\quad = (a^2\cos^2\theta + a^2\sin^2\theta)\, d\theta^2 = a^2\, d\theta^2$;

$\quad ds = a\, d\theta$;

$$s = \int_0^\pi a\, d\theta = \pi a.$$

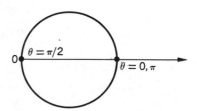

*Problems*

In Problems 1 to 18 find the length of arc between the indicated points.

*1.* $y = \log(1 - x^2)$;   $x = 0$ to $x = \frac{3}{4}$          $[\log\sqrt{7} - \frac{3}{4}]$

*2.* $y^2 = 2px$;   $x = 0$ to $x = \dfrac{p}{2}$          $\left[\dfrac{p}{2}[\sqrt{2} + \log(1 + \sqrt{2})]\right]$

   $[\text{Use } ds^2 = (1 + x'^2)\, dy^2.]$

*3.* The catenary $y = \dfrac{a}{2}(e^{x/a} + e^{-x/a})$;   $(0, a)$ to $(x, y)$          $\left[\dfrac{a}{2}(e^{x/a} - e^{-x/a})\right]$

*4.* $x = 4t^3$, $y = 2t^2$;   $t = 0$ to $t = 1$          $[\frac{4}{27}(10\sqrt{10} - 1)]$

*5.* The whole hypocycloid $x^{2/3} + y^{2/3} = a^{2/3}$          $[6a]$

6. The cardioid $r = a(1 - \cos\theta)$                 [8a]

7. One arch of the cycloid $x = a(\theta - \sin\theta),\ y = a(1 - \cos\theta)$       [8a]

8. $y = \log x;\quad x = 1$ to $x = e$       $\left[\sqrt{e^2+1} - \sqrt{2} + \log\dfrac{e(1+\sqrt{2})}{1+\sqrt{1+e^2}}\right]$

9. The loop of the curve $3ay^2 = x(a-x)^2$       $[4a/\sqrt{3}]$

10. The circle $x = a\cos\theta,\ y = a\sin\theta$

11. $y = \log\sec x;\quad x = 0$ to $x = \pi/4$       $[\log(1+\sqrt{2})]$

12. $y = \log\dfrac{e^x+1}{e^x-1};\quad x = 1$ to $x = 2$       $[\log(e^2+1)-1]$

13. The spiral of Archimedes $r = a\theta;\quad \theta = 0$ to $\theta = \pi$

$$\left[\frac{\pi a}{2}\sqrt{1+\pi^2} + \frac{a}{2}\log(\pi + \sqrt{1+\pi^2})\right]$$

14. The equiangular spiral $r = e^{a\theta};$    (a) $\theta = 0$ to $\theta = \theta_1;$
      (b) $\theta = -\infty$ to $\theta = \theta_1$

$$\left[(b)\ \frac{e^{a\theta_1}\sqrt{1+a^2}}{a}\right]$$

15. The spiral $r\theta = a;\quad \theta = \frac{1}{2}$ to $\theta = 1$       $\left[a\left(\sqrt{5} - \sqrt{2} + \log\dfrac{2+2\sqrt{2}}{1+\sqrt{5}}\right)\right]$

16. The whole curve $r = a\cos^3(\theta/3)$       $[3\pi a/2]$

17. The curve $x = a(\cos\theta + \theta\sin\theta),\ y = a(\sin\theta - \theta\cos\theta);\quad \theta = 0$ to $\theta = \theta_0$
      $[a\theta_0^2/2]$

18. $x = a\cos^3\theta,\ y = b\sin^3\theta;\quad \theta = 0$ to $\theta = \pi/2$       $\left[\dfrac{a^2+ab+b^2}{a+b}\right]$

19. *Set up an integral for the length of the ellipse* $x = a\cos\theta,\ y = b\sin\theta$. *The integral is not elementary; it is called an elliptic integral.*

$$\left[4\int_0^{\pi/2}\sqrt{b^2 + (a^2-b^2)\sin^2\theta}\ d\theta\right]$$

20. *Set up an integral for the length of the lemniscate* $r^2 = a^2\cos 2\theta$.

$$\left[4a\int_0^{\pi/4}\sqrt{\sec 2\theta}\ d\theta\right]$$

8  *Area of Surfaces of Revolution*

Areas of arbitrary surfaces must be deferred until Chapter 15 where double integrals are discussed. However, at this point we can, with single integrals, find areas of surfaces of revolution.

Consider the surface obtained by revolving the curve

$$y = f(x), \qquad a \leq x \leq b,$$

about the $x$-axis. With the subdivision

$$a = x_0 < x_1 < \cdots < x_n = b,$$

the segment of length $\Delta l_i = \sqrt{\Delta x_i^2 + \Delta y_i^2}$ of the figure generates a frustum of a cone, whose lateral area is*

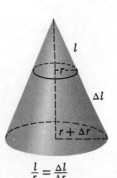

$$\Delta S_i = \pi(y_{i-1} + y_{i-1} + \Delta y_i) \, \Delta l_i$$

$$= \pi(2y_{i-1} + \Delta y_i) \sqrt{\Delta x_i^2 + \Delta y_i^2}$$

$$= 2\pi \left( y_{i-1} + \frac{\Delta y_i}{2} \right) \sqrt{1 + \left( \frac{\Delta y_i}{\Delta x_i} \right)^2} \, \Delta x_i$$

$$= 2\pi \left( y_{i-1} + \frac{\Delta y_i}{2} \right) \sqrt{1 + [f'(\xi_i)]^2} \, \Delta x_i, \qquad x_{i-1} < \xi_i < x_i,$$

where the last equality follows from the MVT of differential calculus. Therefore, an approximation to the area of the surface will be given by

$$(1) \qquad \sum_{i=1}^{n} \Delta S_i = 2\pi \sum_{i=1}^{n} \bar{y}_i \sqrt{1 + [f'(\xi_i)]^2} \, \Delta x_i,$$

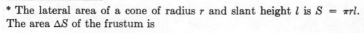

---

* The lateral area of a cone of radius $r$ and slant height $l$ is $S = \pi r l$. The area $\Delta S$ of the frustum is

$$\Delta S = \pi(r + \Delta r)(l + \Delta l) - \pi r l$$

$$= \pi(r + (r + \Delta r)) \, \Delta l,$$

because $l/r = \Delta l/\Delta r$.

where $\bar{y}_i = y_{i-1} + (\Delta y_i)/2 = f(\eta_i)$, for some $\eta_i$ with $x_{i-1} \leq \eta_i \leq x_i$. The limit of (1) as the mesh of the subdivision goes to zero is an integral* that will be the area $S$ of the surface.

$$S = 2\pi \int_a^b y\sqrt{1 + y'^2}\,dx = 2\pi \int_a^b y\,ds.$$

Note that the differential element of surface area is

$$dS = 2\pi y\,ds,$$

the area swept out by the differential of arc.

If the rotation is about the $y$-axis, then the differential element of surface is

$$dS = 2\pi x\,ds.$$

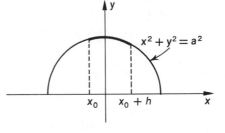

$dS = 2\pi y\,ds$

In any case, the formula for $dS$ can be in terms of any of $x$ or $y$ or some parameter or polar coordinates.

**Example 1** Find the area of a zone of a sphere of radius $a$.

Consider the zone generated by revolving the arc between $x_0$ and $x_0 + h$ of the semicircle $x^2 + y^2 = a^2$, $y \geq 0$, about the $x$-axis.

$$ds = \sqrt{1 + y'^2}\,dx = \sqrt{1 + \left(\frac{-x}{y}\right)^2}\,dx$$

$$= \frac{a}{y}\,dx \quad \text{(because } y > 0\text{)},$$

so

$$S = 2\pi \int_{x_0}^{x_0+h} y\,ds = 2\pi \int_{x_0}^{x_0+h} a\,dx = 2\pi ah.$$

---

* The sum (1) is not in exactly the correct form to invoke the definition of the definite integral. This is the case because $\bar{y}_i = f(\eta_i)$ and, in general, $\eta_i \neq \xi_i$. That the limit of (1) is precisely the integral can be established by Bliss's theorem, which states that

$$\int_a^b f(x)g(x)\,dx = \lim \sum_{i=1}^n f(\xi_i')g(\xi_i'')\,\Delta x_i$$

if $f$ and $g$ are continuous, where $\xi_i'$, $\xi_i''$ are in the $i$th interval.

A careful development of the volume formula, using shells, of Section 6 would also require Bliss's theorem.

**Example 2**   The limaçon $r = 2 + \cos \theta$ is revolved about the polar axis. Find the area of the surface generated. We use $dS = 2\pi y \, ds$, where $y = r \sin \theta = (2 + \cos \theta) \sin \theta$, and $ds = \sqrt{r^2 + \dot{r}^2} \, d\theta$.

$$r = 2 + \cos \theta, \qquad \frac{dr}{d\theta} = -\sin \theta;$$

$$ds = \sqrt{(2 + \cos \theta)^2 + \sin^2 \theta} \, d\theta$$

$$= \sqrt{5 + 4 \cos \theta} \, d\theta.$$

$$S = 2\pi \int_0^\pi (2 + \cos \theta) \sin \theta \, (5 + 4 \cos \theta)^{1/2} \, d\theta$$

$$= 2\pi \left[ -\tfrac{1}{6}(2 + \cos \theta)(5 + 4 \cos \theta)^{3/2} + \tfrac{1}{60}(5 + 4 \cos \theta)^{5/2} \right] \Big|_0^\pi$$

$$= 93\pi/5.$$

$\theta = \pi$

$\theta = 0, 2\pi$

$r = 2 + \cos \theta$

## Problems

In Problems 1 to 19 an arc is described and a line about which it is rotated. Find the area of the surface generated.

1.   $y = x^2, 0 \leq x \leq 1$;   about $y = 0$

$$\left[ \frac{\pi}{32} \left[ 18\sqrt{5} - \log \, (2 + \sqrt{5}) \right] \right]$$

2.   $y = x^2, 0 \leq x \leq 1$;   about $x = 0$

$$\left[ \frac{\pi}{6} (5\sqrt{5} - 1) \right]$$

3. ✓ $y = \dfrac{x^3}{3} + \dfrac{1}{4x}, 1 \leq x \leq 2$;   about $x = 0$

$$\left[ \frac{\pi}{2} (15 + \log 2) \right]$$

4.   $y = \dfrac{x^3}{3} + \dfrac{1}{4x}, 1 \leq x \leq 2$;   about $y = 0$

$$\left[ \frac{515\pi}{64} \right]$$

5.   $x = \sqrt{2}t, y = \dfrac{t^2}{\sqrt{2}}, 0 \leq t \leq 2$;   about $y = 0$ $\left[ \dfrac{\pi}{4} \left[ 18\sqrt{5} - \log \, (2 + \sqrt{5}) \right] \right]$

6.   $x = \sqrt{2}t, y = \dfrac{t^2}{\sqrt{2}}, 0 \leq t \leq 2$;   about $x = 0$

$$\left[ \frac{4\pi}{3} (5\sqrt{5} - 1) \right]$$

7. ✓ $x^2 + y^2 = a^2$;   about $y = a$

$$[4\pi^2 a^2]$$

8.  $x^2 + y^2 = a^2$; about $y = b > a$     $[4\pi^2 ab]$

9.  $x = a \cos^3 \theta, y = a \sin^3 \theta$; about $y = 0$     $[12\pi a^2/5]$

10. $y = \dfrac{a}{h} x, 0 \leq x \leq h$; about $y = 0$     $[\pi a \sqrt{a^2 + h^2}]$

11. One arch of the cycloid $x = a(\theta - \sin \theta), y = a(1 - \cos \theta)$; about $y = 0$
$[\tfrac{64}{3}\pi a^2]$

12. One arch of the cycloid $x = a(\theta - \sin \theta), y = a(1 - \cos \theta)$; about $x = 0$
$[16\pi^2 a^2]$

13. $y = e^x, -\infty < x \leq 0$; about $y = 0$     $[\pi[\sqrt{2} + \log (1 + \sqrt{2})]]$

14. The loop of $3ay^2 = x(a - x)^2$; about the $y$-axis     $[\tfrac{56}{45}\sqrt{3}\pi a^2]$

15. One arch of $y = \sin x$; about $y = 0$     $[2\pi[\sqrt{2} + \log (1 + \sqrt{2})]]$

16. The ellipse $b^2 x^2 + a^2 y^2 = a^2 b^2, a > b$; about $y = 0$
$$\left[ 2\pi \left( b^2 + \frac{ab}{e} \text{Arc sin } e \right), \quad e = \text{eccentricity} \right]$$

17. The ellipse of Problem 16 about $x = 0$     $\left[ 2\pi \left( a^2 + \dfrac{b^2}{2e} \log \dfrac{1+e}{1-e} \right), e = \text{eccentricity} \right]$

18. $r^2 = a^2 \cos 2\theta$; about $\theta = \pi/2$     $[2\sqrt{2}\pi a^2]$

19. $r^2 = a^2 \cos 2\theta$; about $\theta = 0$     $[2\pi a^2 (2 - \sqrt{2})]$

## 9   Force and Work

We have but scratched the surface of applications of the definite integral. Whenever the desired quantity can be expressed as a limit of Riemann sums, we are then in business. The applications to physics are particularly abundant. Examples will illustrate a few cases. In each we will set up the differential element, and so the integral, without justifying our procedure by inequalities on Riemann sums.

**FORCE DUE TO FLUID PRESSURE**    If a container is filled with liquid to a depth $h$ (see the figure), the force due to fluid pressure on the horizontal *bottom* of the container is equal to the pressure at the bottom times the area of the bottom: $F = pA$. The pressure at the bottom is equal to the weight $\rho$ per unit volume of the liquid, times the depth $h$: $p = \rho h$. (In the case of water, $\rho = 62.4$ lbs/cu ft.) Thus, the force on the bottom is $F = \rho h A$.

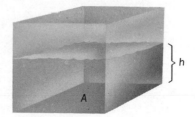

$f = pA, \quad p = \rho h$

It is somewhat more complicated to compute the force on a side* of a container because the pressure varies with the depth. A method for finding the force on a side is to subdivide the side into thin horizontal strips and to compute the approximate force on each strip. Addition of these approximate forces and a passage to the limit will yield a definite integral for the force. Example 1 will illustrate the method.

**Example 1**   A circular window in the wall of a tank of water is of radius 1 ft with its center 4 ft below the surface. Find the force on the window due to the water pressure. (Water weighs 62.4 lb/cu ft.)

Choose a coordinate system, for algebraic convenience, with its origin at the center of the circle, as in the figure.

The differential strip being approximately all at the depth $h$ has on it a force

$$dF = \text{pressure} \times \text{area}$$
$$= (62.4)h\,(l\,dy)$$
$$= (62.4)\,(4 - y)\,(2\sqrt{1 - y^2})\,dy,$$

because $h = 4 - y$ and $l = 2x = 2\sqrt{1 - y^2}$. "Summing" these differential forces over the window gives the force on the window:

$$F = \int_{-1}^{1} 2\,(62.4)\,(4 - y)\sqrt{1 - y^2}\,dy$$

$$\left\{ \begin{array}{l} \text{the odd part gives zero,} \\[4pt] \displaystyle\int_{-1}^{1} y\sqrt{1 - y^2}\,dy = 0 \end{array} \right\}$$

$$= 2\,(62.4)\int_{-1}^{1} 4\sqrt{1 - y^2}\,dy$$

$$= 249.6\pi \approx 784\ \text{lb}.$$

The axes can be chosen in many ways. Had they been chosen as in this figure, we would have for the force

$$F = \int_{3}^{5} (62.4)\,(y)2\sqrt{1 - (y - 4)^2}\,dy$$

$$= 249.6\pi\ \text{lb}.$$

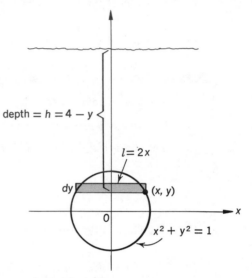

depth $= h = 4 - y$

$l = 2x$

$dy$   $(x, y)$

$0$

$x^2 + y^2 = 1$

Area strip $= l\,dy$
pressure on the strip $= 62.4h$ lb/sq ft

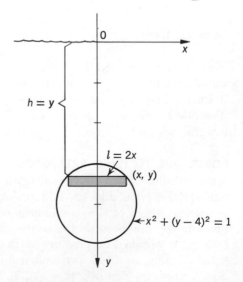

$0$

$x$

$h = y$

$l = 2x$

$(x, y)$

$x^2 + (y - 4)^2 = 1$

$y$

---

* Pressure at a given depth is exerted in all directions equally. This is Pascal's principle. Therefore, the pressure on a small plate *at a given depth* is the same regardless of the way the plate is tilted.

**WORK**   If a *constant* force $F$ (in pounds) acts along a line for a distance $d$ (in feet), then the work done is $W = Fd$ (in foot-pounds). Suppose, however, that the force is a continuous function $F(x)$ of the coordinate $x$ along a line, and that the force acts between $x = a$ and $x = b$. If we subdivide the interval $[a, b]$ as usual, then the work $\Delta W_i$ done in the $i$th interval satisfies the inequality:

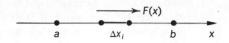

$$F_{\min} \Delta x_i \leqq \Delta W_i \leqq F_{\max} \Delta x_i,$$

where $F_{\min}$ and $F_{\max}$ are the minimum and the maximum values of $F$ in the interval. Then $F_{\min} = F(\xi_i')$ and $F_{\max} = F(\xi_i'')$ for some $\xi_i'$, $\xi_i''$ in the $i$th interval. If we sum over all the subintervals, we get, for the work $W$,

(1) $$\sum F(\xi_i') \Delta x_i \leqq W \leqq \sum F(\xi_i'') \Delta x_i.$$

The two sums in (1) are Riemann sums for the same definite integral. As the mesh approaches zero, the two sums squeeze down on $W$. Thus,

$$W = \text{Work} = \int_a^b F(x) \, dx.$$

Observe that the differential element of work is $dW = F(x) \, dx$.

*Example 2*   To find the work done in stretching a perfectly elastic spring we must know the restoring force $F$. According to Hooke's Law, $F$ is proportional to the displacement $x$ of the spring.

$$F = kx, \qquad k = \text{the spring constant.}$$

Then the work done *on* the spring in stretching it an amount $x_0$ is

$$\text{Work} = W = \int_0^{x_0} F(x) \, dx$$

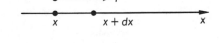

$$= \int_0^{x_0} kx \, dx = \tfrac{1}{2} k x_0^2.$$

This is precisely the potential energy stored in the spring.

If a particular spring exerts a restoring force of 4 lb when stretched 6 in. $(= \tfrac{1}{2}$ ft), the spring constant can be found:

$$4 = k(\tfrac{1}{2}),$$

whence   $k = 8$ lb per ft.

Then the work done in stretching the spring $x_0$ feet is $4x_0^2$.

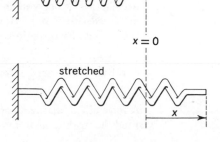

*Example 3*     A hemispherical tank of diameter 10 ft is full of water. If the water in the tank is pumped out to a height 8 ft above the surface, find the work done.

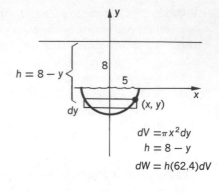

We compute the total work by finding the work required to lift a thin horizontal slab of water to the required level and then "summing" over all thin slabs.

If we choose horizontal and vertical axes as shown, then the element of volume is

$$dV = \pi x^2\, dy = \pi (25 - y^2)\, dy,$$

$$-5 \leqq y \leqq 0.$$

The weight of this slab is $62.4\, dV$, and to pump it out it must be moved $8 - y$ ft. Thus, the differential work required to raise this slab $8 - y$ ft is

$$dW = (8 - y)(62.4)(\pi)(25 - y^2)\, dy.$$

The total work is

$$W = 62.4\pi \int_{-5}^{0} (25 - y^2)(8 - y)\, dy$$

$$\approx 161{,}300 \text{ ft-lb.}$$

If the axes were selected as suggested at the right the integral for the work would be

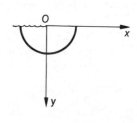

$$W = 62.4\pi \int_{0}^{5} (25 - y^2)(8 + y)\, dy \approx 161{,}300 \text{ ft-lb.}$$

## Problems

1.  Find the force due to water pressure on a rectangular window 12 in. wide and 6 in. high in the side of a tank if the top of the window is 6 ft below the surface.
    [195 lb]

2.  An elliptical window in an oil tank has its major axis horizontal. The major and minor axes are 2 and 1 ft, respectively, and the top of the window is 10 ft below the surface. Find the force on the window if the oil weighs 50 lb/cu ft.
    [$525\pi/2$]

3.  A rubber band exerts a force of 8 oz when stretched 8 in. How much work was done?
    [$\frac{1}{6}$ ft-lb]

4.  A force of $f$ lb acting along the $x$-axis varies according to $f(x) = -x^2 + x - 1$. Find the work done in moving from $x = -1$ to $x = 1$.
    [$-\frac{8}{3}$]

5. A force of attraction along a line is toward the origin and is proportional to the directed distance from the origin. What is the work done in moving from $-a$ to $a$?
[0]

6. A conical reservoir (vertex down) has diameter 10 ft and height 10 ft. (a) Find the work needed to pump it out of the top if it is full of water. (b) How much work would be needed to fill the reservoir from a pond at the base if the water is pumped to the top?          [(a) $13000\pi$ ft-lb, (b) $52000\pi$ ft-lb]

7. A conical reservoir has diameter 10 ft and height 10 ft with the vertex up. Find the work necessary to pump it out of the top.          [$39000\pi$ ft-lb]

8. A river dam has the shape of a trapezoid 50 ft across the top, 40 ft across the bottom, and 20 ft high. What is the force on the dam when the water is 15 ft deep?          [298,350 lb]

9. A vertical dam has a parabolic shape and dimensions at water level as shown. Find the force on the dam in tons.          [37,440]

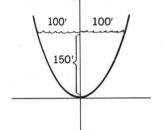

10. A tank consists of a slender cylinder 8 ft tall and 2 ft in diameter, surmounting a flat cylinder 2 ft tall and 6 ft in diameter. If it is filled with water to a depth of 8 ft, find the work necessary to pump it out of the top.

[11,981$\pi$ ft-lb]

11. A swimming pool is 30 ft × 15 ft. The bottom slopes from a depth of 3 ft at one end to 9 ft at the other. What is the total force on the bottom?          [168,500 lb]

12. A cylindrical tank 8 ft in diameter and 8 ft high, with axis vertical, has 6 ft of water in it. What work is required to pump out half the water to a height 8 ft above the top of the tank?          [108,200 ft-lb]

13. A water trough 15 ft long has a cross section that is an isosceles right triangle 4 ft across the top. What is the force on one of the long sides?          [36,504 lb]

14. How much work is required to pump oil (50 lb/cu ft) from the full lower hemispherical tank to the upper tank if (a) the pipe is at the bottom of the top tank? (b) at the top of the top tank?     [(a) 86,400π ft-lb, (b) 102,600π ft-lb]

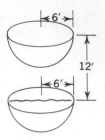

15. The force of gravity (outside the earth) attracting a mass $m$ is proportional to the mass $m$ and inversely proportional to the square of the distance from the center of the earth. Find the work done in moving a mass that weighs 1 lb at the surface of the earth to 10 mi above the surface. Assume the radius of the earth is 4000 mi. At the earth's surface, a force of 1 lb, is required to lift a mass of 1 lb.
     [52,668 ft-lb]

16. Interior to the earth the gravitational attraction varies directly as the distance from the center of the earth. What is the work necessary to lift a mass, weighing 1 lb at the surface, to the surface from the center?     [10,560,000 ft-lb]

17. A cable over a pulley weighs 2 lb/ft. If the weight being raised weighs 200 lb, what work is needed to raise it 100 ft?
     [30,000 ft-lb]

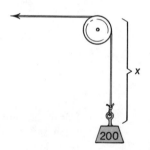

18. A trough $\frac{3}{4}$ full of water has the semi-elliptical end shown. Find the force on one end.

     [83.9 lb]

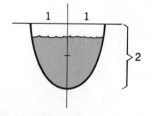

19. A cylindrical oil tank with axis horizontal and diameter 8 ft is full of oil weighing 50 lb/cu ft. Find the force on one end.     [3200π lb]

20. Find the work done against gravity to lift a payload of one ton to a height of 400 mi above the earth (see Problem 15).     [3,840,000,000 ft-lb]

21. When a gas expands a small amount, a differential element of volume, the work done *by* the gas is $p\,dV$, where $p$ is the gas pressure.

Find the work done in compressing the gas from $V_1$ to $V_2$, if the compression follows the formula (isothermal)
$$pV = \text{constant} = C.$$
$$[C \log (V_2/V_1)]$$

22. Find the work done in compressing a gas from $V_1$ to $V_2$ (see Problem 21) if the compression follows the formula $pV^\gamma = \text{constant} = C$, where $\gamma$ is a constant $> 1$.
$$\left[ \frac{1}{\gamma - 1} (p_1 V_1 - p_2 V_2) \right]$$

23. Use the result of Problem 22 to find the work done by a gas expanding from a pressure of 30 lb/sq in. and volume 5 cu ft to a pressure of 10 lb/sq in., if $\gamma = 1.4$.
$$[14{,}550 \text{ ft-lb}]$$

24. A spherical reservoir of diameter 6 ft is full of oil weighing 50 lb/cu ft. Find the work to pump the oil to a height 20 ft above the top of the reservoir.
$$[41{,}400\pi \text{ ft-lb}]$$

25. A reservoir has the shape shown, with the water level half way up the cylindrical part. Find the work done in pumping the water out to the top of the tank. $\quad [118{,}250\pi \text{ ft-lb}]$

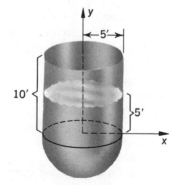

## CHAPTER 13

# THEORY OF INFINITE SERIES

## 1  Convergence and Divergence of Sequences

In Chapter 8 we used power series to compute values of some of the elementary functions. In those examples we added finitely many terms of the series and then observed that the addition of more terms affected the sum but little.

**Example 1**    For the series

$$\sum_{n=1}^{\infty} \frac{1}{n!} = 1 + \frac{1}{2!} + \frac{1}{3!} + \cdots + \frac{1}{n!} + \cdots$$

the first term is:                     $s_1 = 1,$
the sum of the first two terms is:     $s_2 = 1 + 1/2! = 1.5,$
the sum of the first three terms is:   $s_3 = 1 + 1/2! + 1/3! = 1.666\ldots,$
the sum of the first $n$ terms is:     $s_n = 1 + 1/2! + 1/3! + \cdots + 1/n!$

Reference to the table of factorials on page 202 shows that the difference $s_n - s_{n-1}$ is very small when $n$ is large. This suggests that the numbers $s_n$ approach a limit. It seems natural to define the "sum of the series" as the limit of the succession of numbers $s_1, s_2, s_3, \ldots, s_n, \ldots$. This succession of numbers is an example of an *infinite sequence*.

**Example 2**    For the series (called the harmonic series)

$$\sum_{n=1}^{\infty} \frac{1}{n} = 1 + \frac{1}{2} + \frac{1}{3} + \cdots + \frac{1}{n} + \cdots$$

the first term is:                                  $s_1 = 1,$
the sum of the first two terms is:    $s_2 = 1.5,$
the sum of the first three terms is:  $s_3 = 1.833\ldots,$
the sum of the first $n$ terms is:       $s_n = 1 + \frac{1}{2} + \cdots + 1/n.$

At first glance it may appear that here too the sequence of numbers $s_n$ has a limit because $s_n - s_{n-1} = 1/n$ is small when $n$ is large. But we shall see later that this is *not* the case; the numbers $s_n$ get arbitrarily large!

These examples show that we need to state clearly what we mean by an infinite sequence and what is meant by the limit of a sequence.

**DEFINITION 1**     An *infinite sequence* of real numbers is a real-valued function whose domain is the set of positive integers. It is usual to use subscripts instead of functional notation:

Functional notation:   $s(1), s(2), \ldots, s(n), \ldots$
Subscript notation:    $s_1, s_2, \ldots, s_n, \ldots$

The sequence is often denoted by $\{s_n\}$.

*Example 3*     Arithmetic and geometric progressions are examples of infinite sequences.
     The arithmetic progression $\{3, 7, 11, 15, \ldots\}$ is a sequence with $s_1 = 3$, $s_2 = 7$, $s_3 = 11, \ldots.$ The terms of the sequence can be described by a *recursion formula*:

$$s_1 = 3, \quad s_n = s_{n-1} + 4, \qquad \text{for} \qquad n = 2, 3, 4, \ldots.$$

The geometric progression $\{5, \frac{5}{2}, \frac{5}{4}, \frac{5}{8}, \ldots\}$ is a sequence with $s_1 = 5$, $s_2 = \frac{5}{2}, s_3 = \frac{5}{4}, \ldots.$ The terms of the sequence can be described by a recursion formula:

$$s_1 = 5, \quad s_n = \tfrac{1}{2}s_{n-1}, \qquad \text{for} \qquad n = 2, 3, 4, \ldots.$$

The sequence can also be denoted by $\{5/2^n\}$.

*Example 4*     The sequence whose $n$th term is given by $s_n = 1 + 1/n, n = 1, 2,$ $3, \ldots,$ is neither an arithmetic nor a geometric sequence. Here $s_1 = 2, s_2 = \frac{3}{2},$ $s_3 = \frac{4}{3}, s_4 = \frac{5}{4}, \ldots.$ The sequence can also be denoted by $\{1 + 1/n\}$.

**DEFINITION 2**     The sequence $\{s_n\}$ is said to *converge to the limit s* if, for each positive number $\epsilon$, there is an integer $N$ (depending on $\epsilon$) such that

(1)                          $| s_n - s | < \epsilon \qquad \text{if} \qquad n \geqq N.$

The limit of the sequence is denoted by

$$\lim_{n \to \infty} s_n = s.$$

If such a number $s$ does not exist, the sequence is said to *diverge*.

*Remark.* To use the definition to prove that a particular sequence converges one must first have *guessed s*. We shall see that the definition provides a precise *arithmetic* criterion, namely (1), which can be used in proofs of general theorems.

**Example 5**    If $s_n = 1 + 1/n$, then, obviously,

$$s = \lim_{n \to \infty} s_n = \lim_{n \to \infty} \left(1 + \frac{1}{n}\right) = 1.$$

To show that (1) is satisfied, suppose a positive number $\epsilon$ is given. We are to show that

$$|s_n - s| = \left|\left(1 + \frac{1}{n}\right) - 1\right| = \frac{1}{n} < \epsilon,$$

if $n$ is large enough. If we choose an integer $N > 1/\epsilon$, then

$$|s_n - s| = \frac{1}{n} < \epsilon \qquad \text{if} \qquad n \geqq N.$$

**Example 6**    If $s_n = (-1)^n$, $n = 1, 2, \ldots$, then $\lim_{n \to \infty} s_n$ does not exist. Suppose that $\epsilon < 1$ is given; then, for any number $s$, $|s_n - s| \geqq \epsilon$ for either $n$ odd or $n$ even. Thus, the sequence $-1, 1, -1, 1, \ldots$ diverges.

**Example 7**    If $s_n = \sqrt{n}$, $n = 1, 2, \ldots$, then $\lim_{n \to \infty} s_n$ does not exist. This is so because, if $\epsilon = 1$, then for any number $s$ the difference $|\sqrt{n} - s| > 1$ for all $n$ large enough. Thus, the sequence $\{\sqrt{n}\}$ diverges.

In this example the terms of the sequence become arbitrarily large. To denote this behavior of a sequence we sometimes write

$$\lim_{n \to \infty} s_n = \infty, \qquad \text{or, also,} \qquad s_n \longrightarrow \infty,$$

and say that the sequence *diverges to infinity.*

As was remarked above, one must find, or guess, the number $s$ in order to apply Definition 2. Although it often happens that the limit of a sequence is evident, that is not always the case. Sometimes l'Hospital's rule can be applied.

**Example 8**    If $s_n = (1 + \log n)/\sqrt{n}$, both numerator and denominator become infinite as $n \to \infty$, and so $\lim_{n \to \infty} s_n$ is an indeterminate form of the type $\infty/\infty$. l'Hospital's rule applies, and we have

$$\lim_{n \to \infty} s_n = \lim_{n \to \infty} \frac{1 + \log n}{\sqrt{n}} = \lim_{n \to \infty} \frac{1/n}{(\frac{1}{2})n^{-1/2}} = 0.$$

As was observed in the above remark, condition (1) requires that $s$ be known. What one would prefer is a condition on the terms of the sequence alone. Such a condition is given in the following theorem (due to Cauchy), which we state without proof (see Appendix C). The condition of the theorem is seldom useful for particular sequences, but it is frequently useful in proving general theorems.

**THEOREM 1 (THE CAUCHY CRITERION)** *A sequence $\{s_n\}$ converges if and only if to each positive number $\epsilon$ there is a positive integer $N$ such that, if $n \geq N$, then*

(2) $$|s_{n+p} - s_n| < \epsilon \qquad \text{for all positive integers } p.$$

**COROLLARY** *If the sequence $s_n$ converges, then*

$$\lim_{n \to \infty} (s_{n+1} - s_n) = 0.$$

### Problems

In Problems 1 to 8 a few terms of sequences with simple formulas are given. **Guess** the general term $s_n$.

1. $a, 2a^2, 3a^3, \ldots$ $\qquad\qquad\qquad\qquad\qquad\qquad\qquad [s_n = na^n]$

2. $x, \dfrac{x^2}{4}, \dfrac{x^3}{9}, \dfrac{x^4}{16}, \ldots$

3. $6, -3, \frac{3}{2}, -\frac{3}{4}, \ldots$ $\qquad\qquad\qquad\qquad\qquad\qquad [s_n = 6(-\frac{1}{2})^{n-1}]$

4. $\dfrac{2}{1\cdot 3}, \dfrac{2}{3\cdot 5}, \dfrac{2}{5\cdot 7}, \ldots$ $\qquad \left[ s_n = \dfrac{1}{2n-1} - \dfrac{1}{2n+1} = \dfrac{2}{(2n-1)(2n+1)} \right]$

5. $\dfrac{1}{1}, \dfrac{-1}{2}, \dfrac{1}{6}, \dfrac{-1}{24}, \ldots$

6. $\dfrac{1}{4}, \dfrac{-4}{9}, \dfrac{9}{16}, \dfrac{-16}{25}, \ldots$

7. $\dfrac{1}{2\cdot 3}, \dfrac{1\cdot 3}{2\cdot 4\cdot 5}, \dfrac{1\cdot 3\cdot 5}{2\cdot 4\cdot 6\cdot 7}, \ldots$ $\qquad \left[ s_n = \dfrac{1\cdot 3 \cdots (2n-1)}{2\cdot 4 \cdots (2n)(2n+1)} \right]$

8. $\dfrac{3}{2}, \dfrac{5}{5}, \dfrac{7}{10}, \dfrac{9}{17}, \ldots$

In Problems 9 to 18 the nth term of a sequence is given. Find the limit of the sequence if it converges.

9. $\dfrac{1}{1 + \sqrt{n}}$ $\qquad\qquad\qquad$ 11. $\dfrac{\log n}{n}$

10. $\dfrac{n+3}{n^2+1}$ $\qquad\quad$ [0] $\qquad$ 12. $n^2 e^{-n}$

13.  $\dfrac{n^3}{n!}$                                      16.  $(-1)^{n+1}$

14.  $(1 + n)^{1/n}$              [1]           17.  $\sin n\pi/2$              [divergent]

15.  $\left(1 + \dfrac{1}{n}\right)^n$          [e]           18.  $2^n$

19.  Verify condition (1) of Definition 2 for the sequence of Problem 9.
$$[1/(1 + \sqrt{n}) < \epsilon \text{ if } n > 1/\epsilon^2]$$

20.  Verify condition (1) of Definition 2 for the sequence of Problem 10.
$$[(n + 3)/(n^2 + 1) < \epsilon \text{ if } n > 2/\epsilon, \text{ and } n \geq 3]$$

21.  If $s_n < \bar{s}_n$ for $n = 1, 2, 3, \ldots$, can $\lim s_n = \lim \bar{s}_n$? Give an example.

\*22.  Assuming that the sequence $\{s_n\}$, where
$$s_1 = \sqrt{6}, \quad s_2 = \sqrt{6 + s_1}, \quad \ldots, \quad s_n = \sqrt{6 + s_{n-1}}, \quad \ldots,$$
converges, find the limit of the sequence.                    [3]

## 2   *Convergence and Divergence of Series\**

We turn now from sequences to *series*. The symbolic sum

(1)
$$\sum_{n=1}^{\infty} a_n = a_1 + a_2 + \cdots + a_n + \cdots$$

is called an *infinite series*. Because one cannot actually add infinitely many numbers (unless all but a finite number are zero), our first task is to give a definition of what we mean by the sum of the series (1). Examples 1 and 2 of Section 1 show how we associate a certain sequence with a given series.

**DEFINITION 1**      The *partial sums* of the series (1) are the numbers

$$s_1 = a_1$$

$$s_2 = a_1 + a_2 \qquad\qquad = \sum_{k=1}^{2} a_k$$

$$s_3 = a_1 + a_2 + a_3 \qquad\quad = \sum_{k=1}^{3} a_k$$

$$\cdots$$

$$s_n = a_1 + a_2 + \cdots + a_n = \sum_{k=1}^{n} a_k.$$

$$\cdots$$

The sequence $\{s_n\}$ is called the *sequence of partial sums* of the series (1).

---

\* In common English usage the words "series" and "sequence" can have the same meaning. In mathematics they refer to quite different things.

*Notational Remark.* The letter $n$ in $\sum_{n=1}^{\infty} a_n$ is a "dummy variable" because any letter would do as well. Thus,

$$\sum_{n=1}^{\infty} a_n = \sum_{k=1}^{\infty} a_k.$$

However, the letter $N$ in $\sum_{n=1}^{N} a_n$ represents a genuine variable, and the letter $n$ is dummy.

$$s_N = \sum_{n=1}^{N} a_n = \sum_{k=1}^{N} a_k.$$

**DEFINITION 2**     The infinite series

(1)
$$\sum_{n=1}^{\infty} a_n$$

is said to be *convergent,* and to converge to the sum $s$ if the sequence of partial sums converges to $s$:

$$\lim_{n \to \infty} s_n = s.$$

Then we write

$$\sum_{n=1}^{\infty} a_n = s.$$

If the limit does not exist, the series *diverges.*

Definition 2 requires that we examine the partial sums $s_n$. Usually this is no easy task, and the examples of Chapter 8 show that one may have to perform computations. Occasionally $s_n$ is given by a nice formula, and the limit is clear.

*Example 1*     For the series

$$\sum_{n=1}^{\infty} \frac{1}{n(n+1)} = \frac{1}{1 \cdot 2} + \frac{1}{2 \cdot 3} + \cdots + \frac{1}{n(n+1)} + \cdots$$

$$= \left(1 - \frac{1}{2}\right) + \left(\frac{1}{2} - \frac{1}{3}\right) + \cdots + \left(\frac{1}{n} - \frac{1}{n+1}\right) + \cdots,$$

the partial sums are

$$s_1 = 1 - \tfrac{1}{2}$$
$$s_2 = (1 - \tfrac{1}{2}) + (\tfrac{1}{2} - \tfrac{1}{3}) = 1 - \tfrac{1}{3}$$
$$\cdots$$
$$s_n = \left(1 - \frac{1}{2}\right) + \left(\frac{1}{2} - \frac{1}{3}\right) + \cdots + \left(\frac{1}{n} - \frac{1}{n+1}\right) = 1 - \frac{1}{n+1}$$
$$\cdots$$

because of "telescoping." Then

$$s = \sum_{n=1}^{\infty} \frac{1}{n(n+1)} = \lim_{n \to \infty} s_n = \lim_{n \to \infty} \left(1 - \frac{1}{n+1}\right) = 1.$$

*Example 2*    One of the most important series is the *geometric series*:

$$\sum_{n=0}^{\infty} ax^n = a + ax + ax^2 + \cdots + ax^n + \cdots.$$

For this series the partial sums are (see Appendix A):

$$s_1 = a$$

$$s_2 = a + ax = a(1 + x) = a\frac{1 - x^2}{1 - x}$$

$$s_3 = a + ax + ax^2 = a(1 + x + x^2) = a\frac{1 - x^3}{1 - x}$$

$$\cdots$$

$$s_n = a + ax + \cdots + ax^{n-1} = a\frac{1 - x^n}{1 - x}.$$

$$\cdots$$

Then

$$\sum_{n=0}^{\infty} ax^n = s = \lim_{n\to\infty} s_n$$

$$= \lim_{n\to\infty} a\frac{1 - x^n}{1 - x} = \frac{a}{1 - x}, \quad \text{if } |x| < 1.$$

The series converges if $|x| < 1$ to the sum $a/(1 - x)$. The series diverges if $|x| \geq 1$.

The Cauchy criterion (Theorem 1 of Section 1) for convergence of sequences provides the following useful criterion (in proofs, though not so often in examples) for the convergence of series.

**THEOREM 1**    *The series $\sum_{n=1}^{\infty} a_n$ converges if and only if to each positive number $\epsilon$ there is an integer $N$ such that, if $n \geq N$, then*

$$(2) \qquad |a_{n+1} + a_{n+2} + \cdots + a_{n+p}| < \epsilon$$

*for all positive integers $p$.*

**PROOF**    The series converges if and only if the sequence of partial sums converges. The sequence $\{s_n\}$ of partial sums converges if and only if, to each positive number $\epsilon$, there is an integer $N$ such that if $n \geq N$

$$(3) \qquad |s_{n+p} - s_n| < \epsilon \quad \text{for all positive integers } p.$$

But

$$s_{n+p} = a_1 + \cdots + a_n + a_{n+1} + \cdots + a_{n+p}$$

$$s_n = a_1 + \cdots + a_n$$

and so

$$s_{n+p} - s_n = a_{n+1} + \cdots + a_{n+p}.$$

Then inequality (3) becomes inequality (2), and the proof is complete.

The following theorems are easily proved by using the Cauchy criterion.

**THEOREM 2**    *If* $\sum_{1=n}^{\infty} a_n$ *is convergent, then*

$$\lim_{n \to \infty} a_n = 0.$$

**PROOF**    $a_n = s_n - s_{n-1}$. Then by the Corollary to Theorem 1 of Section 1,

$$\lim_{n \to \infty} a_n = \lim_{n \to \infty} (s_n - s_{n-1}) = 0.$$

**THEOREM 3**    *The convergence or divergence of* $\sum_{n=1}^{\infty} a_n$ *is not affected by the deletion of a finite number of terms.* (Of course, the sum is changed.)

**PROOF**    Suppose the first $k$ terms are deleted. Then, if $n > k$,

$$| a_{n+1} + \cdots + a_{n+p} |$$

is the same for both the given series and the series with $k$ terms deleted. Therefore, the criterion (2) of Theorem 1 is either satisfied for both series or fails for both series.

The following theorem can be proved directly from the definition of convergence of series. We shall omit the proof.

**THEOREM 4**    *If* * $\sum a_n$ *and* $\sum b_n$ *converge,* $\sum a_n = a$ *and* $\sum b_n = b$, *then*

(i) $$\sum (a_n + b_n) = a + b,$$

(ii) $$\sum c a_n = c \sum a_n = ca.$$

### Problems

In Problems 1 to 8 obtain a formula for $s_n$ and hence obtain the sum of the infinite series. All sums are from 1 to infinity.

1.  $\sum \dfrac{1}{2^{n-1}}$ $\qquad\qquad\qquad\qquad\qquad$ $\left[ s_n = 2 - \dfrac{1}{2^{n-1}} \right]$

2.  $\sum \dfrac{1}{(n+2)(n+3)}$ $\qquad\qquad$ $\left[ s_n = \dfrac{1}{3} - \dfrac{1}{n+3} \right]$

---

* When the range on the subscript $n$ is clear from context, say from 1 to $\infty$, we shall write for brevity $\sum a_n$ in place of $\sum_{n=1}^{\infty} a_n$.

3.  $\displaystyle\sum \frac{1}{n(n+2)}$ $\qquad\left[ s_n = \frac{1}{2}\left(1 - \frac{1}{n+2}\right)\right]$

4.  $\displaystyle\sum \frac{1}{3^n}$ $\qquad\left[ s_n = \frac{1 - 3^{-n}}{2}\right]$

5.  $\displaystyle\sum \frac{2}{n(n+1)(n+2)}$ $\qquad\left[ s_n = \frac{1}{2} - \frac{1}{n+1} + \frac{1}{n+2}\right]$

6.  $\displaystyle\sum \frac{1}{e^{2n}}$

7.  $\displaystyle\sum 100\,(0.9)^n$

8.  $\displaystyle\sum 2^n/10^{10}$

9.  If $\sum a_n$ diverges and $\sum b_n$ diverges, must $\sum (a_n + b_n)$ diverge? Give an example.

10.  Give an example of an infinite series whose sum is $\sqrt{2}$.

11.  Use the geometric series to show that the repeating decimal $0.999\ldots = \sum 9\left(\frac{1}{10}\right)^k = 1$.

12.  Use the geometric series to express the repeating decimal $0.027027\ldots$ as a quotient of integers.

## 3  *Series of Non-negative Terms*

It is often difficult to determine whether a given series converges or diverges by *direct* application of the definition of convergence (Definition 2 of Section 2). So one seeks tests for convergence or divergence that will apply to large classes of series. This section and the next develop several such tests.

When all the terms of the series are non-negative, then there are simple tests for convergence and divergence. The basis for these tests is the following theorem.

**THEOREM 1**  *If $a_n \geq 0$ for all $n$, and if there is a number $B$ such that for the partial sums $s_n$, $s_n \leq B$ for all $n$, then $\sum a_n$ converges to a sum\* $s \leq B$.*

---

\* If the partial sums have the property $s_n \leq B$ for all $n$, they are said to be *bounded above*.

**PROOF**   Because $a_n \geq 0$ for all $n$, the partial sums are non-decreasing and bounded above by $B$:

$$s_1 \leq s_2 \leq \cdots \leq s_n \leq s_{n+1} \leq \cdots \leq B.$$

It is a fundamental property of the real number system that such a sequence converges (see Appendix C).

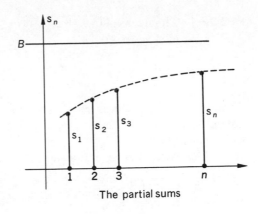

The partial sums

*Remark 1.* It suffices to have $a_n \geq 0$ for all $n$ sufficiently large, say greater than $n_0$, and then to have bounded partial sums.

*Remark 2.* If there is no such bound $B$ as required by the theorem, then clearly $\lim s_n = \infty$, and the given series diverges. In this case we say that the series *diverges to infinity* and write $\sum a_n = \infty$.

**THEOREM 2 (THE COMPARISON TESTS)**   *Suppose that* $\sum a_n$ *and* $\sum b_n$ *are series of non-negative terms.*

(1)   *If* $a_n \leq b_n$ *for all $n$, and* $\sum b_n$ *converges, then* $\sum a_n$ *converges.*

(2)   *If* $a_n \geq b_n$ *for all $n$, and* $\sum b_n$ *diverges, then* $\sum a_n$ *diverges.*

**PROOF**   In case (1) the partial sums of $\sum a_n$ are no larger than the partial sums of $\sum b_n$,

$$a_1 + a_2 + \cdots + a_n \leq b_1 + b_2 + \cdots + b_n,$$

and the latter are bounded because $\sum b_n$ converges. By Theorem 1 the series $\sum a_n$ converges.

In case (2) the partial sums of $\sum a_n$ are at least as large as those of $\sum b_n$, and the latter are unbounded.

In order to use this theorem it is clear that one needs a variety of series to use for comparison. The integral test (which applies only to series of positive terms) will give us many examples.

**THEOREM 3 (THE INTEGRAL TEST)**   *Suppose $f$ is a continuous, positive, and decreasing function on the interval $[1, \infty)$ and $f(n) = a_n$. Then either*

$$\sum_{n=1}^{\infty} a_n \qquad and \qquad \int_1^{\infty} f(x)\, dx$$

*both converge, or they both diverge.*

**PROOF**    The figure shows that

$$s_n = a_1 + a_2 + \cdots + a_n$$

$$\leq a_1 + \int_1^n f(x)\, dx.$$

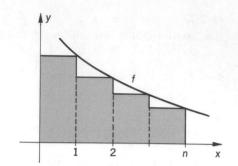

Therefore, if the integral converges, the partial sums are bounded above, and the series converges.

The second figure shows that

$$s_n \geq \int_1^{n+1} f(x)\, dx.$$

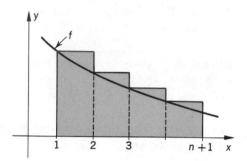

Therefore, if the series is convergent, then the integral is convergent. Hence, the convergence of one implies the convergence of the other. From this it follows that, if one diverges, so does the other.

*Remark 3.* To apply the test one can neglect a finite number of terms and

consider only    $\int_{n_0}^{\infty} f(x)\, dx.$

*Example 1*    The series

$$\sum_{n=1}^{\infty} \frac{1}{n^p} = 1 + \frac{1}{2^p} + \frac{1}{3^p} + \cdots + \frac{1}{n^p} + \cdots, \qquad (p > 0)$$

is called a *p-series.* We shall use the integral test to show that the series converges if $p > 1$ and diverges if $p \leq 1$.

Let $f(x) = 1/x^p$. Then $f(n) = 1/n^p$, and for $1 \leq x < \infty$, $f(x) > 0$, and $f$ is decreasing [because $f'(x) = -p/x^{p+1} < 0$]. Therefore, the integral test may be used.

$$\int_1^{\infty} \frac{dx}{x^p} = \lim_{b \to \infty} \int_1^b \frac{dx}{x^p} = \lim_{b \to \infty} \left[ \frac{b^{1-p}}{1-p} + \frac{1}{1-p} \right] = \begin{cases} \infty & \text{if } p < 1, \\[2ex] \dfrac{1}{1-p} & \text{if } p > 1. \end{cases}$$

Thus, $\sum 1/n^p$ converges if $p > 1$ and diverges to infinity if $p < 1$.

For the case $p = 1$ we have

$$\int_1^{\infty} \frac{dx}{x} = \lim_{b \to \infty} \log b = \infty.$$

Therefore, the series

$$\sum \frac{1}{n} = 1 + \frac{1}{2} + \frac{1}{3} + \cdots + \frac{1}{n} + \cdots \quad \{\text{Harmonic series}$$

diverges to infinity. This important series is called the *harmonic series*. Observe that for this series the terms decrease and approach 0, and yet the series diverges.

**Example 2**    The series

$$\sum_{n=2}^{\infty} \frac{1}{n \log n} = \frac{1}{2 \log 2} + \frac{1}{3 \log 3} + \cdots + \frac{1}{n \log n} + \cdots$$

diverges to infinity because

$$\int_{2}^{\infty} \frac{dx}{x \log x} = \lim_{b \to \infty} \left[ \log (\log b) - \log (\log 2) \right] = \infty.$$

Here we have used in the integral test $f(x) = 1/x \log x$ for $x \geqq 2$. Then $f(x) > 0$, and $f$ is decreasing.

**Example 3**    The $p$-series can be used with the comparison test. Thus, the series $\sum 1/(2 + n^2)$ converges because

$$\frac{1}{2 + n^2} < \frac{1}{n^2} \qquad \text{for all } n,$$

and $\sum 1/n^2$ is the $p$-series with $p = 2$.

### Problems

In Problems 1 to 20 determine whether the series converge or diverge. All sums are from 1 to $\infty$ unless noted.

1. $\sum \dfrac{2}{1 + n\sqrt{n}}$    [converges]

2. $\sum \dfrac{1}{(2n + 1)(2n + 2)}$

3. $\sum \dfrac{1}{2n + 1}$

4. $\sum \dfrac{1}{2n}$

5. $\sum \dfrac{1}{n\sqrt{n}}$

6. $\sum \dfrac{1}{n\sqrt[3]{n}}$

7. $\displaystyle\sum_{n=1000}^{\infty} \dfrac{1}{n (\log n) \log (\log n)}$    [diverges]

8. $\sum \dfrac{1}{(n + 1) \log^2 (n + 1)}$    [converges]

9. $\sum \dfrac{n}{e^n}$    [converges]

10. $\sum \dfrac{n^2}{e^n}$

*11. $\sum \dfrac{(1.1)^n}{n^{1.1}}$ [diverges] 16. $\sum \dfrac{1}{\sqrt{n^2+1}}$ [diverges]

12. $\sum \dfrac{(1.01)^n}{10{,}000}$ 17. $\sum \dfrac{1}{\sqrt{n^3+1}}$ [converges]

13. $\sum \dfrac{1}{99n+5}$ 18. $\sum \dfrac{\log n}{n^2}$ [converges]

14. $\sum \dfrac{n^2+3n+4}{n^4}$ 19. $\sum a^n, \quad |a|<1$

15. $\sum \dfrac{\frac{1}{2}n-71}{n^2}$ [diverges] 20. $\sum \dfrac{2n-1}{(n+1)^3}$

*21. Show that the harmonic series is divergent by showing that $s_{2^n} \to \infty$ as follows:

$$s_1 = 1, \quad s_2 \geqq s_1 + \tfrac{1}{2}, \quad s_4 \geqq s_2 + \tfrac{1}{2}, \quad \dots, \quad s_{2^n} \geqq s_{2^{n-1}} + \tfrac{1}{2},$$

because
$$s_{2^n} = s_{2^{n-1}} + \frac{1}{2^{n-1}+1} + \cdots + \frac{1}{2^n}$$

$$> s_{2^{n-1}} + 2^{n-1}\left(\frac{1}{2^n}\right) = s_{2^{n-1}} + \tfrac{1}{2}.$$

Therefore, $\quad s_{2^n} > (n-1)/2.$

## 4  Absolute Convergence

Section 2 was concerned with series whose terms were of one sign. Some series converge because the positive and the negative terms are "suitably balanced."

**Example 1**   The *alternating harmonic series*

$$\sum (-1)^{n+1}\frac{1}{n} = 1 - \frac{1}{2} + \frac{1}{3} - \frac{1}{4} + \cdots + (-1)^{n+1}\frac{1}{n} + \cdots$$

converges because the partial sums alternately increase and decrease, and $1/n \to 0$.

A proof of convergence is as follows. We have, for all $n$,

$$s_2 < s_4 < \cdots < s_{2n} < s_{2n+2} < \cdots$$
$$< s_{2n+1} < s_{2n-1} < \cdots < s_1.$$

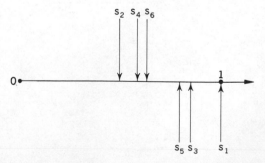

The even partial sums increase and are bounded above by $s_1$ ($s_{2n} < s_1$). The odd partial sums decrease and are bounded below by $s_2$ ($s_2 < s_{2n+1}$). Therefore, both

$$\lim_{n \to \infty} s_{2n} \qquad \text{and} \qquad \lim_{n \to \infty} s_{2n+1}$$

exist. (See Appendix C, page 532.) Because $s_{2n+1} - s_{2n} = 1/(2n+1)$, the two limits are equal, and the series converges.

This example is interesting because both the positive terms alone, and the negative terms alone, diverge:

$$\sum_{n=1}^{\infty} \frac{1}{(2n-1)} = 1 + \frac{1}{3} + \frac{1}{5} + \cdots + \frac{1}{2n+1} + \cdots = \infty,$$

$$\sum_{n=1}^{\infty} \frac{-1}{2n} = -\frac{1}{2} - \frac{1}{4} - \cdots - \frac{1}{2n} - \cdots = -\infty.$$

Example 1 is representative of a class of series whose convergence is easy to establish. A series $\sum a_n$ is said to be *alternating* if successive terms differ in sign: $\text{sign } a_{n+1} = -\text{sign } a_n$ for all $n$.

**THEOREM 1**    *If $\sum a_n$ is an alternating series, and if*

(i) $$|a_{n+1}| \leq |a_n| \qquad \text{for all } n,$$

(ii) $$\lim_{n \to \infty} a_n = 0,$$

*then the series converges.*

**PROOF**    A proof is analogous to the proof of Example 1.

*Example 2*    To apply Theorem 1 we must verify conditions (i) and (ii). Thus, for the alternating series

(1) $$\sum_{n=2}^{\infty} (-1)^n \frac{\log n}{n}$$

we have:

(i) $$\frac{\log(n+1)}{n+1} < \frac{\log n}{n} \qquad \text{for } n \geq 3,$$

because

$$\frac{d}{dx} \frac{\log x}{x} = \frac{1 - \log x}{x^2} < 0 \qquad \text{if } x > e.$$

(ii) $$\lim_{n \to \infty} \frac{\log n}{n} = \lim_{n \to \infty} \frac{1/n}{1} = 0$$

by using l'Hospital's rule. Therefore the series (1) converges.

Other series converge because the terms are "small enough" regardless of their sign. These are the absolutely convergent series.

**DEFINITION**     The series $\sum a_n$ is *absolutely convergent* if $\sum |a_n|$ converges. A series that is convergent, but not absolutely convergent, is called *conditionally convergent*.

Observe that the definition does not claim that an absolutely convergent series is convergent. That this is the case is given by the following theorem.

**THEOREM 2**     *An absolutely convergent series is convergent.*

**PROOF**     We apply the Cauchy criterion (2) of Section 2. First we observe that

(2)     $$|a_{n+1} + \cdots + a_{n+p}| \leqq |a_{n+1}| + |a_{n+2}| + \cdots + |a_{n+p}|.$$

Now, given $\epsilon > 0$, choose $N$ so that if $n \geqq N$, then

$$|a_{n+1}| + \cdots + |a_{n+p}| < \epsilon \qquad \text{for all positive integers } p.$$

This is possible because of the Cauchy criterion and the absolute convergence of the series. Therefore the left member of (2) is also small: If $n \geqq N$, then

$$|a_{n+1} + \cdots + a_{n+p}| < \epsilon \qquad \text{for all positive integers } p.$$

By the Cauchy Criterion the series $\sum a_n$ converges.

A commonly used test for convergence is the ratio test.

**THEOREM 3 (THE RATIO TEST)**     *For the series $\sum a_n$ suppose that*

$$\lim_{n \to \infty} \left| \frac{a_{n+1}}{a_n} \right|$$

*exists or is infinite. Then the series* $\sum a_n$

(i)     *converges absolutely if*  $\displaystyle \lim_{n \to \infty} \left| \frac{a_{n+1}}{a_n} \right| < 1,$

(ii)     *diverges if*  $\displaystyle \lim_{n \to \infty} \left| \frac{a_{n+1}}{a_n} \right| > 1$  *or is* $\infty.$

*If* $\lim_{n \to \infty} |a_{n+1}/a_n| = 1$, *the test gives no information.*\*

---

\* We will use the theorem as it is stated. However, a stronger theorem has the same proof: *If there is an integer N such that*

$$\left| \frac{a_{n+1}}{a_n} \right| \leqq p < 1 \qquad \text{for all } n \geqq N,$$

*then the series converges absolutely.*

**PROOF SKETCH**     Suppose

$$\lim_{n \to \infty} \left| \frac{a_{n+1}}{a_n} \right| = p < 1.$$

Choose $p_0$ such that $p < p_0 < 1$. Then there is an integer $N$ such that

$$\left| \frac{a_{n+1}}{a_n} \right| < p_0 \qquad \text{if } n \geq N.$$

Then
$$| a_{N+1} | \leq | a_N | \, p_0$$
$$| a_{N+2} | \leq | a_{N+1} | \, p_0 \leq | a_N | \, p_0^2$$

$$\cdots$$

(3)
$$| a_{N+k} | \leq | a_N | \, p_0^k.$$

$$\cdots$$

Therefore, after the first $N$ terms we have

$$\sum_{k=1}^{\infty} | a_{N+k} | \leq | a_N | \sum_{k=1}^{\infty} p_0^k.$$

This last series is a geometric series and converges because $p_0 < 1$. By comparison, the series $\sum_{k=1}^{\infty} | a_{N+k} |$ converges, and so the given series converges absolutely.

If $p > 1$ or is $\infty$, then inequalities similar to (3) show that the terms do not approach zero, and so the series diverges.

**Example 3**     The ratio test applies to these series:

The geometric series $\sum ax^n$ converges if $| x | < 1$ and diverges if $| x | > 1$ because:

$$\lim \left| \frac{a_{n+1}}{a_n} \right| = \lim \left| \frac{ax^{n+1}}{ax^n} \right| = | x |.$$

The series $\sum \dfrac{n!}{1 \cdot 3 \cdot 5 \cdots (2n - 1)}$ converges because:

$$\lim \left| \frac{a_{n+1}}{a_n} \right| = \lim \left( \frac{(n + 1)!}{1 \cdot 3 \cdots (2n + 1)} \right) \left( \frac{1 \cdot 3 \cdots (2n - 1)}{n!} \right)$$

$$= \lim \frac{n + 1}{2n + 1} = \frac{1}{2}.$$

The series $\sum \dfrac{(2n)!}{(n!)^2}$ diverges because:

$$\lim \left| \frac{a_{n+1}}{a_n} \right| = \lim \frac{(2n + 2)!(n!)^2}{[(n + 1)!]^2 (2n)!} = \lim \frac{(2n + 2)(2n + 1)}{(n + 1)^2} = 4.$$

*Example 4*    The ratio test fails in these cases:

The *p*-series $\sum \frac{1}{n^p}$ : $\lim \left| \frac{a_{n+1}}{a_n} \right| = \lim \frac{n^p}{(n+1)^p} = 1.$

Yet the series converges if $p > 1$, and diverges if $p \leq 1$.

The series $\sum \frac{2 + (-1)^n}{n}$ : $\lim \left| \frac{a_{n+1}}{a_n} \right| = \lim \frac{2 + (-1)^{n+1}}{2 + (-1)^n} \frac{n}{n+1}.$

This limit does not exist. Yet the series diverges by comparison with the harmonic series.

*Example 5*    The following series are conditionally convergent:

The alternating harmonic series: $\sum \frac{(-1)^{n+1}}{n}$ .

The series $\sum (-1)^n \frac{\log n}{n}$ .

This last series is convergent by Example 2, but it is not absolutely convergent by comparison with the harmonic series; that is, $\log n/n > 1/n$ for $n > 2$.

*Remark.* With a given series to be tested for convergence it is usually a good plan first to try the ratio test. Then, if that fails try comparison, or the integral test, or try to be ingenious.

### Problems

In Problems 1 to 18 test for absolute or conditional convergence. All sums are from 1 to ∞.

1. $\sum \frac{n}{2^n}$    [abs. conv.]

5. $\sum \frac{(-1)^n 2n}{n^2 + 1}$    [cond. conv.]

2. $\sum \frac{n}{2n + 1}$    [div.]

6. $\sum \frac{(-1)^n n}{2n + 3}$    [div.]

3. $\sum \frac{a^n}{n!}$    [abs. conv.]

7. $\sum \frac{(n + 3)2^n}{n(n + 2)^2}$    [div.]

4. $\sum \frac{n!}{a^n}$    [div.]

8. $\sum \frac{2 + \cos n\pi}{n^2 + 1}$    [abs. conv.]

9. $\sum n^2[\log 2]^n$ [abs. conv.]

10. $\sum \dfrac{(-1)^{n+1}}{\sqrt{n}}$ [cond. conv.]

11. $\sum \dfrac{\sin (n + \frac{1}{2})\pi}{\log (n + 1)}$ [cond. conv.]

12. $\sum \dfrac{(-1)^{n+1}(n + 1)}{2n}$ [div.]

13. $\sum \dfrac{(-1)^n}{(2n)n!}$ [abs. conv.]

14. $\sum (-1)^n\left(1 + \dfrac{1}{n}\right)^n$ [div.]

15. $\sum \dfrac{(2n - 1)}{2^n}$ [abs. conv.]

16. $\sum \dfrac{(-1)^{n+1}(3n^2 - 3n + 5)}{n^3 + n^2 + n + 1}$ [cond. conv.]

17. $\sum \dfrac{(n!)^2(-3)^n}{(2n)!}$ [abs. conv.]

18. $\sum (-1)^n n^2 \left(\dfrac{a}{a + 1}\right)^n$, $a > 0$ [abs. conv.]

*19. Prove Theorem 1.

*20. The terms of the alternating harmonic series can be rearranged to give any desired sum. Describe how to rearrange the terms of the series to obtain 10 for the sum of the series. *Hint*: Both the positive terms and the negative terms diverge. First take just enough of the positive terms to obtain a sum $\geq 10$. (why is this possible?) Then take just enough of the negative terms to give a total sum $< 10$. Then take just enough of the remaining positive terms to obtain a total sum $\geq 10$. Then continue with the negative terms as before. Continuing with alternate use of the positive terms and the negative terms, all terms will be used (why?) and give a series whose sum is 10 (why?).

Any conditionally convergent series has the same property.

*21. The *root test* of Cauchy also tests for absolute convergence: *If $\sqrt[n]{|a_n|} \leq p < 1$ for all n, then the series $\sum a_n$ is absolutely convergent.* Prove the root test.

5   *The Convergence of Power Series*

In Chapter 8 we obtained the Taylor series for some of the elementary functions. At that time Taylor's formula was used:

If $f$ has $n + 1$ continuous derivatives in an interval containing $a$, then

$$(1) \qquad f(x) = f(a) + f'(a)(x - a) + \cdots + \frac{f^{(n)}(a)}{n!}(x - a)^n + R_n(x)$$

where

$$R_n(x) = \frac{f^{(n+1)}(\xi)}{(n + 1)!}(x - a)^{n+1},$$

and $\xi$ is between $a$ and $x$. Then, if $f$ is infinitely differentiable and if

$$\lim_{n \to \infty} R_n(x) = 0,$$

the infinite Taylor series

$$\sum_{n=0}^{\infty} \frac{f^{(n)}(a)}{n!}(x - a)^n = f(a) + f'(a)(x - a) + \cdots + \frac{f^{(n)}(a)}{n!}(x - a)^n + \cdots$$

converges to $f(x)$.

In this chapter, however, our point of view is different. We *start* with a power series:

$$(2) \qquad \sum_{n=0}^{\infty} a_n(x - a)^n = a_0 + a_1(x - a) + \cdots + a_n(x - a)^n + \cdots$$

and ask about properties of a function *defined* by the power series (2). For convenience we shall suppose that $a = 0$. The series (2) can be reduced to this by the substitution $t = x - a$.

Those values of $x$ for which the power series converges constitute the *convergence set* for the series. The sum of the series *defines* the value at $x$ of a function $f$,

$$(3) \qquad f(x) = \sum_{n=0}^{\infty} a_n x^n = a_0 + a_1 x + \cdots + a_n x^n + \cdots$$

if the series converges. We therefore need to know something about where the series (3) converges. The following theorem is basic.

**THEOREM 1**   *If the series $\sum a_n x^n$ converges for $x_0 \neq 0$, then the series converges absolutely for all $x$ such that $|x| < |x_0|$.*

**PROOF**   Because $\sum a_n x_0^n$ converges, the terms $a_n x_0^n$ approach $0$ as $n \to \infty$. Consequently there is a positive number $M$ such that

$$(4) \qquad |a_n x_0^n| \leq M \qquad \text{for all } n = 0, 1, 2, \ldots.$$

Now consider, for $|x| < |x_0|$, the series $\sum |a_n x^n|$.

(5)
$$\sum_{n=0}^{\infty} |a_n x^n| = \sum_{n=0}^{\infty} |a_n x_0^n| \left|\frac{x}{x_0}\right|^n \leq M \sum_{n=0}^{\infty} \left|\frac{x}{x_0}\right|^n,$$

because of (4). The last series in (5) is a geometric series with ratio $|x/x_0| < 1$ and, therefore, convergent. By the comparison test, $\sum |a_n x^n|$ converges, and so $\sum a_n x^n$ converges absolutely.

**COROLLARY**    *Either the power series $\sum a_n x^n$ converges for all $x$, or converges only for $x = 0$, or there is a positive real number $R$ such that the series converges for $|x| < R$ and diverges for $|x| > R$.*

**\*PROOF**    Suppose that the series converges at $x_0 \neq 0$, and does not converge for all $x$. Let $x_1$ be such that the series does not converge at $x_1$. Then $|x_0| \leq |x_1|$ (why?) so that $-|x_1| \leq x_0 \leq |x_1|$. Therefore, the convergence set is bounded. Let $R$ be the least upper bound of the convergence set (see Appendix C.) Then the series converges absolutely for $|x| < R$ and diverges for $|x| > R$.

The number $R$ is called the *radius\** *of convergence* of the series. It gives an interval on the real line

$$-R < x < R,$$

called the *interval of convergence*. (The series may or may not converge at $x = \pm R$.) If the series converges only for $x = 0$, we set $R = 0$. If the series converges for all $x$, we set $R = \infty$.

The Corollary assures us that a radius of convergence exists but does not provide a technique for calculating it. In most problems in this text the ratio test will be applicable and will provide a simple method for finding $R$. Examples will illustrate.

*Example 1*    The exponential series $\sum x^n/n!$ converges for all $x$ because

$$\lim_{n \to \infty} \left|\frac{x^{n+1}}{(n+1)!} \frac{n!}{x^n}\right| = \lim_{n \to \infty} \frac{|x|}{n+1} = 0 < 1 \qquad \text{for all } x.$$

Therefore, $R = \infty$.

*Example 2*    Some series converge only at $x = 0$. The series $\sum n! \, x^n$ is such a one because

$$\lim_{n \to \infty} \left|\frac{(n+1)! \, x^{n+1}}{n! \, x^n}\right| = \infty \qquad \text{if } x \neq 0.$$

Therefore $R = 0$.

---

\* The word radius is used because in the complex plane there is a circle of radius $R$ interior to which the series converges. See Section 8.

*Example 3*    The binomial series

$$1 + \sum_{n=1}^{\infty} \frac{m(m-1)\cdots(m-n+1)}{n!} x^n,$$

converges for $|x| < 1$ and diverges for $|x| > 1$ because

$$\lim_{n\to\infty} \left| \frac{m(m-1)\cdots(m-n+1)(m-n)x^{n+1}n!}{m(m-1)\cdots(m-n+1)x^n(n+1)!} \right| = \lim_{n\to\infty} \left| \frac{m-n}{n+1} x \right| = |x|.$$

Therefore, $R = 1$.

*Example 4*    That a series can converge or diverge at the ends of the interval of convergence is shown by the series $\sum x^n/n$. This series converges for $|x| < 1$ and diverges for $|x| > 1$, by the ratio test. At $x = 1$ the series becomes the harmonic series and diverges. At $x = -1$ the series becomes a convergent alternating series.

*Example 5*    Find the interval of convergence of the series

$$\sum_{n=0}^{\infty} \frac{n(x - \frac{3}{2})^n}{3^n}.$$

From the ratio test

$$\lim_{n\to\infty} \left| \frac{(n+1)(x-\frac{3}{2})^{n+1}}{3^{n+1}} \frac{3^n}{n(x-\frac{3}{2})^n} \right| = \frac{|x - \frac{3}{2}|}{3}.$$

Therefore, the series converges for $|x - \frac{3}{2}|/3 < 1$, or for $|x - \frac{3}{2}| < 3$ or

$$-3 < x - \tfrac{3}{2} < 3 \quad \text{or} \quad -\tfrac{3}{2} < x < \tfrac{9}{2}.$$

## Problems

In Problems 1 to 12 find the radius of convergence and an interval of convergence. Disregard the ends of the interval. All sums are from $n = 1$ to $\infty$.

1.  $\sum \dfrac{x^n}{2n!}$                                          $[-\infty < x < \infty]$

2.  $\sum \dfrac{(-1)^n (x-2)^n}{n}$                                 $[1 < x < 3]$

3.  $\sum \dfrac{n!(x + \frac{1}{2})^n}{2 \cdot 1 \cdot 3 \cdots (2n-1)}$     $[-2\tfrac{1}{2} < x < 1\tfrac{1}{2}]$

4.  $\sum 2n(2n-1)\cdots(n+1)x^n$                                    $[R = 0]$

5.  $\sum \dfrac{\log(n+1)2^n(1-x)^n}{n+1}$                          $[\tfrac{1}{2} < x < \tfrac{3}{2}]$

6.  $\displaystyle\sum \frac{1\cdot 3\cdot 5\cdots (2n-1)(x+1)^n}{2^n n!}$    $[-2 < x < 0]$

7.  $\displaystyle\sum \frac{(10^{10}x)^n}{n!}$    $[-\infty < x < \infty]$

8.  $\displaystyle\sum \frac{(-1)^n[x-(\pi/2)]^n}{n\sqrt{n}}$    $\left[\dfrac{\pi}{2}-1 < x < \dfrac{\pi}{2}+1\right]$

9.  $\displaystyle\sum (-1)^n \frac{1\cdot 3\cdots (2n-1)}{2\cdot 4\cdots (2n)}\, x^n$

10.  $\displaystyle\sum \frac{(x+a)^n}{nb^{n-1}}$

11.  $\displaystyle\sum \frac{\sin\,(n-\frac{1}{2})\pi}{(2n-1)3^n}\,(x-2)^n$

12.  $\displaystyle\sum \frac{\log^2 n}{n}\,(x+\tfrac{1}{2})^n$

## *6    Operations on Power Series†

The operations of differentiation, integration, addition, subtraction, multiplication, and division are familiar when applied to functions. Because power series represent functions in their intervals of convergence, the same operations should apply to power series and yield series with known radii of convergence.

With respect to differentiation and integration consider first an example.

*Example 1*    The geometric series $\sum x^n$ converges for $|x| < 1$ and diverges for $|x| > 1$ because

$$\lim \left|\frac{x^{n+1}}{x^n}\right| = \lim_{n\to\infty} |x| = |x|.$$

Moreover, the so-called *derived* series, obtained by differentiating term by term,

$$\sum nx^{n-1} = 1 + 2x + 3x^2 + \cdots + nx^{n-1} + \cdots$$

also converges for $|x| < 1$ because

$$\lim_{n\to\infty} \left|\frac{(n+1)x^{n+1}}{nx^n}\right| = \lim_{n\to\infty} \left|\frac{n+1}{n}\,x\right| = |x|.$$

---

† Sections 6, 7, 8, and 9 are concerned with analysis, and the ideas in these sections are often first encountered in more advanced courses. Yet the topics in these sections follow naturally from our previous considerations, and the student is urged to at least read through these sections.

The following theorem shows that this example is typical.

**THEOREM 1**   *If $\sum a_n x^n$ has radius of convergence $R$, then the derived series obtained by differentiating term by term, and the integrated series obtained by integrating term by term,*

$$\sum n a_n x^{n-1} \quad and \quad \sum \frac{a_n}{n+1} x^{n+1},$$

*also have radius of convergence $R$.*

**PROOF SKETCH**   If the ratio test can be applied to $\sum a_n x^n$, then a proof is easy and is left to the reader. See Problem 19.

In general, suppose that $\sum a_n x_0^n$ converges and $|x| < |x_0|$. In Theorem 1 of Section 5 we proved that $\sum a_n x^n$ is convergent by comparing it with the convergent geometric series: $M \sum |x/x_0|^n$. According to Example 1, the derived series $M \sum n |x/x_0|^{n-1}$ is also convergent. Therefore, by comparison the series $\sum n |a_n x^{n-1}|$ is convergent. This proves that the derived series $\sum n a_n x^{n-1}$ has radius of convergence at least as large as that of $\sum a_n x^n$.

Now consider the integrated series. Its radius of convergence is also at least equal to the radius of convergence of the given series, because the terms are smaller in absolute value than those of the given series.

These two statements imply that the three series have the same radius of convergence. For, let $R$ be the radius of convergence of the given series, $R_d$ be the radius of convergence of the derived series, and $R_i$ be the radius of convergence of the integrated series. We have seen that

$$R_d \geq R \quad and \quad R_i \geq R.$$

However, the given series is the derived series of the integrated series, whence $R \geq R_i$. Analogously, the given series is the integrated series of the derived series, whence $R \geq R_d$. Therefore, $R_d = R = R_i$.

**THEOREM 2**   *If $\sum a_n x^n$ and $\sum b_n x^n$ converge for $|x| < R$, then*

$$\sum (a_n + b_n) x^n \quad and \quad \sum c a_n x^n$$

*also converge for $|x| < R$.*

Theorem 2 shows that addition of convergent series and multiplication by constants yield convergent series. We omit the proof.

To consider multiplication and division of series we need to define what we mean.

**DEFINITION 1**   If $\sum a_n x^n$ and $\sum b_n x^n$ are any series, their *Cauchy product* is the series $\sum c_n x^n$ where $c_n = a_0 b_n + a_1 b_{n-1} + \cdots + a_n b_0$:

$$\left(\sum a_n x^n\right)\left(\sum b_n x^n\right) = \sum c_n x^n = \sum (a_0 b_n + a_1 b_{n-1} + \cdots + a_n b_0) x^n.$$

*Example 2* The product of the sine series and the cosine series is

$$\left(\sum (-1)^n \frac{x^{2n+1}}{(2n+1)!}\right)\left(\sum (-1)^n \frac{x^{2n}}{(2n)!}\right)$$

$$= \left(x - \frac{x^3}{3!} + \frac{x^5}{5!} - \cdots\right)\left(1 - \frac{x^2}{2!} + \frac{x^4}{4!} - \cdots\right) = x - \frac{2x^3}{3} + \frac{2x^5}{15} - \frac{4x^7}{315} + \cdots.$$

Though hardly obvious as it stands, this is the Maclaurin series for $\frac{1}{2}\sin 2x$.

**DEFINITION 2** The quotient of $\sum a_n x^n$ by $\sum b_n x^n$, where $b_0 \neq 0$, is the power series $\sum c_n x^n$ such that

$$\left(\sum c_n x^n\right)\left(\sum b_n x^n\right) = \sum a_n x^n,$$

where the left side is the Cauchy product.

There are two ways of computing the coefficients of the quotient series. One way is to compute successively the $c_k$ by the infinite sequence of recurrence relations:

$$c_0 b_0 = a_0 \quad \text{(Since } b_0 \neq 0\text{, this gives } c_0\text{.)}$$

$$c_0 b_1 + c_1 b_0 = a_1 \quad \text{(This gives } c_1 \text{ in terms of } c_0, a_1, b_0, b_1\text{.)}$$

$$c_0 b_2 + c_1 b_1 + c_2 b_0 = a_2 \quad \text{(This gives } c_2 \text{ in terms of } c_0, c_1, a_2, b_0, b_1, b_2\text{.)}$$

$$\cdots$$

$$c_0 b_n + c_1 b_{n-1} + \cdots + c_n b_0 = a_n \quad \text{(This gives } c_n \text{ in terms of } c_0, \ldots, c_{n-1}, a_n,$$
$$b_0, \ldots, b_n\text{.)}$$

$$\cdots$$

The quotient series can also be obtained simply by the familiar long division algorithm of algebra, as Example 3 below will illustrate.

*Example 3* The quotient of the sine series by the cosine series is obtained by division. We obtain in this way the Maclaurin series for $\tan x$.

$$x + \frac{x^3}{3} + \frac{16x^5}{5!} + \cdots$$

$$1 - \frac{x^2}{2!} + \frac{x^4}{4!} - \cdots \div x - \frac{x^3}{3!} + \frac{x^5}{5!} - \cdots$$

$$x - \frac{x^3}{2!} + \frac{x^5}{4!} - \cdots$$

$$\frac{x^3}{3} - \frac{4x^5}{5!} + \cdots$$

$$\frac{x^3}{3} - \frac{x^5}{6} + \cdots$$

$$\frac{16x^5}{5!} - \cdots$$

The following theorem which we state without proof gives information about the convergence of the product and quotient series.

**THEOREM 3**   *If* $\sum a_n x^n$ *and* $\sum b_n x^n$ *converge for* $|x| < R$, *then*

(1)  *the product series converges for* $|x| < R$;

(2)  *the quotient series converges in some interval,* $|x| < R_1$, *if* $b_0 \neq 0$.

*Problems*

In Problems 1 to 4 find the derived series and the integrated series and verify Theorem 1 using the ratio test. All sums are from $n = 1$ to $n = \infty$.

1.  $\sum (-1)^{n+1} \dfrac{(2x)^{2n+1}}{(2n+1)!}$

2.  $\sum (-1)^n \dfrac{x^{n+1}}{n+1}$

3.  $\sum \dfrac{1 \cdot 3 \cdot 5 \cdots (2n-1)}{2 \cdot 4 \cdot 6 \cdots (2n)} x^n$

4.  $\sum (-1)^{n+1} \dfrac{\log (n+1)}{(n+1)^2} x^{2n}$

In Problems 5 to 15 obtain the indicated sum, difference, product, or quotient power series. All sums are from $n = 0$ to $n = \infty$.

5.  $\frac{1}{2} \sum (-1)^n x^n + \frac{1}{2} \sum x^n$ $\qquad\qquad\qquad\qquad$ $\left[ \sum x^{2n} \right]$

6.  $\sum x^n - \sum x^{n+1}$

7.  $\sum \dfrac{x^n}{n!} + \sum \dfrac{(-x)^n}{n!}$ $\qquad\qquad\qquad\qquad$ $\left[ \sum \dfrac{2x^{2n}}{(2n)!} \right]$

8.  $(\sum x^{2n})(1 - x^2)$ $\qquad\qquad\qquad\qquad\qquad\qquad$ $[1]$

9.  $(\sum x^n)[\sum (-x)^n]$ $\qquad\qquad\qquad\qquad$ $\left[ \sum (-1)^n x^{2n} \right]$

10.  $[\sum (-x)^n](1 + x)$

11.  $\left( \sum (-1)^n \dfrac{x^n}{n!} \right)\left( \sum \dfrac{x^n}{n!} \right)$ $\qquad\qquad\qquad\qquad$ $[1]$

12.  $\dfrac{1}{\sum (x^n/n!)}$

13.  $(1 + x^2)/(1 + x^4)$ $\qquad\qquad$ $[1 + x^2 - x^4 - x^6 + x^8 + x^{10} - \cdots]$

14.    $(1 - x + x^2)/(1 + x^3)$                            $[\sum (-x)^n]$

*15.    $\dfrac{\sum (-1)^n (2x)^{2n+1}/(2n+1)!}{\sum (-1)^n x^{2n}/(2n)!}$        $\left[ 2 \sum \dfrac{(-1)^n x^{2n+1}}{(2n+1)!} \right]$

16.    In Problem 6 what are the radii of convergence of the given series? What minimum radius of convergence can you *predict* for their sum series?

17.    In Problems 8 and 11 what are the radii of convergence of the series in each product? What minimum can you *predict* for the radius of convergence of the product series?

18.    In Problems 12, 14, and 15 what are the radii of convergence of the series in numerator and denominator? What can you *predict* about the radius of convergence of the quotient series?

19.    Prove Theorem 1, assuming that the ratio test is applicable to the given series.

*20.    Prove Theorem 2.

## *7   Functions Represented by Power Series

Theorem 1 of Section 6 states that a power series, its derived series, and its integrated series all have the same radius of convergence. This fact *suggests* that if a function is represented by a power series then its derived and integrated series represent the function's derivative and integral respectively:

If
$$f(x) = \sum_{n=0}^{\infty} a_n x^n \qquad \text{for} \qquad |x| < R,$$

then
$$f'(x) = \frac{d}{dx} \sum_{n=0}^{\infty} a_n x^n = \sum_{n=0}^{\infty} \frac{d}{dx} a_n x^n = \sum_{n=0}^{\infty} n a_n x^{n-1}, \qquad |x| < R,$$

that is, the derivative of the sum of the series is equal to the sum of the derivatives of the terms*; and

$$\int f(x) \, dx = \int \left( \sum_{n=0}^{\infty} a_n x^n \right) dx = \sum_{n=0}^{\infty} \int a_n x^n \, dx = \sum_{n=0}^{\infty} \frac{a_n x^{n+1}}{n+1} + C,$$

that is, the integral of the sum of the series is equal to the sum of the integrals of the terms.*

---

* These statements appear obvious to the beginner, but they are far from trivial and are in general false for series other than power series. The problem is one of interchange of order in taking limits. The sum of the series is a limit, and both the derivative and the integral are limits. Thus, the problem, expressed symbolically, is to show that $D \sum = \sum D$ and $\int \sum = \sum \int$.

Now we shall prove these assertions. First, in order to make our proof later, we prove that functions represented by power series are continuous functions. This, too, is an obvious appearing, but non-trivial, theorem.

**THEOREM 1**   *If $f(x) = \sum a_n x^n$, for $-R < x < R$, then $f$ is continuous on the interval $(-R, R)$.*

**PROOF**   To show that $f$ is continuous at $x_0$ in $(-R, R)$ we must show (see Appendix C) that for any positive number $\epsilon$ there is a positive number $\delta$ such that $|f(x) - f(x_0)| < \epsilon$ if $x$ is in $(-R, R)$ and $|x - x_0| < \delta$.

Let

$$P_N(x) = \sum_{n=0}^{N} a_n x_n, \qquad R_N(x) = \sum_{n=N+1}^{\infty} a_n x^n.$$

Then

$$(1) \qquad |f(x) - f(x_0)| \leq |P_N(x) - P_N(x_0)| + |R_N(x) - R_N(x_0)|.$$

we will first choose $N$ so the terms $R_N(x)$ and $R_N(x_0)$ are both small. Choose $x_1$ so that $|x_0| < |x_1| < R$.

Then

$$|R_N(x)| = |\sum_{n=N+1}^{\infty} a_n x^n| \leq \sum_{n=N+1}^{\infty} |a_n x^n| \leq \sum_{n=N+1}^{\infty} |a_n x_1^n|$$

for all $x$ such that $|x| \leq |x_1|$. The series converges absolutely at $x_1$ so

$$\sum_{n=N+1}^{\infty} |a_n x_1^n|$$

can be kept as small as desired by choosing $N$ large. We choose $N$ so that this sum is less than $\epsilon/3$. Then

$$(2) \qquad |R_N(x)| < \epsilon/3 \qquad \text{and} \qquad |R_N(x_0)| < \epsilon/3.$$

Now $P_N$ is a polynomial function and so continuous. Choose $\delta > 0$ such that if $|x - x_0| < \delta$, and $|x| \leq |x_1|$, then

$$(3) \qquad |P_N(x) - P_N(x_0)| < \epsilon/3.$$

Combining equations (1), (2), and (3), we get

$$|f(x) - f(x_0)| < \frac{\epsilon}{3} + \frac{\epsilon}{3} + \frac{\epsilon}{3} = \epsilon$$

if $|x| < |x_1|$ and $|x - x_0| < \delta$.

**THEOREM 2***    *If $f(x) = \sum a_n x^n$ for $|x| < R$, then*

(i)   *f is differentiable and*

$$f'(x) = \sum n a_n x^{n-1} \qquad \text{for } |x| < R;$$

(ii)   *the integral of f is*

$$F(x) = \int_0^x f(t)\, dt = \sum \frac{a_n}{n+1} x^{n+1} \qquad \text{for } |x| < R.$$

**PROOF**    We prove part (ii) first. By Theorem 1 $f$ is continuous, so the integral exists. Choose $x_1$ so that $|x| < |x_1| < R$ and $N$ so large that

$$\sum_{n=N+1}^{\infty} |a_n x_1^n|$$

is small, say less than $\epsilon$. Then with the notation of the proof of Theorem 1,

$$f(x) = P_N(x) + R_N(x) \qquad \text{with} \qquad |R_N(x)| < \epsilon,$$

$$\int_0^x f(t)\, dt - \int_0^x P_N(t)\, dt = \int_0^x R_N(t)\, dt$$

and

(4)
$$\left| F(x) - \sum_{n=0}^{N} \frac{a_n}{n+1} x^{n+1} \right| = \left| \int_0^x R_n(t)\, dt \right|.$$

The last integral is small.

(5)
$$\left| \int_0^x R_N(t)\, dt \right| \leq \left| \int_0^x |R_N(t)|\, dt \right| \leq \left| \int_0^x \epsilon\, dt \right| = \epsilon |x|.$$

Combining (4) and (5) we get

$$\left| F(x) - \sum_{n=0}^{N} \frac{a_n}{n+1} x^{n+1} \right| < \epsilon |x|.$$

Now, as $N \to \infty$, $\epsilon \to 0$. Thus, the partial sums of the integrated series converge to $F$. This proves (ii).

Turning now to (i), we observe that the derived series determines a continuous function $f_1$,

$$f_1(x) = \sum n a_n x^{n-1} \qquad \text{for } |x| < R.$$

---

* **Historical Note.** The early analysts were attracted to the use of series because of the extreme simplicity exhibited by Theorem 2.

By what has just been proved, we can integrate $f_1$ term by term and get

$$F_1(x) = \int_0^x f_1(t)\,dt = \sum_{n=1}^{\infty} a_n x^n = f(x) - a_0.$$

By the fundamental theorem of calculus $F_1$ is differentiable and $F_1'(x) = f_1(x)$. Thus, $f$ is differentiable, and $f'(x) = f_1(x)$, which was to be shown.

**COROLLARY 1**    *Functions represented by power series have derivatives of all orders.*

**COROLLARY 2**    *The coefficients in the power series are the Taylor coefficients:*

$$a_n = \frac{f^{(n)}(0)}{n!}.$$

**PROOF**    If

$$f(x) = \sum_{n=0}^{\infty} a_n x^n,$$

for $|x| < R$ and $R > 0$, then by Theorem 2 $f$ has derivatives of all orders, and these derivatives are given by the successive derived series.

$$f'(x) = \sum_{n=0}^{\infty} n a_n x^{n-1}, \qquad\qquad |x| < R;$$

$$f''(x) = \sum_{n=0}^{\infty} n(n-1) a_n x^{n-2}, \qquad\qquad |x| < R;$$

$$\cdots$$

$$f^{(N)}(x) = \sum_{n=0}^{\infty} n(n-1)\cdots(n-N+1) a_n x^{n-N}, \quad |x| < R.$$

$$\cdots$$

Then for $x = 0$, we get $f^{(N)}(0) = N!\, a_N$.

Theorem 2 provides information about functions that are *defined* by power series. This is in contrast to what was done in Chapter 8. In that chapter we began with a function (an elementary function) and inquired about conditions under which it is represented by its Taylor series. We then had Taylor's formula (see page 206) for series in powers of $x$:

$$(6) \qquad\qquad f(x) = \sum_{n=0}^{N} \frac{f^{(n)}(0)}{n!}\, x^n + R_N(x)$$

with the remainder $R_N(x)$ given by

(7) $$R_N(x) = \frac{f^{(N+1)}(\xi)}{(N+1)!} x^{N+1},$$

with $\xi$ between 0 and $x$. The sum in (6) is the $(N+1)$th partial sum of the Taylor series for $f$ in powers of $x$,

$$s_N(x) = \sum_{n=0}^{N} \frac{f^{(n)}(0)}{n!} x^n,$$

and is the $N$th Taylor polynomial. Consequently the Taylor series converges to $f(x)$ if and only if

$$\lim_{N \to \infty} s_N(x) = f(x).$$

This is the case if and only if

$$\lim_{N \to \infty} R_N(x) = 0.$$

Corollary 2 asserts that, if a function is represented by a power series, then the series must be the Taylor series of the function. The following corollary is therefore clear.

**COROLLARY 3**    *The power series representing a function is unique. That is, if*

$$f(x) = \sum a_n x^n = \sum b_n x^n \qquad for \; |x| < R,$$

*then $a_n = b_n$ for all $n$.*

*Example 1*    From Theorem 2 we conclude that, because

$$\frac{1}{1+x} = 1 - x + x^2 - \cdots \qquad \text{for } |x| < 1,$$

then

$$\frac{d}{dx}\frac{1}{1+x} = \frac{-1}{(1+x)^2} = -1 + 2x - 3x^2 + \cdots \quad \text{for } |x| < 1,$$

and

$$\int_0^x \frac{dt}{1+t} = \log(1+x) = x - \frac{x^2}{2} + \frac{x^3}{3} - \cdots \quad \text{for } |x| < 1.$$

Theorems 2 and 3 of Section 6 state that addition and multiplication of power series in $x$, convergent for $|x| < R$, result in power series that are also convergent for $|x| < R$. It is reasonable to expect that for two functions represented by power series their sum and their product will be represented by the sum and the product series. That this is the case is part of the following theorem which we state without proof.

**THEOREM 3**   *If*      $f(x) = \sum\limits_{n=0}^{\infty} a_n x^n$      *for* $|x| < R$,

*and*      $g(x) = \sum\limits_{n=0}^{\infty} b_n x^n$      *for* $|x| < R$,

*then*:      $f(x) + g(x) = \sum\limits_{n=0}^{\infty} (a_n + b_n) x^n$      *for* $|x| < R$,

*and*      $f(x)g(x) = \left( \sum\limits_{n=0}^{\infty} a_n x^n \right)\left( \sum\limits_{n=0}^{\infty} b_n x^n \right)$

$$= \sum_{n=0}^{\infty} c_n x^n \qquad \textit{for } |x| < R. \quad \{\textit{the Cauchy product.}$$

*Moreover, if $b_0 \neq 0$ then the quotient*

$$\frac{f(x)}{g(x)} = \frac{\sum\limits_{n=0}^{\infty} a_n x^n}{\sum\limits_{n=0}^{\infty} b_n x^n}$$

*is given by the quotient series in some interval containing the origin.*

**Example 2**   Because

$$\frac{1}{1-x} = \sum_{n=0}^{\infty} x^n, \qquad |x| < 1$$

$$\frac{1}{1+x} = \sum_{n=0}^{\infty} (-x)^n, \qquad |x| < 1$$

we have

$$\frac{1}{1-x}\frac{1}{1+x} = \frac{1}{1-x^2} = \left(\sum_{n=0}^{\infty} x^n\right)\left(\sum_{n=0}^{\infty} (-x)^n\right) = \sum_{n=0}^{\infty} x^{2n}, \qquad |x| < 1.$$

Another example is furnished by $\sin x/(1+x)$. We can either get the desired series by division:

$$\frac{\sin x}{1+x} = \frac{\sum \dfrac{(-1)^n x^{2n+1}}{(2n+1)!}}{1+x}$$

$$= x + x^2 - \frac{7}{3!}x^3 + \frac{7}{3!}x^4 - \frac{139}{5!}x^5 + \frac{139}{5!}x^6 - \cdots,$$

valid in some interval, or by multiplication:

$$\frac{\sin x}{1 + x} = \left( \sum \frac{(-1)^n x^{2n+1}}{(2n + 1)!} \right) \left( \sum (-x)^n \right)$$

$$= x + x^2 - \frac{7}{3!} x^3 + \frac{7}{3!} x^4 - \cdots,$$

valid for $|x| < 1$.

Finally, to conclude the discussion of the representation of functions by power series we derive an alternative form for the remainder $R_N$ in Taylor's formula, a form often easier to use than Lagrange's form, given by (7).

**THEOREM 4 (THE INTEGRAL FORM FOR THE REMAINDER)**    *If $f$ has continuous derivatives up to order $N + 1$, and if*

$$f(x) = \sum_{n=0}^{N} \frac{f^{(n)}(a)}{n!} (x - a)^n + R_N(x)$$

*then*

$$R_N(x) = \int_a^x \frac{(x - t)^N}{N!} f^{(N+1)}(t) \, dt.$$

**PROOF SKETCH**    The proof is by induction and repeated integration by parts. For $N = 0$ we have

$$f(x) = f(a) + \int_a^x f'(t) \, dt.$$

Now integrate by parts with $dv = dt$ and $v = -(x - t)$:

$$f(x) = f(a) + \left[ -f'(t)(x - t) \Big|_a^x + \int_a^x (x - t) f''(t) \, dt \right]$$

$$= f(a) + f'(a)(x - a) + \int_a^x (x - t) f''(t) \, dt, \qquad (N = 1),$$

which establishes the theorem for $N = 1$.

To complete the proof by induction we must use integration by parts on

$$(8) \qquad f(x) = \sum_{n=0}^{N} \frac{f^{(n)}(a)}{n!} (x - a)^n + \int_a^x \frac{(x - t)^N}{N!} f^{(N+1)}(t) \, dt,$$

where the integral is $R_N$, to obtain $R_{N+1}$. The details are left to the reader. See Problem 12.

*Example 3*     The binomial series offers an instance where the integral form of the remainder is more tractable* than Lagrange's form. We have

$$f(x) = (1+x)^m, \qquad f^{(n)}(x) = m(m-1)\cdots(m-n+1)(1+x)^{m-n},$$

so

$$(1+x)^m = 1 + mx + \cdots + \frac{m(m-1)\cdots(m-n+1)}{n!} x^n + R_n(x)$$

with

$$R_n(x) = \frac{m(m-1)\cdots(m-n)}{n!} \int_0^x (x-t)^n (1+t)^{m-n-1}\, dt$$

$$= \frac{m(m-1)\cdots(m-n)}{n!} \int_0^x \left(\frac{x-t}{1+t}\right)^n (1+t)^{m-1}\, dt.$$

Now it can be shown (see Problem 13) that

$$(9) \qquad \left| \frac{x-t}{1+t} \right| \leq |x| \qquad \text{if } -1 < x < 1$$

and $t$ is between 0 and $x$. Therefore, the remainder can be estimated:

$$|R_n(x)| \leq \begin{cases} \dfrac{m(m-1)\cdots(m-n)}{n!} \displaystyle\int_0^x |x|^n (1+t)^{m-1}\, dt & \text{if } 0 \leq x < 1, \\[3ex] \dfrac{m(m-1)\cdots(m-n)}{n!} \displaystyle\int_x^0 |x|^n (1+t)^{m-1}\, dt & \text{if } -1 < x < 0. \end{cases}$$

In either case there is a number $M$ such that

$$(10) \qquad |R_n(x)| \leq M \frac{m(m-1)\cdots(m-n)}{n!} |x|^n.$$

The limit of $R_n(x)$ as $n \to \infty$ can be shown to be zero. To do this consider the terms in (10) as terms of a series and apply the ratio test (see Problem 14). Therefore, the binomial series converges if $|x| < 1$ to the value $(1+x)^m$.

*Example 4*     The Maclaurin series for $e^{-x^2}$ can be obtained from the series for

$$e^u = \sum_{n=0}^{\infty} \frac{u^n}{n!},$$

whence, with $u = -x^2$, we have

$$(11) \qquad e^{-x^2} = \sum_{n=0}^{\infty} \frac{(-x^2)^n}{n!} = \sum_{n=0}^{\infty} (-1)^n \frac{x^{2n}}{n!}.$$

---

* Lagrange's form (7) requires that the $(N+1)$th derivative be estimated at the point $\xi$. With the integral form this task is evaded.

The remainder after $n$ terms in this series is most easily obtained from the remainder for the series for $e^u$. If

$$e^u = \sum_{n=0}^{N} \frac{u^n}{n!} + R_N(u),$$

then by Theorem 4

$$R_N(u) = \int_0^u \frac{(u-t)^N}{N!} e^t \, dt.$$

Now, if $u$ is negative (and it is in $e^{-x^2}$), then $e^t \leq 1$ in the above integral, and

$$|R_N(u)| \leq \left| \int_0^u \frac{(u-t)^N}{N!} \, dt \right| = \frac{|u|^{N+1}}{(N+1)!}.$$

Then the remainder after $x^{2N}$ in series (11) is less than or equal to $(x^2)^{N+1}/(N+1)!$. If $N = 4$, the remainder is less than $x^{10}/5!$. The series (11) and this estimate of the remainder can be used to compute

$$\int_0^{0.5} e^{-x^2} \, dx$$

with known error (see Problem 2).

### Problems

1. Write the binomial series for $(1 + x)^{-3/2}$ and compute $(0.98)^{-3/2}$ to four decimal places.

2. Compute $\int_0^{0.5} e^{-x^2} \, dx$ to four decimal places. See Example 4.  [0.4613]

3. Find five terms of the Maclaurin expansion of

$$F(x) = \int_0^x \frac{dt}{\sqrt[3]{1+t^2}}.$$

For what $x$ is the expansion valid?

4. Expand $e^x (1 - x)^{-1/2}$ in powers of $x$. Find five terms. For what values of $x$ is the expansion valid?

5. Show that $D \sin = \cos$ by using the known series expansions of sine and cosine. What theorems justify this procedure?

6. Obtain the Maclaurin series for $\log (1 + x^2)$ by integration. Where is the expansion valid and why?

7. Obtain the Maclaurin series for Arc tan $x$ by integration. Where is the expansion valid and why?

8.   Obtain five terms in the Maclaurin expansion of $\log (1 + x)/\sqrt{1 + (x/2)}$. Where is the expansion valid and why?

9.   Obtain four terms in the Maclaurin expansion of $\tan x$. Where is the expansion valid?

10.   Obtain four terms in the Maclaurin expansion of $(1 - x)^{1/2}(1 + x)^{-1/2}$. Where is the expansion valid and why?

11.   Obtain the Maclaurin series for $\sqrt{1 + x^2}/\sqrt{1 - x^2}$ in two different ways. Where is the expansion valid?

12.   Complete the proof of Theorem 4 by integrating by parts in equation (8). Use $v = -(x - t)$.

13.   Prove equation (9) of Example 3.

   *Hint:* Consider two cases—(a) $0 \leqq t \leqq x < 1$, (b) $-1 < x \leqq t \leqq 0$. In case (a) $\left| (x - t)/(1 + t) \right|$ is a decreasing function of $t$, and in case (b) it is an increasing function.

14.   Show that the remainder (10) in Example 3 approaches 0 as $n$ becomes infinite if $|x| < 1$ by showing that the series

$$\sum \frac{m(m - 1)\cdots(m - n)}{n!} x^n$$

is convergent. Why does this do the job?

## *8   Complex Power Series

Until now we have dealt exclusively with real-valued functions of real variables. Now, through use of power series we can extend the domains of definition of many of the elementary functions. Suppose

$$f(x) = \sum a_n x^n \qquad \text{for } -R < x < R.$$

Its partial sums are polynomials,

$$s_N(x) = \sum_{n=0}^{N} a_n x^n,$$

which are defined for all *complex* numbers $x$. The question, then, is whether the limit of the partial sums exists when $x$ is a complex number.

Examination of the proofs of convergence of power series given in Section 5 will show that they are independent of whether the variables are real or complex numbers. All that matters is that the absolute value of $x$ is sufficiently small. For example, Theorem 1 of Section 5 then becomes:

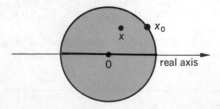

*If the series $\sum a_n x^n$ converges at $x_0$, then the series converges absolutely for all complex numbers $x$ such that $|x| < |x_0|$.* In this circle the series represents a function of a complex variable—and indeed an analytic one.*

In this way many of the elementary functions have vastly expanded domains. To conform to a frequently used notation we shall use $z$ for the complex variable. Thus, the exponential function and the sine and cosine functions, $e^z$, $\sin z$, $\cos z$, are *defined* by their series.

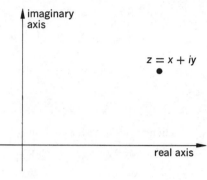

$$e^z = 1 + z + \frac{z^2}{2} + \cdots + \frac{z^n}{n!} + \cdots,$$

$$\sin z = z - \frac{z^3}{3!} + \cdots + (-1)^n \frac{z^{2n+1}}{(2n+1)!} + \cdots,$$

$$\cos z = 1 - \frac{z^2}{2!} + \cdots + (-1)^n \frac{z^{2n}}{(2n)!} + \cdots,$$

where the series converge for all complex numbers $z$.

**Example 1**   If $\theta$ is real, then

(1)     $$e^{i\theta} = 1 + i\theta + \frac{(i\theta)^2}{2!} + \frac{(i\theta)^3}{3!} + \cdots$$

(2)     $$= \left(1 - \frac{\theta^2}{2!} + \frac{\theta^4}{4!} - \cdots\right) + i\left(\theta - \frac{\theta^3}{3!} + \frac{\theta^5}{5!} - \cdots\right)$$

because the partial sums of (1) are the sums of the appropriate partial sums of (2). But the series in (2) are familiar. Thus we obtain

$$e^{i\theta} = \cos\theta + i\sin\theta, \quad \{\text{Euler's formula}$$

an at first startling relation, called *Euler's formula*.†

---

* The theory of analytic functions of a complex variable is essentially the single-handed invention of Cauchy. The important basic theorems were discovered by him and he first saw the relevance of the theory for all of analysis.

† This is but one of the numerous formulas attributed to the great Leonard Euler (1707–1783). He was the most prolific of mathematicians and blessed with a fertile imagination. He computed with series with complete abandon—just as if they were polynomials!

*Example 2*    There are two real functions that are related to the sine and cosine. Because the sine series and cosine series comprise, except for sign, the terms of the exponential series, one is led to define new functions:

(3)
$$\sinh z = z + \frac{z^3}{3!} + \frac{z^5}{5!} + \cdots, \quad \{\text{Hyperbolic sine}$$

$$\cosh z = 1 + \frac{z^2}{2!} + \frac{z^4}{4!} + \cdots. \quad \{\text{Hyperbolic cosine}$$

Then    $e^z = \cosh z + \sinh z,$

$e^{-z} = \cosh z - \sinh z;$

(4)
$$\cosh z = \frac{e^z + e^{-z}}{2},$$

$$\sinh z = \frac{e^z - e^{-z}}{2}.$$

Frequently used as the definitions of $\sinh z$ and $\cosh z$

If $z = i\theta$ in (3), then

(5)
$$\sinh i\theta = i\left(\theta - \frac{\theta^3}{3!} + \frac{\theta^5}{5!} - \cdots\right) = i \sin \theta,$$

$$\cosh i\theta = 1 - \frac{\theta^2}{2!} + \frac{\theta^4}{4!} - \cdots = \cos \theta,$$

which shows the intimate relation between the hyperbolic functions and the circular functions. From equations (4) it is easy to derive the analogue of a familiar identity for sine and cosine:

(6)
$$\cosh^2 z - \sinh^2 z = 1.$$

Other hyperbolic functions are defined in terms of sinh and cosh analogously to the circular functions. For example, the hyperbolic tangent is defined as:

$$\tanh z = \frac{\sinh z}{\cosh z}.$$

The graphs of sinh $x$, cosh $x$, and tanh $x$ for real $x$ are given below.

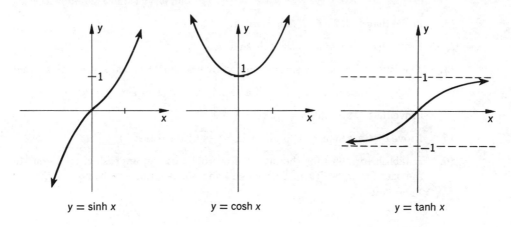

$$y = \sinh x \qquad\qquad y = \cosh x \qquad\qquad y = \tanh x$$

## Problems

1. For what *complex* numbers is the function $f$ given by $f(z) = 1/(1 + z^2)$ defined? Give the power series, in powers of $z$, which represents this function near $z = 0$. What is the radius of convergence of the series?

2. Show by multiplying series that $e^{z_1}e^{z_2} = e^{z_1+z_2}$. In other words one of the laws of exponents holds for complex exponents.

3. Use Problem 2 to show that $e^z$ is never zero.

   *Hint:* Let $z = x + iy$ and find the real and imaginary parts of $e^z$.

4. Use equations (4) to prove that

   $$\sinh (x \pm y) = \sinh x \cosh y \pm \cosh x \sinh y$$

   $$\cosh (x \pm y) = \cosh x \cosh y \pm \sinh x \sinh y.$$

5. Find the real and imaginary parts of $\sinh (x + iy)$ where $x$ and $y$ are real numbers.

6. Prove that $De^z = e^z$.

7. Obtain formulas for $D \sinh z$ and $D \cosh z$.

8. Define the hyperbolic cotangent, secant, and cosecant: $\coth z$, $\operatorname{sech} z$, and $\operatorname{csch} z$.

9. Prove that $\qquad \tanh (x \pm y) = \dfrac{\tanh x \pm \tanh y}{1 \pm \tanh x \tanh y}$ .

10. The Maclaurin series for $\sqrt{1 - z}$, obtained for real $z$, converges for what complex $z$? Use four terms to approximate $\sqrt{(4 - i)/4}$. Check by squaring.

    $$[1.008 - 0.126i]$$

11.    Find the real and imaginary parts of $\sin i\theta$ and $\cos i\theta$, for real $\theta$.
$$[\sin i\theta = i \sinh \theta,\ \cos i\theta = \cosh \theta]$$

12.    Prove that $\sinh 2x = 2 \sinh x \cosh x$.

13.    Prove that $\cosh 2x = \cosh^2 x + \sinh^2 x$.

14.    The inverse hyperbolic sine is denoted by arg sinh. For real $x$ it is unique, and

$$\text{arg sinh } x = y \qquad \text{if and only if} \qquad x = \sinh y.$$

Prove that $D \text{ arg sinh } x = 1/\sqrt{x^2 + 1}$.

15.    Prove that $\text{arg sinh } x = \log (x + \sqrt{x^2 + 1})$ for real $x$.

16.    Establish formulas for the inverse hyperbolic cosine, arg cosh $x$, analogous to the ones for arg sinh $x$ in Problems 14 and 15. Restrict $y$ to be greater than or equal to 0.


## *9    Power Series and Differential Equations

One of the earliest uses of power series was to solve differential equations. To early writers the method appeared very general, and even Lagrange thought that most functions could be represented by power series. We shall see that, if the functions occurring in certain special (but important) differential equations are represented by power series, then the solution of the differential equation is also represented by a power series. These special differential equations are the *linear* differential equations, of the first and second orders (analogous statements are valid for higher order equations):

(1)
$$\frac{dy}{dx} + a(x)y = f(x), \quad \{\text{Linear first order}$$

(2)
$$\frac{d^2y}{dx^2} + a(x)\frac{dy}{dx} + b(x)y = f(x). \quad \{\text{Linear second order}$$

The following theorem, which we state without proof, assures us that the technique we use in the examples will be successful.

**THEOREM 1**    *If $a$ and $f$ are functions represented by convergent power series, in powers of $x - x_0$, in an interval $I$ where $|x - x_0| < R$, and if $y_0$ is any given number, then there is a unique solution $y$, also represented by a power series in $I$, of the equation*

$$\frac{dy}{dx} + a(x)y = f(x)$$

*and satisfying the initial condition $y(x_0) = y_0$.*

**THEOREM 2**    *If a, b, and f are functions represented by power series in an interval I where $|x - x_0| < R$, and if $y_0$ and $y_1$ are any given numbers, then there is a unique solution y, also represented by a power series in I, of the equation*

$$\frac{d^2y}{dx^2} + a(x)\,\frac{dy}{dx} + b(x)y = f(x)$$

*and satisfying the initial conditions $y(x_0) = y_0$, $y'(x_0) = y_1$.*

**Example 1**    We solve, using series, the equation $dy/dx - y = 0$ with initial conditions $x_0 = 0$ and $y_0 = 1$. (Naturally, we could solve the equation more easily by separating variables.) Observe that the equation has the form of a linear equation (1), with $a = -1$ and $f = 0$. The functions $a$ and $f$, being constants, are represented by power series; and so by Theorem 1, the solution of the differential equation is also represented by a power series convergent for all $x$:

$$y(x) = \sum_{n=0}^{\infty} c_n x^n.$$

Then $\qquad y'(x) = \sum nc_n x^{n-1}$, {Theorem 2 of Section 7

and

$$y'(x) - y(x) = \sum nc_n x^{n-1} - \sum c_n x^n$$

(3) $$\qquad = \sum_{n=0}^{\infty} ((n+1)c_{n+1} - c_n)x^n \quad \text{\{Theorem 3 of Section 7.}$$

This last series represents the function identically zero because $y$ solves the differential equation $y' - y = 0$. Therefore, by Corollary 3 to Theorem 2 of Section 7, the uniqueness of power series representation, all the coefficients in equation (3) must be zero*:

(4) $$\qquad (n+1)c_{n+1} - c_n = 0, \qquad n = 0, 1, 2, \ldots.$$

Because $y(0) = 1$, we get $c_0 = y(0) = 1$. Now, using (4) for successive values of $n$, we get:

$$n = 0: \quad c_1 - c_0 = 0 \qquad \text{so} \qquad c_1 = c_0 = 1,$$

$$n = 1: \quad 2c_2 - c_1 = 0 \qquad \text{so} \qquad c_2 = \tfrac{1}{2}c_1 = \tfrac{1}{2},$$

$$n = 2: \quad 3c_3 - c_2 = 0 \qquad \text{so} \qquad c_3 = \frac{1}{3}c_2 = \frac{1}{3!}.$$

---

* Equation (4) is another example of a *recurrence relation*, or *recursion formula*. The coefficients can be found one after the other. This one is very simple and easy to solve.

In general, $nc_n - c_{n-1} = 0$ and $c_n = 1/n!$. Thus,

$$y(x) = \sum_{n=0}^{\infty} \frac{x^n}{n!},$$

which we recognize as the exponential function $e^x$.

If we do not specify the initial condition $y(0) = 1$, the coefficient $c_0$ is arbitrary. Then all the other coefficients are expressed in terms of $c_0$ (the constant of integration): $c_n = c_0/n!$.

*Example 2*    With second order equations the recurrence relation can be harder to solve. We shall select only simple ones with simple coefficient functions. To solve by series the equation

$$(5) \qquad y'' + \frac{x}{1 + x^2} y' + \frac{1}{1 + x^2} y = 0,$$

it is much more convenient to work with the equivalent equation

$$(6) \qquad (1 + x^2)y'' + xy' + y = 0.$$

Because the coefficient functions in (5) are given by power series in $x$ for $|x| < 1$, we can expect the series solution for $y$ to be valid for $|x| < 1$.

$$y(x) = \sum_{n=0}^{\infty} c_n x^n,$$

$$y'(x) = \sum nc_n x^{n-1},$$

$$y''(x) = \sum n(n-1)c_n x^{n-2}.$$

Substitution in (6) yields

$$\sum n(n-1)c_n x^{n-2} + \sum n(n-1)c_n x^n + \sum nc_n x^n + \sum c_n x^n = 0.$$

The coefficient of $x^n$ in this last equation must be 0. It is

$$(n+2)(n+1)c_{n+2} + n(n-1)c_n + nc_n + c_n = 0$$

or

$$(7) \qquad (n+2)(n+1)c_{n+2} + (n^2+1)c_n = 0.$$

If we allow $c_0$ and $c_1$ to be arbitrary (these are the initial values, $y(0) = c_0$ and $y'(0) = c_1$), then we compute the remaining coefficients from (7)

$$n = 0: \qquad 2c_2 + c_0 = 0 \qquad \text{so} \qquad c_2 = -\tfrac{1}{2}c_0$$

$$n = 1: \quad 3 \cdot 2 \cdot c_3 + 2c_1 = 0 \qquad \text{so} \qquad c_3 = -\frac{2}{3!}c_1$$

$$n = 2: \quad 4 \cdot 3 \cdot c_4 + 5c_2 = 0 \qquad \text{so} \qquad c_4 = \frac{5}{4!}c_0$$

$$n = 3: \quad 5 \cdot 4 \cdot c_5 + 10c_3 = 0 \qquad \text{so} \qquad c_5 = \frac{2(10)}{5!}c_1$$

$$\cdots$$

whence

$$y(x) = c_0\left(1 - \frac{1}{2!}x^2 + \frac{1\cdot 5}{4!}x^4 - \cdots + \frac{(-1)^n 1\cdot 5\cdots (1+(2n-2)^2)}{(2n)!}x^{2n} + \cdots\right)$$

$$+ c_1\left(x - \frac{2}{3!}x^3 + \frac{2\cdot 10}{5!}x^5 - \cdots + \frac{(-1)^n 2\cdot 10\cdots (1+(2n-1)^2)}{(2n+1)!}x^{2n+1} + \cdots\right).$$

### Problems

In Problems 1 to 8 find power series in x that give solutions to the differential equations and satisfy the initial conditions given. When possible give the general coefficient in the series; otherwise give at least four terms.

1.   $(1+x)y' + y = 0;\quad y(0) = 1$         $\left[\sum (-x)^n\right]$

2.   $(1+x)y' + y = 0;\quad y(0) = 0$         $[0]$

3.   $(1+x)y' + y = x;\quad y(0) = 0$         $\left[\dfrac{x^2}{2}\sum(-x)^n\right]$

4.   $(1+x^2)y' + xy = 0;\quad y(0) = y_0$

$$\left[y_0\left(1 + \sum_{n=1}^{\infty} \frac{(-1)^n\cdot 1\cdot 3\cdots(2n-1)}{2\cdot 4\cdots(2n)}x^{2n}\right)\right]$$

5.   $(1+x^2)y' + xy = x;\quad y(0) = y_0$

$$\left[1 + (y_0-1)\left(1 + \sum_{n=1}^{\infty} \frac{(-1)^n\cdot 1\cdot 3\cdots(2n-1)}{2\cdot 4\cdots(2n)}x^{2n}\right)\right]$$

6.   $(1+x)y'' + y' = 0;\quad y(0) = 0, y'(0) = y_1$

$$\left[y_1\sum_{n=1}^{\infty} \frac{(-1)^{n+1}}{n}x^n = y_1\log(1+x)\right]$$

7.   $(1+x)y'' + y' = 0;\quad y(0) = y_0, y'(0) = 0$         $[y_0]$

8.   $(1+x^2)y'' + xy' - y = 0;\quad y(0) = y_0, y'(0) = y_1$

$$\left[y_0\left(1 + \sum_{n=1}^{\infty} \frac{(-1)^{n+1}\cdot 1\cdot 3\cdots(2n-3)}{2\cdot 4\cdots(2n)}x^{2n}\right) + y_1 x\right]$$

9.   Apply Theorem 1 to Problems 1 to 5 to predict an interval in which the power series converges to the solution.

10.   Apply Theorem 2 to Problems 6, 7, and 8 to predict an interval in which the power series converges to the solution.

## PART III

MULTIVARIATE CALCULUS

Parts I and II of this text deal with the calculus of functions of *one* variable. We now turn to functions of two or more variables. Such functions arise naturally in mathematics and its applications. For example, the volume $V$ of a right circular cylinder is a function of two variables, the radius of the base $r$, and the height $h$: $V = \pi r^2 h$. This relationship can be expressed symbolically by $V = f(r, h)$. Other examples are:

The internal energy $E$ of a gas is a function of the pressure $p$ and the temperature $T$: $E = f(p, T)$.

The current $i$ flowing in a simple series circuit $t$ seconds after a switch is closed is a function of $t$, the resistance $R$, the capacitance $C$, the inductance $L$, and the impressed voltage $E$: $i = f(t, R, C, L, E)$.

Conditions at a point in space are functions of the coordinates of the point and the time. The behavior of a satellite after launch is a complicated function of the inherent characteristics of the hardware, the weather, the launch position on the earth, and control parameters that direct the path.

Indeed, the more precisely that one tries to describe real phenomena, the more numerous are the variables that one has to take into consideration.

In this text we restrict our attention to functions of two and three variables, for in these cases there are simple geometric interpretations in the plane or in space. Yet what is done here has natural extensions to functions of more than three variables. When more variables occur, we carry over the language (and definitions) that are used in dimensions two and three to the description of events in dimensions 4, 5, ..., $n$. Thus, we speak of points $(x_1, x_2, ..., x_n)$ in $n$-dimensional space with coordinates $x_1, x_2, ..., x_n$; or of vectors in $n$-space, surfaces and curves in $n$-space, etc. There is then a geometry of higher dimensional spaces that depends much on the more elementary beginnings that are taken up here.

## 1 Surfaces

If for each point $P = (x, y)$ of a region $R$ of the $xy$-plane there is associated a real number $z$, we say that $z$ *is a function $f$ of the two variables $x$ and $y$*, and we write

$$z = f(P) \qquad \text{or} \qquad z = f(x, y).$$

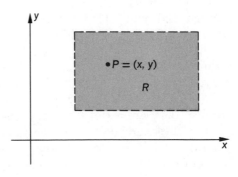

The figures suggest how to picture the function.*

We shall be concerned mainly with functions $f$ that are given by formulas in $x$ and $y$ involving the elementary functions. The domain $R$ of the function will be presumed to be evident from the formula; that is, $R$ will consist of all points $(x, y)$ for which the formula is defined.

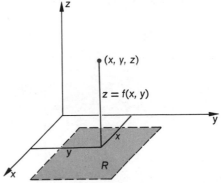

---

* See Appendix A, page 507 for a review of Cartesian coordinates in space.

*Example 1*    Some functions of two variables are:

$$z = x^2 - 2xy - y^2 - xy^3;$$

$$z = \text{Arc tan } (x + 2y^2);$$

$$z = e^x \sin y + e^y \sin x + y^2;$$

$$z = \sqrt{4 - x^2 - y^2}. \quad \{R \text{ is the disc } x^2 + y^2 \leq 4$$

The graph of $z = f(x, y)$ is a surface in space. See the figure.

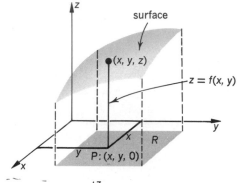

Although surfaces could be sketched by simply plotting points, a more helpful device is to draw some of the curves in which the surface intersects planes parallel to the coordinate planes. The middle figure on the right shows a curve $C$, the curve of intersection of the surface $z = f(x, y)$, and the plane $x = x_0 = $ constant. The equation of $C$ in the plane $x = x_0$, is $z = f(x_0, y)$.

In a similar way, curves on the surface and parallel to the $xz$-plane have equations in a plane $y = y_0 = $ constant, given by $z = f(x, y_0)$. The figure at the left below shows such a curve $C$.

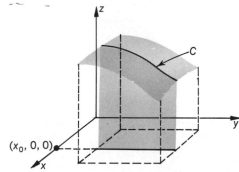

Finally, we have curves on the surface that are parallel to the $xy$-plane. These are curves of constant height, $z = z_0 = $ constant. In the plane $z = z_0$ they have the equation $z_0 = f(x, y)$. The graphs of these equations in the $xy$-plane are called *level curves*.

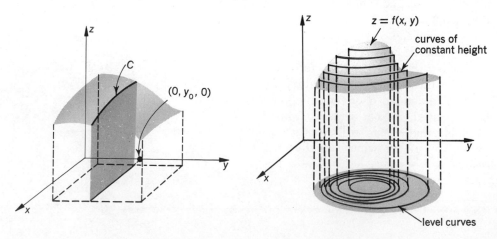

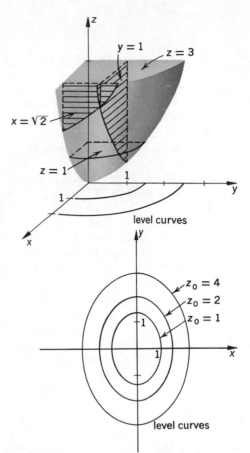

**Example 2** Sketch the part of the surface $z = x^2 + \frac{1}{2}y^2$ that is in the first octant, where $x \geqq 0$, $y \geqq 0$, and $z \geqq 0$.

Curves on the surface parallel to the $yz$-plane are parabolas $z = \frac{1}{2}y^2 + x_0^2$, with $x_0$ constant. Curves parallel to the $xz$-plane are parabolas $z = x^2 + \frac{1}{2}y_0^2$, with $y_0$ constant. Curves of constant height $z_0$ are ellipses $x^2 + \frac{1}{2}y^2 = z_0$. These are the level curves. The figure shows the surface with some cross sections parallel to coordinate planes.

**Example 3** Sketch the surface $z = x^2$.

Here all cross sections by planes parallel to the $xz$-plane are the same, namely the parabola $z = x^2$. The surface is called a *parabolic* cylinder. The term cylinder is used because if $(x, 0, z)$ is on the surface, so is $(x, y, z)$ for any $y$. A part of the surface for $y \geqq 0$ is shown in the figure.

The level curves ($z_0 = x^2$, $z_0 =$ constant) are straight lines parallel to the $y$-axis.

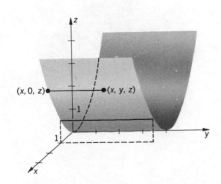

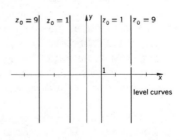

***Example 4***     Sketch the surface $z = y^2 - x^2$.

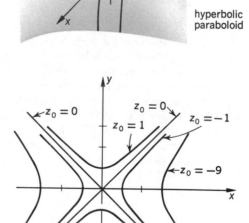

hyperbolic paraboloid

    Cross sections on the surface parallel to the $xz$-plane are parabolas $z = -x^2 + y_0^2$, with $y_0$ constant. Cross sections parallel to the $yz$-plane are parabolas $z = y^2 - x_0^2$, with $x_0$ constant. Cross sections parallel to the $xy$-plane are hyperbolas $z_0 = y^2 - x^2$, with $z_0$ constant. The figure shows part of the surface with some cross sections parallel to the $xy$- and $xz$-planes. The surface has the appearance of a saddle. It is called a *hyperbolic paraboloid*.

    The level curves are the hyperbolas $z_0 = y^2 - x^2$, with $z_0$ constant, in the $xy$-plane. Observe that, when $z_0 = 0$, the level curve is a pair of straight lines $y = x$ and $y = -x$. See the figure.

level curves

***Example 5***     Sketch the surface $y = 1 - z^2 - x^2$.

    Here $y$ is given as a function of $x$ and $z$. Thus, one may think of the surface as extending above (or below) the $xz$-plane. Cross sections parallel to the $yz$-plane are parabolas $y = 1 - z^2 - x_0^2$, with $x_0$ constant. Cross sections parallel to the $xy$-plane are parabolas $y = 1 - z_0^2 - x^2$, with $z_0$ constant. The level curves are circles in the $xz$-plane $y_0 = 1 - z^2 - x^2$, with $y_0$ constant.

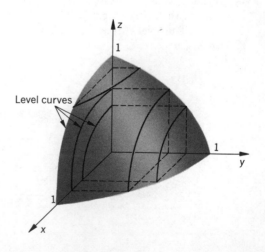

Level curves

    The figure shows the portion of the surface in the first octant. Parts of three level curves are shown.

*Example 6*    The set of points $(x, y, z)$ where $x$, $y$, and $z$ satisfy the equation $y^2 - x^2 - z^2 = 1$ is also a surface. Cross sections parallel to the $xz$-plane are circles $y_0^2 - x^2 - z^2 = 1$, with $y_0$ constant, where $y_0 \geqq 1$. Cross sections parallel to the other coordinate planes are hyperbolas. The figure shows the surface, a *hyperboloid of two sheets*, with the cross section $z = 0$, $y^2 - x^2 = 1$ shaded.

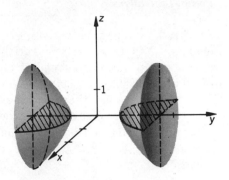

Observe that from the equation, $y^2 - x^2 - z^2 = 1$, we can define $z$ as one or more functions of $x$ and $y$. In this case we say that $z$ is defined *implicitly* as a function of $x$ and $y$.

## Problems

In Problems 1 to 16 sketch the surfaces. Sketch several cross sections parallel to coordinate planes. Also sketch level curves in the appropriate plane.

1.  $z = \sqrt{4 - x^2 - y^2}$

2.  $z = 2 - y^2$

3.  $z = 6 - \frac{3}{2}x - 2y$

4.  $z = \sin x$

5.  $z = \sin (x + y)$

6.  $z = y^2 - x$

7.  $z = y$

8.  $z = y - 1$

9.  $x = a - z$

10.  $z = \sqrt{x^2 + y^2}$

11.  $z = x + y$

12.  $y = x^2 + z^2$

13.  $y = x + z^2$

14.  $x = \sqrt{4 - y^2 - z^2}$

15.  $x = 1 - z^2$

16.  $z = \sqrt{4 + x^2 + y^2}$

In Problems 17 to 20 sketch the surfaces. Show enough of the surface to indicate how it goes, and also show several cross sections.

17.  $x^2 + y^2 - z^2 = 1$

18.  $z^2 = x^2 + y^2$

19.  $x^2 + y^2 + z^2 = 0$

20.  $x^2 - y^2 + 2z^2 = 4$

## 2   *Partial Derivatives*

In Part I we found derivatives, or rates of change, of functions of one variable. We now compute rates of change of functions of several variables.

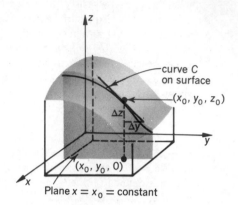

curve C on surface

$(x_0, y_0, z_0)$

$(x_0, y_0, 0)$

Plane $x = x_0 =$ constant

Consider a function $f$ and the surface $z = f(x, y)$. The figure shows the cross section $C$ of the surface and the plane $x = x_0$. The curve $C$ is given by the equation

$$z = f(x_0, y)$$

in the plane $x = x_0$. The slope $m$ of $C$ at the point $(x_0, y_0, z_0)$ is given by

$$(1) \qquad m = \lim_{\Delta y \to 0} \frac{\Delta z}{\Delta y} = \lim_{\Delta y \to 0} \frac{f(x_0, y_0 + \Delta y) - f(x_0, y_0)}{\Delta y}.$$

This slope $m$ is the rate of change of $f$ in the "$y$-direction" at $(x_0, y_0, z_0)$. It is also called the *partial derivative of $f$ with respect to $y$ at* $(x_0, y_0)$.

If the curve $C$ is the intersection of the surface with the plane $y = y_0$, its equation in the plane $y = y_0$ is

$$z = f(x, y_0).$$

The slope $m$ of the curve $C$ at the point $(x_0, y_0, z_0)$ is

$$(2) \quad m = \lim_{\Delta x \to 0} \frac{\Delta z}{\Delta x}$$

$$= \lim_{\Delta x \to 0} \frac{f(x_0 + \Delta x, y_0) - f(x_0, y_0)}{\Delta x}.$$

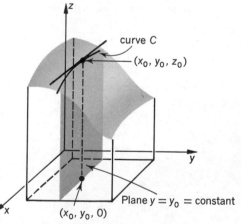

curve C

$(x_0, y_0, z_0)$

Plane $y = y_0 =$ constant

$(x_0, y_0, 0)$

This slope $m$ is the rate of change of $f$ in the "$x$-direction" at $(x_0, y_0, z_0)$. It is called the *partial derivative of $f$ with respect to $x$ at* $(x_0, y_0)$.

There is no need to keep the subscripts on $x$ and $y$ because $x_0$ and $y_0$ can be any numbers (such that $(x_0, y_0)$ is in the domain of $f$). We summarize the above considerations with a definition.

**DEFINITION**   If $f$ is a function of the variables $x$ and $y$ in a region of the $xy$-plane, the *partial derivative of $f$ with respect to $x$ at the point $(x, y)$* is

$$\frac{\partial f}{\partial x} = \lim_{\Delta x \to 0} \frac{f(x + \Delta x, y) - f(x, y)}{\Delta x}.$$

The *partial derivative of f with respect to y at* $(x, y)$ is

$$\frac{\partial f}{\partial y} = \lim_{\Delta y \to 0} \frac{f(x, y + \Delta y) - f(x, y)}{\Delta y}.$$

If $f$ is given in terms of elementary functions, then no new differentiation formulas are needed in order to compute partial derivatives. Thus, for example, to find $\partial f/\partial x$ one simply regards $y$ as a constant in $f(x, y)$ and differentiates with respect to $x$.

The notations for partial derivatives are even more abundant than for ordinary derivatives. Here are some for $\partial f/\partial x$, if we write $z = f(x, y)$:

$$\frac{\partial z}{\partial x}, \ \frac{\partial f}{\partial x}, \ \frac{\partial f(x, y)}{\partial x}, \ f_x(x, y), \ D_x f, \ z_x, \ f_x$$

with obvious modifications for partials with respect to $y$.

**Example 1**    If $f(x, y) = e^x \sin y + e^y \sin x + y^2$, then

$$\frac{\partial f}{\partial x} = e^x \sin y + e^y \cos x \qquad \text{and} \qquad \frac{\partial f}{\partial y} = e^x \cos y + e^y \sin x + 2y.$$

**Example 2**    Find $\partial f/\partial x$ and $\partial f/\partial y$ at the point $(1, \frac{3}{2})$ if $f(x, y) = \sqrt{4 - x^2 - y^2}$.

We first compute the partial derivatives at an arbitrary point $(x, y)$, and then substitute $x = 1$, $y = \frac{3}{2}$.

$$\frac{\partial f}{\partial x} = \frac{-x}{\sqrt{4 - x^2 - y^2}}, \qquad \frac{\partial f}{\partial y} = \frac{-y}{\sqrt{4 - x^2 - y^2}}.$$

At the point $(1, \frac{3}{2})$,

$$\frac{\partial f}{\partial x} = -\frac{2}{\sqrt{3}} \qquad \text{and} \qquad \frac{\partial f}{\partial y} = -\frac{3}{\sqrt{3}}.$$

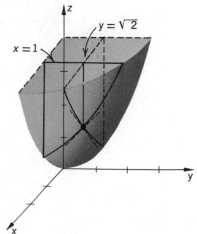

**Example 3**    Sketch the part in the first octant of the surface $z = x^2 + \frac{1}{2}y^2$. Sketch the curves where the surface meets the planes $x = 1$ and $y = \sqrt{2}$. Find the slopes of these curves at the point $(1, \sqrt{2}, 2)$.

The surface was sketched in Example 2, Section 1. The partial derivatives are

$$\frac{\partial z}{\partial x} = 2x, \qquad \frac{\partial z}{\partial y} = y.$$

Thus, the slopes at $(1, \sqrt{2}, 2)$ are 2 and $\sqrt{2}$.

**Example 4**    Sketch the part of the surface $z = -\frac{2}{3}x - y + 3$ that lies in the first octant. Find $z_x$ and $z_y$ and interpret geometrically.

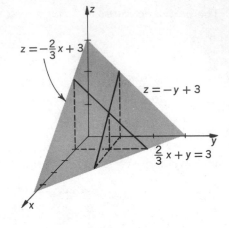

The surface is a plane cutting across the first octant as shown. It is easily sketched by drawing the lines of intersection with the coordinate planes. Thus, in the $yz$-plane the curve of intersection is the line $z = -y + 3$.

The partial derivatives are

$$z_x = -\frac{2}{3} \quad \text{and} \quad z_y = -1.$$

Cross sections by planes parallel to the coordinate planes are straight lines. Because $z_x = -\frac{2}{3} = $ constant, all cross sections parallel to the $xz$-plane have the same slope. A similar statement holds for $z_y$.

Partial derivatives can also be defined for functions of three or more variables. For example, if $u$ is a function of $x$, $y$, and $z$ in a region of space: $u = f(x, y, z)$, then

$$\frac{\partial u}{\partial x} = \lim_{\Delta x \to 0} \frac{f(x + \Delta x, y, z) - f(x, y, z)}{\Delta x},$$

with similar definitions for $\partial u/\partial y$ and $\partial u/\partial z$.

**Example 5**    Find $u_x$, $u_y$, and $u_z$ if $u = x^2 y + \text{Arc tan } xz$.

$$u_x = 2xy + \frac{z}{1 + x^2 z^2}, \quad u_y = x^2, \quad u_z = \frac{x}{1 + x^2 z^2}.$$

**Example 6**    Partial derivatives can also be found when a function is defined implicitly. For example, find $z_x$, $z_y$ if

$$(3) \qquad\qquad x^2 + y^2 + z^2 + \log(xz) = 0.$$

Here we assume that there is a function $z = z(x, y)$ defined implicitly by (3). Then, to find $z_x$, we hold $y$ constant and differentiate (3) implicitly with respect to the variable $x$:

$$2x + 2zz_x + \frac{xz_x + z}{xz} = 0.$$

Thus,

$$z_x = -\frac{z}{x} \frac{2x^2 + 1}{2z^2 + 1}.$$

Similarly, to find $z_y$, we hold $x$ constant and differentiate with respect to $y$:

$$2y + 2zz_y + \frac{xz_y}{xz} = 0.$$

Thus,

$$z_y = -\frac{2yz}{2z^2 + 1}.$$

## Problems

In Problems 1 to 10 compute the indicated partial derivatives.

1. $z = k\sqrt{x^2 + y^2}$   ; $\partial z/\partial x$   $[kx/\sqrt{x^2 + y^2}]$

2. $z = x/(x - y)$   ; $\partial z/\partial y$   $[x/(x - y)^2]$

3. $u = x^3 - xy^2 + 2y^3$   ; $u_x, u_y$   $[3x^2 - y^2, -2xy + 6y^2]$

4. $v = \text{Arc tan } x/y$   ; $v_x, v_y$   $[y/(x^2 + y^2), -x/(x^2 + y^2)]$

5. $v = (x - y)(y - x)$   ; $v_x, v_y$   $[v_x = -2x + 2y = -v_y]$

6. $r = \sqrt{x^2 + y^2 + z^2}$   ; $\partial r/\partial x$   $[x/r]$

7. $u = x^y$   ; $\partial u/\partial x, \partial u/\partial y$   $[u_x = yx^{y-1}, u_y = x^y \log x]$

8. $\theta = \text{Arc sin } \dfrac{x + y}{z}$   ; $\theta_x, \theta_z$   $\left[ \begin{array}{c} \theta_x = \dfrac{|z|}{z\sqrt{z^2 - (x + y)^2}}, \\[3mm] \theta_z = \dfrac{-(x + y)}{|z|\sqrt{z^2 - (x + y)^2}} \end{array} \right]$

9. $w = \log \dfrac{xy}{x^2 + y^2}$   ; $\dfrac{\partial w}{\partial x}$   $\left[ \dfrac{y^2 - x^2}{x(x^2 + y^2)} \right]$

10. $\varphi = \text{Arc cos } \dfrac{z}{\sqrt{x^2 + y^2 + z^2}}$ ; $\varphi_x, \varphi_z$   $\left[ \begin{array}{c} \varphi_x = \dfrac{xz}{(x^2 + y^2 + z^2)\sqrt{x^2 + y^2}}, \\[3mm] \varphi_z = -\dfrac{\sqrt{x^2 + y^2}}{x^2 + y^2 + z^2} \end{array} \right]$

In Problems 11 to 23 sketch the surface, usually the part in the first octant will suffice. Draw the curves of intersection of the surface with planes parallel to the xz- and yz-planes and through the indicated point, or points. Find the slopes of these curves at those points and draw a small portion of the tangent lines.

11. $z = 2 - x$   ; $(1, 2, 1)$   $[z_x = -1, z_y = 0]$

12. $z = 6 - \frac{3}{2}x - 2y$   ; $(1, 1, \frac{5}{2})$   $[z_x = -\frac{3}{2}, z_y = -2]$

13. $3z + 2y = 6$   ; $(0, 1, \frac{4}{3})$   $[z_x = 0, z_y = -\frac{2}{3}]$

14. $z = \frac{1}{2}x - y$   ; $(2, 0, 1), (2, 1, 0)$   $[z_x = \frac{1}{2}, z_y = -1]$

15. $z = 2 - y^2$   ; $(0, 1, 1)$   $[z_x = 0, z_y = -2]$

16. $z = 2 - x^2 - y^2$   ; $(1/\sqrt{2}, 1/\sqrt{2}, 1)$   $[z_x = -\sqrt{2} = z_y]$

17. $z = \sqrt{a^2 - x^2 - y^2}$ ; $(0, 0, a)$   $[z_x = 0 = z_y]$

18. $z = \sqrt{a^2 - x^2 - y^2}$ ; $(a/2, 0, \sqrt{3}a/2)$     $[z_x = -1/\sqrt{3}, z_y = 0]$

19. $z = \sqrt{a^2 - x^2 - y^2}$ ; $(a/2, a/2, a/\sqrt{2})$     $[z_x = -1/\sqrt{2} = z_y]$

20. $3x^2 + 2y^2 + 4z^2 = 12$; $(\sqrt{2}, 1, 1)$     $[z_x = -3\sqrt{2}/4, z_y = -\tfrac{1}{2}]$

21. $\dfrac{z^2}{4} - \dfrac{x^2}{4} - \dfrac{y^2}{2} = 1$  ; $(1, 1, \sqrt{7})$     $\left[ z_x = \dfrac{1}{\sqrt{7}}, z_y = \dfrac{2}{\sqrt{7}} \right]$

22. $z^2 = 4 + y^2 - x^2$  ; $(0, 0, 2), (0, 1, \sqrt{5}), (2, 1, 1)$   $\left[ \begin{array}{l} z_x = 0 = z_y \\[4pt] z_x = 0, z_y = 1/\sqrt{5} \\[4pt] z_x = -2, z_y = 1 \end{array} \right]$

23. $z = y^2 - x^2$         ; $(1, 1, 0)$     $[z_x = -2, z_y = 2]$

In Problems 24 to 33 show that the given function satisfies the equation (a partial differential equation) involving the function and its partial derivatives.

24. $z = \log (e^x + e^y)$         ; $z_x + z_y = 1$

25. $z = \sin^2 (x + y)$         ; $z_x = z_y$

26. $z = \dfrac{a}{2}\left( x + \dfrac{y}{a} \right)^2$         ; $z_x z_y - \dfrac{1}{a} y z_x - a x z_y = 0$

27. $u = r + \log (r - s)$         ; $u_r + u_s = 1$

28. $u = xy/(x + y)$         ; $x^2 u_x - y^2 u_y = 0$

29. $u^2 = xy + \tan (x/y)$         ; $x u u_x + y u u_y = xy$

30. $\varphi (x, y, z) = xy^2 + yz^2 + zx^2$ ; $\varphi_x + \varphi_y + \varphi_z = (x + y + z)^2$.
Also show that $x\varphi_x + y\varphi_y + z\varphi_z = 3\varphi$.

31. $\dfrac{1}{z} = \dfrac{1}{y} - \log \left( \dfrac{1}{x} - \dfrac{1}{y} \right)$     ; $x^2 z_x + y^2 z_y = z^2$

32. $f(x, y) = x^3 + x^2 y + xy^2 + y^3$; $x\dfrac{\partial f}{\partial x} + y\dfrac{\partial f}{\partial y} = 3f$

33. $f(x, y) = x^n \cos (y/x)$         ; $x f_x + y f_y = nf$

## 3   *The Total Differential; the Fundamental Lemma*

In Chapter 1 we observed that for differentiable functions $f$ of one variable

(1)             $\Delta y = f(x + \Delta x) - f(x) = f'(x) \, \Delta x + \eta \, \Delta x,$

where $\eta$ is a function of $\Delta x$ and $\eta \to 0$ as $\Delta x \to 0$. This led naturally to the definition of the differential,

$$dy = f'(x) \, dx.$$

A similar situation prevails for functions of several variables. The following theorem is basic to the development of this chapter.

**THE FUNDAMENTAL LEMMA**  *If $f_x$ and $f_y$ are continuous, then*

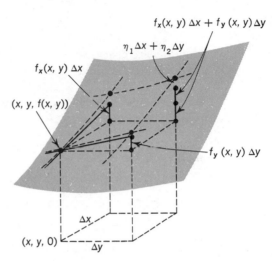

(2)  $\Delta z = f(x + \Delta x, y + \Delta y) - f(x, y)$

$\quad\quad = f_x(x, y) \, \Delta x + f_y(x, y) \, \Delta y$

$\quad\quad\quad + \eta_1 \, \Delta x + \eta_2 \, \Delta y.$

*where $\eta_1$ and $\eta_2$ are functions of $\Delta x$ and $\Delta y$ and approach 0 as $\Delta x$ and $\Delta y \to 0$.*

**PROOF**  We first rewrite $\Delta z$ in a judicious way,

$$\Delta z = [f(x + \Delta x, y + \Delta y) - f(x, y + \Delta y)] + [f(x, y + \Delta y) - f(x, y)]$$

Then, by the MVT, with $0 < \theta_1 < 1$ and $0 < \theta_2 < 1$, we have

(3)  $$\Delta z = f_x(x + \theta_1 \, \Delta x, y + \Delta y) \, \Delta x + f_y(x, y + \theta_2 \, \Delta y) \, \Delta y.$$

Define functions $\eta_1$ and $\eta_2$ by

(4)  $$\begin{cases} \eta_1 = f_x(x + \theta_1 \, \Delta x, y + \Delta y) - f_x(x, y), \\ \\ \eta_2 = f_y(x, y + \theta_2 \, \Delta y) - f_y(x, y). \end{cases}$$

Because $f_x$ and $f_y$ are continuous at $(x, y)$, we have that $\eta_1 \to 0$ and $\eta_2 \to 0$ as $\Delta x$ and $\Delta y \to 0$. Substitution of equations (4) in (3) yields equation (2).

A similar theorem is valid for functions of more than two variables: *If $z = f(x_1, \ldots, x_n)$, and $f$ has continuous partial derivatives, then*

$$\Delta z = \frac{\partial f}{\partial x_1} \, \Delta x_1 + \cdots + \frac{\partial f}{\partial x_n} \, \Delta x_n + \eta_1 \, \Delta x_1 + \cdots + \eta_n \, \Delta x_n,$$

*where $\eta_1, \ldots, \eta_n \to 0$ as $\Delta x_1, \ldots, \Delta x_n \to 0$.*

When $\Delta x$ and $\Delta y$ are small, then $\eta_1 \, \Delta x$ and $\eta_2 \, \Delta y$ are "small of second order."

Therefore, the approximation

$$\Delta z \approx f_x \, \Delta x + f_y \, \Delta y$$

should be quite good, and suggests the following definition.

**DEFINITION**   The *total differential* of $z = f(x, y)$ is

$$dz = f_x(x, y) \, dx + f_y(x, y) \, dy,$$

where $dx$ and $dy$ are arbitrary numbers. Observe that $dz$ is a linear function of $dx$ and $dy$. There is an analogous definition for a function of more than two variables.

The differential is a good approximation to the change in the function and differs from $\Delta z$ by $\eta_1 \, \Delta x + \eta_2 \, \Delta y$.

*Example 1*   If $w = xy - y^2$, then

$$dw = y \, dx + (x - 2y) \, dy$$

$$\Delta w = (x + \Delta x)(y + \Delta y) - (y + \Delta y)^2 - xy + y^2$$

$$= y \, \Delta x + x \, \Delta y - 2y \, \Delta y - \Delta y^2.$$

So, with $dx = \Delta x$, $dy = \Delta y$,

$$\Delta w = dw - \Delta y^2.$$

Thus, $$\eta_1 = 0, \qquad \eta_2 = -\Delta y.$$

*Example 2*   If a box, with lid, with inner dimensions 2, 3, and 4 ft is to be made of sheet metal $\frac{1}{16}$ in. thick the *approximate* volume of metal can be found using differentials.

If the dimensions of the box are $x$, $y$, and $z$, the volume of the box is

$$V(x, y, z) = xyz.$$

With $x = 2$, $y = 3$, $z = 4$, $dx = dy = dz = (\frac{1}{16})(\frac{1}{12})(2) = \frac{1}{96}$ ft (two thicknesses for each dimension), the *exact* volume of metal is

$$V(x + dx, y + dy, z + dz) - V(x, y, z) = \Delta V.$$

But $$\Delta V \approx dV = yz \, dx + xz \, dy + xy \, dz.$$

An easy calculation gives the *approximate* volume of metal.

$$dV = 3(4)\tfrac{1}{96} + 2(4)\tfrac{1}{96} + 2(3)\tfrac{1}{96} = \tfrac{13}{48} \text{ cu ft.}$$

If sheet metal weighs 480 lb/cu ft, the box would weigh about 130 lb.

*Example 3*   If $\theta = \text{Arc tan } y/x$, use differentials to find an approximate value for $\theta$ when $x = 0.95$ and $y = 1.05$.

With $x = 1, y = 1, dx = -0.05, dy = 0.05$, we seek $\theta(x + dx, y + dy) =$ Arc tan $1.05/0.95$. But

$$\theta(x + dx, y + dy) \approx \theta(x, y) + d\theta,$$

and
$$d\theta = -\frac{y}{x^2 + y^2}\, dx + \frac{x}{x^2 + y^2}\, dy.$$

Therefore,

$$\text{Arc tan } \frac{1.05}{0.95} \approx \text{Arc tan } \frac{1}{1} + \left(-\frac{1}{2}\right)(-0.05) + \left(\frac{1}{2}\right)(0.05)$$

$$= \frac{\pi}{4} + 0.05 \approx 0.735.$$

### Problems

**In Problems 1 to 12 write the total differential.**

1.  $f(x, y) = x^3 - xy^2 + x^2 y - y^3$

2.  $\varphi(r, s) = e^{rs} + e^{r-s}$

3.  $F(x, y, z) = x^2 + y^2 - z^2 - xy + yz - 2$

4.  $G(x, y, z) = z$ Arc tan $\dfrac{y}{x}$  $\qquad \left[ dG = \dfrac{-yz}{x^2 + y^2}\, dx + \dfrac{xz}{x^2 + y^2}\, dy + \dfrac{G}{z}\, dz \right]$

5.  $F(x, y) = \sin x \cos y + \cos x \sin y - \cos(x + y) \tan(x + y)$  $\qquad [dF = 0]$

6.  $\varphi(x_1, \ldots, x_n) = (x_1 + \cdots + x_n)^2$

7.  $\varphi(x_1, \ldots, x_n) = x_1^2 + \cdots + x_n^2$

8.  $V = \pi r^2 h$  $\qquad$ (Interpret geometrically.)

9.  $z = \log \sqrt{x^2 + y^2}$

10. $u = $ Arc tan $\dfrac{z}{\sqrt{x^2 + y^2}}$

$$\left[ du = \frac{1}{(x^2 + y^2 + z^2)\sqrt{x^2 + y^2}} \{-xz\, dx - yz\, dy + (x^2 + y^2)\, dz\} \right]$$

11. $U = \dfrac{1}{r}, r = \sqrt{x^2 + y^2 + z^2}$  $\qquad \left[ dU = -\dfrac{1}{r^3}(x\, dx + y\, dy + z\, dz) \right]$

12. $u = x^3 + y^3 + 3axy$

13. Compute approximately using differentials $\sqrt{(2.02)^2 + 4(1.97)^2 - (1.98)^2}$.
$\qquad [3.96]$

14. Find, approximately, the volume of a right circular cylinder of height 10.1 cm and radius of base 8.9 cm. $[800.1\pi]$

15. The *relative error* in a function $u = f(x, y, z)$ is defined to be $du/u$. Find the relative error in $u = xyz$ due to to errors $dx$, $dy$, $dz$ in $x$, $y$, $z$.

16. What is the relative error in $\log (x + y + z)$ due to errors in $x$, $y$, and $z$?

17. What is the relative error in $\exp (x_1^2 + \cdots + x_n^2)$ due to errors in $x_1, \ldots, x_n$?

18. A right circular cone has height $= 10.00 \pm 0.05$ cm and radius of base $= 4.00 \pm 0.02$ cm. What is the maximum error in the volume? the relative error? the percentage error?

19. Pressure $p$, volume $v$, and absolute temperature $T$ of a perfect gas obey the equation of state: $pv = RT$ where $R$ is a constant. If the volume is 120 cu ft when the pressure is 14.7 lb/in.$^2$ and $T = 295°$ absolute, use differentials to find the approximate pressure of the gas when the volume is 121 cu ft and $T = 300°$. $[p = 14.83]$

20. What is the percentage error in the lateral surface area of a right circular cylinder if the radius is increased by 1 per cent and the height is decreased by 1 per cent?

21. How much will the angle $\theta$ possibly change if $a$, $b$ are each possibly in error by 1 per cent? $[\text{at most } 0.02ab/(a^2 + b^2)]$

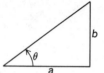

22. The period of a simple pendulum of length $l$ is $p = 2\pi\sqrt{l/g}$ where $g$ is the acceleration of gravity. If $g = 980 \pm 1$ cm/sec$^2$ and $l = 100 \pm 0.1$, what is the possible error in the period? $[\text{about } 0.002 \text{ sec}]$

23. Compute $\Delta z$ and $dz$ if $z = x^2 - xy + y^2$ and $x = 1$, $y = 1$, $\Delta x = dx = 0.1$, and $\Delta y = dy = -0.1$. $[\Delta z = 0.03, dz = 0]$

In Problems 24 and 25 compute $\Delta z$ and $dz$. Then find $\eta_1$ and $\eta_2$ of the Fundamental Lemma and verify that $\eta_1$ and $\eta_2 \to 0$ as $\Delta x$, $\Delta y \to 0$. (Use $\Delta x = dx$, $\Delta y = dy$.)

24. $z = x^2 - xy + y^2$ $\qquad [\Delta z - dz = dx^2 - dx\,dy + dy^2$, whence either $\eta_1 = dx - dy, \eta_2 = dy$ or $\eta_1 = dx, \eta_2 = -dx + dy]$

25. $z = x^3 + y^3 - 3xy$

## 4 The Chain Rule

The Fundamental Lemma of Section 3 has numerous applications. One of these is the chain rule.

**CHAIN RULE** *If $u = f(x, y)$ and $x = x(r, s)$ and $y = y(r, s)$ have continuous partial derivatives, then $u$ is a differentiable function of $r$ and $s$, and*

$$(1) \quad \begin{bmatrix} \dfrac{\partial u}{\partial r} = \dfrac{\partial u}{\partial x}\dfrac{\partial x}{\partial r} + \dfrac{\partial u}{\partial y}\dfrac{\partial y}{\partial r} \\[4mm] \dfrac{\partial u}{\partial s} = \dfrac{\partial u}{\partial x}\dfrac{\partial x}{\partial s} + \dfrac{\partial u}{\partial y}\dfrac{\partial y}{\partial s} \end{bmatrix}$$

**PROOF**   We shall prove the formula for $\partial u/\partial r$. From the Fundamental Lemma

$$(2) \quad \Delta u = \frac{\partial u}{\partial x}\,\Delta x + \frac{\partial u}{\partial y}\,\Delta y + \eta_1\,\Delta x + \eta_2\,\Delta y.$$

If we put

$$\Delta x = x(r + \Delta r, s) - x(r, s),$$
$$\Delta y = y(r + \Delta r, s) - y(r, s)$$

in equation (2) and divide both sides of (2) by $\Delta r$, then

$$\frac{\Delta u}{\Delta r} = \frac{\partial u}{\partial x}\frac{\Delta x}{\Delta r} + \frac{\partial u}{\partial y}\frac{\Delta y}{\Delta r} + \eta_1\frac{\Delta x}{\Delta r} + \eta_2\frac{\Delta y}{\Delta r}.$$

Now, as $\Delta r \to 0$, $\Delta x$ and $\Delta y \to 0$; hence, $\eta_1$ and $\eta_2 \to 0$, and the difference quotients approach the partial derivatives, whence

$$\frac{\partial u}{\partial r} = \frac{\partial u}{\partial x}\frac{\partial x}{\partial r} + \frac{\partial u}{\partial y}\frac{\partial y}{\partial r}.$$

The other partial derivative is obtained analogously.

The chain rule is not limited to functions of two variables. One could have $u = f(x, y, z, \ldots)$ and then $x, y, z, \ldots$ functions of one or more variables. Of course, equation (1) then becomes altered in ways that should be obvious from the Fundamental Lemma.

*Example 1*   If $u = y + \log xy$ and $x = \sin(r^2 + s)$, $y = \sin(r + s^2)$, find $\partial u/\partial r$ in terms of $r$ and $s$.

$$\frac{\partial u}{\partial r} = \frac{\partial u}{\partial x}\frac{\partial x}{\partial r} + \frac{\partial u}{\partial y}\frac{\partial y}{\partial r} = \left(\frac{1}{x}\right)(2r\cos(r^2 + s)) + \left(1 + \frac{1}{y}\right)\cos(r + s^2)$$

$$= \frac{2r\cos(r^2 + s)}{\sin(r^2 + s)} + \cos(r + s^2) + \frac{\cos(r + s^2)}{\sin(r + s^2)}.$$

*Example 2*   If $z = F(t)$ and $t = x^2 + e^{xy}$, then $z$ is a function of $x$ and $y$: $z = F(x^2 + e^{xy})$. Find $\partial z/\partial x$ and $\partial z/\partial y$.

From the chain rule,

$$\frac{\partial z}{\partial x} = \frac{dF}{dt}\frac{\partial t}{\partial x} = (2x + ye^{xy})F'(t)$$

$$\frac{\partial z}{\partial y} = \frac{dF}{dt}\frac{\partial t}{\partial y} = xe^{xy}F'(t).$$

**Example 3**  If we change to polar coordinates, $x = r\cos\theta$, $y = r\sin\theta$, then $u = u(x, y)$ becomes a function of $r$ and $\theta$, and

$$\frac{\partial u}{\partial r} = \frac{\partial u}{\partial x}\cos\theta + \frac{\partial u}{\partial y}\sin\theta,$$

$$\frac{\partial u}{\partial \theta} = -\frac{\partial u}{\partial x}r\sin\theta + \frac{\partial u}{\partial y}r\cos\theta.$$

An important special case of the chain rule occurs when $x$ and $y$ are functions of a single variable $t$. In applications $t$ is frequently a measure of time. Then $u$ is a function of $t$, and the partial derivative becomes an ordinary derivative.

$$(3) \qquad \frac{du}{dt} = \frac{\partial u}{\partial x}\frac{dx}{dt} + \frac{\partial u}{\partial y}\frac{dy}{dt}.$$

An even more special case is that for which $x = t$. Then $y = y(x)$ and equation (3) becomes

$$\frac{du}{dx} = \frac{\partial u}{\partial x} + \frac{\partial u}{\partial y}\frac{dy}{dx}.$$

Observe that $du/dx$ and $\partial u/\partial x$ are different.

**Example 4**  The sides $a$ and $b$ and included angle $\theta$ of a triangle are changing at the rates

$$\dot{a} = 0.1\ \text{cm/sec}, \quad \dot{b} = -0.05\ \text{cm/sec},$$

$$\dot{\theta} = 1\ \text{deg/sec} = \pi/180\ \text{rad/sec}.$$

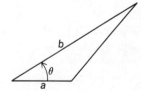

How fast is the area changing at the instant when $\theta = \pi/3$, $a = 5$, $b = 20$?

$$\text{Area} = A = \tfrac{1}{2}ab\sin\theta;$$

$$\frac{dA}{dt} = \frac{1}{2}b\sin\theta\frac{da}{dt} + \frac{1}{2}a\sin\theta\frac{db}{dt} + \frac{1}{2}ab\cos\theta\frac{d\theta}{dt}.$$

Whence, at the instant of concern

$$\frac{dA}{dt} = \frac{1}{2}(20)\frac{\sqrt{3}}{2}(0.1) - \frac{1}{2}(5)\frac{\sqrt{3}}{2}(0.05) + \frac{1}{2}(5)(20)\frac{1}{2}\frac{\pi}{180}$$

$$= 1.19\ \text{cm}^2/\text{sec}.$$

## Problems

1. If $w = u^2 - v^2$ and $u = s + t, v = s - t$, compute $\partial w/\partial s$ and $\partial w/\partial t$ in two ways.
$$[\partial w/\partial s = 2(u - v) = 4t, \; \partial w/\partial t = 2(u + v) = 4s]$$

2. If $w = \log x + e^y + xy$ and $x = \cos t, y = \log \sin t$, find $dw/dt$.
$$[\dot{w} = -\tan t + \cos t + \csc t - \sin t - \sin t \log \sin t]$$

3. Find $du/dt$ if $u = \text{Arc} \sin(x + y)$ and $x = 2t, y = t^2$, with $t > 0$.
$$[2/\sqrt{1 - 2t - t^2}]$$

4. Find $du/dx$ if $u = e^x(y + z)$ and $y = \sin x, z = \cos x$.  $\qquad [2e^x \cos x]$

5. If $w = r^2 + s^2 + rs$ and $r = \log t, s = e^t$, find $\dot{w}$.  $\quad [(2r + s)t^{-1} + (2s + r)e^t]$

6. The height of a right circular cone is 10 cm and is increasing at the rate of $\frac{1}{2}$ cm/sec. The radius of the base is 6 cm and is decreasing at the rate of $\frac{1}{4}$ cm/sec. How fast is the volume changing?  $\qquad$ [decreasing at $4\pi$ cm³/sec]

7. The sides of a rectangle at a certain instant are 8 and 12 in. If the shorter side is decreasing at the rate of $a$ in/min, and the longer side is increasing at the same rate, how fast is the area changing?  $\qquad$ [decreasing at $4a$ sq in/min]

8. For a perfect gas the equation of state is $pv = RT$, where $R$ is a constant and $p$, $v$, and $T$ are pressure, volume, and absolute temperature. At a given instant $p = 20$ lb/in², $v = 10$ cu ft, and $T = 300°$. If the gas is being compressed at the rate $\dot{v} = 0.5$ cu ft/min, and the pressure is rising at the rate of 2 lb per in²/min, what is the rate of change of the temperature?

$\qquad$ [increasing 15°/min]

9. The altitude of a right circular cylinder is increasing at a certain constant rate. The radius of the base is decreasing at the same rate. What are the proportions of the cylinder when the volume is not changing?  $\qquad$ [radius = 2 × altitude]

10. If $w = f(x, y)$ and $x = e^u \sin v, y = e^u \cos v$, compute $w_u^2 + w_v^2$.
$$[e^{2u}(w_x^2 + w_y^2)]$$

11. If $u = f(x - at)$, compute $au_x + u_t$.  $\qquad\qquad\qquad\qquad$ [0]

12. If $z = f(x + y)$, compute $z_x - z_y$.

*13. If $z = F\left(\dfrac{y + \log(z - x)}{z}\right)$,

show that $\qquad (z - x)\dfrac{\partial z}{\partial x} + \dfrac{\partial z}{\partial y} = 0.$

14. Draw the part in the first octant of the curve of intersection of $z = \sqrt{a^2 - x^2 - y^2}$ and $y = x^2$. Find $dz/dx$ along this curve.
$$\left[\frac{dz}{dx} = \frac{-x(1 + 2x^2)}{\sqrt{a^2 - x^2 - x^4}}\right]$$

15. On the curve of intersection of $y = x^2 + z^2$ and $x^2 + y^2 = 4$, find $dz/dx$ at the point where $x = 1$ and $z > 0$. Sketch the curve. [$-1.5$]

16. If $f(x, y) = x^n \varphi(y/x)$, show that $xf_x + yf_y = nf$.

17. A function $f$ of two variables is said to be *homogeneous of degree n* if

   (1) $$f(tx, ty) = t^n f(x, y)$$

   identically in $t$ for all $x, y$. Derive *Euler's formula* for homogeneous functions:

   $$xf_x + yf_y = nf$$

   by differentiating (1) with respect to $t$ by the chain rule and setting $t = 1$.

18. Test Arc tan $y/x$ for homogeneity according to the definition in Problem 17. Then verify Euler's formula.

19. Test $r = \sqrt{x^2 + y^2 + z^2}$ for homogeneity. (Here one restricts $t$ to be positive.) Then verify Euler's formula.

20. Compute $xf_x + yf_y + zf_z$ if

   $$f(x, y, z) = \frac{x + y + z}{\sqrt{x^2 + y^2 + z^2}}.$$ [$0$]

21. If $u = f(x, y)$ and $x = r \cos \theta$, $y = r \sin \theta$ compute $u_r^2 + \dfrac{1}{r^2} u_\theta^2$. [$u_x^2 + u_y^2$]

22. Now do Problem 21 the other way round. $u = F(r, \theta)$ and $r = \sqrt{x^2 + y^2}$, $\theta = $ Arc tan $(y/x)$. Compute $u_x^2 + u_y^2$ in terms of $u_r$ and $u_\theta$.

*23. If $u = f(x, y, z)$, express $u_x^2 + u_y^2 + u_z^2$ in terms of derivatives with respect to spherical coordinates $x = r \cos \theta \sin \varphi$, $y = r \sin \varphi \sin \theta$, $z = r \cos \varphi$.

   $$\left[ u_r^2 + \frac{1}{r^2 \sin^2 \varphi} u_\theta^2 + \frac{1}{r^2} u_\varphi^2 \right]$$

## 5 Directional Derivative

If $u$ is a function of $x$, $y$, and $z$: $u = u(x, y, z)$, then $u_x$, $u_y$, and $u_z$ are the rates of change of $u$ in the $x$, $y$, and $z$ directions, respectively. In this section we calculate (as an application of the chain rule) the rate of change of $u$ in an arbitrary direction. This rate of change is called a *directional derivative*. The reader may wish to consult Appendix A, page 507, for a resumé of the pertinent analytic geometry.

Suppose a direction in space is specified by the arrow (directed line segment) from $P_0 = (x_0, y_0, z_0)$ to $P = (x, y, z)$. The direction of the arrow is determined by its *direction angles* $\alpha, \beta, \gamma$. Observe that $\alpha, \beta, \gamma$ are always non-negative and cannot exceed $\pi: 0 \leq \alpha, \beta, \gamma \leq \pi$. If $s$ is the length of the arrow, then we have, because the triangles shown in the figure are right triangles,

$$\cos \alpha = (x - x_0)/s,$$

(1) $$\cos \beta = (y - y_0)/s,$$

$$\cos \gamma = (z - z_0)/s.$$

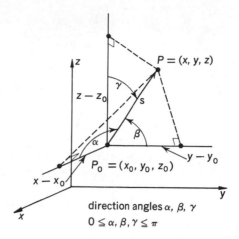

direction angles $\alpha, \beta, \gamma$
$0 \leq \alpha, \beta, \gamma \leq \pi$

The numbers $\cos \alpha$, $\cos \beta$, and $\cos \gamma$ are the *direction cosines* of the directed line segment. Observe that

$$\cos^2 \alpha + \cos^2 \beta + \cos^2 \gamma = 1.$$

Any arrow in the same direction as the given arrow will have the same direction angles, and cosines. Thus, the direction cosines determine the direction.

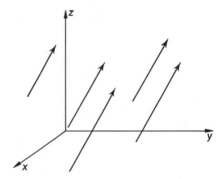

**Example 1** Find the direction cosines of the arrow (a) from $(0,0,0)$ to $(1, -2, \frac{3}{2})$ and (b) from $(-2, \frac{1}{2}, -1)$ to $(0, -\frac{7}{2}, 2)$.

(a) The distance between the points is

$$s = \sqrt{(1-0)^2 + (-2-0)^2 + (\tfrac{3}{2}-0)^2} = \sqrt{29}/2.$$

Thus,

$$\cos \alpha = \frac{1-0}{\sqrt{29}/2} = \frac{2}{\sqrt{29}},$$

$$\cos \beta = \frac{-2-0}{\sqrt{29}/2} = \frac{-4}{\sqrt{29}},$$

$$\cos \gamma = \frac{\tfrac{3}{2}-0}{\sqrt{29}/2} = \frac{3}{\sqrt{29}}.$$

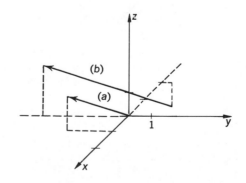

(b) The distance between the points is

$$s = \sqrt{(0+2)^2 + (-\tfrac{7}{2} - \tfrac{1}{2})^2 + (2+1)^2} = \sqrt{29}.$$

Thus,    $\cos \alpha = 2/\sqrt{29}, \quad \cos \beta = -4/\sqrt{29}, \quad \cos \gamma = 3/\sqrt{29}.$

The directions in (a) and (b) are the same.

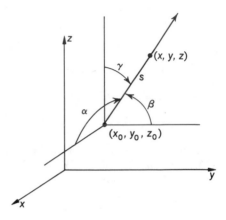

Suppose that we wish to find the rate of change of $u = u(x, y, z)$ at the point $(x_0, y_0, z_0)$ in the direction with direction angles $\alpha, \beta, \gamma$. That is, we want the derivative of $u$ with respect to distance in the given direction. If $s$ is the distance from $(x_0, y_0, z_0)$ to a point $(x, y, z)$ on the given ray then, from equations (1),

$$(2) \quad \begin{cases} x = x_0 + s \cos \alpha, \\ y = y_0 + s \cos \beta, \\ z = z_0 + s \cos \gamma. \end{cases}$$

These equations are parametric equations of the line through $(x_0, y_0, z_0)$ and inclined as shown. (The other ray on the line is obtained for $s < 0$.)

Now, from equations (2), $du/ds$ is easily computed from the chain rule:

$$\frac{du}{ds} = \frac{\partial u}{\partial x}\frac{dx}{ds} + \frac{\partial u}{\partial y}\frac{dy}{ds} + \frac{\partial u}{\partial z}\frac{dz}{ds}$$

$$(3) \qquad = \frac{\partial u}{\partial x}\cos \alpha + \frac{\partial u}{\partial y}\cos \beta + \frac{\partial u}{\partial z}\cos \gamma,$$

where the partial derivatives are evaluated at $(x_0, y_0, z_0)$. The number $du/ds$ is called the *directional derivative* of $u$ in the direction $\alpha, \beta, \gamma$.

*Remark.* The directional derivative in the direction of the positive $x$-axis is $du/ds = \partial u/\partial x$ because in that direction, $\alpha = 0, \beta = \pi/2$, and $\gamma = \pi/2$.

***Example 2***    Find the directional derivative of $u = \log \sqrt{x^2 + y^2 + z^2}$ at the point $(1, 0, 1)$ in the direction toward the point $(2, 4, 3)$.

From equation (3) we have

$$\frac{du}{ds} = \frac{x}{x^2 + y^2 + z^2}\cos \alpha + \frac{y}{x^2 + y^2 + z^2}\cos \beta + \frac{z}{x^2 + y^2 + z^2}\cos \gamma,$$

where $(x, y, z) = (1, 0, 1)$, and

$$\cos \alpha = 1/\sqrt{21}, \quad \cos \beta = 4/\sqrt{21}, \quad \cos \gamma = 2/\sqrt{21}.$$

Thus, $\qquad \dfrac{du}{ds} = \left(\dfrac{1}{2}\right)\left(\dfrac{1}{\sqrt{21}}\right) + 0 + \dfrac{1}{2}\left(\dfrac{2}{\sqrt{21}}\right) = \dfrac{3}{2\sqrt{21}}\,.$

In case $u$ is a function of but two variables, say $u = u(x, y)$, one can consider directions in the $xy$-plane. Then $\gamma = \pi/2$ and $\cos \gamma = 0$ if we consider the $xy$-plane in coordinate 3-space. In this case $\cos^2 \alpha + \cos^2 \beta = 1$. Equation (3) then gives us for the directional derivative in the plane:

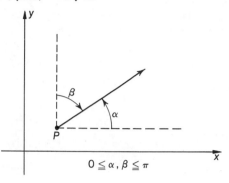

$0 \leq \alpha, \beta \leq \pi$

(4) $\qquad \dfrac{du}{ds} = \dfrac{\partial u}{\partial x} \cos \alpha + \dfrac{\partial u}{\partial y} \cos \beta.$

**Example 3**   Find the directional derivative of $u = 2x^2 + 3y^2$ in a direction tangent to any level curve at the point of tangency.

Level curves are ellipses $2x^2 + 3y^2 = u_0 = $ constant. At any point $(x_0, y_0)$ on a level curve the tangent has slope

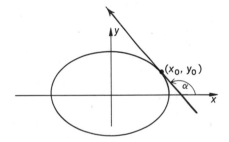

$$dy/dx = -2x_0/3y_0 = \tan \alpha,$$

where $\alpha$ is the inclination of the tangent— and is the direction angle of the upward direction on the tangent line, shown in red in the figure.

To compute the direction cosines we plot the point $(-3y_0, 2x_0)$ obtained from the numerator and denominator of $\tan \alpha$ and read from the figure $\cos \alpha$ and $\cos \beta$. The oppositely directed ray will have cosines opposite in sign.

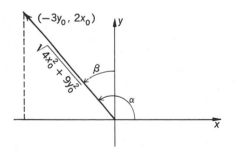

$$\cos \alpha = \dfrac{\mp 3y_0}{\sqrt{4x_0^2 + 9y_0^2}}\,,$$

$$\cos \beta = \dfrac{\pm 2x_0}{\sqrt{4x_0^2 + 9y_0^2}}\,.$$

For either direction the directional derivative at $(x_0, y_0)$ is

$$\dfrac{du}{ds} = 4x_0 \cos \alpha + 6y_0 \cos \beta$$

$$= 4x_0\left(\dfrac{\mp 3y_0}{\sqrt{4x_0^2 + 9y_0^2}}\right) + 6y_0\left(\dfrac{\pm 2x_0}{\sqrt{4x_0^2 + 9y_0^2}}\right) = 0.$$

The rate of change of $u$ in a direction tangent to a level curve is zero.

*Problems*

In Problems 1 to 6 find the directional derivative at the given point in the direction given by the direction cosines.

1.  $u = xy + yz + zx$;  $(0, 0, 0)$, direction: $(-\frac{1}{3}, \frac{2}{3}, \frac{2}{3})$  $[du/ds = 0]$

2.  $u = e^x \cos y + e^x \cos z$;  $(0, \pi/2, -\pi/2)$, direction: $(0, -1/\sqrt{2}, 1/\sqrt{2})$  $[dy/ds = \sqrt{2}]$

3.  $u = x^2 + y^2$;  $(1, 2, 0)$, direction: $(-1/\sqrt{5}, -2/\sqrt{5}, 0)$  $[du/ds = -2\sqrt{5}]$

4.  $u = x^2 + y^2$;  $(1, 2, 0)$, direction: $(-2/\sqrt{5}, 1/\sqrt{5}, 0)$  $[du/ds = 0]$

5.  $u = x + y + z$;  $(0, 1, 0)$, direction: $(1/2, -1/2, 1/\sqrt{2})$  $[du/ds = 1/\sqrt{2}]$

6.  $u = ax + by + cz$;  $(0, 0, 0)$, direction:
$$\left( \frac{-b}{\sqrt{a^2 + b^2}}, \frac{a}{\sqrt{a^2 + b^2}}, 0 \right)$$
$[du/ds = 0]$

In Problems 7 to 12 find the directional derivative of *u* at the first point in the direction of the second point.

7.  $u = x^2 + y^2 + z^2$ at $(1, 2, 1)$ toward $(0, 4, 0)$  $[4/\sqrt{6}]$

8.  $u = x^2 + 2y^2 + 7z^2$ at $(1, 2, 1)$ toward $(0, 4, 0)$  $[0]$

9.  $u = \log \sqrt{x^2 + y^2}$ at $(1, 0)$ toward $(2, 4)$  $[1/\sqrt{17}]$

10.  $u = \text{Arc tan} \frac{y}{x}$ at $(1, -1)$ toward $(3, 0)$  $\left[ \frac{3}{2\sqrt{5}} \right]$

11.  $u = \frac{z}{r}, r = \sqrt{x^2 + y^2 + z^2}$ at $(0, 1, 2)$ toward $(1, 2, 0)$  $\left[ \frac{-2}{75}\sqrt{30} \right]$

12.  $u = ax + by + cz$ at $(1, 1, 1)$ toward $(0, 0, 0)$  $\left[ \frac{-a - b - c}{\sqrt{3}} \right]$

13.  Find the directional derivative of $u = 1/\sqrt{x^2 + y^2 + z^2}$ at $(x, y, z)$ in (a) the direction from $(0, 0, 0)$ to $(x, y, z)$ and (b) the direction from $(x, y, z)$ to $(0, 0, 0)$.

14.  Find the directional derivative of $u = \log 1/\sqrt{x^2 + y^2}$ at $(x, y)$ in (a) the direction from $(0, 0)$ to $(x, y)$, (b) the direction from $(x, y)$ to $(0, 0)$. $\left[ \text{(a)} - \frac{1}{\sqrt{x^2 + y^2}} \right]$

15.  Find the directional derivative of $u = 2x^2 + 3y^2$ in a direction normal to any level curve.  $[\pm 2\sqrt{4x_0^2 + 9y_0^2}]$

16. Find the directional derivative of $u = x^2 + y^2$ in a direction tangent to a level curve. [0]

17. Find the directional derivative of $u = 2x^2 - y^2$ in a direction tangent to a level curve.

## 6   *The Gradient*

The directional derivative of $u$, in the direction with direction angles $\alpha$, $\beta$, $\gamma$, is

$$\frac{du}{ds} = u_x \cos \alpha + u_y \cos \beta + u_z \cos \gamma.$$

One might wonder about the possible geometric significance of the triple of numbers $u_x, u_y, u_z$. We shall see in this section that this triple of numbers determines a vector, called the *gradient* of $u$, whose length is the greatest space rate of change of $u$. In other words the direction of this vector is that direction for which $du/ds$ is a maximum, and the length of this vector is equal to this maximum.

**DEFINITION**   The triple of numbers $(u_x, u_y, u_z)$, at a point $(x, y, z)$, is called the *gradient* of $u$ and is denoted by grad $u$:

$$\text{grad } u = (u_x, u_y, u_z).$$

If not all three partial derivatives are zero, grad $u$ determines an arrow, or vector,* as follows: The length, or magnitude of the vector is

$$|\text{ grad } u | = \sqrt{u_x^2 + u_y^2 + u_z^2}.$$

The direction of the vector is given by the direction cosines

$$\frac{u_x}{|\text{ grad } u |}, \quad \frac{u_y}{|\text{ grad } u |}, \quad \frac{u_z}{|\text{ grad } u |}.$$

Now suppose $\alpha$, $\beta$, $\gamma$ is any set of direction angles. The triple of numbers $\cos \alpha$, $\cos \beta$, $\cos \gamma$ determine an arrow of unit length in this direction. The two vectors grad $u$, and $(\cos \alpha, \cos \beta, \cos \gamma)$ determine a triangle in space with included angle $\theta$. From the law of cosines we have

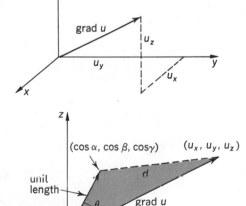

(1)   $d^2 = 1^2 + |\text{ grad } u |^2 - 2(1) |\text{ grad } u | \cos \theta.$

* For the reader familiar with the usual vector notation, grad $u = u_x\mathbf{i} + u_y\mathbf{j} + u_z\mathbf{k}$.

But we also have, from the coordinates,

$$d^2 = (u_x - \cos \alpha)^2 + (u_y - \cos \beta)^2 + (u_z - \cos \gamma)^2.$$

Substitution of this last for $d^2$ in equation (1) gives, after some simplification,

(2) $$u_x \cos \alpha + u_y \cos \beta + u_z \cos \gamma = |\operatorname{grad} u| \cos \theta.$$

But the left member of (2) is the directional derivative $du/ds$ in the direction $\alpha, \beta, \gamma^*$;

(3) $$\frac{du}{ds} = |\operatorname{grad} u| \cos \theta.$$

Equation (3) shows that the directional derivative is the projection of gradient $u$ on the given direction.

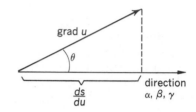

From equation (3) we see that $du/ds$ is a maximum when $\theta = 0$ ($\cos \theta = 1$). This proves the following theorem.

**THEOREM**    *The magnitude of the gradient of u is the maximum of the directional derivative of u. The gradient points in the direction of this maximum space rate of change.*

If $\theta = \pi/2$ in equation (3), then $du/ds = 0$. Thus, in directions perpendicular to the gradient the rate of change of $u$ is zero. Conversely, if $du/ds = 0$ in some direction (and $|\operatorname{grad} u| \neq 0$), then $\cos \theta = 0$, and the direction is perpendicular to the gradient.

*Example*    Find the gradient of $u = k/r$, where $r = \sqrt{x^2 + y^2 + z^2}$, and $k =$ constant. Find the direction of maximum $du/ds$.

The gradient is

$$\operatorname{grad} u: \quad \frac{\partial u}{\partial x} = \frac{kx}{r^3}, \quad \frac{\partial u}{\partial y} = \frac{ky}{r^3}, \quad \frac{\partial u}{\partial z} = \frac{kz}{r^3}.$$

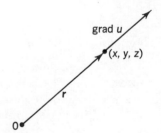

The magnitude of the gradient is

$$|\operatorname{grad} u| = \sqrt{\frac{k^2 x^2 + k^2 y^2 + k^2 z^2}{r^6}} = \frac{|k|}{r^2}.$$

Because $u_x$, $u_y$, and $u_z$ are proportional to $x$, $y$, and $z$, respectively, the direction of grad $u$ is either the same as that of the arrow from $(0, 0, 0)$ to $(x, y, z)$ or the opposite direction, depending on the sign of $k$.

---

\* For the reader familiar with vector notation,

$$\frac{du}{ds} = (u_x i + u_y j + u_z k) \cdot (\cos \alpha i + \cos \beta j + \cos \gamma k).$$

The function $u$, for a suitable $k$, is the gravitational potential at $(x, y, z)$ due to a mass at the origin. Then grad $u$ is the gravitational attraction exerted on a unit mass at $(x, y, z)$.

## Problems

In Problems 1 to 4 find the magnitude of the greatest space rate of change of $u$ at any point.

1.   $u = \log \sqrt{x^2 + y^2}$                         $[1/\sqrt{x^2 + y^2}]$

2.   $u = \text{Arc tan } y/x$                           $[1/\sqrt{x^2 + y^2}]$

3.   $u = \sqrt{x^2 + y^2 + z^2}$                        $[1]$

4.   $u = ax + by + cz$                                  $[\sqrt{a^2 + b^2 + c^2}]$

In Problems 5 to 10, draw the surface $u = $ constant through the given point. Then find and draw the gradient vector at that point.

5.   $u = x + y$;   $(1, 1, 0)$                 8.   $u = x^2 + y^2$;   $(x_0, y_0, 0)$

6.   $u = ax + by + cz$;   $(1, 1, 1)$          9.   $u = x^2 + 4y^2$;   $(-1, \sqrt{3}/2, 0)$

7.   $u = x^2 + y^2 + z$;   $(1, 1, 1)$         10.   $u = \text{Arc tan } (y/x)$;   $(x_0, y_0, 0)$

11.   What are the level curves of $u = ay/(x^2 + y^2)$? Show that grad $u$ is perpendicular to them.

12.   Show that the circle $x^2 + y^2 = 1$ is a level curve of $u = ay - ay/(x^2 + y^2)$. Show that on this circle the gradient is perpendicular to the circle.

13.   Obtain equation (2) from equation (1) on pages 411–412.

14.   A plate with vertices at $(0, 0)$, $(1, 0)$, $(1, 1)$, and $(0, 1)$ is heated, so that the temperature $T$ at a point $(x, y)$ is given by $T = \sin (\pi x/2) \sin (\pi y/2)$. A remarkably intelligent bug is at $(\frac{1}{2}, \frac{1}{3})$. In what direction should it move to cool off as rapidly as possible?

15.   The gravitational potential at $(x, y, z)$ due to a long wire along the $z$-axis is given by $u = k \log r$, where $r = \sqrt{x^2 + y^2}$.

   Show that the gravitational attraction is inversely proportional to $r$.

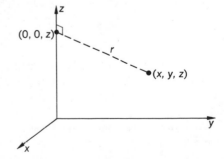

## 7 Curves in Space

A convenient way of representing curves in space is by parametric equations (see Chapter 6). The set of points $(x, y, z)$ such that

(1) $$x = \varphi(t), \quad y = \psi(t), \quad z = \chi(t); \qquad a \leq t \leq b,$$

where $\varphi$, $\psi$, $\chi$ are differentiable functions, constitutes a *smooth curve** in space. Equations (1) are parametric equations with parameter $t$.

The parameter $t$ may have physical or geometric significance. There are two special cases that are important. In the first of these $t$ is the time measured from some instant. Then the derivatives of the coordinate functions in (1) give the components of the velocity, just as was the case in Chapter 6 in the plane. For any parameter the derivatives are the components of a vector tangent to the curve.

If $P = (\varphi(t), \psi(t), \chi(t))$ is a point on the curve (1) and $P' = (\varphi(t + \Delta t),$ $\psi(t + \Delta t), \chi(t + \Delta t))$ is a nearby point, then

$$\Delta x = \varphi(t + \Delta t) - \varphi(t)$$

$$\Delta y = \psi(t + \Delta t) - \psi(t)$$

$$\Delta z = \chi(t + \Delta t) - \chi(t)$$

are components of the arrow (vector) from $P$ to $P'$, and so

(2) $$\frac{\Delta x}{\Delta t}, \quad \frac{\Delta y}{\Delta t}, \quad \frac{\Delta z}{\Delta t}$$

are components of a vector in the same direction (if $\Delta t$ is positive).

Now let $\Delta t \to 0$. The line through $P$ and $P'$ will approach a limiting position that is the *tangent line* to the curve. The limits of the quotients (2) are

(3) $$\dot{x} = \frac{dx}{dt}, \quad \dot{y} = \frac{dy}{dt}, \quad \dot{z} = \frac{dz}{dt}$$

and will (if they do not all vanish) be components of a vector pointing along the tangent line. Direction cosines of this vector are

(4) $$\cos \alpha = \frac{\dot{x}}{\sqrt{\dot{x}^2 + \dot{y}^2 + \dot{z}^2}}, \quad \cos \beta = \frac{\dot{y}}{\sqrt{\dot{x}^2 + \dot{y}^2 + \dot{z}^2}}, \quad \cos \gamma = \frac{\dot{z}}{\sqrt{\dot{x}^2 + \dot{y}^2 + \dot{z}^2}}.$$

If the parameter $t$ is a measure of time the derivatives, (3) are components of velocity $v_x = \dot{x}$, $v_y = \dot{y}$, and $v_z = \dot{z}$. The speed, which is the magnitude of the velocity, is $v = \sqrt{v_x^2 + v_y^2 + v_z^2}$.

---

* To be precise we should require that not all three derivatives $\varphi'$, $\psi'$, $\chi'$ vanish at the same point. See Problem 13.

*Example 1*    Find the direction cosines of the tangent line to the curve $x = t \cos t$, $y = t \sin t$, $z = \frac{1}{2}t$ at the point where $t = 1$.

The curve spirals out, and up, from the origin and lies on the cone $z^2 = \frac{1}{4}(x^2 + y^2)$.

$$\frac{dx}{dt} = \cos t - t \sin t,$$

$$\frac{dy}{dt} = \sin t + t \cos t,$$

$$\frac{dz}{dt} = \frac{1}{2}.$$

At $t = 1$ we have

$$x = 0.540, \quad y = 0.841, \quad z = 0.500,$$

and $$\dot{x} = -0.301, \quad \dot{y} = 1.382, \quad \dot{z} = 0.500.$$

The magnitude of this vector is $\sqrt{\dot{x}^2 + \dot{y}^2 + \dot{z}^2} = 1.50$, whence the direction cosines are

$$\cos \alpha = -0.202, \quad \cos \beta = 0.921, \quad \cos \gamma = 0.333.$$

The second special parameter is $s$, the arc length. The simplest example is provided by parametric equations for the line through $(x_0, y_0, z_0)$ with direction angles $\alpha, \beta, \gamma$ [see Section 5, equation (2)]:

(5)    $$x = x_0 + s \cos \alpha, \quad y = y_0 + s \cos \beta, \quad z = z_0 + s \cos \gamma.$$

Observe that for the line (5),

$$\frac{dx}{ds} = \cos \alpha, \quad \frac{dy}{ds} = \cos \beta, \quad \frac{dz}{ds} = \cos \gamma.$$

We shall see that these equations are quite generally true. Toward this end we define the length of a curve between two points to be the limit of lengths of inscribed polygons. Then the derivation of a formula for this length is completely analogous to what was done in Chapter 6 for plane curves. We content ourselves with giving the result:

*The length of the curve (1) between $P_0 = (\varphi(t_0), \psi(t_0), \chi(t_0))$ and $P_1 = (\varphi(t_1), \psi(t_1), \chi(t_1))$ is*

(6)    $$s = \int_{t_0}^{t_1} \sqrt{\left(\frac{dx}{dt}\right)^2 + \left(\frac{dy}{dt}\right)^2 + \left(\frac{dz}{dt}\right)^2}\, dt.$$

As a corollary to equation (6) we have the expected formula for the differential of arc:

$$ds^2 = dx^2 + dy^2 + dz^2,$$

so that

$$\left(\frac{dx}{ds}\right)^2 + \left(\frac{dy}{ds}\right)^2 + \left(\frac{dz}{ds}\right)^2 = 1,$$

from which we deduce from equations (4) that

(7) $$\frac{dx}{ds} = \cos\alpha, \quad \frac{dy}{ds} = \cos\beta, \quad \frac{dz}{ds} = \cos\gamma,$$

as was predicted.

*Example 2*   Find the length of the curve of Example 1 between $t = 0$ and $t = 1$.

$$\left(\frac{ds}{dt}\right)^2 = (\cos t - t \sin t)^2 + (\sin t + t \cos t)^2 + (\tfrac{1}{2})^2 = \tfrac{5}{4} + t^2.$$

So $$s = \int_0^1 \sqrt{\tfrac{5}{4} + t^2}\, dt = \frac{t}{2}\sqrt{\tfrac{5}{4} + t^2} + \tfrac{5}{8}\log\left| t + \sqrt{\tfrac{5}{4} + t^2}\right|\ \Big|_0^1$$

$$= \tfrac{3}{4} + \tfrac{5}{8}\log\sqrt{5} \approx 1.253.$$

*Example 3*   Often a curve is given as the intersection of two surfaces. In such cases it may be advantageous to use one of the coordinates as the parameter. Thus, the intersection of the parabolic cylinder $x^2 + z = 4$, and the plane $y = z$ is given by the equations

$$x = x, \quad y = 4 - x^2, \quad z = 4 - x^2.$$

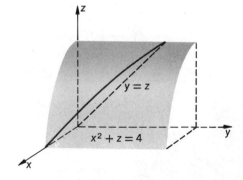

The figure shows the part in the first octant.

The length of the curve between $x = 0$ and $x = x_1$ is

$$s = \int_0^{x_1} \sqrt{1 + (-2x)^2 + (-2x)^2}\, dx = \int_0^{x_1} \sqrt{1 + 8x^2}\, dx.$$

### Problems

In Problems 1 to 4 find direction cosines of the tangent line at the point indicated. Draw the tangent line.

1.   $x = t,\ y = t^2,\ z = t^3;\quad t = 1$     $[\pm 1/\sqrt{14},\ \pm 2/\sqrt{14},\ \pm 3/\sqrt{14}]$

2.   $x = \cos\theta,\ y = \sin\theta,\ z = \theta;\quad \theta = \pi/4$     $[\mp\tfrac{1}{2},\ \pm\tfrac{1}{2},\ \pm 1/\sqrt{2}]$

3.  $x = t, y = e^t, z = e^{-t};\quad t = 0$ $\qquad\qquad$ $[\pm 1/\sqrt{3}, \pm 1/\sqrt{3}, \mp 1/\sqrt{3}]$

4.  $x = t^2, y = 1 - t^2, z = t^3;\quad t = 1$ $\qquad$ $[\pm 2/\sqrt{17}, \mp 2/\sqrt{17}, \pm 3/\sqrt{17}]$

5.  Set up an integral for the arc length in Problem 1 between $t = 0$ and $t = 1$.

6.  Set up, for Problem 3, an integral for the arc length between 0 and $t$.

7.  Find the length of the helix of Problem 2 between $\theta = 0$ and $\theta = 2\pi$. Show that the curve lies on a cylinder. $\hspace{4cm}$ $[2\sqrt{2}\pi]$

8.  Show that the curve of Problem 4 lies in a plane. Find the length of the curve between $t = 0$ and $t = 1$. $\hspace{3cm}$ $[(17\sqrt{17} - 16\sqrt{2})/27]$

9.  Find the direction cosines of the tangent line to the curve of intersection of $x^2 + y^2 + z^2 = 4$ and $x^2 + y^2 - 2y = 0$ at the point $(1, 1, \sqrt{2})$. Draw a figure.
$$[0, \pm \sqrt{\tfrac{2}{3}}, \mp 1/\sqrt{3}]$$

10. Show that the curve $x = t$, $y = t^{3/2}$, $z = t^2$ is part of the intersection of the cylinders $x^2 = z$ and $x^3 = y^2$. Set up an integral for the length from $t = 0$ to $t = t_1$.

11. Show that the length of the curve $x = t + (t^3/3)$, $y = t^2$, $z = t - (t^3/3)$ from 0 to $t$ is equal to $\sqrt{2}x$.

12. Find direction cosines of the tangent to the curve of intersection of $y = x^2 - x$ and $z = 2x - 2$ at the point $(1, 0, 0)$. Draw a figure.
$$[\pm 1/\sqrt{6}, \pm 1/\sqrt{6}, \pm 2/\sqrt{6}]$$

13. Does the curve $x = t^3$, $y = |t|^3$, $z = 0$ have a tangent line at $(0, 0, 0)$?
$$[\text{No, there is a corner and } \dot{x}(0) = \dot{y}(0) = \dot{z}(0) = 0]$$

## 8   Tangent Planes to Surfaces

Suppose a surface is given by

$$f(x, y, z) = 0,$$

where $f$ has continuous partial derivatives. At any point $P_0$ on the surface there should be a *tangent plane*—and it seems reasonable that the tangent plane should have the property that it contains the tangent line to any curve through $P_0$ on the surface.

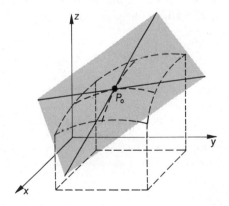

Suppose that a curve $C$ is on the surface and is given parametrically in terms of the arc length from $P_0 = (x_0, y_0, z_0)$:

$$x = x(s), \quad y = y(s), \quad z = z(s).$$

Then for all $s$

$$f[x(s), y(s), z(s)] = 0.$$

Whence, by the chain rule

$$(1) \qquad \frac{df}{ds} = \frac{\partial f}{\partial x}\frac{dx}{ds} + \frac{\partial f}{\partial y}\frac{dy}{ds} + \frac{\partial f}{\partial z}\frac{dz}{ds} = 0.$$

But at the point $P_0$

$$\frac{dx}{ds} = \cos \alpha, \quad \frac{dy}{ds} = \cos \beta, \quad \frac{dz}{ds} = \cos \gamma$$

are direction cosines of a vector along the tangent line. From Section 6

$$(2) \qquad \frac{df}{ds} = |\operatorname{grad} f| \cos \theta,$$

where $\theta$ is the angle between the gradient and the vector. Therefore, from equations (1) and (2) we conclude that (assuming that grad $f$ is not zero at $P_0$) the gradient of $f$ at $P_0$,

$$(3) \qquad \left.\frac{\partial f}{\partial x}\right|_{P_0}, \quad \left.\frac{\partial f}{\partial y}\right|_{P_0}, \quad \left.\frac{\partial f}{\partial z}\right|_{P_0}$$

is a vector perpendicular to the tangent line to $C$ for *any* curve $C$ through $P_0$ and on the surface. In other words the plane perpendicular to the gradient at $P_0$ contains the tangent lines to all curves $C$ through $P_0$ and on the surface. *We define this plane as the tangent plane at $P_0$.*

It remains now to find an equation of the tangent plane. Suppose that $P = (x, y, z)$ is any point on the tangent plane at a distance $\lambda \neq 0$ from $P_0$. Then, if $\alpha$, $\beta$, $\gamma$ are direction angles of the ray from $P_0$ to $P$

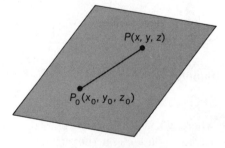

$$(x - x_0)/\lambda = \cos \alpha,$$

$$(y - y_0)/\lambda = \cos \beta,$$

$$(z - z_0)/\lambda = \cos \gamma.$$

Now $\cos \alpha$, $\cos \beta$, $\cos \gamma$ are also direction cosines of a tangent to a curve on the surface. Therefore, from equation (1)

$$(4) \qquad \left.\frac{\partial f}{\partial x}\right|_{P_0}(x - x_0) + \left.\frac{\partial f}{\partial y}\right|_{P_0}(y - y_0) + \left.\frac{\partial f}{\partial z}\right|_{P_0}(z - z_0) = 0.$$

Conversely, if equation (4) is satisfied by $(x, y, z) = P$ then the ray from $P_0$ to $P$ must be perpendicular to the gradient and $P$ must be on the plane. Therefore, equation (4) is an equation of the tangent plane.

*Example 1*    Consider the plane $ax + by + cz = d$, as the graph of

$$f(x, y, z) = ax + by + cz = \text{constant},$$

where the constant is $d$.

The gradient of $f$ is

$$\text{grad } f = (f_x, f_y, f_z) = (a, b, c).$$

The tangent plane through the point $(x_0, y_0, z_0)$ is, therefore, by equation (4),

$$a(x - x_0) + b(y - y_0) + c(z - z_0) = 0,$$

or    $ax + by + cz = ax_0 + by_0 + cz_0 = d.$

The tangent plane to the plane is the plane itself.

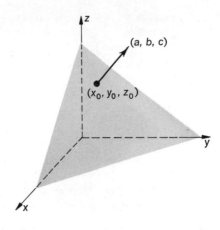

*Example 2*    Find an equation of the plane tangent to the ellipsoid $2x^2 + 3y^2 + 4z^2 = 12$ at the point $P_0 = (\sqrt{3}, 1, \sqrt{3}/2)$.

$$\left.\frac{\partial f}{\partial x}\right|_{P_0} = 4\sqrt{3}, \quad \left.\frac{\partial f}{\partial y}\right|_{P_0} = 6,$$

$$\left.\frac{\partial f}{\partial z}\right|_{P_0} = 4\sqrt{3}.$$

The tangent plane is

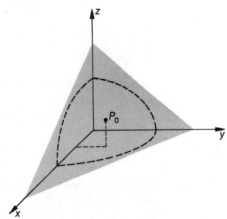

$$4\sqrt{3}(x - \sqrt{3}) + 6(y - 1) + 4\sqrt{3}\left(z - \frac{\sqrt{3}}{2}\right) = 0$$

or    $$2x + \sqrt{3}y + 2z = 4\sqrt{3}.$$

If a surface is given explicitly by $z = F(x, y)$, then we can regard it as the graph of

$$f(x, y, z) = F(x, y) - z = \text{constant},$$

where the constant $= 0$. Then

$$\frac{\partial f}{\partial x} = \frac{\partial F}{\partial x}, \quad \frac{\partial f}{\partial y} = \frac{\partial F}{\partial y}, \quad \frac{\partial f}{\partial z} = -1.$$

Therefore, a vector perpendicular to the surface has components $F_x$, $F_y$, $-1$, and these components at $(x_0, y_0, z_0)$ are used to find the tangent plane.

*Example 3*   Find the tangent plane to the surface $z = x^2 + 2y^2$ at the point $(1, 1, 3)$.

Here we have $F(x, y) = x^2 + 2y^2$, $F_x = 2x$, $F_y = 4y$, and a vector perpendicular to the tangent plane at $(1, 1, 3)$ is $(2, 4, -1)$. The plane is

$$2(x - 1) + 4(y - 1) - (z - 3) = 0$$

or

$$2x + 4y - z = 3.$$

### Problems

1.   Find an equation of the plane tangent to $z = x^2 + y^2$ at the point $(2, 1, 5)$. Sketch the surface and tangent plane.       $[4x + 2y - z = 5]$

2.   Find an equation of the plane tangent to the ellipsoid

$$\frac{x^2}{a^2} + \frac{y^2}{b^2} + \frac{z^2}{c^2} = 1$$

at the point $(x_0, y_0, z_0)$ on it.

$$\left[ \frac{xx_0}{a^2} + \frac{yy_0}{b^2} + \frac{zz_0}{c^2} = 1 \right]$$

3.   Find, using calculus, direction cosines of the line through $(3, 0, -2)$ and perpendicular to the sphere

$$x^2 + y^2 + z^2 - 2x + 4y + 6z + 5 = 0.$$

Then find the center of the sphere and check your answer geometrically.
$$\left[ \pm \tfrac{2}{3}, \pm \tfrac{2}{3}, \pm \tfrac{1}{3} \right]$$

4.   Find an equation of the plane tangent to the surface $x^{1/2} + y^{1/2} = a^{1/2}$ at a point $(x_0, y_0, 0)$ on it.       $[x_0^{-1/2}x + y_0^{-1/2}y = a^{1/2}]$

5.   Show that every plane tangent to the cone $x^2 + y^2 - z^2 = 0$ also passes through the origin.

*6.   If $f$ is a homogeneous function of degree $n$, show that every plane tangent to the surface $f(x, y, z) = 0$ also passes through the origin.

7.   Find an equation of the plane tangent to the cylinder $x^2 + y^2 = a^2$ at a point $(x_0, y_0, z_0)$ on it.       $[x_0 x + y_0 y = a^2]$

8.   Find an equation of the plane perpendicular to the tangent line to the curve $x = t + 1$, $y = t^2 + t$, $z = 2t$, at the point where $t = -1$.
$$[x - y + 2z + 4 = 0]$$

9.   A line has parametric equations $x = x_0 + At$, $y = y_0 + Bt$, $z = z_0 + Ct$. Find an equation of the plane perpendicular to the line at $(x_0, y_0, z_0)$.
$$[A(x - x_0) + B(y - y_0) + C(z - z_0) = 0]$$

10.   Prove that the product of the intercepts on the axes of any tangent plane to $xyz = a^3$ is a constant.       $[\text{constant} = 27a^3]$

11.   Show that the sum of the intercepts on the axes of any tangent plane to $x^{1/2} + y^{1/2} + z^{1/2} = a^{1/2}$ is a constant.                                    [constant $= a$]

12.   Show that the sum of the squares of the intercepts on the axes of any tangent plane to $x^{2/3} + y^{2/3} + z^{2/3} = a^{2/3}$ is a constant.                 [constant $= a^2$]

13.   Two surfaces are *orthogonal* at a point of intersection if the normals to the surfaces are perpendicular at the point. Show that the parabolic cylinders $x^2 = 4a(y + a)$ and $x^2 = 4b(y + b)$, where $a > 0$ and $b < 0$, are orthogonal.

14.   What must be the relation between $r_1, r_2, h, k,$ and $l$ if the spheres $x^2 + y^2 + z^2 = r_1^2$ and $(x - h)^2 + (y - k)^2 + (z - l)^2 = r_2^2$ intersect orthogonally?
                                                                            $[r_1^2 + r_2^2 = h^2 + k^2 + l^2]$

15.   Show that the line through $(1, 1, 1)$ and $(-1, 4, 2)$ is tangent to the surface $xyz = x^2 + 2y - 2$ at the point $(1, 1, 1)$.

Use the following result to find direction cosines of the tangent line to the curve of intersection of the surfaces in Problems 16, 17, 18, at the point indicated. Find another point on the line.

If a curve is given as the intersection of two surfaces, $f(x, y, z) = 0$ and $g(x, y, z) = 0$, then the tangent line at a point $P_0$ on the curve is the intersection of the tangent planes at $P_0$:

$$f_x(x - x_0) + f_y(y - y_0)$$
$$\qquad + f_z(z - z_0) = 0,$$
$$g_x(x - x_0) + g_y(y - y_0)$$
$$\qquad + g_z(z - z_0) = 0,$$

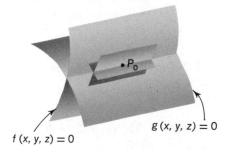

$g(x, y, z) = 0$

$f(x, y, z) = 0$

where the partial derivatives are evaluated at $P_0$.

16.   $x^2 + y^2 + z^2 = 6, 2x - y + z = 3;$   $(1, 1, 2)$   $[\pm 1/\sqrt{3}, \pm 1/\sqrt{3}, \mp 1/\sqrt{3}]$

17.   $z = x^2 + y^2, x^2 + z^2 + 2y^2 = 2;$   $(1, 0, 1)$                    $[0, \pm 1, 0]$

18.   $z = xy, x = y;$   $(0, 0, 0)$                                    $[\pm 1/\sqrt{2}, \pm 1/\sqrt{2}, 0]$

## 9   Higher Order Derivatives

The partial derivatives of a function $f$ are also functions, and so may in turn be differentiated. The notations for these second, and higher order, partial derivatives are as follows:

If $u = f(x, y)$, then

$$\frac{\partial}{\partial x}\frac{\partial u}{\partial x} = \frac{\partial^2 u}{\partial x^2} = f_{xx} = u_{xx}, \qquad \frac{\partial}{\partial y}\frac{\partial u}{\partial y} = \frac{\partial^2 u}{\partial y^2} = f_{yy} = u_{yy};$$

$$\frac{\partial}{\partial x}\frac{\partial u}{\partial y} = \frac{\partial^2 u}{\partial x \, \partial y} = f_{yx} = u_{yx}, \qquad \frac{\partial}{\partial y}\frac{\partial u}{\partial x} = \frac{\partial^2 u}{\partial y \, \partial x} = f_{xy} = u_{xy};$$

$$u_{xxx} = \frac{\partial}{\partial x}\frac{\partial^2 u}{\partial x^2} = \frac{\partial^3 u}{\partial x^3}, \qquad \frac{\partial}{\partial x}\frac{\partial^2 u}{\partial y \, \partial x} = \frac{\partial^3 u}{\partial x \, \partial y \, \partial x} = u_{xyx}, \quad \text{etc.}$$

The difference between $u_{xy}$ and $\partial^2 u/\partial x \, \partial y$ should be noted. In $u_{xy}$ the partial derivative with respect to $x$ occurs first. In $\partial^2 u/\partial x \, \partial y$ the partial with respect to $y$ occurs first.

*Example 1*     If $u = e^{ax} \cos by + x^2 y$, then

$$\frac{\partial u}{\partial x} = ae^{ax}\cos by + 2xy, \qquad \frac{\partial u}{\partial y} = -be^{ax}\sin by + x^2;$$

$$\frac{\partial^2 u}{\partial y \, \partial x} = -abe^{ax}\sin by + 2x, \qquad \frac{\partial^2 u}{\partial x \, \partial y} = -abe^{ax}\sin by + 2x.$$

Observe that $u_{xy} = u_{yx}$. The following theorem shows that such is usually the case. The "mixed" partial derivatives are equal. Also observe that

$$u_{xyx} = -a^2 be^{ax}\sin by + 2 = u_{yxx}.$$

**THEOREM**     *If $u = f(x, y)$ and $f$ has continuous partial derivatives of the second order, then*

$$u_{xy} = u_{yx}.$$

***PROOF**     The proof is tricky and requires repeated use of the MVT. We compute the difference

$$\Delta = f(x + \Delta x, y + \Delta y) - f(x + \Delta x, y) - f(x, y + \Delta y) + f(x, y)$$

in two ways.

In the first place,

$$\Delta = [f(x + \Delta x, y + \Delta y) - f(x + \Delta x, y)] - [f(x, y + \Delta y) - f(x, y)]$$
$$= \psi(x + \Delta x) - \psi(x),$$

where $\psi(t) = f(t, y + \Delta y) - f(t, y)$. By the MVT, for some $\theta_1$ with $0 < \theta_1 < 1$,

$$\Delta = \psi'(x + \theta_1 \Delta x)\,\Delta x = [f_x(x + \theta_1 \Delta x, y + \Delta y) - f_x(x + \theta_1 \Delta x, y)]\,\Delta x.$$

Now use the MVT again on $f_x$. We get, for some $\theta_2$ with $0 < \theta_2 < 1$,

(1)                    $$\Delta = f_{xy}(x + \theta_1 \Delta x, y + \theta_2 \Delta y)\,\Delta x \, \Delta y.$$

In the second place

$$\Delta = [f(x + \Delta x, y + \Delta y) - f(x, y + \Delta y)] - [f(x + \Delta x, y) - f(x, y)]$$
$$= \varphi(y + \Delta y) - \varphi(y)$$

where $\varphi(t) = f(x + \Delta x, t) - f(x, t)$. By the MVT, for some $\theta_3$ with $0 < \theta_3 < 1$,

$$\Delta = \varphi'(y + \theta_3 \, \Delta y) \, \Delta y = [f_y(x + \Delta x, y + \theta_3 \, \Delta y) - f_y(x, y + \theta_3 \, \Delta y)] \, \Delta y.$$

Now use the MVT again on $f_y$. We get, for some $\theta_4$ with $0 < \theta_4 < 1$,

(2) $$\Delta = f_{yx}(x + \theta_4 \, \Delta x, y + \theta_3 \, \Delta y) \, \Delta x \, \Delta y.$$

From equations (1) and (2) we obtain

$$\frac{\Delta}{\Delta x \, \Delta y} = f_{xy}(x + \theta_1 \, \Delta x, y + \theta_2 \, \Delta y) = f_{yx}(x + \theta_4 \, \Delta x, y + \theta_3 \, \Delta y).$$

Now let $\Delta x$, $\Delta y \to 0$ in this last equation. Because $f_{xy}$ and $f_{yx}$ are assumed to be continuous the limit exists and

$$f_{xy}(x, y) = f_{yx}(x, y).$$

**Example 2**    Many of the fundamental laws of physics are given by partial differential equations, usually second order differential equations—that is, ones involving second derivatives. For example, Laplace's equation* is satisfied by the gravitational potential $u$ due to any distribution of point masses:

$$\frac{\partial^2 u}{\partial x^2} + \frac{\partial^2 u}{\partial y^2} + \frac{\partial^2 u}{\partial z^2} = 0.$$

If a mass is at the origin the potential at $(x, y, z)$ is

$$u = k/r, \quad k = \text{constant}, \quad r = \sqrt{x^2 + y^2 + z^2}.$$

Then $$u_x = \frac{kx}{r^3}, \quad u_y = \frac{ky}{r^3}, \quad u_z = \frac{kz}{r^3} ;$$

$$u_{xx} = \frac{k(y^2 + z^2 - 2x^2)}{r^5}, \quad u_{yy} = \frac{k(x^2 + z^2 - 2y^2)}{r^5}, \quad u_{zz} = \frac{k(x^2 + y^2 - 2z^2)}{r^5} ;$$

and Laplace's equation is satisfied.

## Problems

**In Problems 1 to 6 compute $u_{xy}$ and $u_{yx}$.**

1.  $u = x^3 - 3xy^2 + x^2 - y^3$ $\hspace{3cm}$ $[-6y]$

2.  $u = x^2y + y^2z + z^2x - 3xyz$ $\hspace{2.5cm}$ $[2x - 3z]$

---

* Pierre-Simon de Laplace (1749–1827), a great analyst, made his major contribution through his book *Méchanique celeste* in which he applied the Newtonian law of gravitation to the solar system. His free use of the phrase, "It is easy to see," made his writing very hard to understand.

3.   $u = \text{Arc tan } \dfrac{y}{x}$

$$\left[\frac{y^2 - x^2}{(x^2 + y^2)^2}\right]$$

4.   $u = \log \dfrac{x}{y} - e^{x+y}$

$$[-e^{x+y}]$$

5.   $u = f(x) + g(y) + h(xy)$

$$[xyh''(xy) + h'(xy)]$$

6.   $u = F(x - y)$

$$[-F''(x - y)]$$

7.   Show that $u = \log \sqrt{x^2 + y^2}$ satisfies Laplace's equation in the plane, $u_{xx} + u_{yy} = 0$.

8.   Show that $u = \text{Arc tan } (y/x)$ satisfies Laplace's equation in the plane.

9.   If $u$ and $v$ are twice continuously differentiable functions of $x$ and $y$, and if $u_x = v_y$ and $u_y = -v_x$, show that both $u$ and $v$ satisfy Laplace's equation in the plane.

10.  Show that $u = 1/r^2$, $r^2 = x_1^2 + x_2^2 + x_3^2 + x_4^2$ satisfies Laplace's equation in four dimensions.

11.  Show that $u = \frac{1}{4}(x^4 - y^4) + y^2$ satisfies $u_{xy} = 0$.

12.  Show that $e^x \sin y$ and $e^x \cos y$ satisfy Laplace's equation in the plane.

13.  Show that $u = A \sin kx \cos kat$ satisfies the "one-dimensional wave equation" $u_{tt} = a^2 u_{xx}$.

14.  Show that $u = A \exp(-a^2 k^2 t) \cos kx$ satisfies the "one-dimensional heat equation" $u_t = a^2 u_{xx}$.

15.  Show that

$$u = \frac{1}{r}(\sin kr)\exp(-a^2 k^2 t) \quad \text{where} \quad r = \sqrt{x^2 + y^2 + z^2},$$

satisfies the heat equation in three dimensions: $u_t = a^2(u_{xx} + u_{yy} + u_{zz})$.

16.  Show that if $P(x, y)\, dx + Q(x, y)\, dy$ is the total differential of a function $u(x, y)$ then $\partial P/\partial y = \partial Q/\partial x$.

17.  Show that (assuming all derivatives are continuous) $u_{xyy} = u_{yxy} = u_{yyx}$, and that $u_{xyz} = u_{xzy} = u_{zyx}$.

18.  Show that, if $u = f(\xi) + g(\eta)$ where $\xi = x - at$ and $\eta = x + at$, then $u_{tt} = a^2 u_{xx}$.

*19.  In thermodynamics one studies various functions of the state of a gas (or any substance). In addition to pressure $p$, volume $v$, and temperature $T$, there are five other quantities, namely the internal energy $u$, the entropy $S$, the enthalpy

$h$, the Helmholtz function $f$, and the Gibbs function $g$. These quantities are related as follows*:

$$\left(\frac{\partial f}{\partial T}\right)_v = -s, \qquad \left(\frac{\partial f}{\partial v}\right)_T = -p,$$

$$\left(\frac{\partial g}{\partial T}\right)_p = -s, \qquad \left(\frac{\partial g}{\partial p}\right)_T = v,$$

$$\left(\frac{\partial h}{\partial s}\right)_p = T, \qquad \left(\frac{\partial h}{\partial p}\right)_s = v,$$

$$\left(\frac{\partial u}{\partial s}\right)_v = T, \qquad \left(\frac{\partial u}{\partial v}\right)_s = -p,$$

where the subscripts show what variable remains constant in the differentiation.
Show that

$$\left(\frac{\partial s}{\partial v}\right)_T = \left(\frac{\partial p}{\partial T}\right)_v.$$

This is the first of four equations called Maxwell's equations. Obtain three others:

$$\left(\frac{\partial s}{\partial p}\right)_T = -\left(\frac{\partial v}{\partial T}\right)_p; \quad \left(\frac{\partial T}{\partial p}\right)_s = \left(\frac{\partial v}{\partial s}\right)_p; \quad \left(\frac{\partial T}{\partial v}\right)_s = -\left(\frac{\partial p}{\partial s}\right)_v.$$

*20.   Change from rectangular to polar coordinates and show that

$$u_x = \frac{x}{r} u_r - \frac{y}{r^2} u_\theta, \qquad u_y = \frac{y}{r} u_r + \frac{x}{r^2} u_\theta.$$

Then show that, in polar coordinates, Laplace's equation becomes

$$u_{rr} + \frac{1}{r} u_r + \frac{1}{r^2} u_{\theta\theta} = 0.$$

*21.   Show that Laplace's equation becomes, in spherical coordinates

$$u_{rr} + \frac{2}{r} u_r + \frac{1}{r^2 \sin^2 \varphi} u_{\theta\theta} + \frac{1}{r^2} u_{\varphi\varphi} + \frac{\cos \varphi}{r^2 \sin \varphi} u_\varphi = 0.$$

See Appendix A for spherical coordinates.

---

* If $z$ is a function of two or more variables, the notation $\partial z/\partial x$ is sometimes inadequate, as it gives no indication of what the other variables are. In physics, for example, the internal energy $u$ of a gas can be given as a function of any two of the variables $p$, $v$, $T$. Thus, there are three different functions that give the internal energy: $u = u_1(p, v)$, $u = u_2(v, T)$, and $u = u_3(p, T)$. The notation $\partial u/\partial p$ does not indicate which function is being differentiated, $u_1$ or $u_3$. The physicist gets around this dilemma by using a subscript. For example, $(\partial u/\partial p)_T$ means that $u$ is a function of $p$ and $T$, and we are differentiating $u_3$ with $T$ being held constant.

## *10   *Maxima and Minima*

**DEFINITION**   If $f$ is a function of $x$ and $y$, then $f$ has a *relative maximum* at the point $(x_0, y_0)$ if there is a square $D$ centered at $(x_0, y_0)$ such that

$$f(x_0, y_0) \geqq f(x, y)$$

for all points $(x, y)$ in $D$. (For a definition of relative minimum reverse the inequality.)

If $f$ has a relative maximum (or minimum) at $(x_0, y_0)$, and if $f$ has partial derivatives, then

(1)   $f_x(x_0, y_0) = 0,$      $f_y(x_0, y_0) = 0.$

This is so because the curves of intersection of the surface $z = f(x, y)$ with the planes $x = x_0$ and $y = y_0$ must have

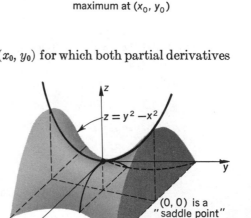

A relative
maximum at $(x_0, y_0)$

maxima (or minima) at $y_0$ or $x_0$. The points $(x_0, y_0)$ for which both partial derivatives vanish are called *critical points*.

Conditions (1) are necessary but not sufficient for a maximum or minimum. Just as with functions of one variable, $f$ need not have a maximum or a minimum at a critical point. The figure shows a surface with a critical point at $(0, 0)$, but the function has neither a maximum nor a minimum there. This example shows that functions of two variables can behave in far more complicated ways near a critical point than can functions of one variable. The reader might try

$z = y^2 - x^2$

$(0, 0)$ is a
"saddle point"
$z_x(0, 0) = z_y(0, 0) = 0$

to imagine a surface with a critical point at $(0, 0)$ such that curves of intersection with several vertical planes through $(0, 0)$ are alternately concave up and concave down.

There is a "second derivative test" for maxima and minima of functions of two variables. It supplies a criterion that, if satisfied, assures one that each cross section by vertical planes through a critical point has the same sense of concavity. We state this test here without proof.

**THEOREM**   *If $f$ has continuous partial derivatives of the second order, and if $(x_0, y_0)$ is a critical point, and if*

$$\Delta = f_{xx}(x_0, y_0)f_{yy}(x_0, y_0) - f_{xy}^2(x_0, y_0),$$

*then*

(a) *there is a minimum at* $(x_0, y_0)$ *if* $\Delta > 0$ *and* $f_{xx}(x_0, y_0) > 0$;

(b) *there is a maximum at* $(x_0, y_0)$ *if* $\Delta > 0$ *and* $f_{xx}(x_0, y_0) < 0$;

(c) *there is neither a maximum nor a minimum if* $\Delta < 0$;

(d) *nothing can be said if* $\Delta = 0$.

*Example*    Examine $f(x, y) = x^2 - xy + y^2 - x - y$ for maxima and minima.
The critical points $(x, y)$ satisfy

$$f_x = 2x - y - 1 = 0, \qquad f_y = -x + 2y - 1 = 0.$$

Solution of these equations yields the critical point $(1, 1)$.

$$f_{xx}(1, 1) = 2, \quad f_{xy}(1, 1) = -1, \quad f_{yy}(1, 1) = 2.$$

$$\Delta = 4 - 1 = 3 \qquad \text{and} \qquad f_{xx} > 0.$$

Therefore, there is a minimum at $(1, 1)$ equal to $-1$.


*Problems*

In Problems 1 to 6 find the critical points and examine them for maxima and minima.

1.  $f(x, y) = x^2 - xy + y^2$                                    [min at $(0, 0)$]

2.  $f(x, y) = xy + x - y - 1$                               [neither at $(1, -1)$]

3.  $f(x, y) = x^3 + y^3 - 3axy, \quad a > 0$    $\begin{bmatrix} \text{neither at } (0, 0), \\ \text{min at } (a, a) \end{bmatrix}$

4.  $u = \dfrac{(x + y + 1)^2}{x^2 + y^2 + 1}$    $\begin{bmatrix} \text{min along the line } x + y = -1, \\ \text{max} = 3 \text{ at } (1, 1) \end{bmatrix}$

5.  $v = -x^2 - 2y^2 + 2x - 4y$                          [max at $(1, -1)$]

6.  $w = a^{2/3} - x^{2/3} - y^{2/3}$                          [max at $(0, 0)$]

7.  Find, using calculus, the point of the plane $x - y + 2z = 6$ at minimum distance from the origin.                    $[(1, -1, 2)]$

8.  What are the dimensions of a box without a top if it is to contain 32 cu ft and require the least amount of sheet metal?                    $[4 \times 4 \times 2]$

9.  Find, using calculus, the point of the sphere $x^2 + y^2 + z^2 = a^2$ that is nearest the point $(2, 1, 3)$.                    $[(2a/\sqrt{14}, a/\sqrt{14}, 3a/\sqrt{14})]$

10.  Find the maximum of $xyz$ if $x + y + z = a = \text{constant} > 0$.                    $[a^3/27]$

11.  Use calculus to find the minimum of the distance from the point $(x_0, y_0, z_0)$ to the plane $Ax + By + Cz + D = 0$.

(*Hint:* It is simplest to consider the equation of the plane as defining $z$ implicitly as a function of $x$ and $y$.)

12.  Show that the function $\cos x + \cos y + \cos (x + y)$ has a maximum at $(0, 0)$, a minimum at $(2\pi/3, 2\pi/3)$, and neither a maximum nor minimum at $(\pi, \pi)$.

*13.  Investigate $f(x, y) = (y^2 - x)(2y^2 - x)$ for maxima and minima. Could you avoid calculus?                                    [no max or min]

14.  Find the maximum of $ax + by + cz$ if $x^2 + y^2 + z^2 = 1$.      $[\sqrt{a^2 + b^2 + c^2}]$

15.  If $f_{xx} > 0$ at a critical point to what extent is there a minimum at the point?

## *11   Functions Defined Implicitly

Suppose that a surface is given by

$$z = f(x, y)$$

where $f$ has continuous partial derivatives in some region of the $xy$-plane.

Often, a plane, $z = z_0$, will meet the surface in a curve given [at least near $(x_0, y_0, z_0)$] by either

$$y = y(x) \qquad \text{or} \qquad x = x(y),$$

or both. The figure shows a curve of intersection in red. The function $y$ is defined implicitly by the equation

(1)                $z_0 = f(x, y)$,

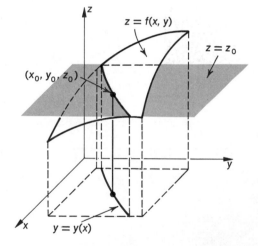

and let us assume that this function $y$ is differentiable. We shall see later a condition that will insure that this is the case.

As we move along, the curve $z$ remains constant, and applying the chain rule to equation (1), we obtain

$$\frac{dz}{dx} = 0 = \frac{\partial f}{\partial x} + \frac{\partial f}{\partial y}\frac{dy}{dx}.$$

Therefore,

(2)                $$\frac{dy}{dx} = -\frac{\partial f/\partial x}{\partial f/\partial y} = -\frac{f_x}{f_y}.$$

Do not commit (2) to memory. Remember the derivation.

*Remark.* Equation (2) makes sense only if $f_y(x_0, y_0) \neq 0$. But this is precisely the condition that insures that equation (1) determines $y$ in terms of $x$. More precisely, the following theorem can be proved.*

**THEOREM**    *If $f_y(x_0, y_0) \neq 0$, then there is an interval $I$ containing $x_0$, and a unique function $y$, differentiable on $I$, such that*

$$y(x_0) = y_0 \qquad and \qquad z_0 = f[x, y(x)]$$

*for $x$ in $I$.*

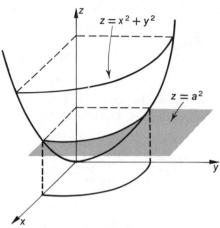

**Example 1**    Consider the equation $x^2 + y^2 = a^2$. Obviously we can solve for $y$ in this example, but let us proceed as in the text development.

If we set $z = x^2 + y^2$, the given equation then represents the intersection of the surface $z = x^2 + y^2$ with the plane $z = a^2$.

Because $\partial z / \partial y = 2y$, we can solve for $y$ in terms of $x$ wherever $y \neq 0$. Then

$$\frac{dy}{dx} = -\frac{\partial z / \partial x}{\partial z / \partial y} = -\frac{2x}{2y} = -\frac{x}{y}.$$

From the Theorem it is clear that if either $f_x \neq 0$ or $f_y \neq 0$, then equation (1), $z_0 = f(x, y)$ defines one of $x$ or $y$ as a function of the other. Points at which *both* $f_x$ and $f_y$ are zero are called *singular points* of the curve given by equation (1). At these points one need not have a unique function defined.

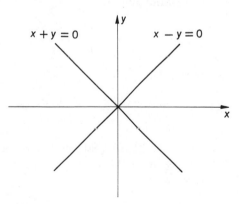

**Example 2**    Find the singular points if any of the curve $x^2 - y^2 = 0$.

Here the curve (two straight lines) is the intersection of the surface $z = x^2 - y^2 = (x - y)(x + y)$ and the plane $z = 0$. Then at a singular point

$$\frac{\partial z}{\partial x} = 2x = 0,$$

$$\frac{\partial z}{\partial y} = -2y = 0.$$

---

* A proof can be found in advanced calculus texts. The hard part of the proof is the *existence* of the function $y$. The existence is geometrically plausible for the following reasons. Since $f_y(x_0, y_0) \neq 0$, the surface is not "horizontal in the $y$-direction." Therefore, near $x_0$ it must lie both above and below the plane $z = z_0$, and so must cross the plane $z = z_0$.

The only solution to these equations that gives a point on the curve is $x = 0$, $y = 0$.

As is seen from the graph, there is no *unique* differentiable function of $x$ (or of $y$) defined near the origin.

The same principles apply to functions of several variables that are defined implicitly. Suppose that $w = f(x, y, z)$ and that $f$ has continuous partial derivatives in a region containing $(x_0, y_0, z_0)$.

Suppose also that

$$w_0 = f(x_0, y_0, z_0)$$

and that

$$\frac{\partial f}{\partial z}(x_0, y_0, z_0) \neq 0.$$

Then it can be shown that in a square centered at $(x_0, y_0)$ there is a unique differentiable function

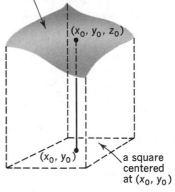

a level surface of $f$: $f(x, y, z) = w_0$

$(x_0, y_0, z_0)$

$(x_0, y_0)$

a square centered at $(x_0, y_0)$

$$z = z(x, y)$$

such that

$$z_0 = z(x_0, y_0)$$

and

(3) $$w_0 \equiv f[x, y, z(x, y)],$$

identically, for all points $(x, y)$ in the square. We shall accept this theorem and proceed to compute $\partial z/\partial x$ and $\partial z/\partial y$. Because $w_0$ is constant in (3), the chain rule gives us

$$0 = \frac{\partial f}{\partial x} + \frac{\partial f}{\partial z}\frac{\partial z}{\partial x} \qquad \text{and} \qquad 0 = \frac{\partial f}{\partial y} + \frac{\partial f}{\partial z}\frac{\partial z}{\partial y}.$$

We obtain then

(4) $$\frac{\partial z}{\partial x} = -\frac{\partial f/\partial x}{\partial f/\partial z} \qquad \text{and} \qquad \frac{\partial z}{\partial y} = -\frac{\partial f/\partial y}{\partial f/\partial z}.$$

Observe that equations (4) require for their validity that $\partial f/\partial z \neq 0$ and that this agrees with the condition we imposed in order that a unique differentiable function exist.

**Example 3**   Find $z_x$ if $e^{x+y} \sin z + e^{x-y} \cos z = 0$. Here $f(x, y, z) = e^{x+y} \sin z + e^{x-y} \cos z$. Then

$$\frac{\partial z}{\partial x} = -\frac{f_x}{f_z} = -\frac{e^{x+y} \sin z + e^{x-y} \cos z}{e^{x+y} \cos z - e^{x-y} \sin z}.$$

Finally we observe that more than one function can be defined implicitly. Suppose that

(5)
$$\begin{cases} u = f(x, y, z, \xi, \eta) = \text{constant}, \\ v = g(x, y, z, \xi, \eta) = \text{constant}, \end{cases}$$

where $f$ and $g$ are functions with continuous partial derivatives, and that one can solve for $\xi$ and $\eta$ in terms of $x$, $y$, and $z$ to obtain differentiable functions:

$$\xi = \xi(x, y, z)$$
$$\eta = \eta(x, y, z).$$

We proceed to compute the partial derivatives of $\xi$ and $\eta$ using the chain rule in equations (5):

$$\frac{\partial u}{\partial x} = 0 = \frac{\partial f}{\partial x} + \frac{\partial f}{\partial \xi}\frac{\partial \xi}{\partial x} + \frac{\partial f}{\partial \eta}\frac{\partial \eta}{\partial x},$$

$$\frac{\partial v}{\partial x} = 0 = \frac{\partial g}{\partial x} + \frac{\partial g}{\partial \xi}\frac{\partial \xi}{\partial x} + \frac{\partial g}{\partial \eta}\frac{\partial \eta}{\partial x}.$$

These are two simultaneous *linear* equations in $\xi_x$ and $\eta_x$. Solving we obtain

(6)
$$\begin{cases} \xi_x = \dfrac{-f_x\, g_\eta + f_\eta\, g_x}{f_\xi\, g_\eta - f_\eta\, g_\xi}, \\ \eta_x = \dfrac{-f_\xi\, g_x + f_x\, g_\xi}{f_\xi\, g_\eta - f_\eta\, g_\xi}. \end{cases}$$

Partial derivatives with respect to $y$ are obtained similarly, by differentiating (5) with respect to $y$. Equations (6) and their analogs for $\xi_y$, $\eta_y$ should not be memorized. The use of the chain rule is to be preferred.

**Example 4**    If $x^2 + y^2 + xy + u^2 - v^2 = 0$ and $x^2 + y^2 - uv + 1 = 0$, define $u$ and $v$ as differentiable functions of $x$ and $y$, find $\partial u/\partial y$ and $\partial v/\partial y$.
    Here we have

$$f(x, y, u, v) = x^2 + y^2 + xy + u^2 - v^2 = 0 = \text{constant},$$

$$g(x, y, u, v) = x^2 + y^2 - uv + 1 = 0 = \text{constant}.$$

From the chain rule,

$$f_y = 0 = 2y + x + 2uu_y - 2vv_y,$$

$$g_y = 0 = 2y - uv_y - vu_y,$$

whence, solving for $u_y$ and $v_y$,

$$u_y = \frac{4yv - (2y + x)u}{2(u^2 + v^2)}, \qquad v_y = \frac{4yu + (2y + x)v}{2(u^2 + v^2)}.$$

### Problems

In Problems 1 to 10 find the indicated derivatives, assuming that differentiable functions are defined by the given equations.

1. $x^2 - xy + y^2 + x - y + 2 = 0$;  $\dfrac{dy}{dx}$   $\left[\dfrac{2x - y + 1}{x - 2y + 1}\right]$

2. $x^2 - xy + y^2 + x - y + 2 = 0$;  $\dfrac{dx}{dy}$   $\left[\dfrac{x - 2y + 1}{2x - y + 1}\right]$

3. $(x^2 + y^2)^2 - a^2(x^2 - y^2) = 0$;  $\dfrac{dy}{dx}$   $\left[\dfrac{x(a^2 - 2(x^2 + y^2))}{y(a^2 + 2(x^2 + y^2))}\right]$

4. $x^{2/3} + y^{2/3} = a^{2/3}$;  $\dfrac{dy}{dx}$   $\left[-\dfrac{y^{1/3}}{x^{1/3}}\right]$

5. $\sqrt{x^2 + y^2} = \operatorname{Arc\,tan}\dfrac{y}{x}$;  $\dfrac{dy}{dx}$   $\left[\dfrac{x\sqrt{x^2 + y^2} + y}{x - y\sqrt{x^2 + y^2}}\right]$

6. $e^{2u} - x^2 - y^2 = 0$;  $\dfrac{\partial u}{\partial x}, \dfrac{\partial u}{\partial y}$   $\left[u_x = \dfrac{x}{x^2 + y^2}, u_y = \dfrac{y}{x^2 + y^2}\right]$

7. $z - xy + \log z = 1$;  $z_x, z_y$   $\left[z_x = \dfrac{yz}{z + 1}, z_y = \dfrac{xz}{z + 1}\right]$

8. $(u^2 - x^2)(x^2 - y^2)(y^2 - u^2) = a^2$;  $\dfrac{\partial u}{\partial y}$   $\left[\dfrac{(u^2 - x^2)(2y^2 - x^2 - u^2)y}{(x^2 - y^2)(y^2 + x^2 - 2u^2)u}\right]$

9. $u^2 + x^2 + y^2 + 2\sin xy = 0$;  $u_y$   $\left[-\dfrac{y + x\cos xy}{u}\right]$

10. $e^{xy} + e^{yz} + e^{zu} + e^{ux} + u = 4$;  $u_z$   $\left[-\dfrac{ye^{yz} + ue^{uz}}{1 + ze^{uz} + xe^{ux}}\right]$

In Problems 11 to 14 assume that the given equations define u and v as differentiable functions of the other variables. Find the indicated derivatives.

11. $u + v + x + y = a$, $u^2 + v^2 + x^2 + y^2 = a^2$;  $u_x, v_x$

$\left[u_x = \dfrac{x - v}{v - u}, v_x = \dfrac{u - x}{v - u}\right]$

12.  $u + v = x + y, u - v = xy; \quad u_y, v_y$

$$\left[ u_y = \frac{1+x}{2}, \ v_y = \frac{1-x}{2} \right]$$

13.  $u^2 + v^2 + x^2 + y^2 + z^2 = a^2, u^3 + v^3 + x^3 + y^3 + z^3 = a^3; \quad \partial v/\partial z$

$$\left[ \frac{z^2 - uz}{uv - v^2} \right]$$

14.  $\sin xu + x^2 + u^2 + v^2 = a^2, u^2 - v^2 + a^2 = 0; \quad du/dx$

$$\left[ -\frac{u \cos xu + 2x}{x \cos xu + 4u} \right]$$

15.  If $f(x, y, z) = 0$ defines each of $x$, $y$, and $z$ as differentiable functions of the other two, show that

$$\frac{\partial x}{\partial y} \frac{\partial y}{\partial z} \frac{\partial z}{\partial x} = -1.$$

16.  The equation of state of a gas has the general form $f(p, v, T) = 0$ where $p, v, T$ are pressure, volume, and temperature. Show that

$$\left( \frac{\partial p}{\partial v} \right)_T = - \frac{(\partial T/\partial v)_p}{(\partial T/\partial p)_v}.$$

17.  Obtain $\partial r/\partial x$ and $\partial r/\partial y$ from the implicit relations connecting rectangular and polar coordinates, $x - r \cos \theta = 0, y - r \sin \theta = 0.$     $[r_x = \cos \theta, r_y = \sin \theta]$

18.  Obtain $\partial r/\partial x$ from the implicit relations connecting rectangular and spherical coordinates, $x - r \sin \varphi \cos \theta = 0, y - r \sin \varphi \sin \theta = 0, z = r \cos \varphi.$

$$[r_x = \sin \varphi \cos \theta]$$

19.  Derive equation (2) directly from the Fundamental Lemma.

20.  If $x = e^\xi \cos \eta, y = e^\xi \sin \eta$, compute $\xi_x \eta_y - \xi_y \eta_x$.

$$\left[ e^{-2\xi} = \frac{1}{x^2 + y^2} \right]$$

21.  If $u = f(x, y), v = g(x, y)$ and one can solve for $x$ and $y$ in terms of $u$ and $v$ and get differentiable functions, find $\partial x/\partial u, \partial y/\partial u$.

$$\left[ \frac{\partial x}{\partial u} = \frac{g_y}{f_x g_y - f_y g_x}, \frac{\partial y}{\partial u} = -\frac{g_x}{f_x g_y - f_y g_x} \right]$$

*22.  The internal energy $u$ of a gas is a function of any two of the variables, pressure $p$, volume $v$, and temperature $T$. Show that

$$\left( \frac{\partial u}{\partial T} \right)_v = \left( \frac{\partial u}{\partial T} \right)_p + \left( \frac{\partial u}{\partial p} \right)_T \left( \frac{\partial p}{\partial T} \right)_v.$$

23.  Find the singular points of the strophoid $y^2(a - x) = x^2(a + x)$. See Appendix A.

$$[(0, 0)]$$

*24.* Find the singular points of the cissoid $y^2(2a - x) = x^3$. See Appendix A.

[ (0, 0) ]

*\*25.* Find the singular points of the conchoid $x^2y^2 = (y + a)^2(b^2 - y^2)$. See Appendix A.

[ (0, -a) ]

## *12   Envelopes

Consider a family of curves given by $f(x, y, c) = 0$. Each value of the parameter* $c$ corresponds to one curve of the family. In some cases each curve of the family is tangent to a curve $E$. The curve $E$ is called an *envelope* of the family.

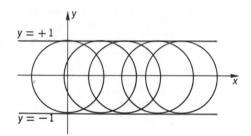

*Example 1*    The family of circles $(x - c)^2 + y^2 = 1$ is shown in the figure. Each of the lines $y = 1$, and $y = -1$ is an envelope of the family.

Consider a family of curves given by

(1)          $f(x, y, c) = 0,$

where $f$ is a differentiable function.

When an envelope exists (and some families do not have envelopes) an equation of the envelope can be found as follows:

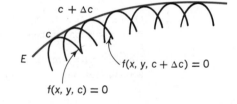

E, the envelope is tangent to each curve of the family

**METHOD**    *An equation of the envelope is obtained by eliminating the parameter $c$ between the equations*

(2)
$$\begin{cases} f(x, y, c) = 0, \\ \dfrac{\partial f}{\partial c}(x, y, c) = 0. \end{cases}$$

*If equations (2) are solved for $x$ and $y$ in terms of $c$, an equation of the envelope is obtained in parametric form.*

---

* The word "parameter" is used in this way as well as in parametric equations. In this case it is a variable that labels, or characterizes, a set of curves.

**PROOF**    Each point on the envelope is the point of
tangency of a particular curve of the family and is
determined by a value of $c$. Therefore, we may suppose
that the envelope $E$ is parameterized by the parameter
$c$,

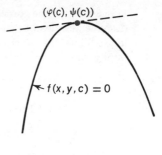

$(\varphi(c), \psi(c))$

$\leftarrow f(x, y, c) = 0$

$$(3) \qquad x = \varphi(c), \qquad y = \psi(c).$$

Then from equations (3), we deduce that, at the point
of tangency

$$(4) \qquad \frac{dy}{dx} = \frac{\psi'(c)}{\varphi'(c)}.$$

But also, since the curve is given by $f(x, y, c) = 0$, its slope at $(\varphi(c), \psi(c))$ is equal
to

$$(5) \qquad \frac{dy}{dx} = -\frac{f_x(\varphi(c), \psi(c), c)}{f_y(\varphi(c), \psi(c), c)}.$$

Thus, from (4) and (5)

$$(6) \qquad f_x \varphi'(c) + f_y \psi'(c) = 0.$$

Now let us return to the envelope. At each point, $f(\varphi(c), \psi(c), c) = 0$,
whence the total derivative with respect to $c$ is zero:

$$(7) \qquad f_x \varphi'(c) + f_y \psi'(c) + f_c = 0.$$

And now if we use equation (6), we obtain

$$f_c(x, y, c) = 0,$$

where the point $(x, y)$ is on the envelope $E$, as well as on a curve $f(x, y, c) = 0$ of
the family.

*Example 2*    Find the envelope of the
family of circles $(x - c)^2 + y^2 = 4c + 4$, where $c > 0$.

The equation of the family, and
the partial derivative with respect to
the parameter are as follows:

$$(8) \qquad \begin{cases} (x - c)^2 + y^2 = 4c + 4, \\ -2(x - c) = 4. \end{cases}$$

Solving for $x$ and $y$ in terms of $c$ gives
the parametric equations of the envelope with parameter $c$,

$$x = c - 2, \qquad y = \pm 2\sqrt{c}.$$

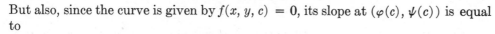

envelope

If we eliminate $c$ between equations (8) we obtain $y^2 = 4(x + 2)$, a parabola.

*Example 3*    The procedure that has been described can lead to curves that are not envelopes. Thus, the process, when applied to the family $(y - c)^2 = x^3$ leads to

$$\begin{cases} (y - c)^2 = x^3, \\ -2(y - c) = 0, \end{cases}$$

whence we get the parametric equations

$$x = 0, \qquad y = c,$$

which represent the $y$-axis.

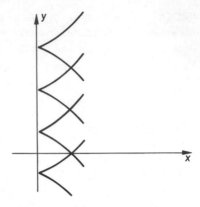

The reason one can get curves other than envelopes by following the method given above, is explained by the following result.

*Suppose that the curves* $f(x, y, c) = 0$ *and* $f(x, y, c + \Delta c) = 0$ *intersect at* $(x_0, y_0)$ *and that as* $\Delta c \to 0$ *the point* $(x_0, y_0)$ *approaches* $(x_1, y_1)$; *then*

(9)          $f_c(x_1, y_1, c) = 0$      *and*      $f(x_1, y_1, c) = 0.$

The proof uses the mean value theorem. We have

$$f(x_0, y_0, c + \Delta c) - f(x_0, y_0, c) = 0 = f_c(x_0, y_0, c + \theta \, \Delta c) \, \Delta c,$$

where $0 < \theta < 1$. Thus,

$$f_c(x_0, y_0, c + \theta \, \Delta c) = 0.$$

Then as $\Delta c \to 0$, $(x_0, y_0)$ approaches the point $(x_1, y_1)$, which is on the curve $f(x, y, c) = 0$.

Re-examination of Example 2 shows that "nearby" curves of the family meet in a point near the line $x = 0$.

As a consequence of this result, it follows that our method for finding the envelope can lead to curves that are not envelopes. Thus, *one must always examine the result of using the method.*

*Example 4*    If we try to find the envelope of the family $y = x^2 + c$, we obtain:

$$y = x^2 + c,$$

$$0 = 1.$$

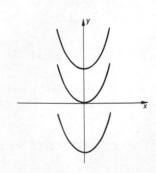

Since these equations are inconsistent, there can be no envelope.

*Problems*

In Problems 1 to 16 find the envelope, if there is one, of the family of curves. Sketch each family and its envelope.

1.  $(x - c)^2 + y^2 = r^2$   (c is the parameter)          $[y = \pm r]$

2.  $(y - c)^2 = 2px$   (c is the parameter)          $[x = 0]$

3.  $(y - c)^2 = 2p(x - c)$   (c is the parameter)          $[y = x + (p/2)]$

4.  $y^2 = (x - c)^3$          $[y = 0]$

5.  $x \cos\theta + y \sin\theta = p$   ($\theta$ is the parameter)          $[x^2 + y^2 = p^2]$

6.  $(x^2/a^2) + a^2y^2 = 1$          $[xy = \pm\tfrac{1}{2}]$

7.  $(x - c)^2 + y^2 = c^2$          [no envelope]

8.  $(x - c)^2 + y^2 = 2c^2$          [no envelope]

9.  $(x - c)^2 + y^2 = \tfrac{1}{2}c^2$          $[y = \pm x]$

10.  $y = cx + c^2$          $[4y = -x^2]$

11.  $y = cx - c^2$          $[4y = x^2]$

12.  $y = cx - (c^2/4)$          $[y = x^2]$

13.  $y = cx + (1/c)$          $[y^2 = 4x]$

14.  $(y - cx)^2 = ac^2 + b$   (c is the parameter)          $[ay^2 + bx^2 = ab]$

15.  $y = cx - ax^2(1 + c^2)$   (c is the parameter)          $[y = (1/4a) - ax^2]$

*16.  $x^2 + y^2 + c^2 = 2cx + 2cy + 2xy$          [the axes]

17.  Find the envelope of the family of straight lines, where $\theta$ is the parameter,

$$y \cos\theta - x \sin\theta = a - a \sin\theta \log \frac{\cos\theta}{1 - \sin\theta}.$$          $\left[ y = \dfrac{a}{2} (e^{x/a} + e^{-x/a}) \right]$

18.  Find the envelope of the family of straight lines the sum of the intercepts of which is a constant $= k$.
$$[x^{1/2} + y^{1/2} = k^{1/2} \text{ or also } x^2 + y^2 + k^2 = 2kx + 2ky + 2xy]$$

10.  Find the envelope of the family of straight lines the sum of the squares of the intercepts of which is a constant $= k^2$.          $[x^{2/3} + y^{2/3} = k^{2/3}]$

20.  Find the envelope of the family of ellipses $b^2x^2 + a^2y^2 = a^2b^2$ if $a + b = $ constant $= k$.          $[x^{2/3} + y^{2/3} = k^{2/3}]$

21.  A family of circles through the origin have their centers on the parabola $y^2 = 2px$. Show that their envelope is the cissoid $x^3 + y^2(2p + x) = 0$.

22.  A straight line passes through the point where a normal to the parabola at a point $P_0$ meets the axis and is parallel to the tangent at $P_0$. Show that the envelope of the family of all such lines is the parabola $y^2 + 2px = 2p^2$.

23. Find the envelope of the family of lines $x \sec^2 \theta + y \csc^2 \theta = a$, where $\theta$ is the parameter. $\qquad [a^2 + x^2 + y^2 = 2ax + 2ay + 2xy]$

24. Show that the envelope of the family of conics ($c$ is the parameter)

$$\frac{x^2}{a^2} + 2c \frac{xy}{ab} + \frac{y^2}{b^2} = 1 - c^2,$$

is the union of the four lines $x = \pm a$, $y = \pm b$.

*25. Show that if the family $f(x, y)c^2 + g(x, y)c + h(x, y) = 0$ has an envelope, it is given by $g^2 - 4fh = 0$. Then interpret this geometrically by considering the number of curves (values of $c$) that pass through a point $(x, y)$.

26. It can be shown that the evolute (see page 175) of a curve is the envelope of the normals to the curve. Use this property to find the evolute of the parabola $y^2 = 2px$.

$$\left[ y^2 = \frac{8}{27p} (x - p)^3 \right]$$

*27. Use the property of Problem 26 to find the evolute of the ellipse $b^2x^2 + a^2y^2 = a^2b^2$. $\qquad [(ax)^{2/3} + (by)^{2/3} = (a^2 - b^2)^{2/3}]$

*28. Show that the family of circles $(x - c)^2 + y^2 = 4$ solves the differential equation, $y'^2 = (4 - y^2)/y^2$, then find the envelope of the family and show that this also solves the differential equation.

*29. Show that the family of lines $y = cx + \frac{1}{4}c^2$ is a solution of the differential equation $y = xy' + \frac{1}{4}y'^2$. Show that the envelope of the family is also a solution. Sketch the family and envelope.

*30. If a family of curves solves a first order differential equation, and if the family has an envelope, give a geometric argument to prove that the envelope also solves the differential equation.

## CHAPTER 15

## 1    *Double Integrals*

The ordinary definite integral is obtained by subdividing the interval and taking a limit of Riemann sums. Double integrals are defined analogously:

(1) Suppose $R$ is a region of the plane, of area $A$, and bounded by a finite number of piecewise smooth curves. See Chapter 6.

(2) Suppose that $f$ is a real-valued function that is continuous on $R$.

(3) Subdivide $R$ into a finite number of subregions $\Delta R_i$, $i = 1, \ldots, n$, (each bounded by a finite number of piecewise smooth curves) of areas $\Delta A_i$, $i = 1, \ldots, n$, such that

$$A = \sum_{i=1}^{n} \Delta A_i.$$

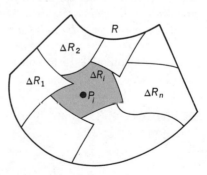

Area $R = A = \sum\limits_{i=1}^{n} \Delta A_i$

The *mesh* of the subdivision is the largest diameter of the regions $\Delta R_i$, where the diameter of a region is the maximum distance between two points of the region.

(4) Choose in each subregion $\Delta R_i$ a point $P_i$ and consider the *Riemann sum*

$$(1) \qquad S_n = \sum_{i=1}^{n} f(P_i)\, \Delta A_i = f(P_1)\, \Delta A_1 + \cdots + f(P_n)\, \Delta A_n.$$

Each term in the sum $S_n$ will be [if $f(P_i) \geqq 0$] the volume of a slender cylinder of base area $\Delta A_i$ and height $f(P_i)$. The sum approximates the volume of the solid under the graph of $f$ and above the region $R$.

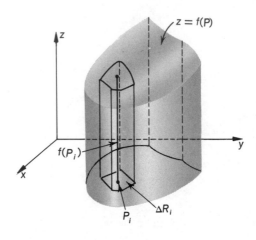

It is plausible that, as the mesh of the subdivision approaches 0 (so necessarily $n \to \infty$), the sums $S_n$ will [for $f(P) \geqq 0$ everywhere] actually approach the volume of the solid under the graph of $f$. We shall *assume* that this is the case and, moreover, that the limit of the Riemann sums exists for any continuous function $f$, whether or not it is always positive. We state this fact, without proof, as a theorem, using the notation we have introduced.

**THEOREM 1**   *If the region $R$ is as described above, and if $f$ is continuous in $R$, then the limit*

$$(2) \qquad \lim \sum f(P_i)\, \Delta A_i$$

*exists as the mesh of the subdivision approaches 0, regardless of how the subdivisions are made or the points $P_i$ are chosen. If $f(P) \geqq 0$ for all points $P$ in $R$, then the above limit is the volume of the solid under the graph of $f$ and over $R$.*

Notation and terminology for the above limit is given by the following definition.

**DEFINITION**   The limit of Theorem 1 is called the *double integral of $f$ over $R$* and is denoted by

$$\int_R \int f(P)\, dA = \lim \sum_{i=1}^{n} f(P_i)\, \Delta A_i.$$

Just as the ordinary definite integral has a few simple and useful general properties, so also does the double integral. The following theorem gives these properties. Their proofs are left as exercises. Observe that they are all rather obvious on the basis of a volume interpretation of the double integral.

**THEOREM 2**    *If the region R is as in Theorem 1, and the functions f and g are continuous in R, then*

(i)    $\displaystyle \int_R \int cf(P)\, dA = c \int_R \int f(P)\, dA,$      $c = constant;$

(ii)    $\displaystyle \int_R \int [f(P) + g(P)]\, dA = \int_R \int f(P)\, dA + \int_R \int g(P)\, dA,$

(iii)    *If* $m \leqq f(P) \leqq M$ *for constants m and M and all P in R, then*

$$mA \leqq \int_R \int f(p)\, dA \leqq MA.$$

(iv) *If R is the union of two subregions R′ and R″, of areas A′ and A″, respectively, such that*

$$A = A' + A'',$$

  *then*

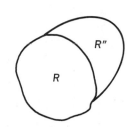

$$\int_R \int f(P)\, dA = \int_{R'} \int f(P)\, dA + \int_{R''} \int f(P)\, dA.$$

For later reference we state without proof a theorem about area that says, in effect, that the boundary of the region has zero area. We use the notation of this section. Let us call a subregion $\Delta R_i$ a *proper* subregion if it does *not* have as part of its boundary, a part of the boundary of R. In the figure the proper subregions are shaded.

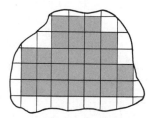

Proper subregions

**THEOREM 3**    *If the sum* $\sum \Delta A_i$ *is taken only over the proper subregions, then the limit of such sums, as the mesh of the subdivision approaches zero, is the area:*

$$\lim \sum \Delta A_i = A.$$

**COROLLARY**    *If the Riemann sums for the double integral of f over the region R are taken only over the proper subregions, then the limit, as the mesh approaches zero, of these restricted Riemann sums is the double integral*

$$\int_R \int f(P)\, dA = \lim \sum f(P_i)\, \Delta A_i. \ \{\textit{The sum over the proper subregions.}$$

The corollary is almost immediate from property (iii) of Theorem 2. It is specially obvious when $f(P)$ is positive everywhere, for then the contribution of the non-proper subregions cannot exceed the maximum value of $f$ times the area of the non-proper subregions, and this is a small number.

***Example***     Suppose $R$ is the square shown and that

$$f(P) = f(x, y) = x + y.$$

We will approximate the double integral of $f$ by using a subdivision of $R$ into the 16 equal squares shown. The point $P_i$ in the $i$th square will be selected as the center of the $i$th square. Then

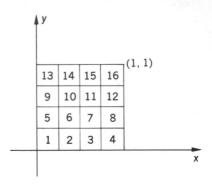

$$\Delta A_i = \tfrac{1}{16} = 0.0625$$

and

$$f(P_1) = 0.25, \quad f(P_2) = 0.50,$$
$$f(P_3) = 0.75, \quad f(P_4) = 1.00,$$
$$f(P_5) = 0.50, \quad f(P_6) = 0.75,$$
$$f(P_7) = 1.00, \quad f(P_8) = 1.25,$$
$$f(P_9) = 0.75, \quad f(P_{10}) = 1.00,$$
$$f(P_{11}) = 1.25, \quad f(P_{12}) = 1.50,$$
$$f(P_{13}) = 1.00, \quad f(P_{14}) = 1.25,$$
$$f(P_{15}) = 1.50, \quad f(P_{16}) = 1.75.$$

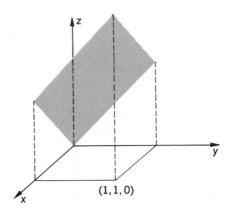

The Riemann sum, approximating the integral is

$$S_{16} = \sum f(P_i) \, \Delta A_i = 1.00,$$

which happens to be exactly equal to the double integral in this case.

## Problems

In Problems 1, 2, and 3 find Riemann sums approximating the double integral of $f$ using the given subdivision and points $P_i$.

1.   The region, function, and subdivision of the example, choosing for $P_i$ the point in $\Delta R_i$ nearest the origin.

2.   The region, function, and subdivision of the example choosing for $P_i$ the point in $\Delta R_i$ farthest from the origin.

*3.* The region and function of the example and the subdivision consisting of the two subregions shown. Choose $P_2$ at the center of $\Delta R_2$ and $P_1$ successively at the points $(0, 0)$, $(0, 1)$, $(1, 1)$. What is the maximum that $f(P_1) \, \Delta A_1$ can be?

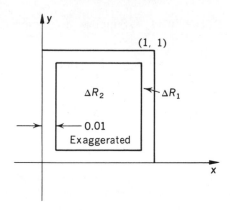

*4.* Draw as large a *quarter* circle as possible on $8\frac{1}{2} \times 11$ squared paper. Count the small squares completely inside the circle to approximate the area. If a function $f$ of maximum value 2 and minimum value $-1$ were integrated over this quarter circle, what effect can there be on the Riemann sums by counting only the squares completely inside the quarter circle?

*5.* Prove part (i) of Theorem 2.

*6.* Prove part (ii) of Theorem 2.

*7.* Prove part (iii) of Theorem 2. (Show that each Riemann sum satisfies the same inequalities.)

*8.* Prove part (iv) of Theorem 2. (Because the Riemann sums approach the integral over $R$ regardless of how the subdivisions are made, as long as the mesh approaches zero, one can consider subdivisions of $R$ that do not overlap on the curve separating $R'$ and $R''$.)

*\*9.* There is a mean value theorem for double integrals (if the region $R$ is connected; that is, in one piece). State and prove a mean value theorem for double integrals.

## 2  *Iterated Integrals*

At this stage evaluation of double integrals requires the calculation of Riemann sums and finding their limit as the mesh approaches zero. As with single integrals, there is a shortcut, an algorithm. We calculate, instead of the double integral, an *iterated integral*, which has the same value as the double integral.

**DEFINITION** Suppose a region $R$ is bounded by curves

$$y = y_1(x) \quad \{\text{Bottom}$$

and $\qquad y = y_2(x) \quad \{\text{Top}$

for $a \leq x \leq b$, and suppose that $f(P) = f(x, y)$ is a continuous function in $R$. Then an *iterated integral of $f$ over $R$* is

$$(1) \quad I = \int_a^b \left[ \int_{y_1(x)}^{y_2(x)} f(x, y) \, dy \right] dx,$$

where in evaluating the inside integral $x$ is held constant.

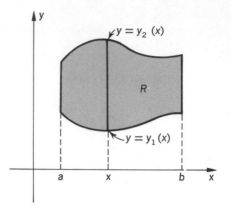

$$(2) \quad \text{If} \qquad\qquad A(x) = \int_{y_1(x)}^{y_2(x)} f(x, y) \, dy$$

with $x = $ constant, then the iterated integral is

$$I = \int_a^b A(x) \, dx.$$

Other notations than that of (1) are used for the iterated integral. One of these simply omits the brackets around the inside integral. Then

$$(3) \qquad\qquad\qquad I = \int_a^b \int_{y_1(x)}^{y_2(x)} f(x, y) \, dy \, dx.$$

The usual notation for the integral, and the one we shall use hereafter, is obtained by placing $dx$ between the integral signs:

$$(4) \qquad\qquad\qquad I = \int_a^b dx \int_{y_1(x)}^{y_2(x)} f(x, y) \, dy.$$

If the region is bounded "left and right" by curves $x = x_1(y)$ and $x = x_2(y)$. respectively, as shown, then the iterated integral is

$$(5) \qquad I = \int_c^d dy \int_{x_1(y)}^{x_2(y)} f(x, y) \, dx.$$

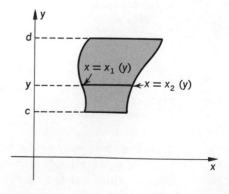

Observe that in both (4) and (5) *the limits on the integrals depend only on the region $R$ and not on the function $f$. The region $R$ is determined by its boundary curves, and it is these curves that give the upper and lower limits of the integral.

**Example 1**    If $R$ is the quarter circle shown, and

$$f(x, y) = xy\sqrt{a^2 - y^2},$$

then the iterated integral (4) is

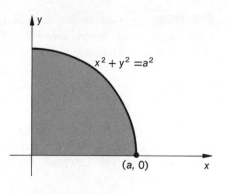

$$I = \int_0^a dx \int_0^{\sqrt{a^2-x^2}} xy\sqrt{a^2 - y^2}\, dy$$

$$= \int_0^a dx \left[ \frac{-x}{3} (a^2 - y^2)^{3/2} \right]_0^{\sqrt{a^2-x^2}}$$

$$= \int_0^a \left( \frac{-x^4}{3} + \frac{a^3x}{3} \right) dx = \frac{a^5}{10}.$$

In this example the region is also bounded left and right. The iterated integral (5) is

$$I = \int_0^a dy \int_0^{\sqrt{a^2-y^2}} xy\sqrt{a^2 - y^2}\, dx$$

$$= \int_0^a dy \left[ \frac{x^2}{2} y\sqrt{a^2 - y^2} \right]_0^{\sqrt{a^2-y^2}}$$

$$= \int_0^a \frac{y}{2}(a^2 - y^2)^{3/2}\, dy = \frac{a^5}{10}.$$

Observe that the two iterated integrals are equal. This is always the case because both are equal to the double integral of $f$ over $R$, as we will see.

**Example 2**    The region $R$, over which the integration takes place, can be found from the form of an iterated integral. Observe that the function $f$ does not enter into the considerations. If

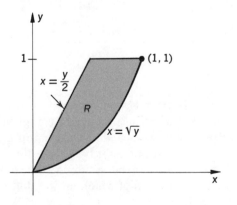

$$I = \int_0^1 dy \int_{y/2}^{\sqrt{y}} e^{\sin(x+y)}\, dx,$$

then $R$ is as shown.

Note that the iterated integral with the integrations performed in reverse order is less convenient because the top curve consists of two pieces, $y = 2x$ over part and $y = 1$ over the remainder.

**THEOREM**    *If the region R is bounded either top and bottom, or left and right by curves so that an iterated integral of f over R exists, then the iterated integral is equal to the double integral,*

$$\int_R \int f(P)\ dA = \int_a^b dx \int_{y_1(x)}^{y_2(x)} f(x, y)\ dy.$$

For a proper proof of this equality the reader is referred to advanced calculus texts. The equality can be made convincing when the function $f$ is positive by observing that both integrals are the volume of the solid under the graph of $f$.

With reference to the figure, the integral

$$A(x) = \int_{y_1(x)}^{y_2(x)} f(x, y)\ dy$$

is simply the area of the cross section of the solid by a plane perpendicular to the $x$-axis. By the methods of Chapter 12 (page 327) the integral

$$\int_a^b A(x)\ dx$$

is the volume, which is equal to the double integral:

$$\text{Volume} = \int_R \int f(P)\ dA = \int_a^b A(x)\ dx$$

$$= \int_a^b dx \int_{y_1(x)}^{y_2(x)} f(x, y)\ dy.$$

In the cases described in this book where the regions are more complicated, so that they are bounded neither top and bottom nor left and right by simple curves, then the region can be subdivided into smaller regions in each of which an iterated integral is defined.

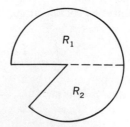

region not bounded top and bottom or left and right by graphs of functions Subregions $R_1$, $R_2$ are so bounded.

We can also regard the iterated integral as being obtained from Riemann sums for the double integral when the subregions $\Delta R_i$ are rectangles. Subdivide the region into rectangles (and parts of rectangles near the boundary) by vertical and horizontal lines:

$$x = x_i, \quad i = 0, \ldots, n;$$

$$y = y_j, \quad j = 0, \ldots, m,$$

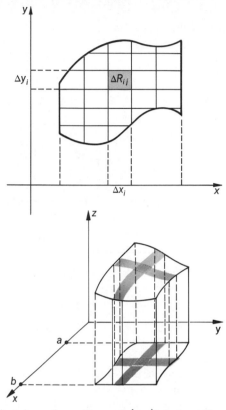

as shown in the figure. The proper regions then can be enumerated using double subscripts: $\Delta R_{ij}$, where the subscript $i$ goes from 0 to $n$, and the subscript $j$ may not always run from 0 to $m$. For the subregions that are rectangles, one has

$$\text{Area} = \Delta R_{ij} = \Delta A_{ij} = \Delta x_i \, \Delta y_j.$$

By Theorem 3 of Section 1, the sum $\sum f(P_{ij}) \, \Delta A_{ij}$, where $P_{ij} = (x_i', y_j')$ is in $\Delta R_{ij}$, taken over the rectangular subregions only, will approach the double integral when the mesh approaches zero. This double sum

$$\sum_i \sum_j f(x_i', y_j') \, \Delta x_i \, \Delta y_j,$$

where for each $i$ the index $j$ runs through the appropriate range, is suggestive of two successive definite integrations, so that in the limit one should have

$$\int_R \int f(P) \, dA = \int_a^b dx \int_{y_1(x)}^{y_2(x)} f(x, y) \, dy.$$

From these considerations the $dA$ in the double integral is often written in terms of the coordinate system:

The differential element of area is

$$dA = dx \, dy = dy \, dx.$$

Then
$$\int_R \int f(P) \, dA = \int_R \int f(x, y) \, dx \, dy$$

$$= \int_R \int f(x, y) \, dy \, dx.$$

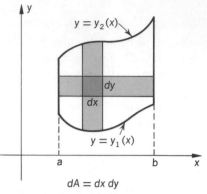

One can think of the iterated integral either as first summing the elements $f(x, y) \, dy$ to get

$$A(x) = \int_{y_1(x)}^{y_2(x)} f(x, y) \, dy,$$

and then summing the elements $A(x) \, dx$,
or, if the region permits, as first summing the elements $f(x, y) \, dx$ to get

$$B(y) = \int_{x_1(y)}^{x_2(y)} f(x, y) \, dx,$$

and then summing the elements $B(y) \, dy$.

Finally, we observe that the area of the region $R$ is the double integral of the constant function, $f(x, y) = 1$.

$$A = \int_R \int dA.$$

**Example 3**   Find the volume under the graph of $z = (a^2 - y^2)^{3/2}$ and above the region $R$, shown in the figure. The volume is the double integral

$$I = \int_R \int (a^2 - y^2)^{3/2} \, dx \, dy.$$

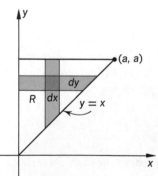

An iterated integral for $I$ is

$$I = \int_0^a dx \int_x^a (a^2 - y^2)^{3/2} \, dy.$$

Although the integrals are elementary, the first integration requires either parts, substitution, or use of tables. If we integrate first with respect to $x$ we get

$$I = \int_0^a dy \int_0^y (a^2 - y^2)^{3/2} \, dx = \int_0^a (a^2 - y^2)^{3/2} y \, dy$$

$$= -\frac{1}{5} (a^2 - y^2)^{5/2} \Big|_0^a = \frac{a^5}{5}.$$

Rather surprisingly, the integrations were trivial this way.

**Example 4**   Find the area of the region under the graph of $y = -\frac{1}{2}x^2 + \frac{1}{2}x + 1$, above the line $y = x$, and between $x = -1$ and $x = 1$.

From the figure we have

$$A = \int_R \int dA = \int_R \int dy\, dx$$

$$= \int_{-1}^{1} dx \int_{x}^{(-x^2+x+2)/2} dy$$

$$= \int_{-1}^{1} (-\tfrac{1}{2}x^2 - \tfrac{1}{2}x + 1)\, dx = \tfrac{5}{3}.$$

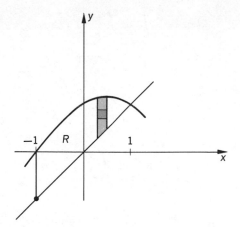

**Example 5**   The reader should be warned that, even when the region is symmetric, the double integral need not be twice the integral over half the region. The function $f$ must also be symmetric. Thus, with the region shown, and $f(x, y) = x + y$,

$$\int_R \int (x + y)\, dA = \int_{-1}^{1} dx \int_{0}^{1-x^2} (x + y)\, dy$$

$$\neq 2 \int_{0}^{1} dx \int_{0}^{1-x^2} (x + y)\, dy.$$

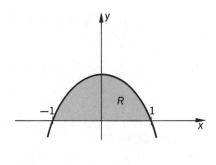

### Problems

In Problems 1 to 8 evaluate the iterated integral and sketch the region $R$ over which the integration takes place.

1. $\displaystyle \int_{0}^{a} dx \int_{0}^{x^2} \sqrt{a^2 - y}\, dy$     $\left[ \left(\dfrac{2}{3} - \dfrac{\pi}{8}\right) a^4 \right]$

2. $\displaystyle \int_{0}^{a} dx \int_{0}^{x^2} e^{y/x}\, dy$     $\left[ e^a (a - 1) - \dfrac{a^2}{2} + 1 \right]$

3. $\displaystyle \int_{a}^{2a} dy \int_{y-a}^{y} xy\, dx$     $\left[ \dfrac{19}{12} a^4 \right]$

4. $\displaystyle \int_{0}^{a} dy \int_{y}^{2y} \sqrt{xy - y^2}\, dx$     $\left[ \dfrac{2a^3}{9} \right]$

5. $\int_{-a}^{a} dx \int_{b-\sqrt{a^2-x^2}}^{b+\sqrt{a^2-x^2}} (x+y)\, dy$        $[a^2 b \pi]$

6. $\int_{0}^{2} dx \int_{0}^{\sqrt{2}\sqrt{2-x}} \dfrac{dy}{\sqrt{2x-x^2}}$        $[4]$

7. $\int_{0}^{a} dy \int_{0}^{\sqrt{a^2-y^2}} \sqrt{a^2-x^2-y^2}\, dx$        $[\frac{4}{3}\pi a^3]$

8. $8\int_{0}^{2} dy \int_{\sqrt{y}}^{2+(y^2/2)} xy\, dx$        $[8]$

In Problems 9 to 12 evaluate the double integrals over the regions R. If the region R permits it, express the double integrals in two ways as iterated integrals.

9. $\displaystyle\int_{R}\!\!\int (xy-y)\, dx\, dy$, where $R$ is bounded by $x=-1$, $x=1$, $y=0$, $y=x+1$

10. $\displaystyle\int_{R}\!\!\int \sqrt{1+y^2}\, dx\, dy$, where $R$ is bounded by $x=1$, $x=y$, $y=-1$

$$[\sqrt{2}+\log{(1+\sqrt{2})}]$$

11. $\displaystyle\int_{R}\!\!\int \dfrac{x}{x^2+y^2}\, dx\, dy$, where $R$ consists of the points $(x,y)$ such that

$$1 \leqq x \leqq \sqrt{2-y^2}, \qquad -1 \leqq y \leqq 1$$

12. $\displaystyle\int_{R}\!\!\int dx\, dy$, where the point $(x,y)$ is in $R$ if and only if

$$0 \leqq x \leqq 1 \qquad \text{and} \qquad x^2 \leqq y \leqq 1$$

In Problems 13 to 18 find, using double integrals, the area of the region R bounded by the given curves.

13. $y=\sin x$, $y=\cos x$, where $\pi/4 \leqq x \leqq 5\pi/4$

14. $y=\sin x$, $y=\cos x$, where $0 \leqq x \leqq \pi/4$

15. $x^{1/2}+y^{1/2}=a^{1/2}$, $x=0$, $y=0$        $[a^2/6]$

16. $x^{1/2}+y^{1/2}=a^{1/2}$, $x+y=a$        $[a^2/3]$

17. $x^2+y^2=2$, $y^2 \leqq x^3$        $[(\pi/2)-\frac{1}{5}]$

18. $x^2-y^2=4$, $x^2+y^2=36$, $x \geqq 2$        $[36\text{ Arc }\sin \frac{2}{3} - 4\log{(2+\sqrt{5})}]$

19. Find the volume in the first octant bounded by the coordinate planes and the plane $bcx+cay+abz=abc$, where $a,b,c>0$.        $[\frac{1}{6}abc]$

20.  Find the volume of the wedge shaped solid bounded by $z = y$, $x^2 + y = 4$, $z = 0$.    $[256/15]$

21.  Find the volume under the plane $x + y + z = 6$, above the plane $z = 0$, and with sides $y = 0$, $y = x$, and $x = 3$.    $[27/2]$

22.  Find the volume of the solid in the first octant bounded by $x = \sin y$, $x^2 + z^2 = 1$, $y = \pi/2$.    $[\frac{1}{16}(4 + \pi^2)]$

23.  Find the volume of the solid above the $xy$-plane and inside the cylinders $y^2 = 4 - 2x$, $x^2 = 4 - z$.    $[1024\sqrt{2}/35]$

24.  The region $R$ is inside the square with opposite vertices at $(0, 0)$ and $(3, 3)$ and outside the square with opposite vertices $(1, 1)$ and $(2, 2)$. Find

$$\int_R \int y \, dx \, dy.$$

25.  Find the volume of the solid in the first octant, inside the cylinder $4x^2 + y^2 = 4$ and below the plane $z = x + y$.    $[2]$

26.  Find the volume in the first octant inside the cylinders $z = 4 - x^2$, $x = 2 - y^2$. Evaluate both iterated integrals, with respect to $x$ and $y$, that give the volume. Also set up the volume as an iterated integral with respect to $x$ and $z$.    $[144\sqrt{2}/35]$

27.  Evaluate the integral

$$\int_1^e dy \int_1^y \frac{x^2}{y} \, dx.$$

Then express it as an iterated integral, integrating first with respect to $y$.

$$\left[\int_1^e dx \int_x^e \frac{x^2}{y} \, dy = \frac{e^3 - 4}{9}\right]$$

28.  Express the iterated integral

$$\int_0^{\sqrt{2}} dx \int_{x^2}^2 e^{x/\sqrt{y}} \, dy,$$

as an iterated integral in the opposite order, and evaluate it.

$$\left[\frac{4\sqrt{2}}{3}(e - 1)\right]$$

29.  The iterated integral

$$\int_0^1 dy \int_0^y (1 - y) \, dx$$

is the volume of what solid? Sketch the solid.

## 3    *Polar Coordinates*

We have, in Section 2, evaluated double integrals as iterated integrals using rectangular coordinates. With rectangular coordinates the area $\Delta A$, of the subregion $\Delta R$ in the double integral, is naturally chosen to be $\Delta x\,\Delta y$.

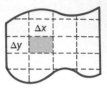

For regions bounded by curves described in polar coordinates another choice for $\Delta A$ is desirable. We usually shall deal with regions bounded by curves given in polar coordinates by

$$
\left.
\begin{aligned}
r &= r_1(\theta) \quad (\text{``inner''}) \\[2mm]
r &= r_2(\theta) \quad (\text{``outer''})
\end{aligned}
\right\} \quad \alpha \leq \theta \leq \beta,
$$

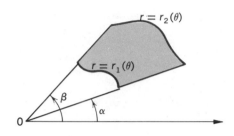

where $0 \leq r_1(\theta) \leq r_2(\theta)$. The subregions $\Delta R_{ij}$ are truncated sectors, or parts of such, with area

$$
\begin{aligned}
\Delta A_{ij} &= \tfrac{1}{2}\big[(r_i + \Delta r_i)^2\,\Delta\theta_j - r_i^2\,\Delta\theta_j\big] \\[2mm]
&= \tfrac{1}{2}(2r_i\,\Delta r_i + \Delta r_i^2)\,\Delta\theta_j \\[2mm]
&= \left(r_i + \frac{\Delta r_i}{2}\right)\Delta r_i\,\Delta\theta_j.
\end{aligned}
$$

If we consider only *proper* subregions $\Delta R_{ij}$, the *limit* of the sum $\sum_{i,j}\Delta A_{ij}$ will be exactly the area $A$ of $R$, because the two boundary curves have zero area, just as was the case with rectangular coordinates.

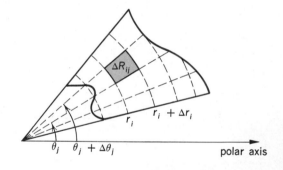

Furthermore, the double integral will be the limit of Riemann sums over these proper subregions. The Riemann sums (over the proper subregions) that approximate the double integral of $f$ over $R$ are then

$$
(1) \qquad \sum_{i,j} f(P_{ij})\,\Delta A_{ij} = \sum_{i,j} f(r_i', \theta_j')\left(r_i + \frac{\Delta r_i}{2}\right)\Delta r_i\,\Delta\theta_j.
$$

These sums approach the double integral

$$
\int_R\!\int f(P)\,dA,
$$

as the mesh approaches zero.

Now we interpret the sum (1) as an approximating Riemann sum for a double integral over a region $\bar{R}$ in the $r\theta$-plane. However, the sum (1) will not have quite the correct form unless

$$r'_i = r_i + \frac{\Delta r_i}{2}.$$

But this choice for $r'_i$ is possible because the point $P_{ij}$ can be chosen anywhere in $\Delta \bar{R}_{ij}$. Then the sum (1) becomes

$$\sum_{i,j} f(r'_i, \theta'_j) r'_i \, \Delta r_i \, \Delta \theta_j,$$

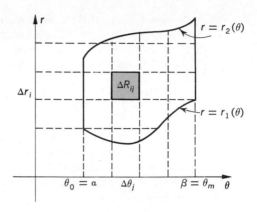

and the limit of this, as the mesh approaches zero, is the double integral

(2)
$$\int_R \int f(r, \theta) r \, dr \, d\theta.$$

Just as with rectangular coordinates we evaluate this as an iterated integral

(3)
$$\int_R \int f(P) \, dA = \int_\alpha^\beta d\theta \int_{r_1(\theta)}^{r_2(\theta)} f(r, \theta) r \, dr.$$

If the region is described by two circular arcs and two curves,

$$\left. \begin{array}{c} \theta = \theta_1(r) \\[2mm] \theta = \theta_2(r) \end{array} \right\} \quad a \leqq r \leqq b$$

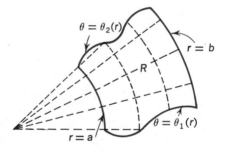

as pictured, then

(4)
$$\int_R \int f(P) \, dA = \int_a^b dr \int_{\theta_1(r)}^{\theta_2(r)} f(r, \theta) r \, d\theta.$$

*Remark.* Observe the factor $r$ in the double integral (2) and in the iterated integrals in (3) and (4). The differential element of area* is

(5)
$$dA = r \, dr \, d\theta$$

The differential rectangle

and is easily remembered by considering the partial sector as a "rectangle" of dimensions $r \, d\theta$ and $dr$.

---

* There is a general theory, for which the reader is referred to an advanced calculus text, that gives the differential area in any coordinate system. If the new coordinates are $u$, $v$ and if $x$ and $y$ are rectangular coordinates then $dA = J \, du \, dv$ where $J$ is the *Jacobian determinant*

$$\begin{vmatrix} \partial x/\partial u & \partial x/\partial v \\[1mm] \partial y/\partial u & \partial y/\partial v \end{vmatrix}.$$

The reader may wish to obtain equation (5) in this way with $u = r$ and $v = \theta$.

*Example 1*    Find the area of the region inside the circle $r = 2a$ and outside the parabola $r = a \sec^2 (\theta/2)$. The region and the differential element of area are shown in the figure.

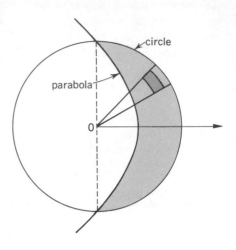

Here the inner boundary of the region is $r = a \sec^2 (\theta/2)$, and the outer boundary is $r = a$. The limits on $\theta$ are $-\pi/2$ and $\pi/2$, but because the region and the integrand are symmetric in $\theta$ we may take twice the integral over half the region. The iterated integral for $A$ is, therefore,

$$A = 2 \int_0^{\pi/2} d\theta \int_{a \sec^2 (\theta/2)}^{2a} r\, dr$$

$$= 2 \int_0^{\pi/2} \left( 2a^2 - \frac{a^2}{2} \sec^4 \frac{\theta}{2} \right) d\theta$$

$$= (2\pi - \tfrac{8}{3})a^2.$$

Had the region been a thin sheet of material of variable density $\rho$, proportional to $r$, so that $\rho = kr$, then the mass of the material is

$$M = \int_R \int kr\, dA = \int_R \int kr\, r\, dr\, d\theta$$

$$= 2 \int_0^{\pi/2} d\theta \int_{a \sec^2 (\theta/2)}^{2a} kr^2\, dr = \frac{ka^3}{3} \left( 8\pi - \frac{112}{15} \right).$$

*Example 2*    Find the volume of the solid inside the cylinder $x^2 + y^2 - 2y = 0$, above the $xy$-plane and under the paraboloid $z = x^2 + y^2$.

The part of the solid in the first octant is shown in the figure. For this solid polar coordinates in the $xy$-plane are convenient (then $r$, $\theta$, $z$ constitute cylindrical coordinates). In these coordinates the equations of the surfaces are $r = 2 \sin \theta$ (cylinder) and $z = r^2$ (paraboloid). The inner boundary of $R$ is $r = 0$. The outer boundary of $R$ is $r = 2 \sin \theta$. The limits on $\theta$ are $\theta = 0$ and $\theta = \pi/2$. Therefore, the total volume (twice that pictured) is

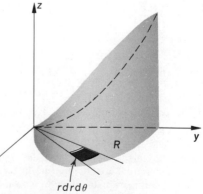

$$V = 2 \int_R \int z\, dA = 2 \int_R \int r^2\, r\, dr\, d\theta$$

$$= 2 \int_0^{\pi/2} d\theta \int_0^{2 \sin \theta} r^3\, dr = 8 \int_0^{\pi/2} \sin^4 \theta\, d\theta = \frac{3\pi}{2}.$$

*Example 3*    Iterated integrals in polar coordinates can be interpreted as double integrals. Thus,

$$\int_{\pi/2}^{\pi} d\theta \int_{a}^{a(1-\cos\theta)} r\cos\theta \, dr$$

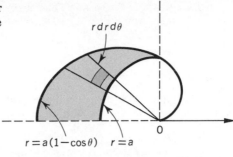

$$= \iint_{R} \cos\theta \, dA$$

where the region $R$ is outside the circle $r = a$ and inside the cardioid $r - a(1 - \cos\theta)$. The function $f$, integrated over $R$ is $f(r, \theta) = \cos\theta$.

### Problems

In Problems 1 to 4 evaluate the iterated integral and express it as a double integral of a function $f$ over a region $R$.

1. $$\int_{-\pi/2}^{\pi/2} d\theta \int_{0}^{2\cos\theta} r^2 \, dr$$

$$\left[\begin{array}{l} 32/9, \ f(r,\theta) = r, \\ R \text{ bounded by } r = 2\cos\theta \end{array}\right]$$

2. $$\int_{0}^{\pi/2} d\theta \int_{0}^{a} \sqrt{a^2 - r^2} \, r \, dr$$

$$\left[\begin{array}{l} f(r,\theta) = \sqrt{a^2 - r^2}, \\ R = \text{quarter circle} \end{array}\right]$$

3. $$\int_{0}^{\pi/4} d\theta \int_{0}^{a\sec\theta} r^{3/2} \sin\theta \, dr$$

$$\left[\begin{array}{l} 4a^{5/2}(\sqrt{2\sqrt{2}} - 1)/15, \\ f(r,\theta) = \sqrt{r}\sin\theta, \\ R = \text{triangle bounded by} \\ \quad x = a, x = y, y = 0 \end{array}\right]$$

4. $$\int_{a}^{2a} dr \int_{r}^{2r} r \, d\theta$$

$$\left[\begin{array}{l} f(r,\theta) = 1, \\ R = \text{region between circles} \\ \quad r = a, r = 2a \text{ and} \\ \quad \text{spirals } \theta = r, \theta = 2r. \end{array}\right]$$

In Problems 5 to 13 find, using double integrals and iterated integrals in polar coordinates, the areas of the regions bounded by the given curves.

5. $r^2 = a^2 \cos 2\theta$ $\qquad\qquad\qquad\qquad\qquad\qquad\qquad [a^2]$

6. $r = a\sec\theta, r = 2a\cos\theta$, where $-\pi/4 \leq \theta \leq \pi/4$ $\qquad [\pi a^2/2]$

7. $r = a, r = 2a\cos\theta$    (Larger part) $\qquad\qquad\qquad [a^2(\pi/3 + \sqrt{3}/2)]$

8. $r = a, r = a(1 - \cos\theta)$    (Inside both) $\qquad\qquad [(5\pi/4 - 2)a^2]$

9. $r = 2a\cos\theta$, polar axis $\qquad\qquad\qquad\qquad\qquad\qquad [\pi a^2/2]$
   (Integrate first with respect to $\theta$.)

10.   $r = a(1 + 2\cos\theta)$ (The larger loop)                    $[a^2(2\pi + 3\sqrt{3}/2)]$

11.   $r = a\cos 3\theta$ (One loop)                    $[\pi a^2/12]$

12.   $y = x^2, y = 2x$                    $[\frac{4}{3}]$

13.   $x^2 + y^2 - 4y = 0$                    $[4\pi]$

14.   A circular disc of radius $a$ has variable density $\rho$ proportional to the radius, so that $\rho = kr$. Find its mass.                    $[2\pi ka^3/3]$

15.   A thin sheet of metal is in the shape of one leaf of the rose $r = a\cos 3\theta$. If the density is equal to the distance from the line of symmetry of the leaf, find its mass.                    $[19a^3/960]$

16.   A thin metal washer of radii $r_1$ and $r_2 > r_1$ has density equal to $k/r$, where $r$ is the distance from the center. Find its mass.                    $[2\pi k(r_2 - r_1)]$

**In Problems 17, 18, and 19 evaluate the iterated integrals by interpreting them as double integrals and by using polar coordinates.**

17.   $\displaystyle\int_0^1 dy \int_{y^2}^y dx$                    $[\frac{1}{6}]$

18.   $\displaystyle\int_0^a dy \int_0^{(a^2-y^2)/2a} dx$                    $[a^2/3]$

19.   $\displaystyle\int_{-a}^a dx \int_{-\sqrt{a^2-x^2}}^{\sqrt{a^2-x^2}} e^{x^2+y^2}\, dy$                    $[\pi(e^{a^2} - 1)]$

(Note the difficulty as it stands.)

20.   Derive from a double integral the formula for area in polar coordinates of Chapter 12,

$$A = \frac{1}{2}\int_\alpha^\beta r^2\, d\theta.$$

**In Problems 21 to 28 find, using polar coordinates, the volume $V$ of the solid bounded as described.**

21.   $z = \sqrt{16 - 4(x^2 + y^2)}, z = 0, y = 0, y = 3x$, in the first octant
                    $[\frac{16}{3}\text{ Arc tan } 3]$

22.   Inside the cone $z = \sqrt{x^2 + y^2}$ and inside the sphere $x^2 + y^2 + z^2 = a^2$
                    $\left[\dfrac{(2 - \sqrt{2})}{3}\pi a^3\right]$

23.   $z = \sqrt{x^2 + y^2}, z = 0, x^2 + y^2 - 2x = 0$                    $[32/9]$

24.    Inside the cylinder $x^2 + y^2 = a^2$ and inside the sphere $x^2 + y^2 + z^2 = b^2$, $b > a$

$$\left[\frac{4\pi}{3}\left(b^3 - (b^2 - a^2)^{3/2}\right)\right]$$

25.    Inside the cylinder $x^2 + y^2 - ay = 0$, and the sphere $x^2 + y^2 + z^2 = a^2$

$$\left[\frac{6\pi - 8}{9}\, a^3\right]$$

26.    The cylinder $r = 1 + \cos\theta$, $z = 0$, $z = y$, and where $y \geq 0$          $\left[\frac{4}{3}\right]$

27.    Outside the hyperboloid $z^2 = x^2 + y^2 - a^2$ and inside the cylinder $x^2 + y^2 = b^2$, $b > a$

$$\left[\frac{4\pi}{3}(b^2 - a^2)^{3/2}\right]$$

28.    $z = \sqrt{x^2 + y^2 + a^2}$, $z = 0$, $x^2 + y^2 = b^2$

$$\left[\frac{2\pi}{3}\left((b^2 + a^2)^{3/2} - a^3\right)\right]$$

29.    Interpret the integral

$$\int_0^{\pi/2} d\theta \int_0^1 (2 - r\cos\theta - 2r\sin\theta)r\, dr$$

as a volume. Sketch the solid.

## 4    First Moments of Plane Regions

Suppose that $(x_1, y_1)$, ..., $(x_n, y_n)$ are $n$ points in the plane, and that associated with each point $(x_i, y_i)$ is a "mass," or "weight," $m_i$. We have, then, a system of mass points, or mass particles.

The *first moments* of the system of mass particles with respect to the $x$- and $y$-axes are *defined* to be:

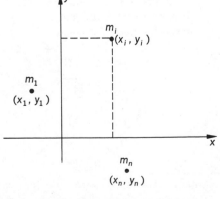

$$
(1)\quad
\begin{cases}
\begin{aligned}
M_x &= m_1 y_1 + \cdots + m_n y_n \\
&= \sum_{i=1}^{n} m_i y_i
\end{aligned}
\left\{
\begin{array}{l}
\text{First moment} \\
\text{with respect to} \\
\text{the } x\text{-axis}
\end{array}
\right. \\[2em]
\begin{aligned}
M_y &= m_1 x_1 + \cdots + m_n x_n \\
&= \sum_{i=1}^{n} m_i x_i
\end{aligned}
\left\{
\begin{array}{l}
\text{First moment} \\
\text{with respect to} \\
\text{the } y\text{-axis}
\end{array}
\right.
\end{cases}
$$

If the total mass

$$M = \sum_{i=1}^{n} m_i$$

were concentrated at one point $(\bar{x}, \bar{y})$ the moments of that mass particle would be $\bar{y}M$ and $\bar{x}M$ with respect to the $x$- and $y$-axes. The point $(\bar{x}, \bar{y})$ such that these moments are equal to the moments $M_x$ and $M_y$ of the *system* of mass particles is called the *center of mass** of the system. Then the center of mass $(\bar{x}, \bar{y})$ has coordinates

(2) $$\bar{x} = M_y/M \qquad \text{and} \qquad \bar{y} = M_x/M.$$

**Example 1**    Masses $m_1 = 1$, $m_2 = 7$, and $m_3 = 4$ are located at the points $(3, -2)$, $(-5, 3)$, and $(\frac{1}{2}, -\frac{1}{4})$, respectively. Find the center of mass of the system.

$$\bar{x} = \frac{(3)(1) + (-5)(7) + (\frac{1}{2})(4)}{1 + 7 + 4} = -\frac{5}{2},$$

$$\bar{y} = \frac{(-2)(1) + (3)(7) + (-\frac{1}{4})(4)}{1 + 7 + 4} = \frac{3}{2}.$$

**Example 2**    The center of mass of a system of particles is independent of the coordinate system used.

We shall show this only for a translation of axes. A proof for a rotation of axes is quite similar.

Suppose that the $xy$- and $x'y'$-axes are as shown so that

$$x' = x - h \qquad \text{and} \qquad y' = y - k.$$

With the system of mass particles in the text we compute the center of mass with respect to the $x'$, $y'$ coordinates

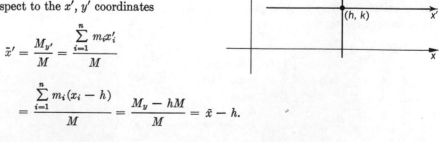

$$\bar{x}' = \frac{M_{y'}}{M} = \frac{\displaystyle\sum_{i=1}^{n} m_i x_i'}{M}$$

$$= \frac{\displaystyle\sum_{i=1}^{n} m_i (x_i - h)}{M} = \frac{M_y - hM}{M} = \bar{x} - h.$$

---

* Center of mass is also called *center of gravity*. The numbers $\bar{x}$, $\bar{y}$ are also called *weighted averages*. First moments are also called *moments of mass*.

A similar computation shows that $\bar{y}' = \bar{y} - k$. Thus, the center of mass is at the same point relative to the system, regardless of the coordinate system. The center of mass is an *invariant*.

Consider a thin material plate of mass density* $\rho = \rho(x, y)$ units/unit area occupying a region $R$ of the plane. If, as in the definition of the double integral, $R$ is subdivided into subregions $\Delta R_i$, then (see the figure) the mass $\Delta M_i$ of the subregion $\Delta R_i$ is approximately $\Delta M_i = \rho(x_i, y_i) \, \Delta A_i$ where $\Delta A_i$ is the area of $\Delta R_i$. If we think of all the mass of the $i$th subregion as being concentrated at the point $(x_i, y_i)$, then the first moments of this system of mass particles are:

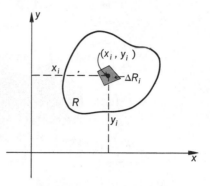

$$(3) \quad \begin{cases} M_x = y_1\rho(x_1, y_1) \, \Delta A_1 + \cdots + y_n\rho(x_n, y_n) \, \Delta A_n = \sum_{i=1}^{n} y_i\rho(x_i, y_i) \, \Delta A_i, \\[2em] M_y = x_1\rho(x_1, y_1) \, \Delta A_1 + \cdots + x_n\rho(x_n, y_n) \, \Delta A_n = \sum_{i=1}^{n} x_i\rho(x_i, y_i) \, \Delta A_i. \end{cases}$$

The total mass is approximately

$$(4) \quad M = \rho(x_1, y_1) \, \Delta A_1 + \cdots + \rho(x_n, y_n) \, \Delta A_n = \sum_{i=1}^{n} \rho(x_i, y_i) \, \Delta A_i.$$

Formulas (3) and (4) are Riemann sums for double integrals. The limit of (4) as the mesh of the subdivision goes to zero is the mass

$$M = \int_R \int \rho(x, y) \, dA.$$

We *define* the limits of the sums in (3) to be the first moments of the plate:

$$M_x = \int_R \int y \, \rho(x, y) \, dA,$$

$$M_y = \int_R \int x \, \rho(x, y) \, dA.$$

Then the center of mass $(\bar{x}, \bar{y})$ of the plate has coordinates

$$\bar{x} = M_y/M \qquad \text{and} \qquad \bar{y} = M_x/M.$$

In the examples and problems that follow we shall abandon the expression "material plate" and simply describe a region with specified density.

---

* If $\rho(x, y)$ is constant, the plate is homogeneous. If $\rho(x, y)$ is not constant, it might be made of varying substances of different densities.

*Example 3*    The density of a semi-circular region of radius $a$ is proportional to the square of the distance from the center of the circle. Find the center of mass.

Choose a coordinate system as shown. Then $\rho(x, y) = k(x^2 + y^2)$, with $k = $ constant

$$M_x = \int_R \int y\rho \, dA = \int_R \int k(x^2 + y^2)y \, dA$$

$$= k \int_{-a}^{a} dx \int_{0}^{\sqrt{a^2 - x^2}} (x^2 + y^2)y \, dy$$

$$= k \int_{-a}^{a} \left( x^2 \frac{(a^2 - x^2)}{2} + \frac{(a^2 - x^2)^2}{4} \right) dx = \frac{2ka^5}{5}.$$

$M_y$ is clearly zero from symmetry (verify this). The mass is

$$M = \int_{-a}^{a} dx \int_{0}^{\sqrt{a^2 - x^2}} k(x^2 + y^2) \, dy$$

$$= k \int_{-a}^{a} \left( x^2\sqrt{a^2 - x^2} + \frac{(a^2 - x^2)^{3/2}}{3} \right) dx$$

$$= \frac{k}{3} \int_{-a}^{a} (a^2 + 2x^2)\sqrt{a^2 - x^2} \, dx = \frac{\pi ka^4}{4},$$

where the last integration is somewhat lengthy. If polar coordinates are used we have, more simply,

$$M_x = \int_{0}^{\pi} d\theta \int_{0}^{a} kr^2(r \sin \theta)r \, dr = \frac{ka^5}{5} \int_{0}^{\pi} \sin \theta \, d\theta = \frac{2ka^5}{5}.$$

$$M = \int_{0}^{\pi} d\theta \int_{0}^{a} k \, r^2 \, r \, dr = \frac{\pi ka^4}{4}.$$

Then for the center of mass we have $\bar{x} = 0$ and

$$\bar{y} = (\tfrac{2}{5}ka^5) / (\tfrac{1}{4}\pi ka^4) = \frac{8a}{5\pi}.$$

If $\rho \equiv 1$ in the region, then

$$M = \int_R \int dA = A,$$

so the mass is numerically equal to the area of the region. In this case (and more generally when $\rho = $ constant) the center of mass is called the *centroid of the region* $R$. When $\rho = 1$, the first moments are called *moments of area*.

In the case of constant density single integrals can be used to compute first moments. Suppose the region is bounded top and bottom by $y = y_2(x)$ and $y = y_1(x)$ and left and right by $x = a$ and $x = b$. Then (5):

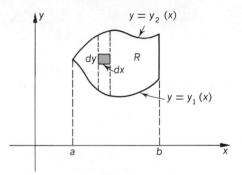

$$M_y = \int_R \int x \, dA = \int_a^b dx \int_{y_1(x)}^{y_2(x)} x \, dy.$$

$$= \int_a^b x[y_2(x) - y_1(x)] \, dx.$$

The integrand in (5) is the distance from the $y$-axis to the centroid of the rectangular area element times the area of the element. Similarly, for $M_x$ we have (6):

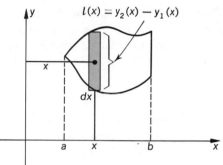

$$M_x = \int_R \int y \, dA = \int_a^b dx \int_{y_1(x)}^{y_2(x)} y \, dy$$

$$= \int_a^b \frac{y_2^2(x) - y_1^2(x)}{2} \, dx$$

$$= \int_a^b \frac{y_2(x) + y_1(x)}{2} [y_2(x) - y_1(x)] \, dx.$$

The integrand in (6) is the distance from the $x$-axis to the centroid of a rectangular area element times the area of the element.

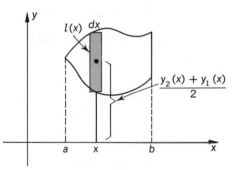

**Example 4**    Find the centroid of the plane region bounded by $y^2 = x$ and $y = x$.
    Here $y_2(x) = \sqrt{x}$, $y_1(x) = x$, $a = 0$, and $b = 1$.

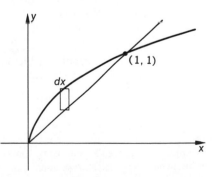

$$M_y = \int_0^1 x(\sqrt{x} - x) \, dx = \frac{1}{15},$$

$$M_x = \int_0^1 \frac{(\sqrt{x} + x)}{2} (\sqrt{x} - x) \, dx = \frac{1}{12},$$

$$M = A = \int_0^1 (\sqrt{x} - x) \, dx = \frac{1}{6}.$$

Hence,

$$\bar{x} = (\tfrac{1}{15})/(\tfrac{1}{6}) = \tfrac{2}{5}, \qquad \bar{y} = (\tfrac{1}{12})/(\tfrac{1}{6}) = \tfrac{1}{2}.$$

There is a connection between the centroid of a plane region and the volume of the solid obtained by revolving the region about a line of the plane.

**THEOREM (PAPPUS' SECOND THEOREM\*)**    *If a plane region be revolved about a line of the plane not meeting the region, except possibly on the boundary, the volume of the solid generated is equal to the area of the region times the circumference of the circle traversed by the centroid.*

**PROOF SKETCH**    Choose the $y$-axis as the axis of rotation, and suppose the $x$-axis is directed so the region is in the first quadrant. If the region is bounded as shown in the figure, then the volume of the solid of revolution is, using the method of shells,

$$V = 2\pi \int_a^b xl(x)\ dx.$$

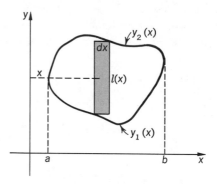

Now, referring back to equation (5), we observe that the above integral is just the first moment $M_y$. Hence,

$$V = 2\pi M_y = 2\pi \bar{x} A,$$

where $A$ is the area of the region.

*Problems*

In Problems 1 to 10 find moments or center of mass, as indicated, of the regions bounded by the given curves and with the given density. Use either rectangular or polar coordinates as seems best.

1.   $x = 0, y = 0, x = a, y = a\ (a > 0);\ \ \rho = k_1 + k_2 x, k_1, k_2$ constants. Find the center of mass.

$$\left[ \bar{x} = \frac{3k_1 a + 2k_2 a^2}{6k_1 + 3k_2 a}, \bar{y} = \frac{a}{2} \right]$$

2.   Same as Problem 1 but with $\rho = k\sqrt{x^2 + y^2}$.

$$\left[ \bar{x} = \bar{y} = \frac{a}{8}\frac{7\sqrt{2} - 2 + 3\log(\sqrt{2} + 1)}{\sqrt{2} + \log(\sqrt{2} + 1)} \right]$$

---

\* Pappus of Alexandria (300 A.D.) was the last of the great Greek geometers. The theorem occurs in his principal work, *Mathematical Collection*. Naturally, our proof is different from his.

3.  $y = \sqrt{2ax - x^2}$, $y = 0$; $\rho = k/\sqrt{x^2 + y^2}$, $k = $ constant. Find $(\bar{x}, \bar{y})$. (Note that in this problem the double integral is an improper one because the density becomes infinite at the origin. However, when polar coordinates are used the iterated integral is no longer improper.)                    $[\bar{x} = 2a/3, \bar{y} = a/3]$

4.  Inside the semicircle $r = 2a \cos \theta$, $0 \leq \theta \leq \pi/2$ and outside the circle $r = a$; $\rho = $ constant. Find $(\bar{x}, \bar{y})$.
$$\left[ \bar{x} = \frac{a(8\pi + 3\sqrt{3})}{2(2\pi + 3\sqrt{3})}, \bar{y} = \frac{11a}{2(2\pi + 3\sqrt{3})} \right]$$

5.  One leaf of the rose $r = a \cos 2\theta$, $-\pi/4 \leq \theta \leq \pi/4$; $\rho$ inversely proportional to the distance from the pole. Find $(\bar{x}, \bar{y})$.                    $[\bar{x} = 4\sqrt{2}a/15, \bar{y} = 0]$

6.  $x = 0$, $y = 0$, $x + y = a$; $\rho = k/\sqrt{x^2 + y^2}$. Find $M_x$ and $M_y$.
$$\left[ M_x = M_y = \frac{ka^2\sqrt{2}}{8} \log (3 + 2\sqrt{2}) \right]$$

7.  One leaf of the lemniscate $r^2 = a^2 \cos 2\theta$; $\rho = $ constant. Find $\bar{x}$.
$$[\bar{x} = \pi a/4\sqrt{2}]$$

8.  Inside $r = a(1 + \cos \theta)$, $0 \leq \theta \leq \pi/2$, and outside $r = a$; $\rho = k \sin \theta$. Find $(\bar{x}, \bar{y})$.
$$\left[ \bar{x} = \frac{39}{40} a, \bar{y} = \frac{a}{2} \left( \frac{17}{15} + \frac{3\pi}{16} \right) \right]$$

9.  $y = ax^2$, $y = abx + 2ab^2$; $\rho = ky$. Find $\bar{x}$.                    $[\bar{x} = \frac{25}{32}b]$

10. Inside the ellipse $4x^2 + 9y^2 = 36$ and outside the circle $x^2 + y^2 - 2x = 0$; $\rho = $ constant. Find $\bar{x}$. Look for a shortcut.                    $[\bar{x} = -\frac{1}{5}]$

11. Find the center of mass of a semicircular region of radius $a$ if the density is proportional to the distance from the straight edge.

**In Problems 12 to 19 use either single or double integrals to find the centroid of the region described.**

12. The region bounded by $y = \sin x$, $y = 0$, $0 \leq x \leq \pi$    $[\bar{x} = \pi/2, \bar{y} = \pi/8]$

13. The region bounded by $y = \cos x$, $y = 0$, $0 \leq x \leq \pi/2$
$$[\bar{x} = (\pi - 2)/2, \bar{y} = \pi/8]$$

14. The triangle with vertices $(0, 0)$, $(a, 0)$, $(b, c)$    $[\bar{x} = \frac{1}{3}(a + b), \bar{y} = \frac{1}{3}c]$

15. The region bounded by $y = x^2$ and $y = 4$    $[\bar{x} = 0, \bar{y} = \frac{12}{5}]$

16. The region bounded by $y = x^2 - 4x + 3$ and $y = 0$    $[\bar{x} = 2, \bar{y} = -\frac{2}{5}]$

17. The region bounded by $x^{1/2} + y^{1/2} = a^{1/2}$ and the axes    $[\bar{x} = \bar{y} = a/5]$

18. The region in the first quadrant and inside the ellipse $b^2x^2 + a^2y^2 = a^2b^2$
$$[\bar{x} = 4a/3\pi, \bar{y} = 4b/3\pi]$$

19. The region pictured
[$4\frac{1}{3}$ and $2\frac{1}{3}$ from the corner]

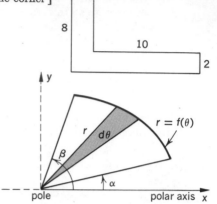

*20. Show that the moments about the x- and y-axes of the polar sector shown are

$$M_x = \tfrac{1}{3}\int_\alpha^\beta r^3 \sin\theta\,d\theta,$$

$$M_y = \tfrac{1}{3}\int_\alpha^\beta r^3 \cos\theta\,d\theta.$$

Thence show that the centroid of the limaçon $r = a(1 + \cos\theta)$ is at $\bar{x} = 5a/6$, $\bar{y} = 0$.

21. Use Pappus' theorem to find the volume of the solid torus shown sliced in half.
[$2\pi^2 a^2 b$]

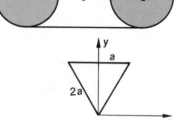

22. Use Pappus' theorem to find the volume of the solid generated by revolving the equilateral triangle of the figure about the x-axis. Check using elementary geometry.
[$4\pi a^3$]

23. Use Pappus' theorem to derive the formula for the volume of a cone.

*24. Show that the center of mass relative to a system of mass particles is unchanged by a rotation of the coordinate system (see Appendix A).

25. In Chapter 12, Section 9, we saw that the force $F$ due to fluid pressure on a vertical plate immersed in a fluid is given by

$$F = \int Wh\,dA = W\int h\,dA,$$

where $W$ is the weight per unit volume of the fluid, and $h$ is the distance below the surface of the element of area $dA$. But $\int h\,dA$ is a first moment and so

$$F = W\int h\,dA = W\bar{h}A,$$

where $\bar{h}$ is the distance from the centroid of the plate to the surface and $A$ is the area of the plate.

Use this result to re-work Problems 1, 2, 8, 9, and 19 on pages 340–342.

*26. Where should the rectangular window be hinged so that it will not tend to turn about the hinge? (This depth is called the center of pressure.)

[$\frac{8}{15}$ ft from the top edge]

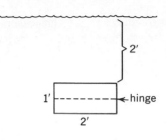

*27. Find the center of pressure (see Problem 26) on the circular window shown.

[$a^2/4\pi b$ below the center]

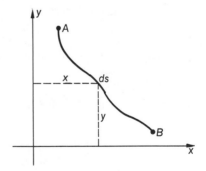

## *5 Centroids of Arcs

In this section we digress from our consideration of double integrals in order to discuss *centroids of arcs* for which single integrals are the natural tool.

Referring to the figure, suppose we have an arc of a curve between $A$ and $B$. If we think of the differential of arc $ds$ as being concentrated at the point $(x, y)$, then the first moments of this element are:

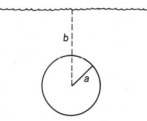

$y\,ds$ = the first moment with respect to the $x$-axis

$x\,ds$ = the first moment with respect to the $y$-axis.

"Summing" these moments over the arc we obtain the first moments $M_x$ and $M_y$ of the arc with respect to the $x$- and $y$-axes, respectively:

(1) $$M_x = \int y\,ds \qquad \text{and} \qquad M_y = \int x\,ds,$$

where the integrations are taken over the arc. The numbers $M_x$ and $M_y$ are *moments of mass* of an arc that has unit mass per unit length. The *centroid* $(\bar{x}, \bar{y})$ of the arc is the point with coordinates given by

(2) $$\bar{x} = M_y/s \qquad \text{and} \qquad \bar{y} = M_x/s,$$

where $s$ is the length of the arc.

*Remark.* If the curve were a material wire of mass density $\rho$ per unit length then formulas (1) and (2) become, for the *center of mass*:

$$M_y = \int x\rho\,ds, \qquad \bar{x} = \frac{M_y}{M},$$

$$M_x = \int y\rho\,ds, \qquad \bar{y} = \frac{M_x}{M},$$

where

$$M = \int \rho\,ds$$

is the mass of the wire.

**Example 1**   Find the centroid of the arc of the polar curve $r = e^\theta$ between $\theta = 0$ and $\theta = \pi/2$.

The length of the arc is

$$s = \int_0^{\pi/2} \sqrt{r^2 + \dot{r}^2}\,d\theta$$

$$= \sqrt{2} \int_0^{\pi/2} e^\theta\,d\theta = \sqrt{2}\,(e^{\pi/2} - 1).$$

$$M_x = \int_0^{\pi/2} y\,ds = \int_0^{\pi/2} r\sin\theta\,\sqrt{2}e^\theta\,d\theta$$

$$= \sqrt{2} \int_0^{\pi/2} e^{2\theta}\sin\theta\,d\theta$$

$$= \frac{\sqrt{2}}{5}\,e^{2\theta}(2\sin\theta - \cos\theta)\,\Big|_0^{\pi/2} = \frac{\sqrt{2}}{5}\,(2e^\pi + 1);$$

$$M_y = \int_0^{\pi/2} x\,ds = \int_0^{\pi/2} r\cos\theta\,\sqrt{2}e^\theta\,d\theta = \sqrt{2}\int_0^{\pi/2} e^{2\theta}\cos\theta\,d\theta$$

$$= \frac{\sqrt{2}}{5}\,e^{2\theta}(\sin\theta + 2\cos\theta)\,\Big|_0^{\pi/2} = \frac{\sqrt{2}}{5}\,(e^\pi - 2).$$

Therefore,

$$\bar{x} = \frac{M_y}{s} = \frac{e^\pi - 2}{5\,(e^{\pi/2} - 1)} \approx 1.2,$$

$$\bar{y} = \frac{M_x}{s} = \frac{2e^\pi + 1}{5\,(e^{\pi/2} - 1)} \approx 2.5.$$

There is a connection between the centroid of a plane arc and the area of the surface of revolution obtained by revolving the arc about a line of the plane. The theorem is called Pappus' First Theorem.

**THEOREM**    *If an arc in a plane be rotated about a line of the plane not cutting the arc then the area of the surface generated is equal to the length of the arc times the circumference of the circle traversed by the centroid.*

**PROOF**    Choose the $x$-axis as the axis of rotation, and the $y$-axis so that on the curve $y \geqq 0$. Then the area of the surface generated is

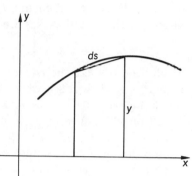

$$S = 2\pi \int y \, ds = 2\pi M_x = 2\pi \bar{y} s,$$

where $s$ is the length of the curve and $\bar{y}$ is the $y$ coordinate of the centroid.

### Problems

1.  Find the centroid (by integration) of the segment between $(0, 0)$ and $(a, b)$; $a, b > 0$.    $[a/2, \, b/2]$

2.  Use Pappus' theorem to find the lateral surface area of the cone formed by revolving the segment of Problem 1 about the $x$-axis.    $[\pi b \sqrt{a^2 + b^2}]$

3.  Find the centroid of one arch of the cycloid $x = a(\theta - \sin\theta)$, $y = a(1 - \cos\theta)$.    $[\bar{x} = \pi a, \, \bar{y} = \frac{4}{3}a]$

4.  Find the centroid of the cardioid $r = a(1 + \cos\theta)$.    $[\bar{x} = 4a/5]$

5.  Find the centroid of the circular arc shown.

$$\left[ \bar{x} = \frac{\sin\alpha}{\alpha} a \right]$$

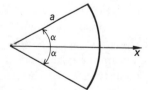

6.  Find the centroid of $r = a\cos\theta$, $0 \leqq \theta \leqq \pi/2$.    $[\bar{y} = a/\pi]$

7.  Find the centroid of the arc in the first quadrant of the hypocycloid $x = a\cos^3\theta$, $y = a\sin^3\theta$.    $[\bar{x} = \bar{y} = \frac{2}{5}a]$

8.  Use Pappus' theorem to find the area of the surface generated when the arc of Problem 7 is rotated about the $x$-axis.    $[6\pi a^2/5]$

*6   *Second Moments of Plane Regions*

Another application of double integrals is to the calculation of *second moments*. These moments arise in physics, statistics, and economics. The terminology that we shall use here conforms to the application to mechanics, where second moments are called *moments of inertia* and occur in problems of rotation of solid bodies.

If a mass particle of mass $m$ is located at $(x, y)$ the second moments, $I_x$ and $I_y$, with respect to the $x$- and $y$-axes are

$$I_x = my^2, \qquad I_y = mx^2.$$

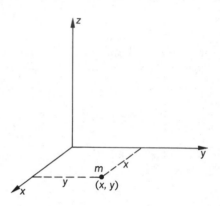

In addition to these there is the polar moment of inertia $I_z$ that arises when the rotation is about the $z$-axis. It is $I_z = m(x^2 + y^2)$.

Consider now a region $R$ of the $xy$-plane with mass density $\rho(x, y)$. Proceeding as in the development of first moments in Section 4, we *define* the second moments of the region $R$ to be:

$$I_x = \int_R \int y^2 \rho(x, y)\, dA, \quad \{\text{moment of inertia about the } x\text{-axis}$$

$$I_y = \int_R \int x^2 \rho(x, y)\, dA, \quad \{\text{moment of inertia about the } y\text{-axis}$$

$$I_z = \int_R \int (x^2 + y^2) \rho(x, y)\, dA. \quad \{\text{polar moment of inertia}$$

Moments of inertia about lines other than the axes are found analogously: The mass element $\rho\, dA$ is multiplied by the square of its distance from the line and the product integrated over $R$.

Associated with moment of inertia is the concept of *radius of gyration* $R_l$, which is a number such that if all the mass $M$ were concentrated at the distance $R_l$ from the line $l$, the same moment of inertia, $I_l$ would be obtained. Thus,

(1)          $$I_l = MR_l^2$$

where $R_l$ = radius of gyration. Usually moments of inertia are written in the form (1) and the square of the radius of gyration is apparent.

*Example*    A thin sheet of constant density $\rho$ is the parabolic segment shown. Find $I_x$ and $I_y$.

$$I_x = \int_0^b dy \int_{-\sqrt{y/a}}^{\sqrt{y/a}} \rho y^2 \, dx$$

$$= \frac{2\rho}{\sqrt{a}} \int_0^b y^{5/2} \, dy = \frac{4\rho b^{7/2}}{7\sqrt{a}}$$

$$I_y = \int_{-\sqrt{b/a}}^{\sqrt{b/a}} dx \int_{ax^2}^b \rho x^2 \, dy = \frac{4\rho b^{5/2}}{15a^{3/2}}.$$

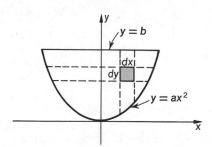

Because

$$M = 2\rho \int_0^b dy \int_0^{\sqrt{y/a}} dx = \frac{4\rho b^{3/2}}{3\sqrt{a}},$$

we can write

$$I_x = \frac{3b^2}{7} M, \qquad I_y = \frac{b}{5a} M.$$

### Problems

In Problems 1 to 12 find the moment of inertia as specified, of the regions bounded by the given curves and with the given densities.

1.   $y = mx$, $x = a$, $y = 0$; $\rho = $ constant. Find $I_x$.    $[\,I_x = M\,(m^2 a^2 / 6)\,]$

2.   Same region and density as in Problem 1. Find the moment of inertia about the line $x = a$.    $[\,I_a = M\,(a^2 / 6)\,]$

3.   $y = ax^2$, $x = 0$, $y = b^2 a$; $\rho = $ constant. Find $I_z$.

4.   $x^2 + y^2 = a^2$; $\rho = $ constant. Find $I_x$.

5.   $x = a$, $x = -a$, $y = a$, $y = -a$; $\rho = $ constant. Find $I_x$.    $[\,M\,(a^2 / 3)\,]$

6.   Same as Problem 5 but find the moment of inertia about the line $x = a$.    $[\,M\,(4a^2 / 3)\,]$

7.   Same as Problem 5 but find the moment of inertia about the line $y = x$.    $[\,M\,(a^2 / 3)\,]$

8.   The ellipse $b^2 x^2 + a^2 y^2 = a^2 b^2$; $\rho = $ constant. Find $I_x$, $I_y$.

9.   $x^2 + y^2 - 2ax = 0$; $\rho = $ constant. Find $I_y$.    $[\,I_y = \tfrac{5}{4} M a^2\,]$

10.   Same as Problem 9 but with $\rho$ proportional to the square of the distance from the center. Find $I_z$.

11.   $r = a(1 + \cos\theta)$; $\rho = $ constant. Find $I_z$.    $[\,I_z = \tfrac{35}{24} M a^2\,]$

12.   $r^2 = a^2 \cos 2\theta$; $\rho = $ constant. Find $I_x$, $I_y$.

13.   Prove that for regions in the $xy$-plane the moments of inertia satisfy the relation $I_z = I_x + I_y$.

*14.* Suppose the region $R$ with density $\rho$ has its center of mass on the $x$-axis. Show that the moment of inertia $I_a$ about the line $y = a$ is given by $I_a = I_x + Ma^2$, where $M$ is the mass. This is called the parallel axis property.

*15.* A region $R$ in the $xy$-plane has density $\rho$. Among all lines parallel to the $x$-axis for which is the moment of inertia least? Among all lines in a given direction for which is the moment of inertia least?

## *7   Surface Area

In Chapter 12 areas of surfaces of revolution were found. Now we consider more general surfaces. The reader should be aware that the concept of surface area is *much* more tricky than one might expect. The "obvious" definitions are not always the "right" ones. Until many examples have been considered an adequate definition of surface area may seem capricious.

The natural procedure, based upon analogy with arc length, is to define the area $S$ of a surface as the limit of the areas of inscribed polyhedra. For a long time, indeed until 1890, this definition was fashionable in elementary text books. In that year H. A. Schwarz showed that even for a circular cylinder one could inscribe polyhedra with arbitrarily small faces and arbitrarily large total surface area. This can be done by making the polyhedron like a very "crinkly" accordion. Each face of the polyhedron is a triangle with its vertices on the cylinder.

Three faces of an accordion-like polyhedron inscribed in a circular cylinder

In the past fifty years surface area has received extensive consideration, and definitions can be given which apply to very general surfaces. The definition adopted here applies to surfaces given by

$$z = f(x, y) \qquad (x, y) \text{ in } R,$$

where $f$ has continuous partial derivatives in a region $R$ of the $xy$-plane.

Suppose $\Delta R$ is a subregion of $R$ and $P = (x, y)$, a point in $\Delta R$. Consider the tangent plane at the point $[x, y, f(x, y)]$ and let $\gamma$ be the acute angle between the upward directed normal at the point and the $z$-axis. The portion of the tangent plane lying over the subregion $\Delta R$ has an area $\Delta S$. This region projects, vertically, onto $\Delta R$, and so, if $\Delta A$ is the area of $\Delta R$,

$$\Delta A = \Delta S \cos \gamma \text{ or } \Delta S = \Delta A \sec \gamma.$$

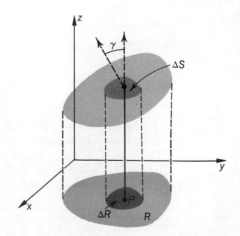

This projection property is valid because lengths in the plane determined by the normal and the vertical project this way, and lengths perpendicular to this plane are preserved. With a subdivision of $R$ into subregions $\Delta R_i$ the sum approximating to the surface area is

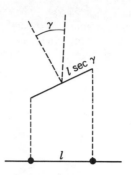

$$\sum \Delta S_i = \sum \sec \gamma_i \, \Delta A_i.$$

The limit of this sum, as the mesh of the subdivision goes to zero, is *by definition* the surface area $S$.

(1)     $$S = \lim \sum \sec \gamma_i \, \Delta A_i = \int_R \int \sec \gamma \, dA.$$

To render the integral (1) a useful formula for the area, $\sec \gamma$ must be expressed in terms of the coordinates of $R$. From the results of Chapter 14 we recall that $\partial f/\partial x$, $\partial f/\partial y$, $-1$ are proportional to the direction cosines of the normal. Therefore, if $\alpha$, $\beta$, $\gamma$ are direction angles of the upward directed normal,

$$\cos \alpha = \frac{-f_x}{\sqrt{f_x^2 + f_y^2 + 1}}, \quad \cos \beta = \frac{-f_y}{\sqrt{f_x^2 + f_y^2 + 1}}, \quad \cos \gamma = \frac{1}{\sqrt{f_x^2 + f_y^2 + 1}}.$$

The positive sign is selected for $\cos \gamma$ because $0 \leq \gamma < \pi/2$. Thus,

$$\sec \gamma = \sqrt{f_x^2 + f_y^2 + 1},$$

and we obtain the desired formula for the area,

(2)     $$S = \int_R \int \sqrt{f_x^2 + f_y^2 + 1} \, dA.$$

*Remark 1.* Should $\gamma$ approach $\pi/2$ anywhere on the surface the integral will be improper. In such cases the double integral will be evaluated as an improper iterated integral.

*Remark 2.* If the surface is given implicitly by $F(x, y, z) = 0$, then the direction cosines of the normal are proportional (see Chapter 14) to $\partial F/\partial x$, $\partial F/\partial y$, $\partial F/\partial z$, and $\sec \gamma$ is given by

$$\sec \gamma = \frac{\sqrt{F_x^2 + F_y^2 + F_z^2}}{|F_z|}.$$

Naturally, in order to compute the integral $\sec \gamma$ must then be expressed in terms of $x$ and $y$.

*Example*  Find the surface area of a sphere of radius $a$. One eighth of the sphere (centered at the origin) lies in the first octant, and we compute the surface area for this part. Its equation is

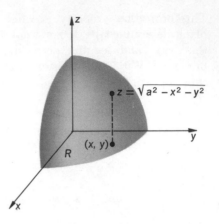

$$z = \sqrt{a^2 - x^2 - y^2}.$$

$$\frac{\partial z}{\partial x} = \frac{-x}{\sqrt{a^2 - x^2 - y^2}},$$

$$\frac{\partial z}{\partial y} = \frac{-y}{\sqrt{a^2 - x^2 - y^2}};$$

$$\sec \gamma = \sqrt{\frac{x^2}{a^2 - x^2 - y^2} + \frac{y^2}{a^2 - x^2 - y^2} + 1}$$

$$= \frac{a}{\sqrt{a^2 - x^2 - y^2}}.$$

Then

$$\frac{1}{8} S = \int_R \int \frac{a}{\sqrt{a^2 - x^2 - y^2}} \, dx \, dy = \int_0^a dy \int_0^{\sqrt{a^2 - y^2}} \frac{a}{\sqrt{a^2 - x^2 - y^2}} \, dx.$$

The integral is improper but causes no difficulty:

$$\frac{1}{8} S = a \int_0^a \left( \text{Arc} \sin \frac{x}{\sqrt{a^2 - y^2}} \, \Big|_0^{\sqrt{a^2 - y^2}} \right) dy = \frac{a^2 \pi}{2},$$

so $S = 4\pi a^2$.

In polar coordinates one would have

$$\sec \gamma = \frac{a}{\sqrt{a^2 - r^2}}$$

$$S = 8 \int_0^{\pi/2} d\theta \int_0^a \frac{a}{\sqrt{a^2 - r^2}} \, r \, dr = 8a \int_0^{\pi/2} a \, d\theta = 4\pi a^2.$$

If the implicit form for the sphere is used, one has $F(x, y, z) = x^2 + y^2 + z^2 - a^2 = 0$;

$$F_x = 2x, \quad F_y = 2y, \quad F_z = 2z;$$

$$\sec \gamma = \frac{\sqrt{4x^2 + 4y^2 + 4z^2}}{|2z|} = \frac{a}{z} = \frac{a}{\sqrt{a^2 - x^2 - y^2}};$$

and the integral is the same as in the explicit case.

### Problems

1. Find the area of the part of the plane $x/a + y/b + z/c = 1$, that lies in the first octant; $a, b, c > 0$. $\quad [\frac{1}{2}\sqrt{b^2c^2 + c^2a^2 + a^2b^2}\,]$

2. Find the area of the part of the surface $2z = x^2 - y^2$ that is inside the cylinder $x^2 + y^2 = a^2$. Use polar coordinates after first finding $\sec \gamma$ in terms of $x$ and $y$.

$$\left[\frac{2\pi}{3}\left[(a^2 + 1)^{3/2} - 1\right]\right]$$

3. Find the area of the part of the hemisphere $z = \sqrt{a^2 - x^2 - y^2}$ that is inside the cylinder $x^2 + y^2 = b^2$, $0 < b < a$. $\quad [2\pi a(a - \sqrt{a^2 - b^2})\,]$

4. Find the area of the part of the sphere $x^2 + y^2 + z^2 = a^2$ that is inside the cylinder $b^2x^2 + a^2y^2 = a^2b^2$, $0 < b < a$. $\quad [8a^2 \text{ Arc sin } (b/a)\,]$

5. Find the area of the part of the cylinder $x^2 + z^2 = a^2$ that is inside the cylinder $x^2 + y^2 = a^2$. $\quad [8a^2]$

6. Find the area of the part of the cone $z = \sqrt{x^2 + y^2}$ that is inside the cylinder $x^2 + y^2 - 2ax = 0$. $\quad [\sqrt{2}\pi a^2]$

7. Find the area of the part of the hemisphere $z = \sqrt{a^2 - x^2 - y^2}$ that is inside the cylinder $x^2 + y^2 - ay = 0$. $\quad [(\pi - 2)a^2]$

8. Find the area of the part of the cylinder $z^2 + x^2 = a^2$ that is above the triangle bounded by $x = 0$, $y = 0$, $x + y = a$.

$$\left[\frac{a^2}{2}(\pi - 2)\right]$$

9. Find the area of the part of the cone $z = \sqrt{x^2 + y^2}$ that is above the lemniscate $r^2 = a^2 \cos 2\theta$ in the $xy$-plane. $\quad [\sqrt{2}a^2]$

10. Find the area of the part of the cone $z = \sqrt{x^2 + y^2}$ that is above the rose $r = a \cos 2\theta$ in the $xy$-plane. $\quad [\pi a^2/\sqrt{2}]$

11. Find the area of the part of the cone $z = a\sqrt{x^2 + y^2}$ that is cut off by the plane $2z = ax + a$. $\quad [(2\pi/3\sqrt{3})\sqrt{a^2 + 1}]$

12. Find the area of the part of the plane $y = mx$, $m > 0$, that lies in the first octant and is cut off by the plane $bx + az = ab$; $a, b > 0$. $\quad [\sqrt{m^2 + 1}\,(ab/2)]$

*13. Find the area of the part of the cylinder $x = y^2$ that lies in the first octant and is cut off by the plane $bx + az = ab$; $a, b > 0$.

$$\left[\frac{b}{32a}(8a - 1)\sqrt{4a^2 + a} + \frac{b}{64a}(16a + 1) \log (2\sqrt{a} + \sqrt{4a + 1})\right]$$

*14. Find the area of the part of the hemisphere $z = \sqrt{a^2 - x^2 - y^2}$ that is inside the cylinder $x^2 + y^2 - 2ay = 0$.

$$\left[a^2\left(\pi - 2\int_0^{\pi/6}\sqrt{1 - 4\sin^2\theta}\,d\theta\right)\right]$$

*15.  The circle $z^2 + y^2 - 2ay = 0$ is revolved about the $z$-axis. Find the area of the surface generated.  $[4\pi^2 a^2]$

*16.  The circle $z^2 + (y - a)^2 = b^2$, $a > b > 0$, is revolved about the $z$-axis. Find the area of the torus generated.  $[4\pi^2 ab]$

*17.  The curve $z = f(y)$, $a \leq y \leq b$ and $f(y) > 0$, is revolved about the $y$-axis. Show, using the double integral formula, that the area of the surface generated is given by the formula of Chapter 12:

$$S = 2\pi \int_a^b z \, ds.$$

## 8  *Triple Integrals*

The definition of a triple integral and its evaluation as an iterated integral follow closely the corresponding definition and theorem for double integrals.

Suppose $R$ is a region of space, a solid, having a volume $V$ and bounded by piecewise smooth surfaces. Suppose $R$ is subdivided into $n$ subregions $\Delta R_i$, with volumes $\Delta V_i$ such that

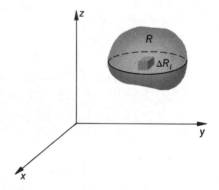

$$V = \sum_{i=1}^{n} \Delta V_i.$$

The *mesh* of the subdivision is the greatest diameter of any of the subregions.

Now choose a point $P_i$ in $\Delta R_i$ and suppose $f$ is a real-valued function continuous in $R$. The *triple integral of $f$ over $R$* is

$$\iiint_R f(P) \, dV = \lim \sum_{i=1}^{n} f(P_i) \, \Delta V_i,$$

where the limit is taken as the mesh of the subdivision approaches zero. It is a theorem of advanced calculus that this limit exists regardless of the sequence of subdivisions chosen and the choices of the points $P_i$, and we shall assume that this is so.

Just as double integrals are evaluated as iterated integrals, so also are triple integrals evaluated. In this section we consider the triple integral in rectangular coordinates. The following theorem, the proof of which must be deferred to a later course, gives the algorithm.

**THEOREM**    *Assume the region $R$ projects into a region $R'$ in the $xy$-plane, and assume $R$ is bounded top and bottom by surfaces*

$$\left.\begin{array}{l} z = z_2(x, y) \\[1.5em] z = z_1(x, y) \end{array}\right\} \quad \text{for } (x, y) \text{ in } R'.$$

*Let $F$ be the function on $R'$ given by*

$$(1) \quad F(x, y) = \int_{z_1(x,y)}^{z_2(x,y)} f(x, y, z)\, dz$$

*where in the integration $x$ and $y$ are constant. Then*

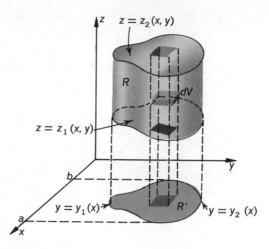

$$(2) \qquad\qquad \iiint_R f(x, y, z)\, dV = \int_{R'} \int F(x, y)\, dA.$$

**COROLLARY**    *If $R'$ is bounded by $y = y_1(x)$ and $y = y_2(x)$ for $a \leq x \leq b$, then*

$$(3) \qquad \iiint_R f(x, y, z)\, dV = \int_a^b dx \int_{y_1(x)}^{y_2(x)} dy \int_{z_1(x,y)}^{z_2(x,y)} f(x, y, z)\, dz,$$

*where in the first integration $x$ and $y$ are constant, and in the second integration $x$ is constant.*

The iterated integral (3) can be visualized as follows. If the subregions $\Delta R$ are small boxes of volume $\Delta x_i\, \Delta y_j\, \Delta z_k$, then the first integral (1) corresponds to a "vertical sum" to obtain an integral over a column of height $z_2 - z_1$ and base $\Delta x_i\, \Delta y_j$. This reduces the triple integral to the double integral (2).

*Remark 1.*    If $R$ is bounded by surfaces $y = y_1(x, z)$ and $y = y_2(x, z)$ over a region $R'$ of the $xz$-plane, then the triple integral can be evaluated by integrating first with respect to $y$, holding $x$ and $z$ constant.

In general, for suitably bounded regions $R$, the triple integral can be evaluated by six different iterated integrals corresponding to different orders of integration with respect to $x$, $y$, and $z$.

*Remark 2.* Integration over more complicated regions $R$ can (in practice) be handled by chopping $R$ into a number of pieces to each of which the theorem applies.

*Example 1*    Evaluate the iterated integral

$$I = \int_0^a dy \int_0^{\sqrt{a^2 - y^2}} dx \int_0^{\sqrt{x^2 + y^2}} (x + xy)\, dz$$

and describe the solid $R$ over which the integral is evaluated. After integrating with respect to $z$ we have

$$I = \int_0^a dy \int_0^{\sqrt{a^2-y^2}} (x+xy)\sqrt{x^2+y^2}\, dx$$

$$= \int_0^a \frac{1+y}{3}\left[(x^2+y^2)^{3/2}\right]\Big|_0^{\sqrt{a^2-y^2}} dy$$

$$= \int_0^a \frac{(1+y)}{3}(a^3-y^3)\, dy$$

$$= \frac{a^4}{4} + \frac{a^5}{5}.$$

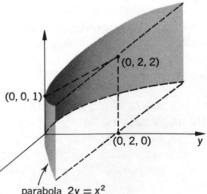

The region $R$ is the solid in the first octant under the cone $z = \sqrt{x^2+y^2}$ and inside the cylinder $x^2+y^2 = a^2$.

In terms of inequalities, $R$ is described by

$$R: \begin{cases} 0 \leq y \leq a, \\ 0 \leq x \leq \sqrt{a^2-y^2}, \\ 0 \leq z \leq \sqrt{x^2+y^2}. \end{cases}$$

**Example 2**    Evaluate

$$I = \iiint_R (x^2+z)\, dV,$$

where $R$ is the region inside the cylinder $2y = x^2$ with $0 \leq y \leq 2$, above the $xy$-plane, and under the plane $z = \frac{1}{2}y + 1$.

The figure shows $R$. It is bounded bottom and top by $z = 0$ and $z = \frac{1}{2}y + 1$, respectively. It projects in the $xy$-plane on a segment of a parabola.

$$I = \int_{-2}^2 dx \int_{x^2/2}^2 dy \int_0^{(y/2)+1} (x^2+z)\, dz$$

$$= \int_{-2}^2 dx \int_{x^2/2}^2 \left[x^2\left(\tfrac{1}{2}y+1\right) + \tfrac{1}{2}\left(\tfrac{1}{2}y+1\right)^2\right] dy$$

$$= \int_{-2}^2 \left(-\frac{13x^6}{3\,(64)} - \frac{13x^4}{3\,(8)} + \frac{8}{3}x^2 + \frac{8}{3}\right) dx = \frac{5286}{315} = 16.8.$$

**Example 3**    *Set up* an integral for the volume of the region in the first octant bounded by the plane $z + 2y = 2$ and the cone $x^2 = y^2 + z^2$.

The region is the solid $OABCD$ shown. In order not to have to cut the region in two, it is best to integrate first with respect to $x$. Then

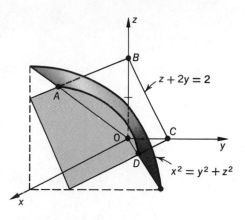

$$V = \iiint\limits_R dV$$

$$= \int_0^2 dz \int_0^{(z-2)/2} dy \int_0^{\sqrt{y^2+z^2}} dx.$$

The integrations are elementary but lengthy and we omit them here.

### Problems

1.  A function $f$ has values between $-1$ and $3$ on a sphere $R$ of radius 2. Give upper and lower bounds for

    $$\iiint\limits_R f(x, y, z)\, dV.$$

    Could your upper and lower bounds be attained for some functions?

2.  On half of a cube $R$ of side of length 2, a function $f$ has values between 1 and 2. On the other half the values are between 2 and $2\frac{1}{2}$. Give upper and lower bounds for the triple integral of $f$ over $R$. Could your upper and lower bounds be obtained for some functions?

3.  State a mean value theorem for triple integrals.

**In Problems 4 to 11 the iterated integrals are equal to triple integrals over regions R. Evaluate the integrals and describe each region R.**

4.  $\displaystyle\int_0^a dy \int_0^{\sqrt{a^2-y^2}} dx \int_0^{\sqrt{a^2-x^2-y^2}} (x+y+z)\, dz$

    [$R$ is the first octant of the sphere of radius $a$ centered at the origin]

5.  $\displaystyle\int_0^a dz \int_0^{\sqrt{a^2-z^2}} dy \int_0^{\sqrt{a^2-y^2-z^2}} (x+y+z)\, dx$

6.  $\displaystyle\int_0^1 dx \int_x^{\sqrt{x}} dy \int_0^{y+x^2} (x^2+y^2)\, dz$

    [$0.092; R: 0 \le z \le y + x^2,\ x \le y \le \sqrt{x},\ 0 \le x \le 1$]

7.  $\displaystyle\int_0^a dz \int_0^{a-z} dx \int_0^{a-x-z} dy$

    [$a^3/6; R$ is the tetrahedron bounded by $x + y + z = a$ and the coordinate planes]

8. $\displaystyle\int_0^2 dz \int_0^{4-z^2} dy \int_0^{4-z^2} (1+z)\, dx$

[$R$ is the region in the first octant inside the cylinders $x = 4 - z^2,\, y = 4 - z^2$]

9. $\displaystyle\int_0^4 dx \int_0^x dy \int_0^{\sqrt{4-x}} dz$  $\qquad\qquad\qquad$ [128/15]

10. $\displaystyle\int_0^1 dx \int_0^x dz \int_0^{x-z} e^{x-y-z}\, dy$

[$e - \frac{5}{2}$; $R$ is the tetrahedron $0 \le y \le x - z,\, 0 \le z \le x,\, 0 \le x \le 1$]

11. $\displaystyle\int_0^1 dy \int_y^{\sqrt{y}} dz \int_0^{\sqrt{y+2z}} xy\, dx$

[5/84; $R: 0 \le x \le \sqrt{y + 2z},\, y \le z \le \sqrt{y},\, 0 \le y \le 1$]

12. Obtain from a triple integral the double integral formula for the volume of a solid $R$ bounded bottom and top by $z = 0$ and $z = f(x, y)$.

13. Find the volume of the solid in the first octant bounded by $z^2 = y - x$, $y + z = 2$, and the coordinate planes.  $\qquad\qquad$ [17/20]

14. Find the volume of the solid in the first octant bounded by $x + y + z = a$, $az = xy$, $z = 0$, $a > 0$.  $\qquad\qquad$ [$a^3 (\frac{17}{12} - 2 \log 2)$]

15. Find the mass of the unit cube, with opposite vertices at $(0, 0, 0)$ and $(1, 1, 1)$ if the density is $\rho = cxyz$, $c = $ constant.  $\qquad\qquad$ [$c/8$]

16. Find the volume of the solid in the first octant bounded by $y + z = 1$ and $x^2 + z^2 = 4$. Integrate first with respect to $y$, then $x$ and $z$. Why not integrate second with respect to $z$? Repeat the problem integrating with respect to $x$, then $y$ and $z$. Repeat again, integrating first with respect to $x$, then $z$ and $y$. Why not integrate first with respect to $y$?
$$\left[ \frac{3\sqrt{3}}{2} - \frac{8}{3} + \frac{\pi}{3} \right]$$

17. Find the volume of the solid in the first octant bounded by the coordinate planes and the cylinders $ax + y^2 = a^2$, $az + y^2 = a^2$. Why is it less convenient to integrate first with respect to $y$ in the iterated integral? Express the volume as six different iterated integrals.  $\qquad\qquad$ [$8a^3/15$]

18. Interpret the integral
$$\int_0^a dx \int_x^a dz \int_x^z z\, dy$$
as a triple integral. Express it as five different iterated integrals. Evaluate it.
$$[a^4/8]$$

### 9  *Cylindrical and Spherical Coordinates*

Cylindrical coordinates $r$, $\theta$, and $z$ and their usual relations to rectangular coordinates are shown in the figure. One has (see Appendix A),

$$x = r \cos \theta, \quad y = r \sin \theta, \quad z = z.$$

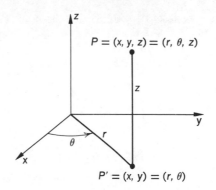

The differential element of volume in cylindrical coordinates is a box-like solid with a base in the $xy$-plane that is the polar coordinate element of area;

$$dV = r \, dr \, d\theta \, dz.$$

Then the triple integral of a function $f$ over the region $R$ bounded top and bottom as in the figure is

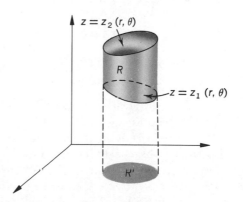

$$(1) \quad \iiint_R f(r, \theta, z) \, dV$$

$$= \int_{R'} \int F(r, \theta) \, dA,$$

where

$$(2) \quad F(r, \theta) = \int_{z_1(r,\theta)}^{z_2(r,\theta)} f(r, \theta, z) \, dz.$$

Therefore, writing the double integral of $F$ over $R'$ as an iterated integral and using (2), we obtain

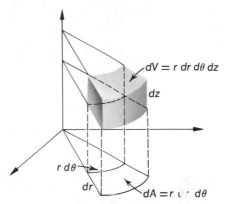

$$\iiint_R f(r, \theta, z) \, dV$$

$$= \int_\alpha^\beta d\theta \int_{r_1(\theta)}^{r_2(\theta)} dr \int_{z_1(r,\theta)}^{z_2(r,\theta)} f(r, \theta, z) r \, dz.$$

*Remark.*  A triple integral in cylindrical coordinates is often written as

$$\iiint_R f(r, \theta, z) r \, dr \, d\theta \, dz,$$

and therefore can be integrated as a triple integral over a region of $r$, $\theta$, $z$-space. As in Section 8 there are (depending on the shape of the region $R$) six possible orders of integration for the triple integral.

*Example 1*    Find the mass of the solid $R$ bounded by $z = 0$, the cone $z = r$, and the cylinder $r = 2a \sin \theta$, if the density $\rho = cz$, where $c = $ constant.

$$\text{Mass} = \iiint_R cz \, dV$$

$$= \iiint_R cz \, r \, dr \, d\theta \, dz$$

$$= \int_0^\pi d\theta \int_0^{2a \sin \theta} dr \int_0^r czr \, dz$$

$$= c \int_0^\pi d\theta \int_0^{2a \sin \theta} \frac{r^3}{2} \, dr$$

$$= c \int_0^\pi (2a^4 \sin^4 \theta) \, d\theta = \frac{3\pi c a^4}{4}.$$

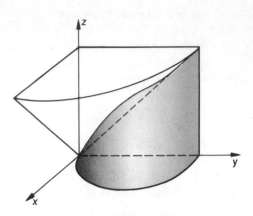

half of the solid

To discuss triple integrals in spherical coordinates we suppose, as usual, that the spherical and rectangular coordinates are related as shown in the figure (see Appendix A):

$$x = r \sin \varphi \cos \theta,$$

$$y = r \sin \varphi \sin \theta,$$

$$z = r \cos \varphi.$$

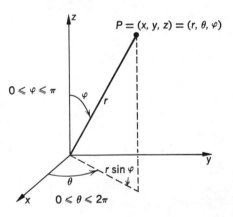

The differential element of volume in spherical coordinates is $dV = r^2 \sin \varphi \, dr \, d\theta \, d\varphi$. This formula is easily recalled by remembering the box-like solid pictured. For a rigorous derivation of the volume $\Delta V$ of a "box" with edges $\Delta r$, $r \, \Delta \varphi$, $r \sin \varphi \, \Delta \theta$, see Problem 15.

If the solid $R$ is subdivided into regions $\Delta R_i$ of volume $\Delta V_i$ of the type pictured, then the Riemann sums

$$\sum f(P_i) \, \Delta V_i$$

approximate the triple integral, which is equal to an iterated integral over $r$, $\theta$, $\varphi$ space.

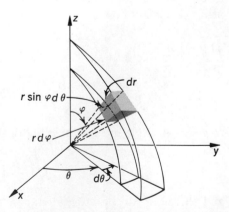

The differential volume of a box-like solid
$dV = r \sin \varphi \, d\theta \cdot dr \cdot r d\varphi$

$$\iiint_R f(r, \theta, \varphi)\, dV = \int_{\theta_1}^{\theta_2} d\theta \int_{\varphi_1(\theta)}^{\varphi_2(\theta)} d\varphi \int_{r_1(\theta,\varphi)}^{r_2(\theta,\varphi)} f(r, \theta, \varphi) r^2 \sin \varphi\, dr.$$

The limits on the integrals depend only on the boundary of the region $R$ and must be determined in each problem from an appropriate figure. Just as with rectangular and cylindrical coordinates, there are six different orders of integration possible, depending on the shape of $R$.

**Example 2**   Find the volume of a right circular cone of height $h$ and base of radius $a$.

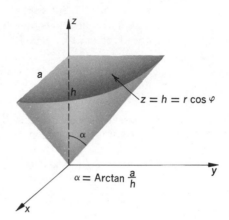

With the vertex at the origin and axis along the $z$-axis, spherical coordinates are convenient. Then

$$V = \int_0^{2\pi} d\theta \int_0^\alpha d\varphi \int_0^{h \sec \varphi} r^2 \sin \varphi\, dr,$$

where $\alpha = \text{Arc tan } (a/h)$ is the semi-vertex angle. Then

$$V = \int_0^{2\pi} d\theta \int_0^\alpha \frac{h^3}{3} \frac{\sin \varphi}{\cos^3 \varphi}\, d\varphi$$

$$= \frac{2\pi h^3}{3} \frac{\cos^{-2} \varphi}{2} \Big|_0^\alpha = \frac{\pi h^3}{3} (\sec^2 \alpha - 1) = \frac{\pi a^2 h}{3}.$$

## Problems

1.  Find, using spherical coordinates the volume of the solid inside the cone $\varphi = \alpha < \pi/2$ and inside the sphere $x^2 + y^2 + z^2 = 2az$.
    $$[4\pi a^3 (1 - \cos^4 \alpha)/3]$$

2.  Find the mass of the solid of Problem 1 if the density is $\rho = cz$, $c = $ constant.
    $$[4\pi a^4 c (1 - \cos^6 \alpha)/3]$$

3.  Find the volume of the solid inside the sphere $z^2 + r^2 = b^2$ and outside the cylinder $r = a < b$ (cylindrical coordinates).    $$[4\pi (b^2 - a^2)^{3/2}/3]$$

4.  Find the mass of the part of the sphere $r = a$ (spherical coordinates) that lies in the first octant if the density is $\rho = cxy$, $c = $ constant.    $$[ca^5/15]$$

5.  Find the volume of the solid in the first octant bounded by $x + y = a$ and $z^2 + y^2 = a^2$. Use cylindrical coordinates $(r, \theta, x)$ with polar coordinates in the $yz$-plane.
    $$\left[a^3 \left(\frac{\pi}{4} - \frac{1}{3}\right)\right]$$

6.   Find the volume of a sphere of radius $a$ using (a) rectangular coordinates, (b) cylindrical coordinates, (c) spherical coordinates.

7.   Find the mass of a sphere of radius $a$ if the density is proportional to the distance from the center.   $[c\pi a^4$, where $\rho = cr]$

*8.   Find the volume of the solid in the first octant bounded by $x = 0$, $x = yz$, and $x + y + z = 1$. Set up the problem, do not integrate, in rectangular coordinates and in cylindrical coordinates $(r, \theta, x)$, with polar coordinates in the $yz$-plane. (See Problem 14, Section 8.)

9.   Find the volume of the solid above the $xy$-plane, below the plane $x + z = a$, and inside the cylinder $r = a \cos 2\theta$ (cylindrical coordinates).   $[\pi a^3/2]$

10.   The spherical shell bounded by $r = a$ and $r = b > a$ has density $\rho = cr^2$, $c =$ constant. Find its mass.   $[\frac{4}{5}\pi c\,(b^5 - a^5)]$

11.   Find the volume of the solid inside the sphere $r^2 + z^2 = 4a^2$ and inside the cylinder $r = 2a \cos \theta$ (cylindrical coordinates).   $\left[16a^3\left(\dfrac{\pi}{3} - \dfrac{4}{9}\right)\right]$

12.   Find the mass of the solid of Problem 11 if the density is $\rho = cr$, $c =$ constant.   $[ca^4\pi^2]$

13.   Find the volume of the solid inside the cone $z^2 = x^2 + y^2$ and inside the sphere $x^2 + y^2 + z^2 = 2az$. Use spherical coordinates.   $[\pi a^3]$

14.   Find the mass of the solid above the $xy$-plane, under the cone $z = r$, and inside the cylinder $r^2 = a^2 \cos 2\theta$ (cylindrical coordinates) if the density is $\rho = cr$, $c =$ constant.   $[c\pi a^4/8]$

*15.   Show that the volume $V$ of the "box" is

$$V = r'r'' \sin \varphi' \, \Delta r \, \Delta \varphi \, \Delta \theta,$$

where $r'$ and $r''$ lie between $r$ and $r + \Delta r$, and $\varphi'$ lies between $\varphi$ and $\varphi + \Delta\varphi$.

   First show that the area $\Delta A$ of a vertical face is $r' \, \Delta r \, \Delta\varphi$. Then use Pappus' theorem for the volume of a solid of revolution, to obtain $\Delta V$ as the product of $\Delta A$ times the distance travelled by the centroid.

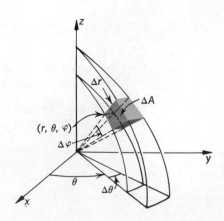

# 10   *Applications of Triple Integrals*

Physical properties, or effects, usually vary from point to point. Therefore, the net property, or effect, is often a "summation" of the contributions of each small portion and is expressed as a triple integral. Such applications of triple integrals are extremely varied, but the principle is the same in all cases. Thus, the contribution due to $f$ of a small portion $\Delta R$ containing a point $P$ will be, approximately, $f(P)\ \Delta V$—the contribution being practically constant over the subregion $\Delta R$. Thus, the differential contribution is $f(P)\ dV$, and integration gives the resultant. We confine ourselves here to the calculation of first and second moments and gravitational attraction.

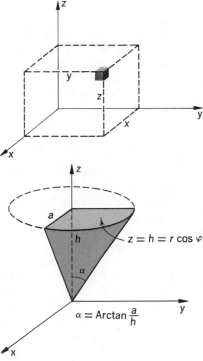

*Example 1*    Find the center of mass, or centroid, of a right circular cone of height $h$ and base of radius $a$, if the density $\rho$ is constant.

Moments of mass of solids are moments with respect to *planes* instead of with respect to lines, as with plane regions. Thus, for example, the contribution of the volume element $dV$ to the moment about the $xz$-plane is $\rho y\ dV$. Then

$$M_{xz} = \iiint\limits_R \rho y\ dV.$$

In the example here, with the axes chosen as pictured the centroid is on the $z$-axis. The moment of mass $M_{xy}$ about the $xy$-plane is

$$M_{xy} = \iiint\limits_R \rho z\ dV.$$

This triple integral is equal to the following iterated integral in spherical coordinates.

$$M_{xy} = \rho \int_0^{2\pi} d\theta \int_0^{\alpha} d\varphi \int_0^{h\sec\varphi} r\cos\varphi\ r^2 \sin\varphi\ dr$$

$$= \frac{1}{4}\rho h^4 \int_0^{2\pi} d\theta \int_0^{\alpha} \cos^{-3}\varphi \sin\varphi\ d\varphi = \frac{1}{4}\rho h^4 \frac{\tan^2\alpha}{2} 2\pi = \frac{\pi\rho h^2 a^2}{4}.$$

Since the mass $M = \frac{1}{3}\pi\rho a^2 h$,

$$\bar{z} = \frac{\pi\rho h^2 a^2}{4}\ \frac{3}{\pi\rho a^2 h} = \frac{3}{4}h.$$

*Example 2*    Find the moment of inertia of the cone of Example 1 about the $z$-axis.

The moment of inertia $I_z$ about the $z$-axis is

$$I_z = \iiint_R \rho\,(x^2 + y^2)\,dV = \iiint_R \rho r^2 \sin^2 \varphi\; r^2 \sin \varphi\, dr\, d\varphi\, d\theta$$

$$= \int_0^{2\pi} d\theta \int_0^\alpha d\varphi \int_0^{h \sec \varphi} \rho r^4 \sin^3 \varphi\, dr = \frac{1}{5}\rho h^5 \int_0^{2\pi} d\theta \int_0^\alpha \sec^5 \varphi \sin^3 \varphi\, d\varphi$$

$$= \frac{1}{5}\rho h^5 \int_0^{2\pi} d\theta \int_0^\alpha \tan^3 \varphi \sec^2 \varphi\, d\varphi = \frac{1}{5}\rho h^5 \frac{\tan^4 \alpha}{4} 2\pi = \frac{1}{10}\pi \rho a^4 h.$$

Since $M = \frac{1}{3}\pi \rho a^2 h$, $I_z = \frac{3}{10}Ma^2$.

*Example 3*    One of the obstacles that supposedly kept Newton from announcing earlier than he did his law of universal gravitation was the determination of the attraction that a homogeneous solid sphere exerted on a unit mass particle at a point outside the sphere.

The gravitational force **F** on a particle of unit mass exerted by a particle of mass $M$ at a distance $R$ from the unit mass is a *vector* of magnitude

$$|\,\mathbf{F}\,| = \frac{GM}{R^2}$$

where $G$ is the universal gravitational constant. The direction of **F** is from the unit mass toward the mass $M$. This law applies also to each small portion of a solid, hence in this example, the components of **F** are integrals over the sphere.

Suppose the sphere is centered at the origin. Then the magnitude of the force at $(0, 0, b)$ due to the differential volume $dV$ is

$$dF = \frac{G\rho\, dV}{R^2}.$$

On integration, the components in the $x$ and $y$ directions are zero by symmetry. The net component in the $z$-direction is

$$F_z = -\iiint_{\text{sphere}} \frac{G\rho}{R^2} \cos \psi\, dV$$

$$= -G\rho \int_0^{2\pi} d\theta \int_0^a dr \int_0^\pi \frac{(b - r \cos \varphi)r^2 \sin \varphi\, d\varphi}{(b^2 + r^2 - 2br \cos \varphi)^{3/2}},$$

$$\cos \psi = \frac{b - r \cos \varphi}{R}$$

$$R^2 = b^2 + r^2 - 2br \cos \varphi$$

where $a$ is the radius of the sphere. This rather awkward integral is best handled by writing it as a sum of two integrals. Let

$$I_1 = \int_0^a dr \int_0^\pi \frac{br^2 \sin \varphi \, d\varphi}{(b^2 + r^2 - 2br \cos \varphi)^{3/2}}$$

$$= \int_0^a \left[ \frac{-r}{(b+r)} + \frac{r}{b-r} \right] dr = \int_0^a \frac{2r^2}{b^2 - r^2} \, dr.$$

The last integral is easy to evaluate, but, as we will see, it will not be necessary to do so.

The other integral is amenable to integration by parts.

$$I_2 = \int_0^a dr \int_0^\pi \frac{-r^3 \cos \varphi \sin \varphi \, d\varphi}{(b^2 + r^2 - 2br \cos \varphi)^{3/2}}$$

$$= \int_0^a \frac{1}{b} \left[ \frac{r^2 \cos \varphi}{(b^2 + r^2 - 2br \cos \varphi)^{1/2}} \Big|_0^\pi + \int_0^\pi \frac{r^2 \sin \varphi \, d\varphi}{(b^2 + r^2 - 2br \cos \varphi)^{1/2}} \right] dr$$

$$= -\int_0^a \frac{2r^2}{b^2 - r^2} \, dr + \int_0^a \frac{r}{b^2} \left[ (b^2 + r^2 - 2br \cos \varphi)^{1/2} \right] \Big|_0^\pi dr$$

$$= -\int_0^a \frac{2r^2}{b^2 - r^2} \, dr + \int_0^a \frac{2r^2}{b^2} \, dr = -\int_0^a \frac{2r^2}{b^2 - r^2} \, dr + \frac{2a^3}{3b^2}.$$

Then
$$F_z = -G\rho 2\pi (I_1 + I_2) = -G\rho 2\pi \frac{2a^3}{3b^2}$$

$$= -\left( \frac{4\pi \rho a^3}{3} \right) \frac{G}{b^2} = -\frac{GM}{b^2},$$

where $M$ is the mass of the sphere. This is the expected result; that is, the gravitational force is directed from the particle at $(0, 0, b)$ toward the center of the sphere and its magnitude is $GM/b^2$.

*Example 4*   First moments of homogeneous solids of revolution can be found using only single integrals. Although the formula can be obtained from physical considerations, we derive it here from the triple integral.

Suppose that the plane region under the graph of $z = f(y)$, $a \leq y \leq b$ is revolved about the $y$-axis. An equation of the surface generated by the curve $z = f(y)$ is:

$$x^2 + z^2 = f^2(y).$$

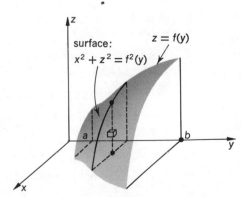

Then

$$M_{xz} = \rho \iiint\limits_{R} y \, dV,$$

where $\rho$ is the constant density and $R$ is the solid of revolution. The integral over the part shown gives $\frac{1}{4}M_{xz}$. Hence

$$M_{xz} = 4\rho \int_{a}^{b} dy \int_{0}^{f(y)} dx \int_{0}^{\sqrt{f^2(y)-x^2}} y \, dz$$

$$= 4\rho \int_{a}^{b} dy \int_{0}^{f(y)} y\sqrt{f^2(y) - x^2} \, dx$$

$$= 4\rho \int_{a}^{b} y \, dy \left[ \frac{x}{2} \sqrt{f^2(y) - x^2} + \frac{f^2}{2} \text{Arc sin } \frac{x}{f} \right] \Big|_{0}^{f}$$

$$= \rho \int_{a}^{b} \pi f^2(y) y \, dy = \rho \int_{a}^{b} \pi z^2 y \, dy.$$

This last integral is just what was to be expected. The integrand $\pi\rho z^2 y \, dy$ is the moment of the thin circular slab of radius $z$ and thickness $dy$, situated at a distance $y$ from the $xz$-plane.

## Problems

1.  Find the centroid of the solid tetrahedron of constant density bounded by the coordinate planes and

    $$\frac{x}{a} + \frac{y}{b} + \frac{z}{c} = 1. \qquad\qquad \left[ \left( \frac{a}{4}, \frac{b}{4}, \frac{c}{4} \right) \right]$$

2.  Find $I_z$ for the tetrahedron of Problem 1.   $[\frac{1}{10}M(a^2 + b^2)]$

3.  A solid tetrahedron of constant density is bounded by $x = 0$, $z = 0$, $x = y$, $x + z = a$. Find $I_x$.   $[\frac{1}{5}Ma^2]$

4.  A cube has constant density and edges of length $a$. Find its moment of inertia about one edge.   $[\frac{2}{3}Ma^2]$

5.  Show, by setting up the integrals, that in Example 1 $M_{xz} = M_{yz} = 0$.

6.  Find the centroid of the hemisphere above the $xy$-plane and bounded by $r = a$ (spherical coordinates).   $[\bar{z} = \frac{3}{8}a]$

7.  Use the formula of Example 4 to do Problem 6.

8.  Find the moment of inertia of a sphere of constant density about a diameter.   $[\frac{2}{5}Ma^2]$

9.  Find $I_z$ for the solid bounded by the cylinder $r^2 = a^2 \cos 2\theta$ (cylindrical coordinates), the cone $z = r$, and the $xy$-plane if the density is $\rho = c/r$.
$$[\pi M a^2/8]$$

10. The centroid of a solid is on the $z$-axis. Show that the moment of inertia about the line parallel to the $z$-axis and through the point $(x_0, y_0, 0)$ is
$$I = M(x_0^2 + y_0^2) + I_z.$$

11. A solid of constant density lies in the first octant and is bounded by the coordinate planes and $z^2 = 1 - y$ and $z^2 = 1 - x$. Find the $z$-coordinate of its centroid.
$$[\bar{z} = \tfrac{5}{16}]$$

12. Find $I_z$ for the solid of Problem 11. $\qquad\qquad$ $[32M/63]$

13. Find the moment of inertia of a solid circular cylinder of radius $a$ and height $h$ about its axis if the density is constant. $\qquad\qquad$ $[\tfrac{1}{2}Ma^2]$

14. Find the moment of inertia of the cylinder of Problem 13 about a line perpendicular to its axis through its centroid.
$$\left[ M \left( \frac{a^2}{4} + \frac{h^2}{12} \right) \right]$$

15. Show that the gravitational force exerted by a solid hemisphere of radius $a$ on a unit mass situated at the center is $GM/\tfrac{2}{3}a^2$, where $M$ is the mass of the hemisphere.

16. Show that the gravitational force exerted by a solid circular cylinder of radius $a$ and height $h$ on a unit mass situated at the center of one end is
$$2\pi G\rho[a - \sqrt{a^2 + h^2} + h].$$

17. Find (by carrying out the integration) the gravitational force exerted by the solid sphere bounded by $x^2 + y^2 + z^2 = 2az$ on a unit mass at the origin.
$$[GM/a^2]$$

18. Find the attraction of the upper half of the solid sphere of Problem 17 on a unit mass at the origin.
$$\left[ \frac{GM(\sqrt{2} - 1)}{a^2} \right]$$

19. Find the attraction of an infinite straight wire of constant linear density $\rho$ on a unit mass at distance $a$ from the wire. $\qquad\qquad$ $[2\rho G/a]$

20. Find the attraction of an infinite plane of constant areal density $\rho$ on a unit mass at distance $a$ from the plane. $\qquad\qquad$ $[2\pi\rho G]$

**In Problems 21 to 24 use the method of Example 4 to find the centroid.**

21. The region bounded by $y^2 = 2px$ and $x = a$ is revolved about the $x$-axis. Find its centroid. $\qquad\qquad$ $[\bar{x} = 2a/3]$

22. The region bounded by $y = \sqrt{2px}$, $x = 0$, and $y = \sqrt{2pa}$ is revolved about the $y$-axis. Find its centroid.
$$[\bar{y} = \tfrac{5}{6}\sqrt{2pa}]$$

**23.**   The region bounded by $y = 0$ and $y = x \log x$ is revolved about the $x$-axis. Find the centroid of the solid.   $[\bar{x} = 27/64]$

**24.**   Find the centroid of the solid generated by revolving half the ellipse $b^2x^2 + a^2y^2 = a^2b^2$, $0 \leqq x \leqq a$, about the $x$-axis.   $[\bar{x} = 3a/8]$

## REMEMBRANCE OF THINGS PAST

Almost all of the material here should be familiar, with the possible exception of part of the analytic geometry of space. By the end of your calculus course most of this material should be at your "mental finger tips," ready for immediate use.

This material is for *reference*, not for learning the first time around. If parts are completely unfamiliar, consult a suitable reference text—one of those listed on page 517, for example.

### THE GREEK ALPHABET

| | | | | | | | | |
|---|---|---|---|---|---|---|---|---|
| A | $\alpha$ | alpha | I | $\iota$ | iota | P | $\rho$ | rho |
| B | $\beta$ | beta | K | $\kappa$ | kappa | $\Sigma$ | $\sigma$ | sigma |
| $\Gamma$ | $\gamma$ | gamma | $\Lambda$ | $\lambda$ | lambda | T | $\tau$ | tau |
| $\Delta$ | $\delta$ | delta | M | $\mu$ | mu | $\Upsilon$ | $\upsilon$ | upsilon |
| E | $\epsilon$ | epsilon | N | $\nu$ | nu | $\Phi$ | $\varphi$ | phi |
| Z | $\zeta$ | zeta | $\Xi$ | $\xi$ | xi | X | $\chi$ | chi |
| H | $\eta$ | eta | O | $o$ | omicron | $\Psi$ | $\psi$ | psi |
| $\Theta$ | $\theta$ | theta | $\Pi$ | $\pi$ | pi | $\Omega$ | $\omega$ | omega |

# 1  *Plane and Space Geometry*

## TRIANGLES

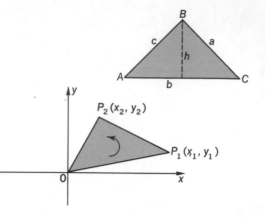

$$\text{Area } \triangle ABC = \tfrac{1}{2}bh = \tfrac{1}{2}bc \sin A$$

$$\text{Area } \triangle OP_1P_2 = \tfrac{1}{2}(x_1y_2 - x_2y_1)$$

(This is positive if the order $OP_1P_2$ is counterclockwise.)

## CIRCLES

$$\text{Arc length } s = r\theta;$$

$$\text{Area of sector} = \tfrac{1}{2}sr = \tfrac{1}{2}r^2\theta,$$

where $\theta$ is in radians.

$$\text{Area of segment} = hr - r^2 \text{ Arc sin } \frac{h}{r}.$$

## PRISMS AND CYLINDERS

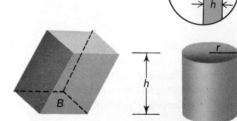

Volume $= Bh$, where $B$ is area of base.

Volume of right circular cylinder $= \pi r^2 h$.

Lateral surface of right circular cylinder $= 2\pi rh$.

## PYRAMIDS AND CONES

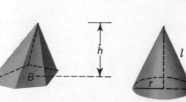

$$\text{Volume} = \tfrac{1}{3}Bh$$

Right circular cone:

$$\text{Volume} = \tfrac{1}{3}\pi r^2 h$$

$$\text{Lateral surface area} = \pi r\sqrt{r^2 + h^2} = \pi rl.$$

$$\text{Lateral surface of a frustum of a cone} = \pi(r_1 + r_2)l.$$

## SPHERES

$$\text{Volume of sphere} = \tfrac{4}{3}\pi R^3$$

$$\text{Surface area of sphere} = 4\pi R^2$$

$$\text{Volume of cap} = \tfrac{1}{3}\pi h^2(3R - h)$$

$$\text{Surface of cap} = 2\pi Rh.$$

$$\text{Volume of segment} = \tfrac{1}{3}\pi h(3R^2 - h^2)$$

$$\text{Surface of zone} = 2\pi Rh$$

## 2  *Algebra; Formulas*

### SOME PRODUCTS AND FACTORS

$$(x \pm y)^3 = x^3 \pm 3x^2y + 3xy^2 \pm y^3$$

$$(x + y + z)^2 = x^2 + y^2 + z^2 + 2xy + 2yz + 2zx$$

$$x^2 - y^2 = (x - y)(x + y)$$

$$x^3 \pm y^3 = (x \pm y)(x^2 \mp xy + y^2)$$

$$x^{2n+1} \pm y^{2n+1} = (x \pm y)(x^{2n} \mp x^{2n-1}y + \cdots + y^{2n})$$

$$x^4 + x^2y^2 + y^4 = (x^2 + xy + y^2)(x^2 - xy + y^2)$$

### VARIATION

If $y$ *varies directly* as $x$, or is proportional to $x$, then there is a constant $k$ such that $y = kx$. Similarly, $y$ *varies inversely* as $x$, if $y = k/x$.

And $z$ *varies jointly* as $x$ and $y$, if $z = kxy$.

### RATIONAL EXPONENTS

If $n$ is a positive integer, then $a^n = a \cdot a \cdots a$ ($n$ factors).

If $n$ is a negative integer, $n = -k$, then $a^n = a^{-k} = 1/a^k$.

If $x$ is a rational number, $x = p/q$, with $p$ and $q$ integers and $q$ positive, then $a^x = \sqrt[q]{a^p}$, where $a$ is restricted to be positive in order to avoid complex numbers.

### IRRATIONAL EXPONENTS

If $x$ is an irrational number, then the definition of $a^x$ given above does not apply. We define

$$a^x = \text{the limit of } a^r \text{ as } r \text{ approaches } x \text{ through rational values,}$$

$$= \lim_{r \to x} a^r, \text{ where } r \text{ is rational.}$$

## THE LAWS OF EXPONENTS

For all real numbers $x, y$,

$$a^x a^y = a^{x+y} \qquad a^x/a^y = a^{x-y} \qquad (a^x)^y = a^{xy}.$$

## SOME NOTATION

The "sigma notation" for sums: $\displaystyle\sum_{i=1}^{n} a_i = a_1 + a_2 + \cdots + a_n$.

The "pi notation" for products: $\displaystyle\prod_{i=1}^{n} a_i = a_1 a_2 \cdots a_n$

$n$-factorial: $n! = n(n-1)(n-2)\cdots(2)(1) = \displaystyle\prod_{k=1}^{n} k,$

0-factorial: $0! = 1$.

## SOME SEQUENCES, OR PROGRESSIONS

*Arithmetic sequence*: $a, a + d, a + 2d, \ldots$

$n$th term $= a + (n-1)d = l$

Sum of $n$ terms $= \dfrac{n}{2}(a + l) = \displaystyle\sum_{k=0}^{n-1}(a + kd)$.

Thus, for example, the sum of the first $n$ positive integers

$$= 1 + 2 + \cdots + n = \frac{n}{2}(n + 1),$$

and the sum of the first $n$ positive odd integers

$$= 1 + 3 + \cdots + (2n - 1) = \frac{n}{2}\big[(2n - 1) + 1\big] = n^2.$$

*Geometric sequence*: $a, ar, ar^2, \ldots$

$n$th term $= ar^{n-1} = l$

Sum of $n$ terms $= \dfrac{a - ar^n}{1 - r} = \displaystyle\sum_{k=0}^{n-1} ar^k$.

A derivation of the formula can be made as follows. Let $s_n$ be the sum of the first $n$ terms. Then

$$s_n = a + ar + \cdots + ar^{n-1}$$

$$s_{n+1} = a + ar + \cdots + ar^{n-1} + ar^n = s_n + ar^n.$$

But also

$$s_{n+1} = a + r(a + ar + \cdots + ar^{n-1}) = a + rs_n,$$

whence, from the last two equations,

$$s_n + ar^n = a + rs_n.$$

Solution of this equation for $s_n$ yields the formula. Thus, for example, the sum

$$2 + 4 + 8 + \cdots + 2^n = \frac{2 - 2 \cdot 2^{n+1}}{1 - 2} = 2^{n+2} - 2.$$

## SUMS OF SQUARES AND CUBES OF INTEGERS

$$1^2 + 2^2 + \cdots + n^2 = \tfrac{1}{6}n(n+1)(2n+1) = \sum_{i=1}^{n} i^2$$

$$1^3 + 2^3 + \cdots + n^3 = \tfrac{1}{4}n^2(n+1)^2 = \sum_{i=1}^{n} i^3.$$

For an inductive proof see mathematical induction page 494.

## PERMUTATIONS AND COMBINATIONS

The number of *permutations* of $n$ things taken $r$ at a time:

$$_nP_r = n(n-1) \cdots (n-r+1) = \frac{n!}{(n-r)!}.$$

The number of *combinations* of $n$ things taken $r$ at a time:

$$_nC_r = \binom{n}{r} = \frac{n(n-1) \cdots (n-r+1)}{r(r-1) \cdots (1)} = \frac{n!}{r!(n-r)!}.$$

## BINOMIAL THEOREM     ($n$ is a positive integer)

$$\frac{n(n-1)(n-2)}{2 \cdot 3} a^{n-3} x^3$$

$$(a + x)^n = a^n + na^{n-1}x + \frac{n(n-1)}{2} a^{n-2}x^2 + \cdots + x^n$$

$$= \sum_{r=0}^{n} {_nC_r}a^{n-r}x^r = \sum_{r=0}^{n} \binom{n}{r} a^{n-r}x^r.$$

Observe that the binomial coefficients $\binom{n}{r}$ are the numbers of combinations of $n$ things $r$ at a time.

## MATHEMATICAL INDUCTION

If a statement $P$ is made for each positive integer $n$, suppose that

(a)   $P$ is true for $n = 1$,
(b)   If $P$ is true for $n = k$, then $P$ is true for $n = k + 1$.

Then $P$ is true for every positive integer.

*Example*    A proof by induction that

$$\sum_{i=1}^{n} i^3 = \frac{n^2}{4}(n+1)^2$$

for all positive integers $n$, can be made as follows:

If $n = 1$, then $1^3 = \frac{1}{4} \cdot 1^2 (1+1)^2$, and the formula is correct when $n = 1$. This proves (a).

Suppose the formula is valid for $n = k$. Then

$$1^3 + 2^3 + \cdots + k^3 = \tfrac{1}{4}k^2(k+1)^2.$$

Therefore, the sum of the first $(k+1)$ cubes is

$$1^3 + 2^3 + \cdots + k^3 + (k+1)^3 = \tfrac{1}{4}k^2(k+1)^2 + (k+1)^3$$

$$= \tfrac{1}{4}(k+1)^2[k^2 + 4(k+1)] = \tfrac{1}{4}(k+1)^2(k+2)^2.$$

But this last expression is the sum by the formula for $n = k + 1$. This proves (b) and completes the proof.

## COMPLEX NUMBERS

Numbers of the form $a + bi$, where $a$ and $b$ are real numbers and $i^2 = -1$ are called complex numbers. They are added and multiplied thus:

$$(a + bi) + (c + di) = (a + c) + (b + d)i,$$

$$(a + bi)(c + di) = (ac - bd) + (ad + bc)i.$$

Like the real numbers the complex numbers form a *field*. One can add, subtract, multiply, and divide (except by 0) with the usual laws of algebra remaining valid.

The *absolute value* of the complex number $a + bi$ is

$$|a + bi| = \sqrt{a^2 + b^2}.$$

## 3  *Algebraic Equations*

An algebraic equation of degree $n$ is obtained by setting a polynomial $f(x)$ of degree $n$ (see Section 5 for information on polynomials) equal to zero:

(1)                $f(x) = a_0 x^n + a_1 x^{n-1} + \cdots + a_n = 0,$

where $a_0 \neq 0$, $n$ is a positive integer, and $a_0, a_1, \ldots, a_n$ are real, or complex, numbers. A *root* of equation (1) is a number $r$ such that $f(r) = 0$.

**THE FACTOR THEOREM**    A number $r$ is a root of (1) if and only if $f(x)$ is divisible by $(x - r)$; that is,

$$f(x) = (x - r)Q(x)$$

where $Q(x)$ is a polynomial of degree $n - 1$.

**THE FUNDAMENTAL THEOREM OF ALGEBRA**    Equation (1) has a root.

**COROLLARY**   There exist numbers (real or complex) $r_1, r_2, \ldots, r_n$ (not necessarily distinct) such that $f(x)$ factors uniquely into linear factors:

(2) $$f(x) = a_0(x - r_1)(x - r_2) \cdots (x - r_n).$$

By possibly counting duplicates among $r_1, \ldots, r_n$ we agree to say that equation (1) has $n$ roots, exactly.

Determination of the roots of (1) and, hence, the factors of $f(x)$, can be difficult. Yet, when the coefficients $a_0, \ldots, a_n$ are integers, rational roots (if any) can be found by trial.

**THEOREM ON RATIONAL ROOTS**   If the coefficients in (1) are integers, and if $r$ is a rational root of (1) with $r = p/q$, where $p$ and $q$ are integers, and if $p/q$ is reduced to lowest terms, then $p$ divides $a_n$ and $q$ divides $a_0$.

For example to find the rational roots, if any, of $2x^3 - x^2 + x + 1 = 0$, we need test only the numbers $\pm 1$ and $\pm \frac{1}{2}$. By trial one root is $-\frac{1}{2}$, so that $x + \frac{1}{2}$ is a factor. Then the equation becomes $2x^3 - x^2 + x + 1 = (2x + 1)(x^2 - x + 1) = 0$. A complete solution is found by solving the quadratic equation $x^2 - x + 1 = 0$.

Although the fundamental theorem of algebra allows the coefficients $a_0, \ldots, a_n$ to be complex numbers, we in this text, are concerned only with real coefficients. In that case the complex roots of equation (1) occur in pairs.

**THEOREM ON COMPLEX ROOTS**   If the coefficients in (1) are real, then the complex roots occur in pairs, each complex root $a + bi$ being paired with its complex conjugate $a - bi$.

**COROLLARY**   Every polynomial with real coefficients can be factored into real linear and quadratic factors. To the real root $r$ corresponds the factor $(x - r)$. To the complex roots $a \pm bi$ corresponds the factor $(x - a - bi)(x - a + bi) = x^2 - 2ax + a^2 + b^2$.

The corollary focuses attention on *quadratic equations*. The quadratic equation $ax^2 + bx + c = 0$ has the roots

$$r_1 = \frac{-b + \sqrt{b^2 - 4ac}}{2a}, \qquad r_2 = \frac{-b - \sqrt{b^2 - 4ac}}{2a}.$$

Thus, the quadratic polynomial factors as:

$$ax^2 + bx + c = a\left(x - \frac{-b + \sqrt{b^2 - 4ac}}{2a}\right)\left(x - \frac{-b - \sqrt{b^2 - 4ac}}{2a}\right).$$

If $a$, $b$, and $c$ are real, then the roots are: (a) real and unequal if $b^2 - 4ac > 0$; (b) real and equal if $b^2 - 4ac = 0$; and (c) complex conjugates if $b^2 - 4ac < 0$. If $b^2 - 4ac \geqq 0$ the factors of $ax^2 + bx + c$ are real. If $b^2 - 4ac < 0$, then $ax^2 + bx + c$ cannot be factored into real factors. It is said to be *irreducible* over the real numbers.

*Example*   The fourth degree polynomial

$$2x^4 - 2x^3 + 3x^2 - x + 1 = (2x^2 + 1)(x^2 - x + 1)$$

has been factored into two real quadratic factors, each of which is irreducible:

$2x^2 + 1$   is irreducible because $b^2 - 4ac = 0^2 - 4(2)(1) < 0$;

$x^2 - x + 1$   is irreducible because $b^2 - 4ac = (-1)^2 - 4(1)(1) < 0$.

## 4  *Functions*

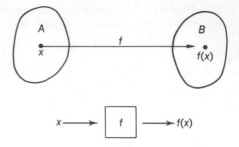

**DEFINITION**    A function $f$ consists of a set $A$, called the *domain* of $f$, and a set $B$ into which $f$ maps $A$. To each element $x$ of $A$, $f$ assigns a unique element $f(x)$ of the set $B$.

   The *range* of $f$ is the set of all elements $f(x)$ in $B$ for $x$ in $A$.

**Examples**

$$f(x) = 1/(x^2 + 1), \quad \text{domain is all real numbers;}$$

$$g(x) = |x| = \left\{ \begin{array}{lll} x & \text{if} & x \geq 0 \\ -x & \text{if} & x < 0 \end{array} \right\}, \quad \text{domain is all real numbers.}$$

$$q(x) = \sqrt{x}, \quad \text{domain is all non-negative real numbers.}$$

Observe that $q(x)$ is never negative. Thus, $q(4) = \sqrt{4} = 2$, and $q(a^2) = \sqrt{a^2} = |a|$, and $q(\sin^2 \theta) = \sqrt{\sin^2 \theta} = |\sin \theta|$.

   In the definition above, $A$ and $B$ can be any sets whatsoever. In most cases we encounter both $A$ and $B$ are sets of real numbers. If $B$ is a set of real numbers, then $f$ is a *real-valued* function. If $A$, too, consists of real values, then $f$ is a real-valued function of a *real variable*. In other examples, notably in Chapters 14 and 15, $A$ consists of a set of points in a plane or in space. In such cases the notation for the value of the function may change. For example, if $A$ is a subset of the plane, and $x$ and $y$ are coordinates in the plane, then, if the point $(x, y)$ is in $A$, the value of $f$ at this point is denoted by $f(x, y)$. In this case $f$ may be described as a function of two variables.

   Functions can be composed or compounded by taking functions of functions.

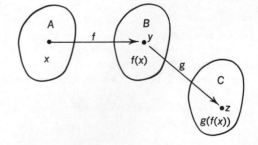

**DEFINITION**    If a function $f$ has domain $A$, and if the domain of a function $g$ contains the range of $f$, then the *composition of $g$ and $f$* has at $x$ the value $g(f(x))$.

   A function $f$ is said to have an inverse if there is a function, denoted by $f^{-1}$, such that $f$ and $f^{-1}$, composed with each other, give identity functions. The following definition gives the details.

**DEFINITION**    If a function $f$ with domain $A$ has the property that $f(x_1) \neq f(x_2)$ whenever $x_1 \neq x_2$, then the correspondence between the domain of $f$ and the range of $f$ is *one-to-one*. In this case there is a unique function $f^{-1}$, called the *inverse* of $f$, whose domain is the range of $f$ and which has the following properties:

$$f^{-1}(f(x)) = x$$

for all $x$ in $A$, the domain of $f$.

$$f(f^{-1}(y)) = y$$

for all $y$ in the domain of $f^{-1}$.

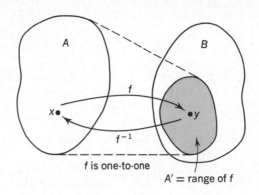

$f$ is one-to-one

$A' = $ range of $f$

For discussion of limits and continuity of real-valued real functions see Appendix C.

## 5  *Polynomial Functions; Rational Functions*

If $a_0, a_1, \ldots, a_n$ are real numbers, then the function $f$ given by

$$f(x) = a_0 x^n + a_1 x^{n-1} + \cdots + a_n, \quad -\infty < x < \infty,$$

where $n$ is a positive integer, is called a *polynomial*. If $a_0 \neq 0$, then the polynomial is of *degree n*. Thus, constant functions, not identically zero, are polynomials of degree zero. The zero polynomial is not assigned a degree.

Two polynomials can be added to obtain a polynomial, and they can be multiplied to obtain a product polynomial. The degree of the product (if the product is not identically zero) is equal to the sum of the degrees of the factors. For division of polynomials the following theorem is basic.

**THE DIVISION ALGORITHM**    Given the polynomials $f$ and $g$ (not zero), there are unique polynomials $Q$ and $R$ such that the degree of $R$ is less than the degree of $g$ and

$$f(x) = g(x)Q(x) + R(x) \qquad \text{for all } x.$$

The polynomial $Q$ is called the *quotient* polynomial of $f$ by $g$, and the polynomial $R$ is called the *remainder*.

The familiar process of long division leads to both quotient $Q$ and remainder $R$. If $R(x)$ is identically zero, then $f(x)$ is *divisible* by $g(x)$, and $f(x)$ is factorable as the product of $g(x)$ and $Q(x)$. In case $g(x) = x - r$, the remainder must be constant, and we get:

**REMAINDER THEOREM**    If $f(x)$ is divided by $x - r$, the remainder is $f(r)$:

$$f(x) = (x - r)Q(x) + f(r).$$

Rational functions are defined to be quotients of two polynomial functions. Thus, $F$ given by

$$F(x) = \frac{a_0 x^n + \cdots + a_n}{b_0 x^m + \cdots + b_m},$$

wherever the denominator $\neq 0$, is a rational function. By the division algorithm, with $g(x) = b_0 x^m + \cdots + b_m$, $F(x)$ is equal to a polynomial function plus a rational function with numerator of lower degree than the denominator.

## 6  *Exponential and Logarithmic Functions*

Suppose that $a$ is positive and not 1; then the function that maps each real number $x \rightarrow a^x$ is called *the exponential function with base a*. It is denoted by $\exp_a$. Thus,

$$\exp_a x = a^x.$$

This function obeys the laws of exponents:

(1)   $a^x a^y = a^{x+y}; \quad a^x/a^y = a^{x-y}; \quad (a^x)^y = a^{xy}.$

The function $\exp_a$ is continuous and is increasing if $a > 1$. It maps the set of all real numbers one-to-one on the set of positive real numbers.

Because the function $\exp_a$ maps the real numbers *one-to-one* on the positive numbers, the function has an inverse. It is called the *logarithm function with base a* and is denoted by

$$\log_a = \exp_a^{-1}.$$

Therefore,

$$\log_a N = x \qquad \text{and} \qquad N = a^x = \exp_a x$$

are equivalent. Each implies the other.

For computational purposes the base $a = 10$ is most suitable. But for calculus another base, $a = e = 2.71828\ldots$, which is defined in Chapter 2, is more convenient. See also Appendix D.

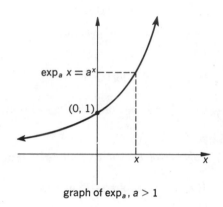

graph of $\exp_a$, $a > 1$

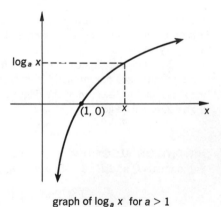

graph of $\log_a x$ for $a > 1$

### LAWS OF LOGARITHMS

The laws of exponents (1) are reflected in corresponding properties of logarithms:

$$\log_a MN = \log_a M + \log_a N \qquad \text{because} \qquad a^x a^y = a^{x+y};$$

$$\log_a \frac{M}{N} = \log_a M - \log_a N \qquad \text{because} \qquad \frac{a^x}{a^y} = a^{x-y};$$

$$\log_a N^y = y \log_a N \qquad \text{because} \qquad (a^x)^y = a^{xy}.$$

If one changes from base $a$ to base $b$, then

$$\log_a N = \log_a b \log_b N.$$

# 7  *The Trigonometric Functions*

For each real number $\theta$, $-\pi < \theta \leq \pi$, there is a unique point $(x, y)$ on the unit circle $x^2 + y^2 = 1$ such that the arc from $(x, y)$ to $(1, 0)$ has length $|\theta|$, and such that for $\theta \geq 0$, $y \geq 0$ and for $\theta < 0$, $y < 0$. In this way the interval $-\pi < \theta \leq \pi$ is "wrapped around" the circle. For values of $\theta$ greater than $\pi$ or less than $-\pi$, this wrapping function is extended in a natural way. With this agreement the trigonometric (or circular) functions are defined as follows (the above notation is shown in the figure):

$$\sin \theta = y, \quad \cos \theta = x, \quad \tan \theta = y/x,$$

$$\csc \theta = 1/y, \quad \sec \theta = 1/x, \quad \cot \theta = x/y.$$

The graphs of the sine, cosine, and tangent functions are shown below. The sine and cosine functions have period $2\pi$. The tangent function has period $\pi$.

$$\sin (\theta + 2\pi) = \sin \theta, \quad \cos (\theta + 2\pi) = \cos \theta, \quad \tan (\theta + \pi) = \tan \theta.$$

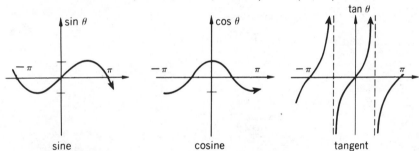

sine          cosine          tangent

The trigonometric functions must have their domains restricted in order that they possess unique inverses. When this restriction is defined properly for the sine, cosine, and tangent, one obtains the *principal values* for the inverse sine, cosine, and tangent. Their graphs are shown below.

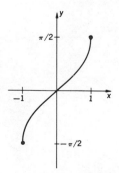

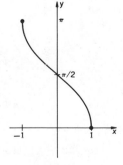

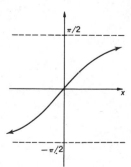

(a) $-\pi/2 \leq \text{Arcsin } x \leq \pi/2$,
   domain: $-1 \leq x \leq 1$

(b) $0 \leq \text{Arc cos } x \leq \pi$,
   domain: $-1 \leq x \leq 1$

(c) $-\pi/2 < \text{Arctan } x < \pi/2$,
   domain: $-\infty < x < \infty$

## 8   *Trigonometric Identities and Formulas*

Numerous identities can be established from the definitions of the functions:

### THE ELEMENTARY IDENTITIES

$$\sin\theta\csc\theta = 1, \quad \cos\theta\sec\theta = 1, \quad \tan\theta\cot\theta = 1,$$

$$\sin^2\theta + \cos^2\theta = 1, \quad 1 + \tan^2\theta = \sec^2\theta, \quad 1 + \cot^2\theta = \csc^2\theta.$$

### THE ADDITION FORMULAS

$$\sin(x \pm y) = \sin x \cos y \pm \cos x \sin y,$$

$$\cos(x \pm y) = \cos x \cos y \mp \sin x \sin y,$$

$$\tan(x \pm y) = \frac{\tan x \pm \tan y}{1 \mp \tan x \tan y}.$$

$$\sin\left(\frac{\pi}{2} \pm \theta\right) = \cos\theta, \quad \cos\left(\frac{\pi}{2} \pm \theta\right) = \mp\sin\theta, \quad \tan\left(\frac{\pi}{2} \pm \theta\right) = \mp\cot\theta.$$

### DOUBLE AND HALF-ANGLE FORMULAS

$$\sin 2\theta = 2\sin\theta\cos\theta,$$

$$\tan 2\theta = \frac{2\tan\theta}{1 - \tan^2\theta},$$

$$\cos 2\theta = \cos^2\theta - \sin^2\theta = 2\cos^2\theta - 1 = 1 - 2\sin^2\theta,$$

$$\sin\frac{\theta}{2} = \pm\sqrt{\frac{1 - \cos\theta}{2}},$$

$$\cos\frac{\theta}{2} = \pm\sqrt{\frac{1 + \cos\theta}{2}},$$

$$\tan\frac{\theta}{2} = \pm\sqrt{\frac{1 - \cos\theta}{1 + \cos\theta}} = \frac{1 - \cos\theta}{\sin\theta} = \frac{\sin\theta}{1 + \cos\theta}.$$

Also, one gets:

$$\sin 3\theta = 3\sin\theta - 4\sin^3\theta, \qquad \cos 3\theta = 4\cos^3\theta - 3\cos\theta.$$

### ADDITIONAL IDENTITIES

$$\sin x + \sin y = 2\sin\tfrac{1}{2}(x + y)\cos\tfrac{1}{2}(x - y),$$

$$\sin x - \sin y = 2\cos\tfrac{1}{2}(x + y)\sin\tfrac{1}{2}(x - y),$$

$$\cos x + \cos y = 2\cos\tfrac{1}{2}(x + y)\cos\tfrac{1}{2}(x - y),$$

$$\cos x - \cos y = -2\sin\tfrac{1}{2}(x + y)\sin\tfrac{1}{2}(x - y),$$

$$\sin\theta\sin\varphi = \tfrac{1}{2}\cos(\theta - \varphi) - \tfrac{1}{2}\cos(\theta + \varphi),$$

$$\cos\theta\cos\varphi = \tfrac{1}{2}\cos(\theta - \varphi) + \tfrac{1}{2}\cos(\theta + \varphi),$$

$$\sin\theta\cos\varphi = \tfrac{1}{2}\sin(\theta - \varphi) + \tfrac{1}{2}\sin(\theta + \varphi).$$

## LAWS OF SINES AND COSINES

$$\frac{\sin A}{a} = \frac{\sin B}{b} = \frac{\sin C}{c},$$

$$c^2 = a^2 + b^2 - 2ab\cos C.$$

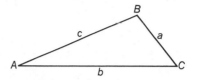

## 9 *Plane Analytic Geometry; Lines*

It is a fundamental property of lines in euclidean geometry that the real numbers can be matched one-to-one with the points of a line so that, if points $P_x$ and $P_y$ correspond to the real numbers $x$ and $y$, then

$$|P_xP_y| = \text{distance from } P_x \text{ to } P_y = |x - y|.$$

Then the order of points on the line is the same as the order of the numbers associated with them. Such a correspondence between the real numbers and the points of a line is called a *coordinatization* of the line.

Inequalities using the absolute value sign can be used to specify intervals on the line.

Open intervals do not contain the end points. Closed intervals do.

Because one can coordinatize lines, it is possible to coordinatize the plane by choosing coordinates on two perpendicular lines with the origin on both at their point of intersection, and with the same unit of length.

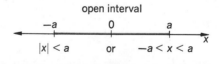

open interval

$|x| < a$    or    $-a < x < a$

closed interval

$|x| \leqq a$    or    $-a \leqq x \leqq a$

open interval

$|x - c| < a$    or    $c - a < x < c + a$

**THE DISTANCE FORMULA,** in rectangular co-ordinates gives the distance from $P_1 = (x_1, y_1)$ to $P_2 = (x_2, y_2)$:

$$|P_1P_2| = \sqrt{(x_1 - x_2)^2 + (y_1 - y_2)^2}.$$

**THE INCLINATION OF A LINE** is the angle $\theta$ between the positive direction of the $x$-axis and the upward direction of the line. Then

$$0 \leqq \theta < \pi.$$

## SLOPE

If $x_1 \neq x_2$, then the slope $m$ of the line through $(x_1, y_1)$ and $(x_2, y_2)$ is

$$m = \frac{y_2 - y_1}{x_2 - x_1} = \tan \theta.$$

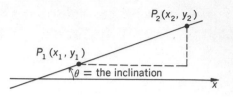

## ANGLES BETWEEN LINES

An angle $\theta$ between line $l_1$ of slope $m_1$ and line $l_2$ of slope $m_2$ is given by

$$\tan \theta = \pm \frac{m_1 - m_2}{1 + m_1 m_2}.$$

The two lines are parallel, $l_1 \parallel l_2$, if $m_1 = m_2$.

The two lines are perpendicular, $l_1 \perp l_2$, if $m_1 m_2 = -1$.

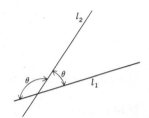

## EQUATIONS OF LINES IN DIFFERENT FORMS

*Point slope:*      $y - y_1 = m(x - x_1).$

*Slope intercept:*   $y = mx + b.$

*Intercept:*      $\dfrac{x}{a} + \dfrac{y}{b} = 1.$

*Normal:*      $x \cos \alpha + y \sin \alpha = p.$

*Parametric:*    $x = x_0 + s \cos \alpha,$

$\qquad\qquad\quad y = y_0 + s \cos \beta,$

where $\alpha$ and $\beta$ are the direction angles, and $s$ is the directed distance from $(x_0, y_0)$.

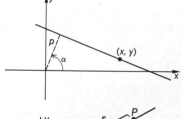

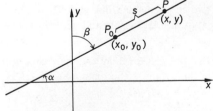

## DISTANCE FROM A LINE TO A POINT

The distance from $(x_0, y_0)$ to the line $ax + by + c = 0$ is

$$d = \frac{|ax_0 + by_0 + c|}{\sqrt{a^2 + b^2}}.$$

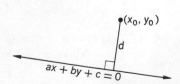

## TRANSLATION AND ROTATION OF AXES

*Translation:*   $x' = x - h,$

$y' = y - k.$

*Rotation:*   $x = x' \cos \theta - y' \sin \theta,$

$y = x' \sin \theta + y' \cos \theta.$

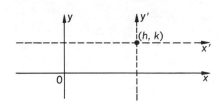

## 10   *The Conics*

### ELLIPSE

The point $P$ is on the ellipse if and only if

$$| PF_1 |/| PM | = e < 1,$$

where $F_1$ and $F_2$ are the two foci of the ellipse, and $| PM |$ is the distance from the point $P$ to the $F_1$ directrix.

If one has
center $(0, 0)$; foci $(\pm c, 0)$; $a^2 = b^2 + c^2$;
eccentricity $= e = c/a$;
directrices: $x = \pm a/e$;
then its standard equation is

$$\frac{x^2}{a^2} + \frac{y^2}{b^2} = 1.$$

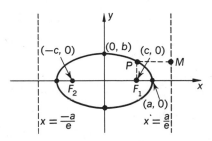

If $P$ is on the ellipse, then $| PF_1 | + | PF_2 | = 2a.$

### PARABOLA

The point $P$ is on the parabola if and only if

$$| PF |/| PM | = e = 1,$$

where $F$ is the focus, and $M$ is the foot of the perpendicular from $P$ to the directrix.

If one has focus $(p/2, 0)$ and directrix $x = -p/2$, then the standard equation is

$$y^2 = 2px.$$

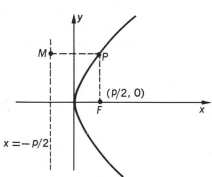

### HYPERBOLA

The point $P$ is on the hyperbola if and only if

$$| PF_1 |/| PM | = e > 1,$$

where $F_1$ and $F_2$ are the two foci of the hyperbola, and $| PM |$ is the distance from the $F_1$ directrix.

If one has
center $(0, 0)$; foci $(\pm c, 0)$;
  $c^2 = a^2 + b^2$;
eccentricity $= e = c/a$;
directrices: $x = \pm a/e$;
asymptotes: $y = \pm (b/a)x$,

then its standard equation is

$$\frac{x^2}{a^2} - \frac{y^2}{b^2} = 1.$$

If $P$ is on the hyperbola, then

$$|\, PF_1 \,| - |\, PF_2 \,| = \pm 2a.$$

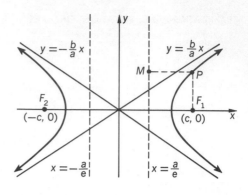

## TEST FOR CONICS

If the second degree equation

$$Ax^2 + 2Bxy + Cy^2 + Dx + Ey + F = 0$$

has a graph consisting of more than one point, and if the left side does not factor into two straight lines, then its graph is

an ellipse if $\qquad B^2 - AC < 0,$
a parabola if $\qquad B^2 - AC = 0,$
a hyperbola if $\qquad B^2 - AC > 0.$

## 11  Polar Coordinates

The pole and the polar axis can be located to please the user. When both polar and rectangular coordinates are involved, one usually places the axes so they are related as shown in the figure. Then

$$x = r \cos \theta, \qquad y = r \sin \theta,$$
$$r = \pm \sqrt{x^2 + y^2}, \qquad \theta = \text{Arc tan } (y/x).$$

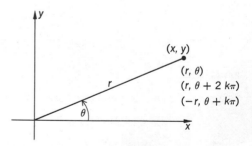

## CONICS

Conics have simple equations in polar coordinates if a focus is at the pole and the directrix is perpendicular to the polar axis. With the directrix as shown an equation of the conic is

$$r = \frac{ep}{1 + e \cos \theta}, \qquad \text{where } e = \text{eccentricity.}$$

For the parabola $e = 1$, this can be changed to

$$r = \frac{p}{2} \sec^2 \frac{\theta}{2}.$$

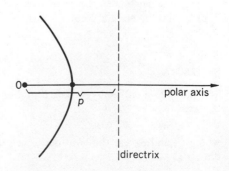

## 12   *Some Plane Curves; Rectangular Coordinates*

**The witch of Agnesi**

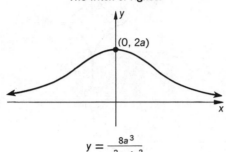

$$y = \frac{8a^3}{x^2 + 4a^2}$$

**The cissoid**

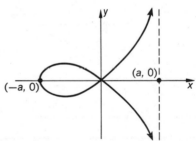

$$y^2 = \frac{x^3}{2a \cdot x}$$

**The folium of Descartes**

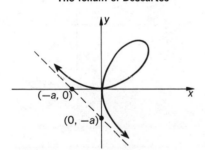

$$x^3 + y^3 - 3axy = 0$$

**The strophoid**

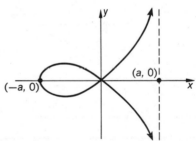

$$y^2 = x^2 \frac{a + x}{a - x}$$

**The conchoid of Nicomedes**

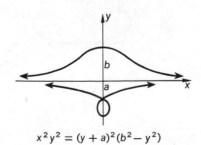

$$x^2 y^2 = (y + a)^2 (b^2 - y^2)$$

**A parabola**

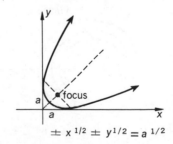

$$\pm x^{1/2} \pm y^{1/2} = a^{1/2}$$

### Cycloid
A circle of radius $a$
rolls on a line

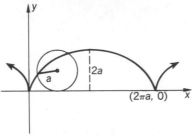

$$x = a\,(\theta - \sin\theta)$$
$$y = a\,(1 - \cos\theta)$$

### Hypocycloid of four cusps
A circle of radius $a/4$ rolls
inside one of radius $a$

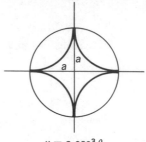

$$x = a\,\cos^3\theta$$
$$y = a\,\sin^3\theta$$

### Catenary
(a hanging chain)

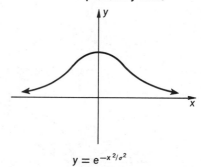

$$y = a\,\frac{e^{x/a} + e^{-x/a}}{2} = a\,\cos h\,\frac{x}{a}$$

### Normal probability curve

$$y = e^{-x^2/\sigma^2}$$

## 13    *Some Plane Curves in Polar Coordinates*

### Cardiods and limacons

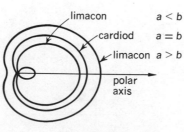

limacon         $a < b$
cardiod         $a = b$
limacon  $a > b$

polar
axis

$$r = a + b\cos\theta$$

### Three-leaved roses

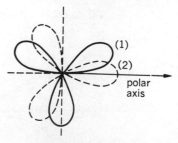

(1)
(2)
polar
axis

**(1)** $r = a\,\sin 3\theta$
**(2)** $r = a\,\cos 3\theta$

**Four-leaved roses**

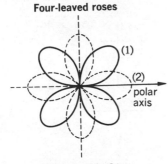

(1) $r = a \sin 2\theta$
(2) $r = a \cos 2\theta$

**Lemniscates**

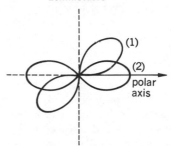

(1) $r^2 = a^2 \sin 2\theta$
(2) $r^2 = a^2 \cos 2\theta$

**Archimedes spiral**

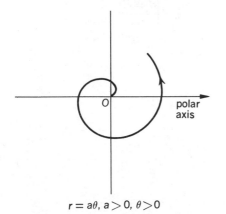

$r = a\theta,\ a > 0,\ \theta > 0$

**Hyperbolic spiral**

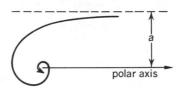

$r\theta = a,\ a > 0$

**Logarithmic spiral**

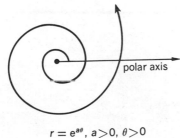

$r = e^{a\theta},\ a > 0,\ \theta > 0$

**Lituus**

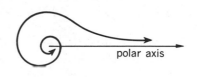

$r^2 \theta = a^2\ \ r > 0$

## 14   *Analytic Geometry of Space; Lines*

To coordinatize Euclidean space choose three mutually perpendicular lines through a point $O$. Coordinatize each line with $O$ as the origin of coordinates and with the same unit of length. Call the three lines the $x$-, $y$-, and $z$-coordinate axes.

The planes containing the axes in pairs are called the coordinate planes, and have the following equations:

$$xy\text{-plane: } z = 0, \quad yz\text{-plane: } x = 0,$$
$$zx\text{-plane: } y = 0.$$

If $P$ is any point the planes through $P$ parallel to the coordinate planes, intersect the axes in points whose coordinates are the rectangular coordinates $x$, $y$, and $z$ of $P$. Three points are shown plotted in the figure.

The coordinate planes separate space into eight parts called octants. The *first octant* is the one where all coordinates are positive.

The distance between the points $P_1 = (x_1, y_1, z_1)$ and $P_2 = (x_2, y_2, z_2)$ is

$$|P_1P_2|$$
$$= \sqrt{(x_1 - x_2)^2 + (y_1 - y_2)^2 + (z_1 - z_2)^2}.$$

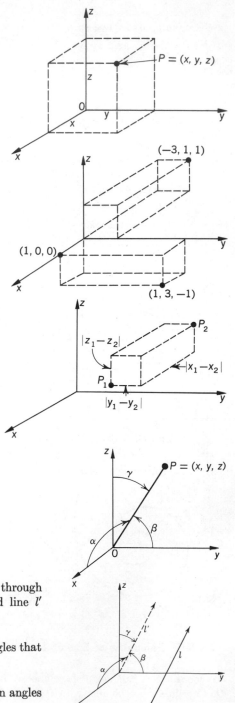

## DIRECTED LINES

The direction of the directed segment (or arrow) from $O$ to $P$ is specified by the angles, called *direction angles*, $\alpha$, $\beta$, and $\gamma$ between the segment and the positive axes. These angles always lie between $0$ and $\pi$: $0 \leqq \alpha, \beta, \gamma \leqq \pi$.

These angles determine a direction on the line through $O$ and $P$.

The direction angles of a directed line $l$ not through $O$ are the same as those of a parallel directed line $l'$ through $O$.

Oppositely directed lines have direction angles that are supplementary in pairs.

The *direction cosines* of a line with direction angles $\alpha$, $\beta$, $\gamma$ are the numbers $\cos \alpha$, $\cos \beta$, $\cos \gamma$.

Referring to the figure, the angles $\angle OXP$, $\angle OYP$, and $\angle OZP$ are right angles. Therefore, we have

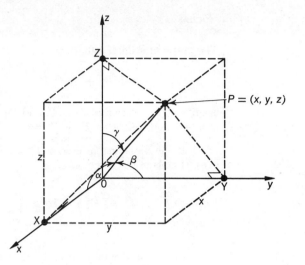

$$(1) \qquad \begin{aligned} x/r &= \cos \alpha, \\ y/r &= \cos \beta, \\ z/r &= \cos \gamma, \end{aligned}$$

where $\alpha$, $\beta$, and $\gamma$ are the direction angles of the ray from $O$ through $P = (x, y, z)$ and

$$r = |\, OP \,| = \sqrt{x^2 + y^2 + z^2}.$$

As a consequence of equations (1) we have

$$\cos^2 \alpha + \cos^2 \beta + \cos^2 \gamma = 1.$$

**PROOF** $\quad \cos^2 \alpha + \cos^2 \beta + \cos^2 \gamma = (x^2/r^2) + (y^2/r^2) + (z^2/r^2) = 1.$

As a further consequence of (1) it follows that, if a line $l$ has direction angles $\alpha$, $\beta$, and $\gamma$ and passes through the point $(x_0, y_0, z_0)$, then the point $(x, y, z)$ is on the line if and only if

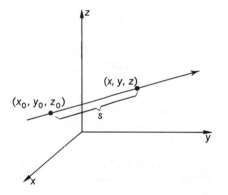

$$x - x_0 = s \cos \alpha, \quad y - y_0 = s \cos \beta,$$

$$z - z_0 = s \cos \gamma,$$

where $s$ is the directed distance from $(x_0, y_0, z_0)$ along the line. Thus, we have the

*Parametric equations*:

$$(2) \qquad \begin{aligned} x &= x_0 + s \cos \alpha, \\ y &= y_0 + s \cos \beta, \\ z &= z_0 + s \cos \gamma. \end{aligned}$$

Numbers proportional to direction cosines of a line are called *direction numbers* of the line. From equations (2) it is apparent that $x - x_0$, $y - y_0$, and $z - z_0$ are direction numbers.

Suppose that $l$ and $l'$ are two directed lines with direction angles $\alpha$, $\beta$, $\gamma$ and $\alpha'$, $\beta'$, $\gamma'$; then the angle $\theta$ between the lines is given by

$$(3) \qquad \cos \theta = \cos \alpha \cos \alpha' + \cos \beta \cos \beta' + \cos \gamma \cos \gamma'.$$

From (3) we conclude that the lines are perpendicular if and only if

$$(4) \qquad \cos \alpha \cos \alpha' + \cos \beta \cos \beta' + \cos \gamma \cos \gamma' = 0.$$

From (4) it follows that, if the lines have direction numbers $A$, $B$, $C$ and $A'$, $B'$, $C'$, then the lines are perpendicular if and only if

$$(5) \qquad AA' + BB' + CC' = 0.$$

## 15   *Planes*

The graph of a linear equation,

$$Ax + By + Cz = D$$

is a plane, providing that $A, B, C$, and $D$ are real numbers and that $A^2 + B^2 + C^2 \neq 0$.

If the plane has intercepts $a$, $b$, $c$ on the axes, the plane has an equation in

*Intercept form*:   $\dfrac{x}{a} + \dfrac{y}{b} + \dfrac{z}{c} = 1.$

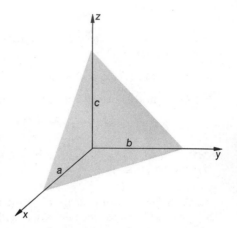

Planes parallel to the coordinate axes have equations with one of the variables absent.

planes parallel to the
   $x$-axis:   $By + Cz = D,$

planes parallel to the
   $y$-axis:   $Ax + Cz = D,$

planes parallel to the
   $z$-axis:   $Ax + By = D.$

Suppose that a line $l$ has direction numbers $A, B$, and $C$, and that a plane is perpendicular to $l$ and passes through the point $(x_0, y_0, z_0)$. Then the plane has an equation

$$(1) \qquad\qquad A(x - x_0) + B(y - y_0) + C(z - z_0) = 0.$$

**PROOF**    The point $(x, y, z)$ is on the plane if and only if the line through $(x, y, z)$ and $(x_0, y_0, z_0)$ is perpendicular to $l$. Equation (1) is then simply equation (5) of Section 14.

## 16   *Quadric Surfaces*

Second degree equations in $x$ and $y$ represent conics in the plane. Second degree equations in $x$, $y$, and $z$ represent surfaces in space and are closely related to the conics. Most of the types that can occur are sketched here, and are so oriented as to have simple equations.

### Paraboloid

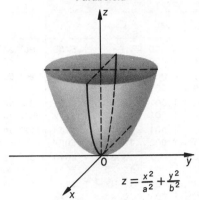

$$z = \frac{x^2}{a^2} + \frac{y^2}{b^2}$$

### Hyperboloid of two sheets

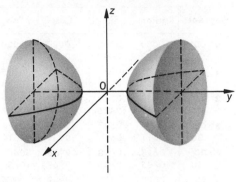

$$\frac{y^2}{b^2} - \frac{x^2}{a^2} - \frac{z^2}{c^2} = 1$$

### Ellipsoid

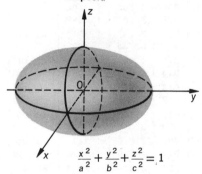

$$\frac{x^2}{a^2} + \frac{y^2}{b^2} + \frac{z^2}{c^2} = 1$$

#### Hyperboloid of one sheet

#### Right elliptic cone

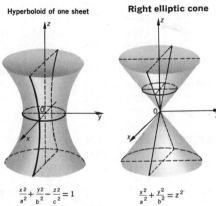

$$\frac{x^2}{a^2} + \frac{y^2}{b^2} - \frac{z^2}{c^2} = 1$$

$$\frac{x^2}{a^2} + \frac{y^2}{b^2} = z^2$$

### Hyperbolic paraboloid

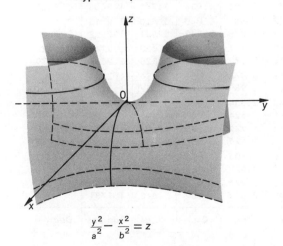

$$\frac{y^2}{a^2} - \frac{x^2}{b^2} = z$$

### Parabolic cylinder

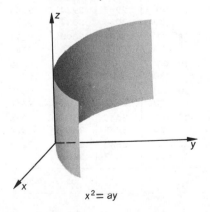

$$x^2 = ay$$

## 17 *Four-Place Values of Trigonometric Ratios; Radians*

| Degrees | Radians | Sin | Cos | Tan | Cot | Sec | Csc | | |
|---|---|---|---|---|---|---|---|---|---|
| **0° 00′** | .0000 | .0000 | 1.0000 | .0000 | — | 1.000 | — | 1.5708 | **90° 00′** |
| 30 | .0087 | .0087 | 1.0000 | .0087 | 114.6 | 1.000 | 114.6 | 1.5621 | 30 |
| **1° 00′** | .0175 | .0175 | .9998 | .0175 | 57.29 | 1.000 | 57.30 | 1.5533 | **89° 00′** |
| 30 | .0262 | .0262 | .9997 | .0262 | 38.19 | 1.000 | 38.20 | 1.5446 | 30 |
| **2° 00′** | .0349 | .0349 | .9994 | .0349 | 28.64 | 1.001 | 28.65 | 1.5359 | **88° 00′** |
| 30 | .0436 | .0436 | .9990 | .0437 | 22.90 | 1.001 | 22.93 | 1.5272 | 30 |
| **3° 00′** | .0524 | .0523 | .9986 | .0524 | 19.08 | 1.001 | 19.11 | 1.5184 | **87° 00′** |
| 30 | .0611 | .0610 | .9981 | .0612 | 16.35 | 1.002 | 16.38 | 1.5097 | 30 |
| **4° 00′** | .0698 | .0698 | .9976 | .0699 | 14.30 | 1.002 | 14.34 | 1.5010 | **86° 00′** |
| 30 | .0785 | .0785 | .9969 | .0787 | 12.71 | 1.003 | 12.75 | 1.4923 | 30 |
| **5° 00′** | .0873 | .0872 | .9962 | .0875 | 11.43 | 1.004 | 11.47 | 1.4835 | **85° 00′** |
| 30 | .0960 | .0958 | .9954 | .0963 | 10.39 | 1.005 | 10.43 | 1.4748 | 30 |
| **6° 00′** | .1047 | .1045 | .9945 | .1051 | 9.514 | 1.006 | 9.567 | 1.4661 | **84° 00′** |
| 30 | .1134 | .1132 | .9936 | .1139 | 8.777 | 1.006 | 8.834 | 1.4573 | 30 |
| **7° 00′** | .1222 | .1219 | .9925 | .1228 | 8.144 | 1.008 | 8.206 | 1.4486 | **83° 00′** |
| 30 | .1309 | .1305 | .9914 | .1317 | 7.596 | 1.009 | 7.661 | 1.4399 | 30 |
| **8° 00′** | .1396 | .1392 | .9903 | .1405 | 7.115 | 1.010 | 7.185 | 1.4312 | **82° 00′** |
| 30 | .1484 | .1478 | .9890 | .1495 | 6.691 | 1.011 | 6.765 | 1.4224 | 30 |
| **9° 00′** | .1571 | .1564 | .9877 | .1584 | 6.314 | 1.012 | 6.392 | 1.4137 | **81° 00′** |
| 30 | .1658 | .1650 | .9863 | .1673 | 5.976 | 1.014 | 6.059 | 1.4050 | 30 |
| **10° 00′** | .1745 | .1736 | .9848 | .1763 | 5.671 | 1.015 | 5.759 | 1.3963 | **80° 00′** |
| 30 | .1833 | .1822 | .9833 | .1853 | 5.396 | 1.017 | 5.487 | 1.3875 | 30 |
| **11° 00′** | .1920 | .1908 | .9816 | .1944 | 5.145 | 1.019 | 5.241 | 1.3788 | **79° 00′** |
| 30 | .2007 | .1994 | .9799 | .2035 | 4.915 | 1.020 | 5.016 | 1.3701 | 30 |
| **12° 00′** | .2094 | .2079 | .9781 | .2126 | 4.705 | 1.022 | 4.810 | 1.3614 | **78° 00′** |
| 30 | .2182 | .2164 | .9763 | .2217 | 4.511 | 1.024 | 4.620 | 1.3526 | 30 |
| **13° 00′** | .2269 | .2250 | .9744 | .2309 | 4.331 | 1.026 | 4.445 | 1.3439 | **77° 00′** |
| 30 | .2356 | .2334 | .9724 | .2401 | 4.165 | 1.028 | 4.284 | 1.3352 | 30 |
| **14° 00′** | .2443 | .2419 | .9703 | .2493 | 4.011 | 1.031 | 4.134 | 1.3265 | **76° 00′** |
| 30 | .2531 | .2504 | .9681 | .2586 | 3.867 | 1.033 | 3.994 | 1.3177 | 30 |
| **15° 00′** | .2618 | .2588 | .9659 | .2679 | 3.732 | 1.035 | 3.864 | 1.3090 | **75° 00′** |
| 30 | .2705 | .2672 | .9636 | .2773 | 3.606 | 1.038 | 3.742 | 1.3003 | 30 |
| **16° 00′** | .2793 | .2756 | .9613 | .2867 | 3.487 | 1.040 | 3.628 | 1.2915 | **74° 00′** |
| 30 | .2880 | .2840 | .9588 | .2962 | 3.376 | 1.043 | 3.521 | 1.2828 | 30 |
| **17° 00′** | .2967 | .2924 | .9563 | .3057 | 3.271 | 1.046 | 3.420 | 1.2741 | **73° 00′** |
| 30 | .3054 | .3007 | .9537 | .3153 | 3.172 | 1.049 | 3.326 | 1.2654 | 30 |
| **18° 00′** | .3142 | .3090 | .9511 | .3249 | 3.078 | 1.051 | 3.236 | 1.2566 | **72° 00′** |
| 30 | .3229 | .3173 | .9483 | .3346 | 2.989 | 1.054 | 3.152 | 1.2479 | 30 |
| **19° 00′** | .3316 | .3256 | .9455 | .3443 | 2.904 | 1.058 | 3.072 | 1.2392 | **71° 00′** |
| 30 | .3403 | .3338 | .9426 | .3541 | 2.824 | 1.061 | 2.996 | 1.2305 | 30 |
| **20° 00′** | .3491 | .3420 | .9397 | .3640 | 2.747 | 1.064 | 2.924 | 1.2217 | **70° 00′** |
| 30 | .3578 | .3502 | .9367 | .3739 | 2.675 | 1.068 | 2.855 | 1.2130 | 30 |
| **21° 00′** | .3665 | .3584 | .9336 | .3839 | 2.605 | 1.071 | 2.790 | 1.2043 | **69° 00′** |
| 30 | .3752 | .3665 | .9304 | .3939 | 2.539 | 1.075 | 2.729 | 1.1956 | 30 |
| **22° 00′** | .3840 | .3746 | .9272 | .4040 | 2.475 | 1.079 | 2.669 | 1.1868 | **68° 00′** |
| 30 | .3927 | .3827 | .9239 | .4142 | 2.414 | 1.082 | 2.613 | 1.1781 | 30 |
| | | Cos | Sin | Cot | Tan | Csc | Sec | Radians | Degrees |

| Degrees | Radians | Sin | Cos | Tan | Cot | Sec | Csc | | |
|---|---|---|---|---|---|---|---|---|---|
| **23° 00′** | .4014 | .3907 | .9205 | .4245 | 2.356 | 1.086 | 2.559 | 1.1694 | **67° 00′** |
| 30 | .4102 | .3987 | .9171 | .4348 | 2.300 | 1.090 | 2.508 | 1.1606 | 30 |
| **24° 00′** | .4189 | .4067 | .9135 | .4452 | 2.246 | 1.095 | 2.459 | 1.1519 | **66° 00′** |
| 30 | .4276 | .4147 | .9100 | .4557 | 2.194 | 1.099 | 2.411 | 1.1432 | 30 |
| **25° 00′** | .4363 | .4226 | .9063 | .4663 | 2.145 | 1.103 | 2.366 | 1.1345 | **65° 00′** |
| 30 | .4451 | .4305 | .9026 | .4770 | 2.097 | 1.108 | 2.323 | 1.1257 | 30 |
| **26° 00′** | .4538 | .4384 | .8988 | .4877 | 2.050 | 1.113 | 2.281 | 1.1170 | **64° 00′** |
| 30 | .4625 | .4462 | .8949 | .4986 | 2.006 | 1.117 | 2.241 | 1.1083 | 30 |
| **27° 00′** | .4712 | .4540 | .8910 | .5095 | 1.963 | 1.122 | 2.203 | 1.0996 | **63° 00′** |
| 30 | .4800 | .4617 | .8870 | .5206 | 1.921 | 1.127 | 2.166 | 1.0908 | 30 |
| **28° 00′** | .4887 | .4695 | .8829 | .5317 | 1.881 | 1.133 | 2.130 | 1.0821 | **62° 00′** |
| 30 | .4974 | .4772 | .8788 | .5430 | 1.842 | 1.138 | 2.096 | 1.0734 | 30 |
| **29° 00′** | .5061 | .4848 | .8746 | .5543 | 1.804 | 1.143 | 2.063 | 1.0647 | **61° 00′** |
| 30 | .5149 | .4924 | .8704 | .5658 | 1.767 | 1.149 | 2.031 | 1.0559 | 30 |
| **30° 00′** | .5236 | .5000 | .8660 | .5774 | 1.732 | 1.155 | 2.000 | 1.0472 | **60° 00′** |
| 30 | .5323 | .5075 | .8616 | .5890 | 1.698 | 1.161 | 1.970 | 1.0385 | 30 |
| **31° 00′** | .5411 | .5150 | .8572 | .6009 | 1.664 | 1.167 | 1.942 | 1.0297 | **59° 00′** |
| 30 | .5498 | .5225 | .8526 | .6128 | 1.632 | 1.173 | 1.914 | 1.0210 | 30 |
| **32° 00′** | .5585 | .5299 | .8480 | .6249 | 1.600 | 1.179 | 1.887 | 1.0123 | **58° 00′** |
| 30 | .5672 | .5373 | .8434 | .6371 | 1.570 | 1.186 | 1.861 | 1.0036 | 30 |
| **33° 00′** | .5760 | .5446 | .8387 | .6494 | 1.540 | 1.192 | 1.836 | .9948 | **57° 00′** |
| 30 | .5847 | .5519 | .8339 | .6619 | 1.511 | 1.199 | 1.812 | .9861 | 30 |
| **34° 00′** | .5934 | .5592 | .8290 | .6745 | 1.483 | 1.206 | 1.788 | .9774 | **56° 00′** |
| 30 | .6021 | .5664 | .8241 | .6873 | 1.455 | 1.213 | 1.766 | .9687 | 30 |
| **35° 00′** | .6109 | .5736 | .8192 | .7002 | 1.428 | 1.221 | 1.743 | .9599 | **55° 00′** |
| 30 | .6196 | .5807 | .8141 | .7133 | 1.402 | 1.228 | 1.722 | .9512 | 30 |
| **36° 00′** | .6283 | .5878 | .8090 | .7265 | 1.376 | 1.236 | 1.701 | .9425 | **54° 00′** |
| 30 | .6370 | .5948 | .8039 | .7400 | 1.351 | 1.244 | 1.681 | .9338 | 30 |
| **37° 00′** | .6458 | .6018 | .7986 | .7536 | 1.327 | 1.252 | 1.662 | .9250 | **53° 00′** |
| 30 | .6545 | .6088 | .7934 | .7673 | 1.303 | 1.260 | 1.643 | .9163 | 30 |
| **38° 00′** | .6632 | .6157 | .7880 | .7813 | 1.280 | 1.269 | 1.624 | .9076 | **52° 00′** |
| 30 | .6720 | .6225 | .7826 | .7954 | 1.257 | 1.278 | 1.606 | .8988 | 30 |
| **39° 00′** | .6807 | .6293 | .7771 | .8098 | 1.235 | 1.287 | 1.589 | .8901 | **51° 00′** |
| 30 | .6894 | .6361 | .7716 | .8243 | 1.213 | 1.296 | 1.572 | .8814 | 30 |
| **40° 00′** | .6981 | .6428 | .7660 | .8391 | 1.192 | 1.305 | 1.556 | .8727 | **50° 00′** |
| 30 | .7069 | .6494 | .7604 | .8541 | 1.171 | 1.315 | 1.540 | .8639 | 30 |
| **41° 00′** | .7156 | .6561 | .7547 | .8693 | 1.150 | 1.325 | 1.524 | .8552 | **49° 00′** |
| 30 | .7243 | .6626 | .7490 | .8847 | 1.130 | 1.335 | 1.509 | .8465 | 30 |
| **42° 00′** | .7330 | .6691 | .7431 | .9004 | 1.111 | 1.346 | 1.494 | .8378 | **48° 00′** |
| 30 | .7418 | .6756 | .7373 | .9163 | 1.091 | 1.356 | 1.480 | .8290 | 30 |
| **43° 00′** | .7505 | .6820 | .7314 | .9325 | 1.072 | 1.367 | 1.466 | .8203 | **47° 00′** |
| 30 | .7592 | .6884 | .7254 | .9490 | 1.054 | 1.379 | 1.453 | .8116 | 30 |
| **44° 00′** | .7679 | .6947 | .7193 | .9657 | 1.036 | 1.390 | 1.440 | .8029 | **46° 00′** |
| 30 | .7767 | .7009 | .7133 | .9827 | 1.018 | 1.402 | 1.427 | .7941 | 30 |
| **45° 00′** | .7854 | .7071 | .7071 | 1.000 | 1.000 | 1.414 | 1.414 | .7854 | **45° 00′** |
| | | Cos | Sin | Cot | Tan | Csc | Sec | Radians | Degrees |

## 18 *Exponential Functions*

| x | $e^x$ | $e^{-x}$ | x | $e^x$ | $e^{-x}$ |
|---|---|---|---|---|---|
| 0.00 | 1.0000 | 1.0000 | 2.5 | 12.182 | 0.0821 |
| 0.05 | 1.0513 | 0.9512 | 2.6 | 13.464 | 0.0743 |
| 0.10 | 1.1052 | 0.9048 | 2.7 | 14.880 | 0.0672 |
| 0.15 | 1.1618 | 0.8607 | 2.8 | 16.445 | 0.0608 |
| 0.20 | 1.2214 | 0.8187 | 2.9 | 18.174 | 0.0550 |
| 0.25 | 1.2840 | 0.7788 | 3.0 | 20.086 | 0.0498 |
| 0.30 | 1.3499 | 0.7408 | 3.1 | 22.198 | 0.0450 |
| 0.35 | 1.4191 | 0.7047 | 3.2 | 24.533 | 0.0408 |
| 0.40 | 1.4918 | 0.6703 | 3.3 | 27.113 | 0.0369 |
| 0.45 | 1.5683 | 0.6376 | 3.4 | 29.964 | 0.0334 |
| 0.50 | 1.6487 | 0.6065 | 3.5 | 33.115 | 0.0302 |
| 0.55 | 1.7333 | 0.5769 | 3.6 | 36.598 | 0.0273 |
| 0.60 | 1.8221 | 0.5488 | 3.7 | 40.447 | 0.0247 |
| 0.65 | 1.9155 | 0.5220 | 3.8 | 44.701 | 0.0224 |
| 0.70 | 2.0138 | 0.4966 | 3.9 | 49.402 | 0.0202 |
| 0.75 | 2.1170 | 0.4724 | 4.0 | 54.598 | 0.0183 |
| 0.80 | 2.2255 | 0.4493 | 4.1 | 60.340 | 0.0166 |
| 0.85 | 2.3396 | 0.4274 | 4.2 | 66.686 | 0.0150 |
| 0.90 | 2.4596 | 0.4066 | 4.3 | 73.700 | 0.0136 |
| 0.95 | 2.5857 | 0.3867 | 4.4 | 81.451 | 0.0123 |
| 1.0 | 2.7183 | 0.3679 | 4.5 | 90.017 | 0.0111 |
| 1.1 | 3.0042 | 0.3329 | 4.6 | 99.484 | 0.0101 |
| 1.2 | 3.3201 | 0.3012 | 4.7 | 109.95 | 0.0091 |
| 1.3 | 3.6693 | 0.2725 | 4.8 | 121.51 | 0.0082 |
| 1.4 | 4.0552 | 0.2466 | 4.9 | 134.29 | 0.0074 |
| 1.5 | 4.4817 | 0.2231 | 5 | 148.41 | 0.0067 |
| 1.6 | 4.9530 | 0.2019 | 6 | 403.43 | 0.0025 |
| 1.7 | 5.4739 | 0.1827 | 7 | 1 096.6 | 0.0009 |
| 1.8 | 6.0496 | 0.1653 | 8 | 2 981.0 | 0.0003 |
| 1.9 | 6.6859 | 0.1496 | 9 | 8 103.1 | 0.0001 |
| 2.0 | 7.3891 | 0.1353 | 10 | 22 026 | 0.00005 |
| 2.1 | 8.1662 | 0.1225 | | | |
| 2.2 | 9.0250 | 0.1108 | | | |
| 2.3 | 9.9742 | 0.1003 | | | |
| 2.4 | 11.023 | 0.0907 | | | |

19  *Natural Logarithms of Numbers*

| n | log<sub>e</sub> n | n | log<sub>e</sub> n | n | log<sub>e</sub> n |
|---|---|---|---|---|---|
| 0.0 |  | 4.5 | 1.5041 | 9.0 | 2.1972 |
| 0.1 | −2.3026 | 4.6 | 1.5261 | 9.1 | 2.2083 |
| 0.2 | −1.6094 | 4.7 | 1.5476 | 9.2 | 2.2192 |
| 0.3 | −1.2040 | 4.8 | 1.5686 | 9.3 | 2.2300 |
| 0.4 | −0.9163 | 4.9 | 1.5892 | 9.4 | 2.2407 |
| 0.5 | −0.6931 | 5.0 | 1.6094 | 9.5 | 2.2513 |
| 0.6 | −0.5108 | 5.1 | 1.6292 | 9.6 | 2.2618 |
| 0.7 | −0.3567 | 5.2 | 1.6487 | 9.7 | 2.2721 |
| 0.8 | −0.2231 | 5.3 | 1.6677 | 9.8 | 2.2824 |
| 0.9 | −0.1054 | 5.4 | 1.6864 | 9.9 | 2.2925 |
| 1.0 | 0.0000 | 5.5 | 1.7047 | 10 | 2.3026 |
| 1.1 | 0.0953 | 5.6 | 1.7228 | 11 | 2.3979 |
| 1.2 | 0.1823 | 5.7 | 1.7405 | 12 | 2.4849 |
| 1.3 | 0.2624 | 5.8 | 1.7579 | 13 | 2.5649 |
| 1.4 | 0.3365 | 5.9 | 1.7750 | 14 | 2.6391 |
| 1.5 | 0.4055 | 6.0 | 1.7918 | 15 | 2.7081 |
| 1.6 | 0.4700 | 6.1 | 1.8083 | 16 | 2.7726 |
| 1.7 | 0.5306 | 6.2 | 1.8245 | 17 | 2.8332 |
| 1.8 | 0.5878 | 6.3 | 1.8405 | 18 | 2.8904 |
| 1.9 | 0.6419 | 6.4 | 1.8563 | 19 | 2.9444 |
| 2.0 | 0.6931 | 6.5 | 1.8718 | 20 | 2.9957 |
| 2.1 | 0.7419 | 6.6 | 1.8871 | 25 | 3.2189 |
| 2.2 | 0.7885 | 6.7 | 1.9021 | 30 | 3.4012 |
| 2.3 | 0.8329 | 6.8 | 1.9169 | 35 | 3.5553 |
| 2.4 | 0.8755 | 6.9 | 1.9315 | 40 | 3.6889 |
| 2.5 | 0.9163 | 7.0 | 1.9459 | 45 | 3.8067 |
| 2.6 | 0.9555 | 7.1 | 1.9601 | 50 | 3.9120 |
| 2.7 | 0.9933 | 7.2 | 1.9741 | 55 | 4.0073 |
| 2.8 | 1.0296 | 7.3 | 1.9879 | 60 | 4.0943 |
| 2.9 | 1.0647 | 7.4 | 2.0015 | 65 | 4.1744 |
| 3.0 | 1.0986 | 7.5 | 2.0149 | 70 | 4.2485 |
| 3.1 | 1.1314 | 7.6 | 2.0281 | 75 | 4.3175 |
| 3.2 | 1.1632 | 7.7 | 2.0412 | 80 | 4.3820 |
| 3.3 | 1.1939 | 7.8 | 2.0541 | 85 | 4.4427 |
| 3.4 | 1.2238 | 7.9 | 2.0669 | 90 | 4.4998 |
| 3.5 | 1.2528 | 8.0 | 2.0794 | 95 | 4.5539 |
| 3.6 | 1.2809 | 8.1 | 2.0919 | 100 | 4.6052 |
| 3.7 | 1.3083 | 8.2 | 2.1041 |  |  |
| 3.8 | 1.3350 | 8.3 | 2.1163 |  |  |
| 3.9 | 1.3610 | 8.4 | 2.1282 |  |  |
| 4.0 | 1.3863 | 8.5 | 2.1401 |  |  |
| 4.1 | 1.4110 | 8.6 | 2.1518 |  |  |
| 4.2 | 1.4351 | 8.7 | 2.1633 |  |  |
| 4.3 | 1.4586 | 8.8 | 2.1748 |  |  |
| 4.4 | 1.4816 | 8.9 | 2.1861 |  |  |

## 20    *Powers and Roots*

| No. | Sq. | Sq. Root | Cube | Cube Root | No. | Sq. | Sq. Root | Cube | Cube Root |
|---|---|---|---|---|---|---|---|---|---|
| 1 | 1 | 1.000 | 1 | 1.000 | 51 | 2,601 | 7.141 | 132,651 | 3.708 |
| 2 | 4 | 1.414 | 8 | 1.260 | 52 | 2,704 | 7.211 | 140,608 | 3.733 |
| 3 | 9 | 1.732 | 27 | 1.442 | 53 | 2,809 | 7.280 | 148,877 | 3.756 |
| 4 | 16 | 2.000 | 64 | 1.587 | 54 | 2,916 | 7.348 | 157,464 | 3.780 |
| 5 | 25 | 2.236 | 125 | 1.710 | 55 | 3,025 | 7.416 | 166,375 | 3.803 |
| 6 | 36 | 2.449 | 216 | 1.817 | 56 | 3,136 | 7.483 | 175,616 | 3.826 |
| 7 | 49 | 2.646 | 343 | 1.913 | 57 | 3,249 | 7.550 | 185,193 | 3.849 |
| 8 | 64 | 2.828 | 512 | 2.000 | 58 | 3,364 | 7.616 | 195,112 | 3.871 |
| 9 | 81 | 3.000 | 729 | 2.080 | 59 | 3,481 | 7.681 | 205,379 | 3.893 |
| 10 | 100 | 3.162 | 1,000 | 2.154 | 60 | 3,600 | 7.746 | 216,000 | 3.915 |
| 11 | 121 | 3.317 | 1,331 | 2.224 | 61 | 3,721 | 7.810 | 226,981 | 3.936 |
| 12 | 144 | 3.464 | 1,728 | 2.289 | 62 | 3,844 | 7.874 | 238,328 | 3.958 |
| 13 | 169 | 3.606 | 2,197 | 2.351 | 63 | 3,969 | 7.937 | 250,047 | 3.979 |
| 14 | 196 | 3.742 | 2,744 | 2.410 | 64 | 4,096 | 8.000 | 262,144 | 4.000 |
| 15 | 225 | 3.873 | 3,375 | 2.466 | 65 | 4,225 | 8.062 | 274,625 | 4.021 |
| 16 | 256 | 4.000 | 4,096 | 2.520 | 66 | 4,356 | 8.124 | 287,496 | 4.041 |
| 17 | 289 | 4.123 | 4,913 | 2.571 | 67 | 4,489 | 8.185 | 300,763 | 4.062 |
| 18 | 324 | 4.243 | 5,832 | 2.621 | 68 | 4,624 | 8.246 | 314,432 | 4.082 |
| 19 | 361 | 4.359 | 6,859 | 2.668 | 69 | 4,761 | 8.307 | 328,509 | 4.102 |
| 20 | 400 | 4.472 | 8,000 | 2.714 | 70 | 4,900 | 8.367 | 343,000 | 4.121 |
| 21 | 441 | 4.583 | 9,261 | 2.759 | 71 | 5,041 | 8.426 | 357,911 | 4.141 |
| 22 | 484 | 4.690 | 10,648 | 2.802 | 72 | 5,184 | 8.485 | 373,248 | 4.160 |
| 23 | 529 | 4.796 | 12,167 | 2.844 | 73 | 5,329 | 8.544 | 389,017 | 4.179 |
| 24 | 576 | 4.899 | 13,824 | 2.884 | 74 | 5,476 | 8.602 | 405,224 | 4.198 |
| 25 | 625 | 5.000 | 15,625 | 2.924 | 75 | 5,625 | 8.660 | 421,875 | 4.217 |
| 26 | 676 | 5.099 | 17,576 | 2.962 | 76 | 5,776 | 8.718 | 438,976 | 4.236 |
| 27 | 729 | 5.196 | 19,683 | 3.000 | 77 | 5,929 | 8.775 | 456,533 | 4.254 |
| 28 | 784 | 5.292 | 21,952 | 3.037 | 78 | 6,084 | 8.832 | 474,552 | 4.273 |
| 29 | 841 | 5.385 | 24,389 | 3.072 | 79 | 6,241 | 8.888 | 493,039 | 4.291 |
| 30 | 900 | 5.477 | 27,000 | 3.107 | 80 | 6,400 | 8.944 | 512,000 | 4.309 |
| 31 | 961 | 5.568 | 29,791 | 3.141 | 81 | 6,561 | 9.000 | 531,441 | 4.327 |
| 32 | 1,024 | 5.657 | 32,768 | 3.175 | 82 | 6,724 | 9.055 | 551,368 | 4.344 |
| 33 | 1,089 | 5.745 | 35,937 | 3.208 | 83 | 6,889 | 9.110 | 571,787 | 4.362 |
| 34 | 1,156 | 5.831 | 39,304 | 3.240 | 84 | 7,056 | 9.165 | 592,704 | 4.380 |
| 35 | 1,225 | 5.916 | 42,875 | 3.271 | 85 | 7,225 | 9.220 | 614,125 | 4.397 |
| 36 | 1,296 | 6.000 | 46,656 | 3.302 | 86 | 7,396 | 9.274 | 636,056 | 4.414 |
| 37 | 1,369 | 6.083 | 50,653 | 3.332 | 87 | 7,569 | 9.327 | 658,503 | 4.431 |
| 38 | 1,444 | 6.164 | 54,872 | 3.362 | 88 | 7,744 | 9.381 | 681,472 | 4.448 |
| 39 | 1,521 | 6.245 | 59,319 | 3.391 | 89 | 7,921 | 9.434 | 704,969 | 4.465 |
| 40 | 1,600 | 6.325 | 64,000 | 3.420 | 90 | 8,100 | 9.487 | 729,000 | 4.481 |
| 41 | 1,681 | 6.403 | 68,921 | 3.448 | 91 | 8,281 | 9.539 | 753,571 | 4.498 |
| 42 | 1,764 | 6.481 | 74,088 | 3.476 | 92 | 8,464 | 9.592 | 778,688 | 4.514 |
| 43 | 1,849 | 6.557 | 79,507 | 3.503 | 93 | 8,649 | 9.644 | 804,357 | 4.531 |
| 44 | 1,936 | 6.633 | 85,184 | 3.530 | 94 | 8,836 | 9.695 | 830,584 | 4.547 |
| 45 | 2,025 | 6.708 | 91,125 | 3.557 | 95 | 9,025 | 9.747 | 857,375 | 4.563 |
| 46 | 2,116 | 6.782 | 97,336 | 3.583 | 96 | 9,216 | 9.798 | 884,736 | 4.579 |
| 47 | 2,209 | 6.856 | 103,823 | 3.609 | 97 | 9,409 | 9.849 | 912,673 | 4.595 |
| 48 | 2,304 | 6.928 | 110,592 | 3.634 | 98 | 9,604 | 9.899 | 941,192 | 4.610 |
| 49 | 2,401 | 7.000 | 117,649 | 3.659 | 99 | 9,801 | 9.950 | 970,299 | 4.626 |
| 50 | 2,500 | 7.071 | 125,000 | 3.684 | 100 | 10,000 | 10.000 | 1,000,000 | 4.642 |

## 21 *References*

Every student should have on his bookshelf, (1) a reference for elementary algebra, (2) a reference for trigonometry, (3) a reference for analytic geometry, (4) a set of mathematical tables and formulas. Appendix A is but a partial substitute for this kind of reference library. The following books are some possible suggestions. Many alternates are available.

> *College Algebra* by M. Richardson, Prentice-Hall, 1966.
>
> *Algebra and Trigonometry* by E. A. Cameron, Holt, Rinehart and Winston, 1965.
>
> *Pre-Calculus Mathematics* by M. Shanks, C. Brumfiel, C. Fleenor, and R. Eicholz, Addison-Wesley, 1965.
>
> *Analytic Geometry* by Middlemiss, Marks, and Smart, McGraw-Hill, 1968.
>
> *Mathematical Tables*, Chemical Rubber Publishing Co., 1967.
>
> *Handbook of Mathematical Tables* by R. S. Burington, Handbook Publishing Co., 1965.

For a view of the period preceding the invention of calculus the reader is urged to consult one of the many good histories of mathematics. A brief treatment of the men of that epoch can be found on pages 281–304 of:

> *An Introduction to the History of Mathematics* by H. Eves, Holt, Rinehart and Winston, 1964.

# A BRIEF TABLE OF INTEGRALS

The symbols $u$ and $v$ are variables and may be functions. The symbols $a$, $b$, $c$, $m$, $n$, and $p$ are constants. All constants of integration have been omitted.

## THE BASIC INTEGRALS

1. $\displaystyle \int (du + dv) = \int du + \int dv$

2. $\displaystyle \int c\,du = c \int du$

3. $\displaystyle \int u\,dv = uv - \int v\,du \quad \{\text{Integration by Parts}$

4. $\displaystyle \int u^n\,du = \frac{1}{n+1} u^{n+1}, \quad n \neq -1$

5. $\displaystyle \int \frac{du}{u} = \log |u|$

6.  $\displaystyle\int e^u \, du = e^u$

7.  $\displaystyle\int a^u \, du = \frac{a^u}{\log a}, \quad a > 0, a \neq 1$

8.  $\displaystyle\int \sin u \, du = -\cos u$

9.  $\displaystyle\int \cos u \, du = \sin u$

10.  $\displaystyle\int \sec^2 u \, du = \tan u$

11.  $\displaystyle\int \csc^2 u \, du = -\cot u$

12.  $\displaystyle\int \sec u \tan u \, du = \sec u$

13.  $\displaystyle\int \csc u \cot u \, du = -\csc u$

14.  $\displaystyle\int \frac{du}{\sqrt{a^2 - u^2}} = \operatorname{Arc\,sin} \frac{u}{a}, \quad a > 0$

15.  $\displaystyle\int \frac{du}{a^2 + u^2} = \frac{1}{a} \operatorname{Arc\,tan} \frac{u}{a}$

16.  $\displaystyle\int \frac{du}{a^2 - u^2} = \frac{1}{2a} \log \left| \frac{a + u}{a - u} \right|$

17.  $\displaystyle\int \frac{du}{\sqrt{u^2 \pm a^2}} = \log \left| u + \sqrt{u^2 \pm a^2} \right|$

18.  $\displaystyle\int \tan u \, du = \log |\sec u| = -\log |\cos u|$

19.  $\displaystyle\int \cot u \, du = \log |\sin u|$

20.  $\displaystyle\int \sec u \, du = \log |\sec u + \tan u|$

21.  $\displaystyle\int \csc u \, du = \log |\csc u - \cot u|$

## SOME RATIONAL INTEGRANDS

The basic technique for rational functions is the method of partial fractions. The following special ones will serve for many problems.

22.  $$\int \frac{du}{a^2 - u^2} = \frac{1}{2a} \log \left| \frac{a + u}{a - u} \right|$$

23.  $$\int \frac{du}{u(a + bu^2)} = \frac{1}{2a} \log \left| \frac{u^2}{a + bu^2} \right|$$

24.  $$\int \frac{du}{u^2(a + bu^2)} = -\frac{1}{au} - \frac{b}{a} \int \frac{du}{a + bu^2}$$

25.  $$\int \frac{du}{(a + bu^2)^2} = \frac{u}{2a(a + bu^2)} + \frac{1}{2a} \int \frac{du}{a + bu^2}$$

26.  $$\int \frac{u^2 \, du}{(a + bu^2)} = \frac{u}{b} - \frac{a}{b} \int \frac{du}{a + bu^2}$$

27.  $$\int \frac{du}{(a + bu^2)^n} = \frac{u}{2(n - 1)a(a + bu^2)^{n-1}} + \frac{2n - 3}{2(n - 1)a} \int \frac{du}{(a + bu^2)^{n-1}}$$

28.  $$\int \frac{u^2 \, du}{(a + bu^2)^n} = \frac{-u}{2b(n - 1)(a + bu^2)^{n-1}} + \frac{1}{2b(n - 1)} \int \frac{du}{(a + bu^2)^{n-1}}$$

29.  $$\int \frac{du}{u^2(a + bu^2)^n} = \frac{1}{a} \int \frac{du}{(a + bu^2)^{n-1}} - \frac{b}{a} \int \frac{du}{(a + bu^2)^n}$$

30.  $$\int u^m (a + bu^n)^p \, du$$

$$= \frac{u^{m+1}(a + bu^n)^p}{np + m + 1} + \frac{anp}{np + m + 1} \int u^m (a + bu^n)^{p-1} \, du$$

$$= \frac{-u^{m+1}(a + bu^n)^{p+1}}{an(p + 1)} + \frac{np + n + m + 1}{an(p + 1)} \int u^m (a + bu^n)^{p+1} \, du$$

$$= \frac{u^{m-n+1}(a + bu^n)^{p+1}}{b(np + m + 1)} - \frac{a(m - n + 1)}{b(np + m + 1)} \int u^{m-n} (a + bu^n)^p \, du$$

$$= \frac{u^{m+1}(a + bu^n)^{p+1}}{a(m + 1)} - \frac{b(np + m + 1)}{a(m + 1)} \int u^{m+n} (a + bu^n)^p \, du$$

## INTEGRANDS THAT ARE RATIONAL FUNCTIONS OF $u$ AND $\sqrt{a + bu}$; $R(u, \sqrt{a + bu})$

These integrands can always be rendered rational by the substitution $\sqrt{a + bu} = v$.

## INTEGRANDS THAT ARE RATIONAL FUNCTIONS OF $u$ AND $\sqrt{a^2 - u^2}$; $R(u, \sqrt{a^2 - u^2})$

For these integrands the trigonometric substitution $u = a \sin \theta$ will often be most convenient. The integrals below are the ones that usually occur. For rational functions of $u$ and $\sqrt{a + bu + cu^2}$ first complete the square under the radical sign. Then make a change of variable.

31. $\displaystyle \int \sqrt{a^2 - u^2} \, du = \frac{u}{2} \sqrt{a^2 - u^2} + \frac{a^2}{2} \operatorname{Arc} \sin \frac{u}{a}, \quad a > 0$

32. $\displaystyle \int u^2 \sqrt{a^2 - u^2} \, du = \frac{u}{8} (2u^2 - a^2) \sqrt{a^2 - u^2} + \frac{a^4}{8} \operatorname{Arc} \sin \frac{u}{a}, \quad a > 0$

33. $\displaystyle \int \frac{u^2 \, du}{\sqrt{a^2 - u^2}} = -\frac{u}{2} \sqrt{a^2 - u^2} + \frac{a^2}{2} \operatorname{Arc} \sin \frac{u}{a}, \quad a > 0$

34. $\displaystyle \int (a^2 - u^2)^{3/2} \, du = \frac{u}{8} (5a^2 - 2u^2) \sqrt{a^2 - u^2} + \frac{3a^4}{8} \operatorname{Arc} \sin \frac{u}{a}, \quad a > 0$

35. $\displaystyle \int \frac{du}{(a^2 - u^2)^{3/2}} = \frac{u}{a^2 \sqrt{a^2 - u^2}}$

36. $\displaystyle \int \frac{u^2 \, du}{(a^2 - u^2)^{3/2}} = \frac{u}{\sqrt{a^2 - u^2}} - \operatorname{Arc} \sin \frac{u}{a}, \quad a > 0$

37. $\displaystyle \int \frac{du}{u \sqrt{a^2 - u^2}} = \frac{1}{a} \log \frac{|u|}{a + \sqrt{a^2 - u^2}}, \quad a > 0$

38. $\displaystyle \int \frac{du}{u^2 \sqrt{a^2 - u^2}} = -\frac{\sqrt{a^2 - u^2}}{a^2 u}$

39. $\displaystyle \int \frac{du}{u^3 \sqrt{a^2 - u^2}} = -\frac{\sqrt{a^2 - u^2}}{2a^2 u^2} + \frac{1}{2a^3} \log \frac{|u|}{a + \sqrt{a^2 - u^2}}, \quad a > 0$

40. $\displaystyle \int \frac{\sqrt{a^2 - u^2}}{u} \, du - \sqrt{a^2 - u^2} + a \log \frac{|u|}{a + \sqrt{a^2 - u^2}}, \quad a > 0$

41. $\displaystyle \int \frac{\sqrt{a^2 - u^2}}{u^2} \, du = -\frac{\sqrt{a^2 - u^2}}{u} - \operatorname{Arc} \sin \frac{u}{a}, \quad a > 0$

42. $\displaystyle \int (a^2 - u^2)^{n/2} \, du = \frac{u (a^2 - u^2)^{n/2}}{n + 1} + \frac{a^2 n}{n + 1} \int (a^2 - u^2)^{(n/2)-1} \, du$

43. $\displaystyle \int \frac{u^m}{\sqrt{a^2 - u^2}} \, du = -\frac{u^{m-1}}{m} \sqrt{a^2 - u^2} + \frac{m - 1}{m} a^2 \int \frac{u^{m-2}}{\sqrt{a^2 - u^2}} \, du$

## INTEGRANDS THAT ARE RATIONAL FUNCTIONS OF $u$ AND $\sqrt{u^2 \pm a^2}$; $R(u, \sqrt{u^2 \pm a^2})$

44. $\displaystyle\int \frac{du}{\sqrt{u^2 \pm a^2}} = \log|u + \sqrt{u^2 \pm a^2}|$

45. $\displaystyle\int \sqrt{u^2 \pm a^2}\, du = \frac{u}{2}\sqrt{u^2 \pm a^2} \pm \frac{a^2}{2}\log|u + \sqrt{u^2 \pm a^2}|$

46. $\displaystyle\int u^2\sqrt{u^2 \pm a^2}\, du = \frac{u}{8}(2u^2 \pm a^2)\sqrt{u^2 \pm a^2} - \frac{a^4}{8}\log|u + \sqrt{u^2 \pm a^2}|$

47. $\displaystyle\int \frac{u^2\, du}{\sqrt{u^2 \pm a^2}} = \frac{u}{2}\sqrt{u^2 \pm a^2} \mp \frac{a^2}{2}\log|u + \sqrt{u^2 \pm a^2}|$

48. $\displaystyle\int (u^2 \pm a^2)^{3/2}\, du = \frac{u}{8}(2u^2 \pm 5a^2)\sqrt{u^2 \pm a^2} + \frac{3a^4}{8}\log|u + \sqrt{u^2 \pm a^2}|$

49. $\displaystyle\int \frac{du}{(u^2 \pm a^2)^{3/2}} = \frac{\pm u}{a^2\sqrt{u^2 \pm a^2}}$

50. $\displaystyle\int \frac{u^2\, du}{(u^2 \pm a^2)^{3/2}} = \frac{-u}{\sqrt{u^2 \pm a^2}} + \log|u + \sqrt{u^2 \pm a^2}|$

51. $\displaystyle\int \frac{du}{u\sqrt{u^2 - a^2}} = \frac{1}{a}\operatorname{Arc\,cos}\frac{a}{u}, \quad u > 0$

52. $\displaystyle\int \frac{du}{u\sqrt{u^2 + a^2}} = \frac{1}{a}\log\left|\frac{\sqrt{u^2 + a^2} - a}{u}\right| = \frac{1}{a}\log\left|\frac{u}{\sqrt{u^2 + a^2} + a}\right|$

53. $\displaystyle\int \frac{du}{u^2\cdot\sqrt{u^2 \pm a^2}} = \mp\frac{\sqrt{u^2 \pm a^2}}{a^2 u}$

54. $\displaystyle\int \frac{du}{u^3\sqrt{u^2 + a^2}} = -\frac{\sqrt{u^2 + a^2}}{2a^2 u^2} - \frac{1}{2a^3}\log\left|\frac{a - \sqrt{u^2 + a^2}}{u}\right|, \quad a > 0$

55. $\displaystyle\int \frac{du}{u^3\sqrt{u^2 - a^2}} = \frac{\sqrt{u^2 - a^2}}{2a^2 u^2} + \frac{1}{a^3}\operatorname{Arc\,cos}\frac{a}{u}, \quad u > 0$

56. $\displaystyle\int \frac{\sqrt{u^2 + a^2}}{u}\, du = \sqrt{u^2 + a^2} + a\log\left|\frac{a - \sqrt{u^2 + a^2}}{u}\right|, \quad a > 0$

57. $\displaystyle\int \frac{\sqrt{u^2 - a^2}}{u}\, du = \sqrt{u^2 - a^2} - a\operatorname{Arc\,cos}\frac{a}{u}, \quad u > 0$

58.  $\displaystyle \int \frac{\sqrt{u^2 \pm a^2}}{u^2}\, du = -\frac{\sqrt{u^2 \pm a^2}}{u} + \log |\, u + \sqrt{u^2 \pm a^2}\,|$

59.  $\displaystyle \int (u^2 \pm a^2)^{n/2}\, du = \frac{u\,(u^2 \pm a^2)^{n/2}}{n+1} \pm a^2\, \frac{n}{n+1} \int (u^2 \pm a^2)^{(n/2)-1}\, du$

## OTHER ALGEBRAIC INTEGRANDS

60.  $\displaystyle \int \sqrt{\frac{a+u}{b+u}}\, du = \sqrt{(a+u)(b+u)} + (a-b) \log \left( \sqrt{a+u} + \sqrt{b+u} \right)$

61.  $\displaystyle \int \sqrt{\frac{a-u}{b+u}}\, du = \sqrt{(a-u)(b+u)} + (a+b)\, \text{Arc sin} \sqrt{\frac{b+u}{a+b}}, \quad a+b>0$

62.  $\displaystyle \int \sqrt{\frac{a+u}{b-u}}\, du = -\sqrt{(a+u)(b-u)} - (a+b)\, \text{Arc sin} \sqrt{\frac{b-u}{a+b}}, \quad a+b>0$

63.  $\displaystyle \int \frac{du}{\sqrt{(a+u)(b+u)}} = \log \left| \frac{a+b}{2} + u + \sqrt{(a+u)(b+u)} \right|$

64.  $\displaystyle \int \frac{du}{\sqrt{(a-u)(b+u)}} = \text{Arc sin} \frac{b-a+2u}{a+b}, \quad a+b>0$

## TRIGONOMETRIC INTEGRANDS

65.  $\displaystyle \int \tan u\, du = \log |\sec u| = -\log |\cos u|$

66.  $\displaystyle \int \cot u\, du = \log |\sin u|$

67.  $\displaystyle \int \sec u\, du = \log |\sec u + \tan u|$

68.  $\displaystyle \int \csc u\, du = \log |\csc u - \cot u|$

69.  $\displaystyle \int \sin^2 u\, du = \tfrac{1}{2}u - \tfrac{1}{4} \sin 2u$

70.  $\displaystyle \int \cos^2 u\, du = \tfrac{1}{2}u + \tfrac{1}{4} \sin 2u$

71.  $\displaystyle \int \sin^n u\, du = -\frac{1}{n} \sin^{n-1} u \cos u + \frac{n-1}{n} \int \sin^{n-2} u\, du$

72. $\displaystyle\int \cos^n u \, du = \frac{1}{n} \cos^{n-1} u \sin u + \frac{n-1}{n} \int \cos^{n-2} u \, du$

73. $\displaystyle\int \tan^n u \, du = \frac{1}{n-1} \tan^{n-1} u - \int \tan^{n-2} u \, du$

74. $\displaystyle\int \cot^n u \, du = \frac{-1}{n-1} \cot^{n-1} u - \int \cot^{n-2} u \, du$

75. $\displaystyle\int \csc^n u \, du = \frac{-1}{n-1} \csc^{n-2} u \cot u + \frac{n-2}{n-1} \int \csc^{n-2} u \, du$

76. $\displaystyle\int \sec^n u \, du = \frac{1}{n-1} \sec^{n-2} u \tan u + \frac{n-2}{n-1} \int \sec^{n-2} u \, du$

77. $\displaystyle\int \cos^m u \sin^n u \, du = \frac{1}{m+n} \cos^{m-1} u \sin^{n+1} u + \frac{m-1}{m+n} \int \cos^{m-2} u \sin^n u \, du$

$\displaystyle = \frac{-1}{m+n} \cos^{m+1} u \sin^{n-1} u + \frac{n-1}{m+n} \int \cos^m u \sin^{n-2} u \, du$

$\displaystyle = \frac{-1}{m+1} \cos^{m+1} u \sin^{n+1} u + \frac{m+n+2}{m+1} \int \cos^{m+2} u \sin^n u \, du$

$\displaystyle = \frac{1}{n+1} \cos^{m+1} u \sin^{n+1} u + \frac{m+n+2}{n+1} \int \cos^m u \sin^{n+2} u \, du$

78. $\displaystyle\int \sin au \sin bu \, du = -\frac{\sin (a+b)u}{2(a+b)} + \frac{\sin (a-b)u}{2(a-b)}$

79. $\displaystyle\int \sin au \cos bu \, du = -\frac{\cos (a+b)u}{2(a+b)} - \frac{\cos (a-b)u}{2(a-b)}$

80. $\displaystyle\int \cos au \cos bu \, du = \frac{\sin (a+b)u}{2(a+b)} + \frac{\sin (a-b)u}{2(a-b)}$

81. $\displaystyle\int \frac{du}{a+b \cos u} = \frac{2}{\sqrt{a^2-b^2}} \operatorname{Arc\,tan} \left( \sqrt{\frac{a-b}{a+b}} \tan \frac{u}{2} \right), \quad |a| > |b|$

$\displaystyle = \frac{1}{\sqrt{b^2-a^2}} \log \left| \frac{\sqrt{a+b} + \sqrt{b-a} \tan (u/2)}{\sqrt{a+b} - \sqrt{b-a} \tan (u/2)} \right|, \quad |a| < |b|$

82. $\displaystyle\int \frac{du}{a + b \sin u} = \frac{2}{\sqrt{a^2 - b^2}} \text{ Arc tan } \frac{a \tan (u/2) + b}{\sqrt{a^2 - b^2}}, \quad |a| > |b|$

$\displaystyle = \frac{1}{\sqrt{b^2 - a^2}} \log \left| \frac{b + a \tan (u/2) - \sqrt{b^2 - a^2}}{b + a \tan (u/2) + \sqrt{b^2 - a^2}} \right|, \quad |b| > |a|$

83. $\displaystyle\int \frac{du}{a^2 \cos^2 u + b^2 \sin^2 u} = \frac{1}{ab} \text{ Arc tan } \frac{b \tan u}{a}$

84. $\displaystyle\int u^n \sin au \, du = -\frac{1}{a} u^n \cos au + \frac{n}{a} \int u^{n-1} \cos au \, du$

85. $\displaystyle\int u^n \cos au \, du = \frac{1}{a} u^n \sin au - \frac{n}{a} \int u^{n-1} \sin au \, du$

## INTEGRANDS INVOLVING EXPONENTIAL AND LOGARITHMIC FUNCTIONS

(Parts should be considered with other forms.)

86. $\displaystyle\int u^n e^{au} \, du = \frac{1}{a} u^n e^{au} - \frac{n}{a} \int u^{n-1} e^{au} \, du$

87. $\displaystyle\int u^n \log u \, du = \frac{u^{n+1} \log u}{n + 1} - \frac{u^{n+1}}{(n + 1)^2}$

88. $\displaystyle\int u^n \log^m u \, du = \frac{u^{n+1}}{n + 1} \log^m u - \frac{m}{n + 1} \int u^n \log^{m-1} u \, du$

89. $\displaystyle\int e^{au} \sin bu \, du = \frac{e^{au} (a \sin bu - b \cos bu)}{a^2 + b^2}$

90. $\displaystyle\int e^{au} \cos bu \, du = \frac{e^{au} (b \sin bu + a \cos bu)}{a^2 + b^2}$

91. $\displaystyle\int \frac{du}{a + be^{cu}} = \frac{1}{ac} (cu - \log |a + be^{cu}|)$

# LIMITS AND CONTINUITY

## Introduction

Precise definitions of limits and of continuity for real functions must avoid vague phrases such as "gets close to," or "approaches," which have a connotation of the passage of time. Real functions map real numbers into real numbers, and so ultimately the definitions and the theorems must be phrased in terms of the arithmetic and order properties of the real numbers. The reader must consult more advanced books for a thorough discussion of the real number system. Here we assume that the arithmetic of the real number system is familiar. Theorems about limits will be based on the following fundamental property of the real number system.

## The Least Upper Bound Property

Suppose $S$ is a non-empty set of real numbers that is bounded above, that is, suppose there is a number $B$ such that

$$x \leqq B \text{ for all numbers } x \text{ in } S.$$

Then there is a *smallest* number $b \leqq B$ such that

$$x \leqq b \text{ for all numbers } x \text{ in } S.$$

The number $b$ is called the *least upper bound* of $S$.

# 1   *Limits of Real Functions*

**DEFINITION**   Assume the function $f$ is defined on an interval $I$, except possibly at the point $x = a$ of $I$ ($a$ may be an endpoint). Then $f$ has the limit $L$ as $x$ approaches $a$,

$$\lim_{x \to a} f(x) = L,$$

if for each positive number $\epsilon$ there is a positive number $\delta$ such that

$$|f(x) - L| < \epsilon$$

if $x$ is in $I$ and

$$0 < |x - a| < \delta.$$

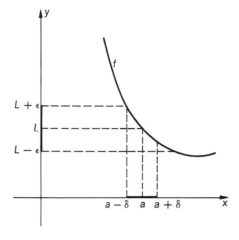

*Remark 1.* The definition does not provide a method for finding $L$. It only provides a criterion for determining whether a number $L$ is the limit of the function.

**Example 1**   To establish that

$$\lim_{x \to 2} (x^2 - x) = 2,$$

we apply the Definition with

$$f(x) = x^2 - x, \qquad a = 2 \quad \text{and} \quad L = 2.$$

**PROOF**   We must show that, given any positive number $\epsilon$, one can find a positive number $\delta$ such that

(1)    $|f(x) - L| = |(x^2 - x) - 2| < \epsilon$    if    $0 < |x - 2| < \delta.$

But $|f(x) - L| = |x^2 - x - 2| = |x - 2| \cdot |x + 1|$ and clearly $|x - 2| \cdot |x + 1|$ will be small when $|x - 2|$ is small. We have but to prescribe a suitable choice for $\delta$. There are obviously many choices because once a suitable $\delta$ has been selected for a given $\epsilon$, then any smaller value for $\delta$ would also serve. Let us choose

$$\delta < \text{minimum of } \epsilon/4 \text{ and } 1.$$

Then, if $0 < |x - 2| < \delta$, we have $-1 < x - 2 < 1$ and $2 < x + 1 < 4$. Whence, from (1)

$$|f(x) - L| = |x^2 - x - 2| = |x - 2||x + 1| < \frac{\epsilon}{4} 4 = \epsilon.$$

This completes the proof.

This example shows that $\delta$ may depend on $\epsilon$ (is a function of $\epsilon$). Indeed, finding this dependence is often the most difficult step in establishing that a limit exists.

The fact that $f(x) = x^2 - x$ is defined at $x = 2$ is irrelevant. Thus, if we had considered the function

$$f(x) = \begin{cases} x^2 - x & \text{if} \quad x \neq 2, \\ 156 & \text{if} \quad x = 2, \end{cases}$$

the result would be the same, namely,

$$\lim_{x \to 2} f(x) = 2,$$

because the value of $f$ at $x = 2$ is never considered in computing the limit.

The difference quotient

$$\frac{f(x + \Delta x) - f(x)}{\Delta x},$$

considered as a function of $\Delta x$ is not defined at $\Delta x = 0$, yet the limit as $\Delta x \to 0$ often exists and is the derivative $f'(x)$.

**Example 2**    The limit of $\sin(1/x)$ as $x \to 0$ does not exist, for rather obviously $\sin(1/x)$ will oscillate between 1 and $-1$ as $x \to 0$.

*Remark 2.* If in the definition of limit $x$ becomes infinite, the definition reads as follows:

$$\lim_{x \to \infty} f(x) = L$$

if for each positive number $\epsilon$ there is a number $N$ such that

$$|f(x) - L| < \epsilon \qquad \text{if} \qquad x > N.$$

In order for the limit to exist, $L$ must be a *number*. Although we do sometimes write

$$\lim_{x \to a} f(x) = \infty$$

this is but a way of saying: For each $M > 0$ there is $\delta > 0$ such that $f(x) > M$ when $0 < |x - a| < \delta$.

Real-valued functions with the same domain can be added, subtracted, multiplied, and divided (if the denominator is not zero). Limits of these new functions are related to limits of the original functions.

**THEOREM 1**    If $f$ and $g$ have the same domain and $\lim f(x) = L$ and $\lim g(x) = M$ as $x \to a$, then, as $x$ approaches $a$,

(a)   $\lim [f \pm g](x) - \lim f(x) \pm \lim g(x) - L \pm M,$

(b)   $\lim [fg](x) = [\lim f(x)][\lim g(x)] = LM,$

(c)   $\lim \dfrac{f}{g}(x) = \dfrac{\lim f(x)}{\lim g(x)} = \dfrac{L}{M} \qquad \text{if} \qquad M \neq 0.$

**PROOF**    We prove part (a) only. The other parts are more troublesome but the principle is the same.

Suppose a positive number $\epsilon$ is given. Choose positive numbers $\delta_1$ and $\delta_2$ so that

$$|f(x) - L| < \epsilon/2 \qquad \text{if} \qquad 0 < |x - a| < \delta_1,$$

$$|g(x) - M| < \epsilon/2 \qquad \text{if} \qquad 0 < |x - a| < \delta_2.$$

This is possible because the limits of $f$ and $g$ are $L$ and $M$. Now let $\delta$ be the lesser of $\delta_1$ and $\delta_2$. Then, if $0 < |x - a| < \delta$,

$$|f(x) + g(x) - (L + M)| = |[f(x) - L] + [g(x) - M]|$$

$$\leqq |f(x) - L| + |g(x) - M|$$

$$< \epsilon/2 + \epsilon/2 = \epsilon.$$

This completes the proof.

*Remark 3.* We have used this theorem again and again in evaluating limits. For example, in deriving the basic differentiation formula for the quotient (see page 35) we had

$$\frac{\Delta y}{\Delta x} = \frac{v(\Delta u/\Delta x) - u(\Delta v/\Delta x)}{(v + \Delta v)v},$$

and to find the limit we used parts (a) and (c) of the theorem.

*Remark 4.* In Example 1, we could have proved separately that

$$\lim_{x \to 2} x^2 = 4 \qquad \text{and} \qquad \lim_{x \to 2} x = 2$$

and then applied part (a) of Theorem 1 to obtain

$$\lim_{x \to 2} (x^2 - x) = 2.$$

The definition of limit provides the ultimate criterion for the existence of a limit of a function, but the following theorem can be useful. It provides an easily verifiable condition that often occurs.

**THEOREM 2**    If $f$ is increasing on an interval $(a, b)$, and if $f$ is bounded above on that interval (that is, there is a number $N$ such that $f(x) \leqq N$, for all $x$ in the interval) then $\lim_{x \to b} f(x)$ exists.

We omit the proof, which is similar to the proof of Theorem 1 of Section 3, this Appendix.

### Problems

By inspection, find the following limits if they exist. Then apply the definition of limit to prove that you are correct.

*1.* $\lim_{x \to 3} (\tfrac{1}{2}x + 2)$

*2.* $\lim_{x \to 3} \dfrac{1}{x}$

*3.* $\lim_{x \to 1} (x^2 + x)$

*4.* $\lim_{x \to -1} (x - 2)$

## 2   Continuity

**DEFINITION**   A function $f$ is *continuous at a point* $x_0$ in its domain if

$$\lim_{x \to x_0} f(x) = f(x_0).$$

A function $f$ is *continuous on an interval* if it is continuous at each point of the interval.

In terms of the $\epsilon$, $\delta$ definition of limit, continuity at $x_0$ becomes: For each positive number $\epsilon$ there is a positive number $\delta$ such that, if $|x - x_0| < \delta$, then

$$|f(x) - f(x_0)| < \epsilon.$$

Continuous functions have a number of rather obvious appearing properties which, nevertheless, are quite subtle. The first of these which we consider here is an immediate consequence of Theorem 1 of Section 1.

**THEOREM 1**   If $f$ and $g$ are continuous at $x_0$, so also are $f \pm g$, $fg$, and $f/g$ if $g(x_0) \neq 0$.

A proof based on Section 1 is left to the reader. Observe that we have frequently used this theorem without specific comment. From Theorem 1 we conclude, starting with the continuity of the identity function, $f(x) = x$, and the constant functions, $f(x) = c$, that rational functions are continuous.

Besides rational functions, the remaining elementary functions are also continuous, their continuity being a consequence of their differentiability. Yet the careful reader may suspect that perhaps we used continuity somewhere in deriving the differentiation formulas. This was indeed the case. On page 42 we used the continuity of the logarithm function, $\log_a$, and on page 53 we assumed as obvious that $\cos \Delta u \to 1$ as $\Delta u \to 0$. The continuity of $\log_a$ and cosine can be established directly from the definitions of these functions. We shall not do so here.

Not only do our differentiation formulas establish the continuity of the simplest elementary functions, but also the chain rule establishes that the composite of two continuous functions is continuous. This fact is not dependent on differentiability, and we state it as a theorem.

**THEOREM 2**   If $f$ and $g$ are continuous, and if the composite $f(g(x))$ is defined for $x$ in some interval, then $f(g)$ is continuous.

**PROOF**   Let $y_0 = g(x_0)$ and $y = g(x)$. If $\epsilon$ is a positive number, choose a positive number $\epsilon_1$ such that

(1) $$|f(y) - f(y_0)| < \epsilon \qquad \text{if} \qquad |y - y_0| < \epsilon_1.$$

Now choose a positive number $\delta$ such that

(2) $$|g(x) - g(x_0)| < \epsilon_1 \qquad \text{if} \qquad |x - x_0| < \delta.$$

Then, combining (1) and (2), we have

$$|f(g(x)) - f(g(x_0))| = |f(y) - f(y_0)| < \epsilon$$

if $|x - x_0| < \delta$.

An important property of continuous real functions is given by the following theorem.

**THEOREM 3**     If $f$ is continuous on an interval $[a, b]$, then $f$ is bounded on the interval. That is, there is a number $M$ such that $-M \leq f(x) \leq M$ for all $x$ in $[a, b]$.

*Comments on a possible proof.* If $f$ were not bounded above, then there would be a sequence $\{x_n\}$, with $a \leq x_n \leq b$, and $f(x_n) > n$. Some subsequence of $\{x_n\}$ (see Section 3) will converge to a point $\bar{x}$ of the interval. One then gets a contradiction with the continuity of $f$ at $\bar{x}$.

The conclusion of the theorem is not valid if the interval is not closed. For example, if $f(x) = 1/x$ for $0 < x \leq 1$, then $f$ is continuous on the half-open interval $(0, 1]$ and is obviously unbounded.

Finally, we prove the theorem on which the proof of Rolle's theorem is based (see page 179).

**THEOREM 4**     If $f$ is continuous on the closed interval $[a, b]$, there are numbers $u$ and $v$, $a \leq u, v \leq b$, such that

$$f(x) \leq f(u) \qquad \text{for all } x \text{ in } [a, b]$$
$$f(v) \leq f(x) \qquad \text{for all } x \text{ in } [a, b].$$

**PROOF**     From Theorem 3, $f$ is bounded. Let $S$ be the set of values, $f(x)$, of $f$, and let $M$ be the least upper bound of $S$. We claim that there is a point $u$ in $[a, b]$ such that $f(u) = M$.

If this were not the case, then $f(x) < M$ for all $x$ in $[a, b]$ and so $M - f(x) > 0$ for $x$ in $[a, b]$. Then, by Theorem 1 the function $g: g(x) = 1/[M - f(x)]$, is continuous in $[a, b]$. By Theorem 3, $g$ is bounded above by some number $L$; that is,

$$0 < \frac{1}{M - f(x)} \leq L$$

for every $x$ in $[a, b]$. Solving this inequality for $f(x)$, we obtain

$$f(x) \leq M - \frac{1}{L}$$

for every $x$ in $[a, b]$. Thus, $f$ has the upper bound $M - (1/L)$, a number less than the least upper bound $M$. This contradiction shows that $f(x)$ must equal $M$ for at least one $x$ in $[a, b]$.

As a final fact about continuous functions we prove another theorem that appeals strongly to the intuition, yet is non-trivial.

**THEOREM 5 (INTERMEDIATE VALUE THEOREM)**     Suppose that $f$ is continuous on the interval $a \leq x \leq b$ and that $f(a) < f(b)$. Let $k$ be any number between $f(a)$ and $f(b)$. Then there is a number $c$ such that $a < c < b$, and $f(c) = k$.

**PROOF SKETCH**    Let $A$ be the set of all numbers $x$ such that $a \leq x < b$ and $f(x) < k$. Then $A$ is not empty (since $a$ is in $A$) and $b$ is an upper bound of $A$. By the least upper bound property there is a number $c$ that is the least upper bound of $A$. We claim that $f(c) = k$. It suffices to prove that it is impossible to have either $f(c) < k$ or $f(c) > k$.

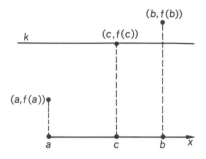

Suppose that $f(c) < k$. Then one can show that there are numbers $x_1 > c$ such that $f(x_1) < k$. Then $x_1$ would be in $A$, and $c$ would not be an upper bound. Now suppose that $f(c) > k$. Then one can show that in some interval $x_1 \leq x \leq c$, $f(x) > k$. Then each number of this interval would also be an upper bound of $A$, and $c$ would not be the *least* upper bound.

## 3    Limits of Sequences

For notation, see Chapter 13. In this section we establish the Cauchy criterion for convergence of sequences. Remember that we must base our proof upon the assumed properties of the real number system (see the introduction to this Appendix). We begin with an easy theorem.

**THEOREM 1**    If $\{x_n\}$ is an increasing sequence that is bounded above; that is,

(i)   $x_n \geq x_m$      if $n > m$,   {Increasing

(ii)   $x_n \leq b$      for all $n$,   {Bounded above

then $\lim_{n \to \infty} x_n$ exists.

**PROOF**    Let $S$ be the set of all the numbers $x_n$, $n = 1, 2, \ldots$. By (ii) above, $S$ is bounded above. Let $L$ be the least upper bound of $S$, which must exist by our basic assumption concerning the real numbers. We claim that

$$\lim_{n \to \infty} x_n = L.$$

Suppose a positive number $\epsilon$ is given. Because $L$ is the least upper bound, there is a positive integer $N$ such that $L - \epsilon < x_N \leq L$. Then if $n \geq N$, we have

$$L - \epsilon < x_N \leq x_n \leq L \qquad \text{and} \qquad |x_n - L| < \epsilon.$$

*Remark.*    A similar result holds for decreasing sequences bounded below. Suppose $\{x_n\}$ is decreasing ($x_n \leq x_m$ if $n \leq m$) and bounded below ($b \leq x_n$ for all $n$). Then $\{-x_n\}$ is increasing and bounded above. By Theorem 1, $\lim (-x_n) = L$ exists, and so $\lim x_n = -L$.

We next prove that bounded sequences have convergent subsequences. A sequence $\{x_n\}$ is bounded if there are numbers $a$ and $b$ such that

$$a \leq x_n \leq b \qquad \text{for all } n = 1, 2, \ldots.$$

The theorem we are interested in concerns subsequences, which are defined as follows:

If $\{x_n\}$ is any infinite sequence, and if $n_0, n_1, \ldots, n_k, \ldots$

is a sequence of positive integers that become infinite, then the sequence $\{x_{n_k}\}$ is called a *subsequence* of $\{x_n\}$.

**THEOREM 2**    Every bounded sequence has a convergent subsequence.

**PROOF SKETCH**    Suppose that $\{x_n\}$ is bounded with $a \leq x_n \leq b$. Denote the interval $[a, b]$ by $I_0$. Choose $n_0$ arbitrarily. Then $x_{n_0}$ is in $I_0$.

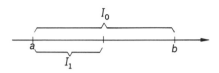

Now bisect $I_0$. Infinitely many $x_n$ must lie in at least one of the two halves. Let $I_1$ be such a half-interval. Choose $n_1 > n_0$ so that $x_{n_1}$ is in $I_1$. We are now in quite the same position with respect to $I_1$ as we were initially with respect to $I_0$.

Continue this selection process. We obtain an infinite sequence of intervals $I_k$ and a subsequence $\{x_{n_k}\}$ of $\{x_n\}$ such that $x_{n_k}$ is in $I_k$, $k = 0, 1, 2, \ldots$, and with the length of $I_k$ equal to $(b - a)/2^k$. Now consider the ends of the intervals $I_k = [a_k, b_k]$ and the sequence $\{a_k\}$. Because the interval $I_k$ contains the interval $I_{k+1}$, we have $a_k \leq a_{k+1} \leq b$. Thus, $\{a_k\}$ is bounded above and increasing. Therefore, by Theorem 1,

$$\lim_{k \to \infty} a_k = L$$

exists. We claim also that

$$\lim_{k \to \infty} x_{n_k} = L.$$

Because

(1)
$$|x_{n_k} - L| \leq |x_{n_k} - a_k| + |a_k - L|,$$

all that needs to be done is to choose $k$ so large that both terms on the right of (1) are small enough. Suppose that a positive number $\epsilon$ is given. Since $a_k \leq x_{n_k} \leq b_k$ and $b_k - a_k = (b - a)/2^k$, choose an integer $N_1$ so that $(b - a)/2^k < \epsilon/2$ if $k \geq N_1$, and choose an integer $N_2$ so that $|a_k - L| < \epsilon/2$ if $k \geq N_2$. Finally, choose $N =$ the maximum of $N_1$ and $N_2$. Then, from (1)

$$|x_{n_k} - L| < \frac{\epsilon}{2} + \frac{\epsilon}{2}, \qquad \text{if } k \geq N.$$

This completes the proof.

We are now able to make a proof of Cauchy's criterion.

**THEOREM 3 (CAUCHY'S CRITERION)**    The sequence $\{x_n\}$ converges if and only if to each positive number $\epsilon$ there is a positive integer $N$ such that, if $n \geq N$,

(2)
$$|x_{n+p} - x_n| < \epsilon \qquad \text{for all positive integers } p.$$

**PROOF**    The "only if" part of the theorem is rather obvious, since, if the sequence converges, say to $L$, then

$$| x_{n+p} - x_n | \leq | x_{n+p} - L | + | L - x_n |.$$

Thus, if $| x_n - L | < \epsilon/2$, for $n \geq N$, then $| x_{n+p} - x_n | < \epsilon$.

To prove the "if" part, we first choose $\epsilon = 1$, and suppose that condition (2) is satisfied. Then there is an integer $N_1$ such that

$$| x_{N_1+p} - x_{N_1} | < 1 \qquad \text{for all positive integers } p.$$

Then, except for the finite set of $N_1 - 1$ numbers $x_1, x_2, \ldots, x_{N_1-1}$, all terms $x_n$ of the sequence lie in the interval $[x_{N_1} - 1, x_{N_1} + 1]$. Therefore, the given sequence is bounded. By Theorem 2 there is a convergent subsequence $\{x_{n_k}\}$ that converges to, say, $L$.

To complete the proof, we must show that the full sequence $\{x_n\}$ also converges to the limit $L$. We shall not fill in the details of such a proof, but be content with the remark that, if $\lim_{n \to \infty} x_n$ were not equal to $L$, then there would be *another* subsequence $x_{m_j}$ converging to a number $M \neq L$. But then $| x_{m_j} - x_{n_k} |$ could not be arbitrarily small for large $m_j$ and $n_k$, as the Cauchy condition (2) requires.

## 1 The Number e

To prove that

$$\lim_{h \to 0} (1 + h)^{1/h}$$

exists, we first prove that the limit exists if $h$ approaches zero through positive values. Then we will prove that the same limit is obtained if $h \to 0$ through negative values.

In the course of our proof we will need the following inequality:

(1) $$(1 + x)^k > 1 + kx \qquad \text{if } x > 0 \text{ and } k > 1.$$

**PROOF**     Let $f(x) = (1 + x)^k - (1 + kx)$. Then $f'(x) = k(1 + x)^{k-1} - k > 0$ and $f(0) = 0$. Thus, $f$ increases, and $f(x) > 0$ for $x > 0$.*

Returning to the main proof, we first prove that $(1 + h)^{1/h}$ increases as $h$ decreases through positive values, or what is the same $(1 + h)^{1/h}$ decreases as $h$ increases. To this end, suppose that $h > 0$, and that $\epsilon$ is any positive number. We wish to show that

(2) $$(1 + h)^{1/h} > (1 + h + \epsilon)^{1/(h+\epsilon)}.$$

---

* The formula for the derivative of the power function with arbitrary real exponent $k$ was obtained in Chapter 2 using logarithmic differentiation. Thus, the proof of (1) above needs modification. The formula for the derivative of the power function with rational exponent $(p/q)$ ($p$ and $q$ are integers) can be obtained from the power function with integer exponent by implicit differentiation. Thus, if $y = x^{p/q}$, then $y^q = x^p$ and $qy^{q-1}y' = px^{p-1}$. Solving for $y'$ we obtain

$$y' = \frac{p}{q} x^{(p/q)-1}$$

Second, an arbitrary real number $k$ can be approximated as closely as one wishes by rational numbers $(p/q)$. Thus, the worst that could happen in (1) is that ">" is replaced by "$\geq$"; this change would not affect the later proof.

Inequality (2) is equivalent to

(3)
$$(1+h)^{(h+\epsilon)/h} > 1+h+\epsilon.$$

Now, if we apply the inequality (1) to $(1+h)^{(h+\epsilon)/h}$, we obtain inequality (3), which implies the desired inequality.

Now, since $(1+h)^{1/h}$ increases as $h$ approaches zero through positive values, it suffices to consider values of $h = 1/n$, where $n$ is a positive integer. Therefore, we consider the sequence

$$a_n = \left(1+\frac{1}{n}\right)^n, \qquad n = 1, 2, \ldots.$$

According to Appendix C, it suffices to prove that the sequence $\{a_n\}$ is bounded because we have already proved that it is increasing. This we do as follows.

By the binomial theorem

$$\left(1+\frac{1}{n}\right)^n = 1 + n\frac{1}{n} + \cdots + \binom{n}{k}\frac{1}{n^k} + \cdots + \frac{1}{n^n}$$

$$= 1 + 1 + \frac{1}{2!}\left(1-\frac{1}{n}\right) + \frac{1}{3!}\left(1-\frac{1}{n}\right)\left(1-\frac{2}{n}\right)$$

$$+ \cdots + \frac{1}{k!}\left(1-\frac{1}{n}\right)\cdots\left(1-\frac{k-1}{n}\right) + \cdots + \frac{1}{n!}\left(1-\frac{1}{n}\right)\cdots\left(1-\frac{n-1}{n}\right)$$

$$< 1 + 1 + \frac{1}{2} + \frac{1}{2^2} + \cdots + \frac{1}{2^{k-1}} + \cdots + \frac{1}{2^{n-1}} < 3.$$

This completes a proof that

$$\lim_{h\to 0+} (1+h)^{1/h}$$

exists.

Now, suppose that $h < 0$, and set $h = -k$. Then

(4)
$$(1+h)^{1/h} = (1-k)^{-1/k} = \left(\frac{1}{1-k}\right)^{1/k}$$

$$= \left(1+\frac{k}{1-k}\right)^{1/k} = \left(1+\frac{k}{1-k}\right)^{(1-k)/k}\left(1+\frac{k}{1-k}\right).$$

Finally, as $h \to 0-$, then $k/(1-k) \to 0+$, and we have

$$\lim_{h\to 0-} (1+h)^{1/h} = \lim_{k\to 0+}\left(1+\frac{k}{1-k}\right)^{(1-k)/k}\left(1+\frac{k}{1-k}\right) = e.$$

This completes the proof of the existence of the limit we call $e$.

**2   $\displaystyle \lim_{\theta \to 0} \frac{\sin \theta}{\theta}$**

To establish the limit, we may confine our attention to positive $\theta$ because $\sin(-\theta)/(-\theta) = \sin\theta/\theta$. The existence of the limit will be based on a simple geometric inequality which we take without proof. Referring to the figure, we have

   Area $\triangle OPQ <$ Area sector $OPR <$ Area $\triangle OPS$.

Using the values of these areas in terms of $\theta$, we have

$$\tfrac{1}{2}\cos\theta\sin\theta < \tfrac{1}{2}\theta < \tfrac{1}{2}\tan\theta,$$

whence, after a step

$$\cos\theta < \frac{\sin\theta}{\theta} < \frac{1}{\cos\theta}.$$

Therefore, as $\theta \to 0$, $\sin\theta/\theta$ is pinched between the limits of $\cos\theta$ and $1/\cos\theta$, namely 1.

$$1 = \lim_{\theta \to 0} \cos\theta \leq \lim_{\theta \to 0} \frac{\sin\theta}{\theta} \leq \lim_{\theta \to 0} \frac{1}{\cos\theta} = 1.$$

This completes the proof.

*Remark.* If we were to use degree measure for $\angle POQ$, then the area of the sector would be $\tfrac{1}{2}\delta\,(\pi/180)$, where $\delta$ is the measure in degrees. In that case, the limit in question would have been $180/\pi$. Thus, if the trigonometric functions were considered as functions of angles measured in degrees, then the differentiation formulas would all have this awkward factor of $180/\pi$. The simplicity of radian measure is evident.

# Index